COMPUTATIONAL
INTEGRATION

COMPUTATIONAL INTEGRATION

ARNOLD R. KROMMER

The Numerical Algorithms Group Ltd.
Oxford, United Kingdom

CHRISTOPH W. UEBERHUBER

Institute for Applied and Numerical Mathematics
Technical University of Vienna
Vienna, Austria

Society for Industrial and Applied Mathematics
Philadelphia

10 9 8 7 6 5 4 3 2

About the cover: The cover was designed by Robert Lettner. He is in charge of the central printing and reprography office at the Academy of Applied Arts in Vienna, Austria.

Library of Congress Cataloging-in-Publication Data

Krommer, Arnold R.
Computational integration / Arnold R. Krommer, Christoph W. Ueberhuber
p. cm.
Includes bibliographical references and indexes.
ISBN 978-0-898713-74-9 (pbk.)
1. Numerical integration. I. Ueberhuber, Christoph W., 1946- II. Title.
QA299.3.K74 1998
515'.624--dc21 98-3274

Contents

Preface

The calculation of integrals is an important task in many fields of scientific computation, ranging from computational statistics to finite element methods. When trying to solve integration problems, one is faced with a vast amount of relevant literature, a bewildering spectrum of computational methods, and a large number of software products.

Literature on computational integration comprises thousands of technical reports and articles in mathematical journals as well as a number of books (for instance, Braß [39], Davis and Rabinowitz [92], Engels [131], Evans [142], Krommer and Ueberhuber [247], Krylov [250], and Stroud [423], [424]). Among the computational methods employed, one can find both symbolic and numerical techniques, the latter based on principles derived from areas as diverse as approximation theory, number theory, statistics, complexity theory, etc. Software for computational integration comprises dozens of published algorithms and programs, special software packages (in particular, QUADPACK [339]), and extensive sections of numerical program libraries such as NAG [311], IMSL [441], etc. Many software products can be obtained free of charge through electronic networks, in particular, the Internet, but are scattered over a number of software repositories and are therefore difficult to find and retrieve.

Scope. This monograph has been written with three major goals in mind. First, it is meant to give a comprehensive yet concise survey of computational integration methods and the fundamental mathematical principles on which they are based. Second, it is intended to cover many recent developments in computational integration, such as parallel integration algorithms, which have not yet been coherently presented in any other textbook. Third, this book attempts to bridge the gap between the theoretical knowledge about and the practical application of computational integration by providing references to and descriptions of numerous relevant software products.

The authors anticipate that three groups of readers will be interested in using this book. The first group is graduate students (and their instructors) who may use this monograph as a textbook in advanced courses on current techniques in scientific computing and numerical analysis. The second group is computational scientists and engineers working on problems for which reliable and efficient methods for calculating integrals are required. For them this volume may serve as a reference book. The third group is researchers in applied numerical analysis and mathematical software development, who will find a well-organized representation of relevant material along with an extensive bibliography that covers many pertinent articles published in the last 10 years.

Synopsis. Part I introduces the topic of computational integration. Chapter 1 concisely outlines the fundamental mathematical concepts of Riemann integration. The practically relevant types of integrals are defined and their most important properties are described. Chapter 2 discusses several prototypical areas of scientific computing in which computational integration plays an important role, thus demonstrating the practical importance of computational integration. Additionally, this chapter clarifies the requirements that computational integration software must meet in order to be of real practical value.

Chapter 3 outlines the fundamentals of computational integration. In particular, it provides a general problem formulation, describes the most frequently occurring integration regions and weight functions, and discusses various problem settings. This chapter also highlights the main features of both numerical and symbolic integration methods, gives advice on how to obtain existing software, and introduces methods of pre- and postprocessing integration problems. The heterogeneity of these topics reflects the broad range of issues one has to deal with when solving integration problems.

Part II is dedicated to symbolic integration methods. Chapter 4 introduces the basics of algebraic computation, sheds light on the inner workings of symbolic integration algorithms, and provides the major theoretical results fundamental to symbolic integration. Special consideration is given to the pitfalls that may arise from the mixed use of algebraic and finite precision calculations.

Part III deals with the theoretical, algorithmic, and computational aspects of numerical integration. In this part of the book, particular emphasis is placed on integration formulas, which are the central part of any numerical method for discretizing and approximating integrals. It is demonstrated that the construction of one-dimensional (quadrature) formulas is far easier than the construction of multidimensional (cubature) formulas. Quadrature formulas are considered in Chapter 5, cubature formulas in Chapter 6.

In practical computations one often has to deal with particular kinds of integrals involving, for instance, oscillatory or singular integrands. This is why substantial efforts have been made to develop tailor-made computational techniques for such special integration problems. Tailor-made numerical integration methods are discussed in Chapter 7.

Chapter 8 considers the construction of numerical integration algorithms. It describes different algorithmic sections—those implementing, for instance, integration formulas, error estimation procedures, or extrapolation schemes—and their incorporation into integration programs.

Chapter 9 deals with the development of numerical integration software for advanced computer systems, in particular, for parallel and distributed computer architectures. This chapter gives a classification of the level and degree of parallelism involved in numerical integration programs and discusses the structure and properties of various parallelization schemes. Particular attention is given to the problem of load distribution, i.e., the problem of allocating tasks to computational resources, which is of crucial importance to the performance of parallel numerical integration programs. Furthermore, parallelization schemes that have already been implemented and tested on advanced computer systems are reviewed.

Criteria for assessing the quality of numerical integration programs such as robustness, efficiency, etc., and for evaluating corresponding quality indices are discussed in Chapter 10.

Acknowledgments. Particular thanks go to Christoph Schmid, who tirelessly and carefully helped us to prepare this book. We would also like to thank Yolanda Goss for helping us express our thoughts as clearly as possible in the English language. Robert Bursill (The University of New South Wales, Sydney, Australia) proofread the manuscript competently.

In addition, we would like to acknowledge the financial support of the Austrian Science Foundation and to express our gratitude to The Numerical Algorithms Group Ltd. for encouraging and supporting us in our work.

ARNOLD R. KROMMER and CHRISTOPH W. UEBERHUBER

Notation

$:=$	Equality by definition	3
$:\Leftrightarrow$	Logical equivalence by definition	95
$\sim$	Asymptotic equality	54
$\approx$	Approximate equality	46
$\equiv$	Identity	5
$n \mid m$	n divides m	95
$\{ x, y, z, \dots \}$	Set of the elements x, y, z, ...	4
$x \in S$	x element of S	3
$\{ x \in S \; : \; p(x) \}$	Set of elements x of S with property p	5
$S_1 \setminus S_2$	Difference of sets S_1 and S_2	12
$S_1 \cap S_2$	Intersection of sets S_1 and S_2	10
$S_1 \cup S_2$	Union of sets S_1 and S_2	12
$S_1 \subset S_2$	S_1 proper subset of S_2	3
$S_1 \subseteq S_2$	S_1 subset of S_2	5
$S_1 \times S_2$	Cartesian product of sets S_1 and S_2	4
S^N	N-fold Cartesian product of set S	4
$S^{N \times N}$	Set of matrices over S	9
$\emptyset$	Empty set	10
$f\colon S_1 \to S_2$	Mapping from S_1 to S_2	3
$x \mapsto y$	x is mapped onto y	6
f'	Derivative of f	7
$f^{(k)}$	kth derivative of f	4
$\overline{f}$	Transformed function	75
$f\|_S$	Restriction of f to S	41
$f_1 \circ f_2$	Composition of functions	8
A^{-1}, f^{-1}	Inverse (of matrix, function)	26
$A^\top$, $\mathbf{x}^\top$	Transpose (of matrix, vector)	4
$\int$	Integral	3
$\sum$	Sum	3
$\sum''$	Sum with first and last term halved	134

$⨍$	Cauchy principal value integral	14
$⨍$	Finite part integral	18
$[a,b]$,$[a,b)$,$(a,b]$,(a,b)	Integration interval	3
$\partial/\partial x$, $\partial^k/\partial x^k$	Partial differential operators	45
$\partial u/\partial \nu$	Normal derivative	45
∇	Gradient operator	39
Δ	Laplacean operator	38
Δ, Δ^k	Difference operator	82
$\lvert\cdot\rvert$	Absolute value	35
$\lvert\cdot\rvert$	Number of elements	180
$\lVert\cdot\rVert$	Arbitrary norm	35
$\lVert\cdot\rVert_1$	L_1 or sum norm	13
$\lVert\cdot\rVert_2$	L_2 or Euclidean norm	45
$\lVert\cdot\rVert_\infty$	L_∞ or maximum norm	63
$\square$	End of proof	115
$\square$	Rounded value	66
∞	Infinity	32
$n!$	Factorial	19
$\{x\}$, $\{\mathbf{x}\}$	Fractional part of real number x or vector $\mathbf{x}$	185
$\lfloor x \rfloor$	Largest integer less than or equal to x	179
$\lceil x \rceil$	Smallest integer greater than or equal to x	19
$\mathbf{0}$	Zero vector	216
$\{\hat{s}_n\}$	Transformed sequence	79
$U^\perp$	Annihilator space	182
X, $\mathbf{X}$	Random variables (scalar, vector)	30
$\binom{n}{k}$	Binomial coefficient	166
$\sqrt{\ }$	Square root	32
$\bar{z}$	Conjugate complex of z	35
(g,v)	Inner product on V	39
$\langle , \rangle$	Inner product	136
$\langle P,Q\rangle_w$	Inner product with respect to w	136
(p,q)	Ordered pair of elements	10
$(x_1,x_2,\ldots,x_N)^\top$	N-dimensional vector with components $x_1,x_2,\ldots,x_N$	4
$\rightarrow$	Convergence	4
$\rightarrow a+0$	Right-hand side limit	14
Γ	Boundary of Ω	38
$\Gamma(x)$	Gamma function	117
$\delta_{m,n}$	Kronecker delta	35
ε	Error requirement	55
ε_{abs}	Absolute error requirement	331

ε_{rel}	Relative error requirement	331
θ	Specific system state	32
Θ	Set of system states	32
$\mu(\mathbf{X})$	Expected (mean) value of **X**	31
π	Pi	32
$\pi(.)$	Permutation	177
$\pi(\theta)$	Prior density	32
$\pi(\theta\|\mathbf{x})$	Posterior density	32
$\rho(L)$	Figure of merit	226
$\sigma(X)$	Standard deviation of X	203
φ_b	Radical-inverse function	192
$\varphi_{d,i}$	Lagrange basis functions	120
Φ	Distribution function of standard normal distr.	77
ψ	Parameterization	24
$\omega(f;\delta)$	Modulus of continuity	115
ω_n	Nodal polynomial	121
Ω	Domain of partial differential equations	38
Ω_k	Finite element	41
arctan	Arcus tangent function	57
$a(m)$, $a(\mathbf{m})$	Fourier coefficients	214
$a(u,v)$	Bilinear form on V	39
$A(E;n)$	Counting function	185
$A(L)$	Abscissa set associated with integration lattice L	217
b	Basis vector of linear space	40
B	Integration region	31
c_i	Integration weight	62
$\mathrm{cl}\,\Omega$	Closure of Ω	38
c_B	Characteristic function	5
$C(\Omega)$	Continuous functions on Ω	38
$C^k(\Omega)$	k-times differentiable functions on Ω	38
$C_0^k(\Omega)$	Functions in $C^k(\Omega)$ vanishing on $\partial\Omega$	39
C_N	Unit cube	51
C_N^+	Positive unit cube	51
C_N^q	Quotient unit cube	217
$\mathbb{C}$	Set of all complex numbers	33
d	Polynomial degree	42
$\deg P$	Degree of polynomial P	94
$\det A$	Determinant of matrix A	9
$\mathrm{diam} S$	Diameter of S	5
$\dim(d,N)$	$\dim \mathbb{P}_d^N$	166
div	Division of whole numbers	363

$d_J \cdots d_0$	Digit expansion	192
D	Degree of accuracy	125
$D_n, D_n^*, D_n^{\mathcal{M}}$	Discrepancy (star, with respect to $\mathcal{M}$)	187
$\mathcal{D}_4$	Dihedral group	177
$e_d(f)$	Error of interpolation polynomial	121
$e_d^*(f)$	Error of best approximation polynomial	126
erf	Error function	107
exp	Exponential function	32
Ei	Exponential integral	107
$E(f)$	Truncation error $\mathrm{Q}f - \mathrm{I}f$	124
$\mathrm{E}(f; B_s)$	Estimate for $\mathrm{Q}(f; B_s) - \mathrm{I}(f; B_s)$	320
E_N^α, $E_N^\alpha(c)$	Korobov spaces	227
f, f_j	Integrand functions	3
F	Transformed function	33
$F(x)$	Field of rational expressions in x over F	95
$F[x]/m(x)$	Field of polynomials over field F modulo irreducible polynomial m	95
$F_{\mathbf{X}}$	Cumulative distribution function	31
$\mathbb{F}$	Set of all floating-point numbers	312
$\mathcal{F}$	Space of integrand functions	33
$g(\mathbf{x}, \theta)$	Joint distribution	33
G_n	Gauss–Legendre formula (n points)	128
G_n^w	Gauss formula with respect to w	138
$\mathcal{G}$	Group of linear transformations	176
$\mathcal{G}(\mathbf{x})$	$\mathcal{G}$-orbit	177
h	Discretization parameter	40
i	Imaginary unit	35
$\mathrm{Im}(z)$	Imaginary part of z	247
$\inf S$	Infimum of set S	126
I	Integral operator	3
$\mathrm{I}(f; B_l)$	Integral of f over $B_l \subset B$	12
$\mathcal{I}^N(\mathbf{x}_1, \ldots, \mathbf{x}_n)$	$\{P \in \mathbb{P}^N \ : \ P(\mathbf{x}_i) = 0, \ i = 1, \ldots, n\}$	182
$\mathcal{I}_d^N(\mathbf{x}_1, \ldots, \mathbf{x}_n)$	$\mathcal{I}^N(\mathbf{x}_1, \ldots, \mathbf{x}_n) \cap \mathbb{P}_d^N$	182
J_n	Bessel function of order n	34
$J(\mathbf{x})$	Jacobian matrix	8
J_n^w, $J_n^{w,R}$, $J_n^{w,L}$	Jacobi matrices of w	137
$k \times \mathrm{Q}$, $k^N \times \mathrm{Q}$	Compound integration formulas	123
$k^N \times \mathrm{Q}$	k^N-copy rule of Q	164
k	Kernel function	50
$K_{f,w}$	Absolute condition number w. r. t. integrand	62
$K_{f,\mathrm{Q}}$	Absolute condition number w. r. t. integrand	62
$K_{f,w}^{\mathrm{rel}}$	Relative condition number w. r. t. integrand	62

$K_{f,B}$	Condition number w. r. t. region	66
$\mathbf{K}$	Structure of invariant integration formula	180
lim	Limit	11
lim sup	Limit superior	193
log	Natural logarithm	57
L	Lattice	216
Li	Logarithmic integral	107
$L^{\perp}$	Dual lattice	221
$f \in L_p[a,b]$	$\|f\|^p$ integrable $(1 \leq p < \infty)$ or f bounded $(p = \infty)$ on $[a,b]$	133
L_∞ norm	Maximum (supremum, uniform) norm	117
m	Number of integrands	50
$\max S = \max_{s\in S} s$	Maximum of elements in S	4
$\min S = \min_{s\in S} s$	Minimum of elements in S	4
mod	Modulo function	362
$m(\mathbf{x})$	Unconditional density function	33
M_h	Dimension of V_h	40
M_k	Bound on absolute value of $f^{(k)}$	54
$M_n f$	Sample mean	203
$\mathcal{M}$	Space of model (approximation) functions or	60
	System of measurable subsets	187
$\mathcal{M}_{\mathbb{P}}$	Space of polynomial model functions	167
n	Number of abscissas	61
$n_{1/2}$	Half-performance length	372
N	Dimension of integration problem	30
N_n	Null rule (n points)	303
$\mathbb{N}$	Set of all natural numbers (positive integers)	51
$\mathbb{N}_0$	$\mathbb{N} \cup \{0\}$ (nonnegative integers)	166
$O(\cdot)$, $o(\cdot)$	Landau symbols	41
P, P_d	Polynomial (of degree d)	119
P	Number of processors	354
$P_{\mathbf{X}}$	Probability distribution of $\mathbf{X}$	30
P_D^*	Best approximation polynomial	126
P_α	Worst-case index	227
$\mathbb{P}^N$	Space of all polynomials in N variables	165
$\mathbb{P}_d, \mathbb{P}_d^N$	Space of all polynomials of maximum degree d	120
$\mathbb{P}_d^N(\mathcal{G})$	$\mathcal{G}$-invariant polynomials of maximum degree d	178
$\mathcal{P}$	Partitioning	4
$Q(D)$	Quotient field of integral domain D	93
$\|\mathrm{Q}\|_1$	Sum norm of Q	65
$\mathrm{Q}f, \mathrm{Q}_n f$	Numerical integration operators	55
$\mathrm{Q}_n(f; B_s)$	Integration formula for f over $B_s \subset B$	301

$Q^{(1)} + \cdots + Q^{(L)}$	Compound integration formula	163
$Q^{(1)} \times Q^{(2)}$	Cartesian product rule	159
$Q^{(1)} \times \cdots \times Q^{(K)}$	Cartesian product rule	160
r_∞	Maximum performance	372
R_n	Riemann sum	3
R_n^l, R_n^r	Particular Riemann sums	114
$R[x]$	Ring of polynomials in x with coefficients in R	93
$\mathbb{R}$	Set of all real numbers	3
$\mathbb{R}_+$	Set of all positive real numbers	14
$\mathbb{R}_0^+$	Set of all nonnegative real numbers	31
$\mathcal{R}(B)$	Set of all Riemann integrable functions on B	5
$\mathrm{Re}(z)$	Real part of z	35
sin	Sine function	59
sinc	Sinc function	248
span S	Linear hull of S	159
sup S	Supremum of set S	42
supp f	Support of f	41
$S(f)$	Sample information (of f)	61
S_N	Unit sphere	51
$S_N^{shell}(K_1, K_2)$	Spherical shell	51
$S_N^{\|\cdot\|}(\mathbf{x}_0; \delta)$	N-dimensional sphere with center $\mathbf{x}_0$ and radius δ	15
S_P	Speed-up	389
t	Run-time	372
tanh	Hyperbolic tangent function	151
T_l	Trapezoidal rule with step length $h = (b-a)/l$	146
$T_n(v)$	Displaced trapezoidal rule	218
T_d	Chebyshev polynomials	134
T_l^k	Romberg formulas (with $l2^k + 1$ points)	153
$T_k(f; y)$	Taylor polynomial of f of degree k at y	19
$\mathcal{T}$	Integral transform	33
$\mathcal{T}_L$	Laplace transform	34
$\mathcal{T}_F$	Fourier transform	34
$\mathcal{T}_{H_n}$	nth Hankel transform	34
U_N	Surface of unit sphere S_N	51
$\mathrm{vol}(B)$	Volume of B	4
$v_{\mathbf{y}}$	Green's function	45
V	Space of solution functions	39
V_h	Finite-dimensional subspace of V	40
$V(f; \mathcal{R}, S)$	Set of indistinguishable functions	299
$V(f; \mathcal{P})$, $V^{(N)}(f; \mathcal{P})$	Discrete variation	187
$V(f)$	Variation in the sense of Hardy and Krause	189
$V^{(N)}(f)$	Variation in the sense of Vitali	189

$V(\mathbf{X})$	Variance-covariance matrix of $\mathbf{X}$	31
w	Weight or density function	30
$w(\mathbf{x}\|\theta)$	Conditional density	32
x_i, $\mathbf{x}_i$	Integration abscissa	62
$\mathbf{x}^{\mathbf{d}}$	Multivariate monomial (of degree $\\|\mathbf{d}\\|_1$)	166
$\mathbf{x}^*$	Linear functional	168
Z_2	Particular transformation group	177
$\mathbb{Z}$	Set of all integers (whole numbers)	36
$\mathbb{Z}_0^+$	Set of all nonnegative integers	31

Part I

Introduction

Chapter 1

Mathematical Foundations of Integration

This chapter gives a concise introduction to the fundamental mathematical concepts underlying integration. Practically relevant types of integrals are defined and their most important properties described. The definitions and properties of integrals provided in this chapter are used throughout the remainder of this book to derive computational methods as well as to characterize their salient features.

1.1. Riemann Integrals

Many mathematical disciplines (such as probability theory, statistics, or functional analysis) rely heavily on the concept of *Lebesgue integration*. However, the definition of Lebesgue integrals is inherently nonconstructive, which is why Lebesgue integration is important only to mathematical theory. The concept relevant to computational practice is the more restricted, but constructive concept of *Riemann integration*.

1.1.1. Univariate Riemann Integrals. Riemann integrals of univariate functions $f\colon [a,b] \subset \mathbb{R} \to \mathbb{R}$,

$$\mathrm{I}f = \int_a^b f(x)\,dx,$$

are defined as follows. For any decomposition of the interval $[a,b]$,

$$a = x_1 < x_2 < \cdots < x_n < x_{n+1} = b,$$

into n subintervals and for any choice of n points

$$\xi_i \in [x_i, x_{i+1}], \quad i = 1, 2, \dots, n,$$

a *Riemann sum* is defined by

$$R_n := \sum_{i=1}^{n} (x_{i+1} - x_i) f(\xi_i). \tag{1.1}$$

If all sequences $\{R_n\}$ of Riemann sums with

$$\Delta_n := \max\{x_2 - x_1, \ldots, x_{n+1} - x_n\} \to 0 \qquad \text{as} \quad n \to \infty$$

approach the same limit R, then f is *Riemann integrable* on $[a, b]$ and

$$\int_a^b f(x)\,dx := R. \tag{1.2}$$

A necessary and sufficient condition for the existence of the Riemann integral of a bounded function f is continuity almost everywhere (at all points except for a set of measure zero) on $[a, b]$ (Trench [433]). In particular, if f is continuous on $[a, b]$ or if f is bounded on $[a, b]$ and continuous except for a finite or countably infinite set of arguments, then the Riemann integral (1.2) exists.

1.1.2. Multivariate Riemann Integrals. Riemann integrals of multivariate functions $f\colon B \subset \mathbb{R}^N \to \mathbb{R}$,

$$\mathrm{I}f = \int_B f(\mathbf{x})\,d\mathbf{x},[1]$$

are defined for N-dimensional cubes

$$B = [a_1, b_1] \times [a_2, b_2] \times \cdots \times [a_N, b_N] \subset \mathbb{R}^N$$

as follows. Let

$$a_k = x_1^{(k)} < x_2^{(k)} < \cdots < x_{n_k}^{(k)} < x_{n_k+1}^{(k)} = b_k, \qquad k = 1, 2, \ldots, N,$$

be partitions of the intervals $[a_k, b_k]$, $k = 1, \ldots, N$, into $n_1, \ldots, n_N$ subintervals, respectively. The set of $n = n_1 n_2 \cdots n_N$ subcubes

$$C_{l_1,\ldots,l_N} = [x_{l_1}^{(1)}, x_{l_1+1}^{(1)}] \times [x_{l_2}^{(2)}, x_{l_2+1}^{(2)}] \times \cdots \times [x_{l_N}^{(N)}, x_{l_N+1}^{(N)}],$$
$$l_k = 1, 2, \ldots, n_k, \quad k = 1, 2, \ldots, N,$$

is a partition $\mathcal{P}$ of B. For any choice of n points

$$\mathbf{x}_{l_1,\ldots,l_N} \in C_{l_1,\ldots,l_N} \subset \mathbb{R}^N, \qquad C_{l_1,\ldots,l_N} \in \mathcal{P},$$

a *Riemann sum* is defined by

$$R_n := \sum_{C_{l_1,\ldots,l_N} \in \mathcal{P}} \mathrm{vol}(C_{l_1,\ldots,l_N}) f(\mathbf{x}_{l_1,\ldots,l_N}). \tag{1.3}$$

[1]From this point on, vectors will appear in boldface type; thus $\mathbf{x}$ denotes the vector $\mathbf{x} = (x_1, \ldots, x_N)^\top$.

If all sequences $\{R_n\}$ of Riemann sums with

$$\Delta_n := \max\{\,\mathrm{diam}(C_{l_1,\dots,l_N}) \;:\; C_{l_1,\dots,l_N} \in \mathcal{P}\,\} \to 0 \qquad \text{as} \quad n \to \infty$$

approach the same limit R, then f is Riemann integrable on B and

$$\int_B f(\mathbf{x})\,d\mathbf{x} := R. \tag{1.4}$$

In general, if the integration region B is bounded, then there exists an N-dimensional cube $[a_1, b_1] \times \cdots \times [a_N, b_N]$ such that

$$B \subseteq [a_1, b_1] \times \cdots \times [a_N, b_N].$$

The Riemann integral of f on B is then defined as the Riemann integral of the function $c_B f$ on $[a_1, b_1] \times \cdots \times [a_N, b_N]$, where c_B denotes the *characteristic function* of B:

$$c_B(\mathbf{x}) := \begin{cases} 1 & \text{for} \quad \mathbf{x} \in B, \\ 0 & \text{for} \quad \mathbf{x} \notin B. \end{cases}$$

If the latter Riemann integral does not exist, then f is not Riemann integrable on B.

In general, practically relevant functions f will not be Riemann integrable on B if the Riemann integral of a constant function $f \equiv c$ does not exist. Because

$$\int_B c\,d\mathbf{x} = c\ \mathrm{vol}(B),$$

a constant function $f \equiv c$ is Riemann integrable on B if and only if the volume of the N-dimensional region B exists. Therefore, the definition of multidimensional Riemann integrals is generally restricted to integration regions whose volume is well defined. Such regions B are generally referred to as *(Jordan) measurable*.

A necessary and sufficient condition for the existence of the Riemann integral of a bounded multivariate function f is continuity almost everywhere (at all points except for a set of measure zero) in B (Trench [433]). In particular, if f is bounded and continuous on B, then the Riemann integral (1.4) exists.

1.1.3. Elementary Properties of Riemann Integrals.

Let $\mathcal{R}(B)$ denote the set of all Riemann integrable functions on a bounded, measurable region $B \subset \mathbb{R}^N$.[2] For any constant $c \in \mathbb{R}$ and for any functions $f, g \in \mathcal{R}(B)$, the functions $c\,f$ and $f + g$, defined by the corresponding pointwise operations,

[2]If B is known to be an interval, $B = [a, b]$, the short form $\mathcal{R}[a, b]$ is generally used instead of $\mathcal{R}([a, b])$.

are again Riemann integrable on B, which shows that the set of functions $\mathcal{R}(B)$ is a vector space over the field of real numbers $\mathbb{R}$. Additionally, since the function $|f|: \mathbf{x} \mapsto |f(\mathbf{x})|$ is continuous at $\mathbf{x} \in B$ whenever the function f is continuous at $\mathbf{x}$, $f \in \mathcal{R}(B)$ implies $|f| \in \mathcal{R}(B)$.

The definition $\mathrm{I}: f \mapsto \mathrm{I}f$ establishes a mapping from the vector space $\mathcal{R}(B)$ to the underlying scalar field $\mathbb{R}$. Such a mapping is generally referred to as a *functional*. The functional I on $\mathcal{R}(B)$ is *linear*,

$$\mathrm{I}(f+g) = \mathrm{I}f + \mathrm{I}g \quad \text{and} \quad \mathrm{I}(c\,f) = c\,\mathrm{I}f,$$

and *positive*,

$$\mathrm{I}f \geq 0 \qquad \text{if} \quad f(\mathbf{x}) \geq 0 \quad \text{for all} \quad \mathbf{x} \in B.$$

The latter property implies the inequality

$$|\mathrm{I}f| \leq \mathrm{I}|f| \qquad \text{for all} \quad f \in \mathcal{R}(B).$$

The linearity of integrals also implies that they are *additive* with respect to any partition of B (the region of integration) into pairwise disjoint subregions $B_1, \ldots, B_L$: a function $f: B \to \mathbb{R}$ is Riemann integrable on a region B if and only if f is Riemann integrable on all subregions $B_1, \ldots, B_L$; the Riemann integral of f on B—if it exists—is given by

$$\int_B f(\mathbf{x})\,d\mathbf{x} = \int_{B_1} f(\mathbf{x})\,d\mathbf{x} + \cdots + \int_{B_L} f(\mathbf{x})\,d\mathbf{x}.$$

In particular,

$$\int_a^b f(x)\,dx = \int_a^c f(x)\,dx + \int_c^b f(x)\,dx \tag{1.5}$$

holds for all $a, b, c \in \mathbb{R}$ with $a \leq c \leq b$ and all $f \in \mathcal{R}[a,b]$. In order for equality (1.5) to be valid for arbitrary values $a, b, c \in \mathbb{R}$ and any function $f \in \mathcal{R}[\min\{a,b,c\}, \max\{a,b,c\}]$, the Riemann integral of f on $[a,b]$, $a > b$, is formally defined by

$$\int_a^b f(x)\,dx := -\int_b^a f(x)\,dx.$$

1.1.4. Riemann Integrals and Differentiation. Let the function f be integrable on $[a,b]$. Then, for any $c \in [a,b]$, the function

$$t \mapsto \int_c^t f(x)\,dx$$

is called an *indefinite integral* of f. The various indefinite integrals obtained by choosing different values $c \in [a,b]$ differ only by additive constants.

If f is *continuous* at a point $t_0 \in [a, b]$, then any indefinite integral of f is differentiable at t_0, and

$$\left(\int_c^t f(x)\,dx\right)'(t_0) = f(t_0).$$

Consequently, if f is continuous on $[a, b]$, $f \in C[a, b]$, then any indefinite integral of f is differentiable on $[a, b]$, and its derivative is equal to f.

In general, a differentiable function $F\colon [a, b] \to \mathbb{R}$ with

$$F'(x) = f(x) \qquad \text{for all} \quad x \in [a, b]$$

is called an *antiderivative* of $f\colon [a, b] \to \mathbb{R}$. For any two antiderivatives F and G of the same function f, there is a constant $c \in \mathbb{R}$ such that

$$F(x) = G(x) + c \qquad \text{for all} \quad x \in [a, b].$$

Using this terminology, any indefinite integral of a continuous function f can be said to be an antiderivative of f. This is why the terms "indefinite integral" and "antiderivative" are often used interchangeably.

Although indefinite integrals of general, possibly discontinuous Riemann integrable functions f are *not* necessarily differentiable, it is possible to generalize this result as follows.

THEOREM 1.1.1 (FUNDAMENTAL THEOREM OF INTEGRAL CALCULUS). *Let $f\colon [a, b] \subset \mathbb{R}$ be a Riemann integrable function, and suppose there exists an antiderivative F of f. Then,*

$$\int_s^t f(x)\,dx = F(t) - F(s) \qquad \textit{for all} \quad s, t \in [a, b];$$

i.e., any indefinite integral of f is differentiable on $[a, b]$ and represents an antiderivative of f.

1.1.5. Transformation of Riemann Integrals. The methods described in this section enable one to convert integrals into mathematically equivalent expressions. The purpose of applying such transformations is usually to reformulate integration problems in such a way that their mathematical analysis and/or numerical computation become easier.

Integration by Parts. The method of integration by parts is mathematically formulated as follows.

THEOREM 1.1.2 (INTEGRATION BY PARTS). *Let f and g be differentiable functions defined on the interval $[a, b]$. If both fg' and $(fg)'$ are Riemann integrable on $[a, b]$, then $f'g$ is also Riemann integrable on $[a, b]$, and*

$$\int_a^b f'(x)g(x)\,dx = f(b)g(b) - f(a)g(a) - \int_a^b f(x)g'(x)\,dx.$$

For integration by parts to be applicable, the integrand must be the product of two functions, of which at least one can be integrated easily.[3] The integrand of the transformed integral then factors into the antiderivative of one and into the derivative of the other factor in the original integrand.

Transformation Rule. The transformation rule for univariate integrals can be mathematically formulated as follows.

THEOREM 1.1.3 (UNIVARIATE TRANSFORMATION RULE). *Let ψ denote a differentiable function defined on the interval $[\overline{a}, \overline{b}]$ with image $[a, b] := \psi([\overline{a}, \overline{b}])$. Suppose the derivative ψ' is Riemann integrable on $[a, b]$ and*

$$\psi'(\overline{x}) \neq 0 \qquad \textit{for all} \quad \overline{x} \in [\overline{a}, \overline{b}].$$

Then, a function $f\colon [a, b] \to \mathbb{R}$ is Riemann integrable on $[a, b]$ if and only if the function $(f \circ \psi)\psi'$ is Riemann integrable on $[\overline{a}, \overline{b}]$. The Riemann integral of f on $[a, b]$—if it exists— is given by

$$\int_{\psi(\overline{a})}^{\psi(\overline{b})} f(x)\,dx = \int_{\overline{a}}^{\overline{b}} f(\psi(\overline{x}))\psi'(\overline{x})\,d\overline{x}. \tag{1.6}$$

If ψ' is continuous on $[\overline{a}, \overline{b}]$ and $\psi'(\overline{x}) \neq 0$ for all $\overline{x} \in [\overline{a}, \overline{b}]$, then ψ' is either strictly positive or strictly negative on $[\overline{a}, \overline{b}]$. If $\psi' > 0$, ψ is strictly increasing:

$$a = \psi(\overline{a}) \qquad \text{and} \qquad b = \psi(\overline{b}).$$

If $\psi' < 0$, ψ is strictly decreasing:

$$a = \psi(\overline{b}) \qquad \text{and} \qquad b = \psi(\overline{a}).$$

Using these relationships, equation (1.6) can be written as

$$\int_a^b f(x)\,dx = \int_{\overline{a}}^{\overline{b}} f(\psi(\overline{x}))|\psi'(\overline{x})|\,d\overline{x}.$$

This form of the univariate transformation rule lends itself to a multivariate generalization. Let the mapping $\psi\colon \overline{B} \subset \mathbb{R}^N \to \mathbb{R}^N$ be differentiable and let $\psi_l\colon \overline{B} \to \mathbb{R}$, $l = 1, \ldots, N$, denote the N components of ψ. The *Jacobian matrix* of ψ is defined by

$$J(\overline{\mathbf{x}}) := \begin{pmatrix} \frac{\partial \psi_1}{\partial \overline{x}_1}(\overline{\mathbf{x}}) & \cdots & \frac{\partial \psi_N}{\partial \overline{x}_1}(\overline{\mathbf{x}}) \\ \vdots & & \vdots \\ \frac{\partial \psi_1}{\partial \overline{x}_N}(\overline{\mathbf{x}}) & \cdots & \frac{\partial \psi_N}{\partial \overline{x}_N}(\overline{\mathbf{x}}) \end{pmatrix}.$$

[3]Since the constant 1 is an implicit factor in all integrands, integration by parts can be applied to virtually all integrands.

THEOREM 1.1.4 (MULTIVARIATE TRANSFORMATION RULE). *Let the function*

$$\psi\colon \overline{B} \subset \mathbb{R}^N \to \mathbb{R}^N$$

be a continuously differentiable, one-to-one mapping of $\overline{B}$ *onto* B, *and assume that the Jacobian matrix of* ψ *is nonsingular on* $\overline{B}$; *i.e.,*

$$\det J(\overline{\mathbf{x}}) \neq 0 \quad \textit{for all} \quad \overline{\mathbf{x}} \in \overline{B}.$$

Then, for any Riemann integrable function $f : B \subset \mathbb{R}^N \to \mathbb{R}$ *the composite mapping* $|\det J|(f \circ \psi)$ *is Riemann integrable on* $\overline{B}$ *and*

$$\mathrm{I}f = \int_B f(\mathbf{x})\, d\mathbf{x} = \int_{\overline{B}} |\det J(\overline{\mathbf{x}})| f(\psi(\overline{\mathbf{x}}))\, d\overline{\mathbf{x}}. \tag{1.7}$$

This rule becomes particularly simple for *affine transformations*

$$\psi(\overline{\mathbf{x}}) = A\overline{\mathbf{x}} + b,$$

where $A \in \mathbb{R}^{N\times N}$ and $b \in \mathbb{R}^N$. In this case the Jacobian matrix is independent of $\overline{\mathbf{x}}$,

$$\det J(\overline{\mathbf{x}}) \equiv \det A.$$

The transformation rule therefore simplifies to

$$\mathrm{I}f = \int_B f(\mathbf{x})\, d\mathbf{x} = |\det A| \int_{\overline{B}} f(A\overline{\mathbf{x}} + b)\, d\overline{\mathbf{x}}.$$

Arguably the most frequently used *nonaffine* transformation ψ converts N-dimensional Cartesian coordinates $\mathbf{x} \in \mathbb{R}^N$ into N*-dimensional spherical coordinates*

$$(r, \varphi_1, \ldots, \varphi_{N-2}, \varphi_{N-1})^\top \in \mathbb{R}_0^+ \times [0,\pi] \times \cdots \times [0,\pi] \times [0, 2\pi]$$

according to

$$\begin{aligned}
x_1 &= r \sin\varphi_1 \sin\varphi_2 \cdots \sin\varphi_{N-3} \sin\varphi_{N-2} \sin\varphi_{N-1},\\
x_2 &= r \sin\varphi_1 \sin\varphi_2 \cdots \sin\varphi_{N-3} \sin\varphi_{N-2} \cos\varphi_{N-1},\\
x_3 &= r \sin\varphi_1 \sin\varphi_2 \cdots \sin\varphi_{N-3} \cos\varphi_{N-2},\\
&\ \vdots\\
x_{N-1} &= r \sin\varphi_1 \cos\varphi_2,\\
x_N &= r \cos\varphi_1.
\end{aligned} \tag{1.8}$$

In this case, the relationship

$$|\det J(r, \varphi_1, \ldots, \varphi_{N-1})| = r^{N-1} q(\varphi_1, \ldots, \varphi_{N-1}) \tag{1.9}$$

holds with

$$q(\varphi_1, \ldots, \varphi_{N-1}) = (\sin\varphi_1)^{N-2} (\sin\varphi_2)^{N-3} \cdots (\sin\varphi_{N-3})^2 \sin\varphi_{N-2}. \tag{1.10}$$

Iteration of Integrals. Using the transformation rule, an integration region B can be transformed into another region $\overline{B}$ which has a different shape, but which has the same dimension as B. In contrast, the iteration of integrals aims at decomposing a given integration problem into integrals of reduced dimensions.

THEOREM 1.1.5 (ITERATED INTEGRATION). *Let f denote a Riemann integrable function $f\colon B \subset \mathbb{R}^N \to \mathbb{R}$, and suppose*

$$N = N_1 + N_2, \qquad i.e., \qquad \mathbb{R}^N = \mathbb{R}^{N_1} \times \mathbb{R}^{N_2}.$$

In addition, let the regions B_1 and $B_2(\mathbf{x}_1)$, $\mathbf{x}_1 \in B_1$, be defined by

$$\begin{aligned} B_1 &:= \left\{ \mathbf{x}_1 \in \mathbb{R}^{N_1} \ : \ \left(\{\mathbf{x}_1\} \times \mathbb{R}^{N_2}\right) \cap B \neq \emptyset \right\} \quad \textit{and} \\ B_2(\mathbf{x}_1) &:= \left\{ \mathbf{x}_2 \in \mathbb{R}^{N_2} \ : \ (\mathbf{x}_1, \mathbf{x}_2)^\top \in B \right\}, \quad \mathbf{x}_1 \in B_1. \end{aligned}$$

Then, the integral If *can be* iterated,

$$\int_B f(\mathbf{x})\, d\mathbf{x} = \int_{B_1} \int_{B_2(\mathbf{x}_1)} f(\mathbf{x}_1, \mathbf{x}_2)\, d\mathbf{x}_2\, d\mathbf{x}_1,$$

provided that all inner integrals

$$\int_{B_2(\mathbf{x}_1)} f(\mathbf{x}_1, \mathbf{x}_2)\, d\mathbf{x}_2, \qquad \mathbf{x}_1 \in B_1,$$

exist.

The iteration process can be repeated for both the outer and inner integrals until If is reduced to a composition of one-dimensional integrals,

$$\begin{aligned} \mathrm{I}f &= \int_B f(\mathbf{x})\, d\mathbf{x} \\ &= \int_{B_1} \int_{B_2(x_1)} \int_{B_3(x_1,x_2)} \cdots \int_{B_N(x_1,\ldots,x_{N-1})} f(x_1, \ldots, x_N)\, dx_N \cdots dx_1. \end{aligned}$$

1.2. Improper Integrals

In the definitions of one- and multidimensional Riemann integrals given in Section 1.1, the integration region B is generally required to be bounded. Moreover, a necessary condition for the existence of the Riemann integral of a function f over B is that f is bounded on B. The Riemann integrals introduced in Section 1.1 are also referred to as *proper* Riemann integrals. In certain circumstances, *improper* Riemann integrals enable the definition of integrals for unbounded regions B and/or unbounded integrands f. Such improper integrals are defined as the limits of certain sequences of (proper) Riemann integrals.[4]

[4]Unless explicitly stated otherwise, the term "Riemann integral" should generally be understood to mean *proper* Riemann integral in the remainder of this book.

1.2.1. Univariate Improper Integrals

Unbounded Region. Suppose the function $f : [a, \infty) \to \mathbb{R}$ is Riemann integrable on all bounded subintervals $[a, \beta] \subset [a, \infty)$. Then, the improper integral of f on $[a, \infty)$ is defined by

$$\int_a^\infty f(x)\,dx := \lim_{\beta\to\infty} \int_a^\beta f(x)\,dx,$$

provided this limit exists. An analogous definition holds for the improper integral

$$\int_{-\infty}^b f(x)\,dx.$$

Example (Integral on a Semi-Infinite Interval). Because

$$\lim_{\beta\to\infty} \int_0^\beta \exp(-x)\,dx = \lim_{\beta\to\infty} (1 - \exp(-\beta)) = 1,$$

the improper integral of $f(x) = \exp(-x)$ over $[0, \infty)$ exists and

$$\int_0^\infty \exp(-x)\,dx = 1.$$

Now suppose the function $f : (-\infty, \infty) \to \mathbb{R}$ is Riemann integrable on all bounded subintervals $[\alpha, \beta] \subset (-\infty, \infty)$. Then, the improper integral of f on $(-\infty, \infty)$ is defined by

$$\int_{-\infty}^\infty f(x)\,dx := \int_{-\infty}^\gamma f(x)\,dx + \int_\gamma^\infty f(x)\,dx, \quad \gamma \in \mathbb{R},$$

provided the two improper integrals over the semi-infinite intervals $(-\infty, \gamma]$ and $[\gamma, \infty)$ exist. It is obvious that this definition of the improper integral of f over $(-\infty, \infty)$ is independent of the choice of $\gamma \in \mathbb{R}$.

Example (Integral on an Infinite Interval). Because

$$\lim_{\beta\to\infty} \int_0^\beta \frac{1}{1+x^2}\,dx = \lim_{\beta\to\infty} \arctan(\beta) = \frac{\pi}{2}$$

and because $f(x) = 1/(1 + x^2)$ is an even function, f is improperly integrable over $(-\infty, \infty)$ and

$$\int_{-\infty}^\infty \frac{1}{1+x^2}\,dx = \pi.$$

Unbounded Integrand. Suppose the function $f : [a, b) \to \mathbb{R}$ is Riemann integrable on all subintervals $[a, \beta] \subset [a, b)$, and assume that f is unbounded at the right-hand interval endpoint b, i.e., that f is unbounded in any neighborhood of b.[5] Then, the improper integral of f over $[a, b)$ is defined by

$$\int_a^b f(x)\,dx := \lim_{\beta\to b-} \int_a^\beta f(x)\,dx,$$

provided this limit exists. An analogous definition holds for the improper integral of f over $(a, b]$ if f is unbounded at a.

Example (Integral of an Unbounded Function). For $\lambda > -1$,

$$\lim_{\beta\to 1-} \int_0^\beta (1-x)^\lambda\,dx = \lim_{\beta\to 1-} \frac{1-(1-\beta)^{1+\lambda}}{1+\lambda} = \frac{1}{1+\lambda}.$$

Therefore, the improper integral of $f(x) = (1-x)^\lambda$, $\lambda > -1$, exists and

$$\int_0^1 (1-x)^\lambda\,dx = \frac{1}{1+\lambda}.$$

If a function $f\colon [a, b] \to \mathbb{R}$ is unbounded at an interior point $c \in (a, b)$, but Riemann integrable on all subintervals $[a, \gamma^-] \subset [a, c)$ and $[\gamma^+, b] \subset (c, b]$, then the improper integral of f over $[a, b]$ is defined by

$$\int_a^b f(x)\,dx := \int_a^c f(x)\,dx + \int_c^b f(x)\,dx,$$

provided the two improper integrals over $[a, c)$ and $(c, b]$ exist.

1.2.2. Multivariate Improper Integrals. Let S_f denote the set of singularities of the function $f : B \subseteq \mathbb{R}^N \to \mathbb{R}$ and suppose f is Riemann integrable over every measurable subset $B_s \subseteq B$ for which the *closure* of B_s, $\operatorname{cl} B_s$, and S_f are disjoint: $(\operatorname{cl} B_s) \cap S_f = \emptyset$. Assume that B is unbounded and/or that f is unbounded, i.e., that S_f is nonempty. Let $B_1 \subset B_2 \subset \cdots$ denote an arbitrary sequence of measurable sets such that

$$(\operatorname{cl} B_l) \cap S_f = \emptyset, \quad l = 1, 2, 3, \ldots,$$

and

$$B \setminus S_f = B_1 \cup B_2 \cup B_3 \cup \cdots.$$

If, for all such sequences $\{B_l\}$, the sequences $\{R_l\}$ defined by

$$R_l := \mathrm{I}(f; B_l) := \int_{B_l} f(\mathbf{x})\,d\mathbf{x}$$

[5]Generally, if a function $f\colon B \subseteq \mathbb{R}^N \to \mathbb{R}$ is unbounded at a limit point $\mathbf{x}_0$ of B, then $\mathbf{x}_0$ is said to be a *singularity* or *singular point* of f.

approach the same limit R, then f is improperly Riemann integrable over B and

$$\int_B f(\mathbf{x})\,d\mathbf{x} := R.$$

If $f \geq 0$ on B, then f can be shown to be improperly Riemann integrable over B if and only if there exists a *single* sequence $B_1, B_2, \ldots$ such that the limit

$$R = \lim_{l \to \infty} \mathrm{I}(f; B_l)$$

exists (Trench [433]).

It should be observed that, for $N = 1$, the definition of general N-dimensional improper Riemann integrals does *not* coincide with—indeed, is more restrictive than—the definition of one-dimensional improper Riemann integrals (see, for instance, Trench [433]).

Example (Multivariate Improper Integrals). Consider the integral

$$\mathrm{I}f = \int_{\mathbb{R}^N} \exp(-\|\mathbf{x}\|_1)\,d\mathbf{x}.$$

The integration region $B = \mathbb{R}^N$ is unbounded. The sequence of subregions

$$B_l := \{\, \mathbf{x} \;:\; \|\mathbf{x}\|_\infty \leq l \,\}$$

satisfies the aforementioned conditions. Because

$$\mathrm{I}(f; B_l) = \int_{B_l} \exp(-\|\mathbf{x}\|_1)\,d\mathbf{x} = \prod_{k=1}^{N} \int_{-l}^{l} \exp(-|x_k|)\,dx_k = 2^N \left(1 - \exp(-l)\right)^N,$$

the limit $R = \lim_{l \to \infty} \mathrm{I}(f; B_l) = 2^N$ can be derived. Because

$$\exp(-\|\mathbf{x}\|_1) > 0 \qquad \text{for all} \quad \mathbf{x} \in \mathbb{R}^N,$$

the improper Riemann integral of $f(\mathbf{x}) = \exp(-\|\mathbf{x}\|_1)$ over $B = \mathbb{R}^N$ exists and

$$\int_{\mathbb{R}^N} \exp(-\|\mathbf{x}\|_1)\,d\mathbf{x} = 2^N.$$

1.2.3. Properties and Transformations of Improper Integrals.

The elementary properties of proper Riemann integrals (see Section 1.1.3), in particular the linearity and positivity of the integral operator, also hold for improper Riemann integrals. Additionally, the transformations applicable to proper Riemann integrals (see Section 1.1.5) also apply to their improper counterparts. The only notable exception to this correspondence is that the existence of the improper Riemann integral of a *univariate* function f does *not* necessarily imply that the absolute value function $|f|$ is improperly Riemann integrable as well.

1.3. Cauchy Principal Value Integrals

A number of applications require the computation of integrals over unbounded integration regions and/or for unbounded integrands which do not exist as proper or improper Riemann integrals, but only more generally as Cauchy principal value integrals. Cauchy principal value integrals occur, for instance, in certain classes of integral equations (see Golberg [177] and Hackbusch [191]) and in certain integral transforms, such as the Hilbert transform (see Jerri [219]).

1.3.1. Univariate Cauchy Principal Value Integrals

Unbounded Integrand. Suppose the function $f\colon [a,b] \to \mathbb{R}$ is unbounded in any neighborhood of some interior point x_0, $a < x_0 < b$. Assume that, for any $\delta \in \mathbb{R}_+$, the function f is Riemann integrable over the subintervals $[a, x_0 - \delta] \subset [a, x_0)$ and $[x_0 + \delta, b] \subset (x_0, b]$. Then the Cauchy principal value integral of f over $[a,b]$ is defined by

$$\text{(1.11)} \qquad ⨍_a^b f(x)\,dx := \lim_{\delta \to 0+} \left(\int_a^{x_0-\delta} f(x)\,dx + \int_{x_0+\delta}^b f(x)\,dx \right),$$

provided that this limit exists. Note that (1.11) is well defined whenever f is improperly Riemann integrable on $[a,b]$. In this case, x_0 is said to be a *weak* singularity of f; otherwise, x_0 is said to be a *strong* singularity of f.

Example (Cauchy Principal Value Integral). Because

$$\lim_{\delta \to 0+} \int_\delta^1 \left(1 + \frac{1}{x}\right) dx = \lim_{\delta \to 0+} (1 - \delta - \log(\delta)) = \infty,$$

$f(x) = 1 + x^{-1}$ is not improperly Riemann integrable over $(0,1]$. Similarly, it can be shown that the improper Riemann integral of f over $[-1,0)$ does not exist. Nevertheless, since $f - 1$ is an odd function, the Cauchy principal value integral of f over $[-1,1]$ is well defined:

$$⨍_{-1}^1 \left(1 + \frac{1}{x}\right) dx = \lim_{\delta \to 0+} \left(\int_{-1}^{-\delta} \left(1 + \frac{1}{x}\right) dx + \int_\delta^1 \left(1 + \frac{1}{x}\right) dx \right) = 2.$$

Unbounded Region. Suppose the function $f\colon \mathbb{R} \to \mathbb{R}$ is Riemann integrable over all finite subintervals $[a,b] \subset \mathbb{R}$. Then the Cauchy principal value integral of f over $\mathbb{R}$ is defined by

$$⨍_{-\infty}^{\infty} f(x)\,dx := \lim_{\beta \to \infty} \int_{-\beta}^{\beta} f(x)\,dx,$$

provided that this limit exists.

Example (Cauchy Principal Value Integral). Because

$$\lim_{\beta\to\infty}\int_0^{\beta}\frac{x}{1+x^2}\,dx=\lim_{\beta\to\infty}\frac{1}{2}\log(1+\beta^2)=\infty,$$

$f(x)=x/(1+x^2)$ is not improperly Riemann integrable over $(-\infty,\infty)$. However, since f is an odd function,

$$⨍_{-\infty}^{\infty}\frac{x}{1+x^2}\,dx=\lim_{\beta\to\infty}\int_{-\beta}^{\beta}\frac{x}{1+x^2}\,dx=\lim_{\beta\to\infty}0=0.$$

Note that by changing the independent variable appropriately, Cauchy principal value integrals over $\mathbb{R}$ can be transformed into Cauchy principal value integrals of unbounded functions on finite intervals.

1.3.2. Multivariate Cauchy Principal Value Integrals. Suppose the function $f: B\subset\mathbb{R}^N\to\mathbb{R}$ is unbounded in any neighborhood of an interior point $\mathbf{x}_0$ of a bounded, measurable region B. Let $S_N^{\|\cdot\|}(\mathbf{x}_0;\delta)$ denote the sphere with center $\mathbf{x}_0$ and radius δ with respect to a norm $\|\cdot\|$ on $\mathbb{R}^N$. Assume that, for any $\delta\in\mathbb{R}_+$, the function f is Riemann integrable over the subregion $B\setminus S_N^{\|\cdot\|}(\mathbf{x}_0;\delta)$. Then the Cauchy principal value integral of f over B with respect to the norm $\|\cdot\|$ is defined by

$$(1.12)\qquad ⨍_B f(\mathbf{x})\,d\mathbf{x}:=\lim_{\delta\to0+}\int_{B\setminus S_N^{\|\cdot\|}(\mathbf{x}_0;\delta)}f(\mathbf{x})\,d\mathbf{x},$$

provided that this limit exists.[6]

For $N=1$ and $B=[a,b]$, this definition of N-dimensional Cauchy principal value integrals coincides with the definition (1.11) of univariate Cauchy principal value integrals: for the absolute-value norm $\|x\|=|x|$ on $\mathbb{R}^1=\mathbb{R}$, the right-hand side of (1.12) is identical to the right-hand side of (1.11) because $S_1^{|\cdot|}(x_0;\delta)=(x_0-\delta,x_0+\delta)$. Since in $\mathbb{R}$ any norm differs from the absolute value norm only by a multiplicative constant, this identity holds for any norm on $\mathbb{R}$.

For $N>1$, it can be shown that the *existence* of the right-hand side limit in (1.12) is *independent* of the particular norm $\|\cdot\|$ chosen, provided that the singularity of f at $\mathbf{x}_0$ is at most of order N (Monegato [306]). The *value* of the Cauchy principal value integral, however, *does* depend on the norm. Most practical Cauchy principal value integrals are calculated with respect to the Euclidean norm $\|\cdot\|_2$ or the maximum norm $\|\cdot\|_\infty$.

[6]There are more general definitions of multivariate Cauchy principal value integrals; see Monegato [306]. For the purposes of this book, however, the definition (1.12) is sufficient.

1.3.3. Properties and Transformations of Cauchy Principal Value Integrals. If a function f is properly or improperly Riemann integrable on a region $B \subseteq \mathbb{R}^N$, then the corresponding Cauchy principal value integral is well defined and equal to the Riemann integral. The converse does not generally hold. Accordingly, Cauchy principal value integrals can be regarded as generalizations of Riemann integrals.

It can easily be seen that the Cauchy principal value integral is, as is the (proper or improper) Riemann integral, a linear, positive functional. Similarly, the Cauchy principal value integral for unbounded functions f is additive with respect to any partition of B into disjoint subregions $B_1, \ldots, B_L$.[7] Furthermore, the method of integration by parts is, in a properly modified form, applicable to univariate Cauchy principal value integrals. However, in order for the univariate transformation rule to hold, more restrictive conditions must be imposed on the transformation function ψ (see, for instance, Hackbusch [191]).

THEOREM 1.3.1 (TRANSFORMATION RULE FOR UNIVARIATE CAUCHY PRINCIPAL VALUE INTEGRALS). *Let ψ be a continuously differentiable function defined on the interval $[\overline{a}, \overline{b}] \subset \mathbb{R}$ with $\psi'(\overline{x}) > 0$ for all $\overline{x} \in [\overline{a}, \overline{b}]$. Suppose the function*

$$f\colon [\psi(\overline{a}), \psi(\overline{b})] \to \mathbb{R}$$

is Riemann integrable on

$$[\psi(\overline{a}), x_0 - \delta] \subset [\psi(\overline{a}), x_0) \qquad \text{and} \qquad [x_0 + \delta, \psi(\overline{b})] \subset (x_0, \psi(\overline{b})]$$

for all $\delta \in \mathbb{R}_+$. Furthermore, assume that the order of singularity in $f(x)$ at x_0 is at most 1, i.e., that the function $f(x)(x - x_0)$ is bounded in a neighborhood of x_0. Then, the Cauchy principal value integral of f exists if and only if the principal value integral of the function $(f \circ \psi)\psi'$ exists. The Cauchy principal value integral of f—if it exists—is given by

$$(1.13) \qquad \fint_{\psi(\overline{a})}^{\psi(\overline{b})} f(x)\, dx = \fint_{\overline{a}}^{\overline{b}} f(\psi(\overline{x}))\psi'(\overline{x})\, d\overline{x}.$$

In practice, most Cauchy principal value integrals are (or can easily be) represented in the form

$$(1.14) \qquad \mathrm{I}_{\mathbf{y}} f := \fint_B \frac{f(\mathbf{x})}{\|\mathbf{x} - \mathbf{y}\|_2^N}\, d\mathbf{x},$$

where $f(\mathbf{y}) \neq 0$, $\mathbf{y} \in B \setminus \partial B$ being the point at which the integrand is singular. In this case, a sufficient (not necessary) condition for the existence of the

[7]It is assumed that the singularity $\mathbf{x}_0$ is located in the interior of one of the subregions.

Cauchy principal value integral (1.14) is that f is *Hölder continuous* at $\mathbf{y}$, i.e., that constants C and $\lambda > 0$ exist such that

$$|f(\mathbf{x}) - f(\mathbf{y})| \leq C\|\mathbf{x} - \mathbf{y}\|_2^\lambda \tag{1.15}$$

for all $\mathbf{x}$ in a neighborhood of $\mathbf{y}$ (Hackbusch [191]). For $N = 1$ and $B = [a, b]$, specifically, the mapping $y \mapsto \mathrm{I}_y f$ is called the *Hilbert transform* of f.

1.4. Hadamard Finite Part Integrals

Hadamard finite part integrals, also known as *hypersingular integrals*, were introduced as early as 1923 by Hadamard [196], and the first applications of hypersingular integrals (in aerodynamics) date back to the early 1950s (Mangler [293]). Nevertheless, widespread interest in hypersingular integrals has arisen only recently with their emergence in boundary integral methods. This is why hypersingular integrals are, in contrast to proper and improper Riemann integrals as well as Cauchy principal value integrals, not part of standard courses in calculus and higher level mathematics. The following section therefore attempts to provide an intuitive introduction to finite part integrals and assumes no prior knowledge of the subject.

1.4.1. Univariate Finite Part Integrals. Consider the integral

$$\mathrm{I}_y f := \int_a^b \frac{f(x)}{(x-y)^{\alpha+1}}\, dx, \qquad y \in [a, b]. \tag{1.16}$$

For the choice $y = a$, $x - y$ is nonnegative on $[a, b]$, and thus the denominator $(x - y)^{-\alpha-1}$ in (1.16) is defined on $[a, b]$ for any $\alpha \in \mathbb{R}$. In contrast, if y is chosen such that $a < y \leq b$ then $x - y$ is negative on $[a, y)$, and thus $(x - y)^{-\alpha-1}$ is well defined only for $\alpha \in \mathbb{N}$. In the following discussion it is tacitly assumed that the parameters y and α are chosen appropriately.

If α is negative, then $\mathrm{I}_y f$ exists as a (possibly improper) Riemann integral provided f is a properly or improperly Riemann integrable function on $[a, b]$ and bounded in a neighborhood of $x = y$. For nonnegative values of α, on the other hand, $\mathrm{I}_y f$ does not generally exist in the Riemann sense for Riemann integrable functions f. Nevertheless, in applications such as boundary integral methods, there is a need to assign a meaning to integrals of type (1.16) for nonnegative values of α, i.e., to define a new kind of integral for which (1.16) exists for nonnegative α. The integral to be defined should be an extension of the usual Riemann integral: whenever the integral (1.16) exists in the Riemann sense, it should also exist in the newly defined sense and the two kinds of integrals should coincide. Furthermore, the integral to be defined is supposed to inherit the fundamental properties, most importantly, additivity and linearity, of Riemann integrals.

Case I: $f \equiv 1$ and $\alpha = 0$. To begin with, consider the simplest integrand $f(x) \equiv 1$. For $\alpha = 0$ and $y = a$, the divergent integral $\mathrm{I}_y f$ can be written as

$$\mathrm{I}_y f = \int_y^b \frac{1}{x-y}\, dx = \lim_{\delta \to 0+} \int_{y+\delta}^b \frac{1}{x-y}\, dx = \log(b-y) - \lim_{\delta \to 0+} \log \delta.$$

Hence, the divergent integral $\mathrm{I}_y f$ can be represented as the sum of the finite term $\log(b-y)$ and the divergent term $\lim_{\delta \to 0+} \log \delta$. The corresponding *finite part integral* is defined by taking only the *finite part* of this sum:

$$\text{(1.17)} \qquad \;\;\;\; \mathchoice{}{}{}{} \int\!\!\!\!\!\!\!=_y^b \frac{1}{x-y}\, dx := \log(b-y).$$

For $y = b$, a similar derivation leads to the definition

$$\text{(1.18)} \qquad \int\!\!\!\!\!\!\!=_a^y \frac{1}{x-y}\, dx := -\log(y-a).$$

In order for the finite part integral to be additive, the finite part integral for $a < y < b$ has to be defined by dividing the interval $[a, b]$ into the subintervals $[a, y]$ and $[y, b]$ and applying (1.17) and (1.18), respectively:

$$\text{(1.19)} \qquad \int\!\!\!\!\!\!\!=_a^b \frac{1}{x-y}\, dx := \log\left(\frac{b-y}{y-a}\right).$$

Case II: $f \equiv 1$ and $\alpha > 0$. For $\alpha > 0$ and $y = a$, the representation

$$\begin{aligned} \mathrm{I}_y f &= \int_y^b \frac{1}{(x-y)^{\alpha+1}}\, dx \\ &= \lim_{\delta \to 0+} \int_{y+\delta}^b \frac{1}{(x-y)^{\alpha+1}}\, dx = -\frac{1}{\alpha(b-y)^\alpha} + \lim_{\delta \to 0+} \frac{1}{\alpha \delta^\alpha} \end{aligned}$$

shows that $\mathrm{I}_y f$ can be split into the finite part $-\alpha^{-1}(b-y)^{-\alpha}$ and the divergent part $\alpha^{-1} \lim_{\delta \to 0+} \delta^{-\alpha}$. Following the same argument as for $\alpha = 0$, the following definition is obtained:

$$\text{(1.20)} \quad \int\!\!\!\!\!\!\!=_a^b \frac{1}{(x-y)^{\alpha+1}}\, dx := \begin{cases} -\dfrac{1}{\alpha(b-y)^\alpha} & \text{for} \quad y = a, \\ \dfrac{1}{\alpha(y-a)^\alpha} & \text{for} \quad y = b, \\ -\dfrac{1}{\alpha}\left(\dfrac{1}{(b-y)^\alpha} - \dfrac{1}{(y-a)^\alpha}\right) & \text{for} \quad a < y < b. \end{cases}$$

Case III: General f. Let f be a general Riemann integrable function defined on $[a, b]$ which is $k := \lceil \alpha \rceil$-times continuously differentiable in a neighborhood of y. If α is an integer, assume additionally that $f^{(k)}$ is Hölder continuous at y. Let $T_k(f; y)$ denote the Taylor polynomial of f of degree k at y:

$$T_k(f; y)(x) := \sum_{l=0}^{k} \frac{f^{(l)}(y)}{l!}(x - y)^l.$$

Since the finite part integral is supposed to be linear, the relationship

$$\int_a^b \frac{f(x)}{(x-y)^{\alpha+1}}\,dx = \int_a^b \frac{f(x) - T_k(f; y)(x)}{(x-y)^{\alpha+1}}\,dx + \sum_{l=0}^{k} \frac{f^{(l)}(y)}{l!} \int_a^b \frac{1}{(x-y)^{\alpha+1-l}}\,dx$$

must be satisfied for any useful definition of the finite part integral. The first integral,

$$\int_a^b \frac{f(x) - T_k(f; y)(x)}{(x-y)^{\alpha+1}}\,dx,$$

on the right-hand side can easily be seen to exist in the Riemann sense, while the remaining finite part integrals,

$$\int_a^b \frac{1}{(x-y)^{\alpha+1-l}}\,dx, \qquad l = 0, 1, \ldots, k,$$

are already well defined by (1.19) and (1.20). Since the finite part integral is supposed to be an extension of the Riemann integral,

$$(1.21) \qquad \int_a^b \frac{f(x)}{(x-y)^{\alpha+1}}\,dx \quad := \quad \int_a^b \frac{f(x) - T_k(f; y)(x)}{(x-y)^{\alpha+1}}\,dx + \sum_{l=0}^{k} \frac{f^{(l)}(y)}{l!} \int_a^b \frac{1}{(x-y)^{\alpha+1-l}}\,dx$$

is the only coherent way of defining the finite part integral for general, sufficiently smooth functions f.

Example (Univariate Finite Part Integral). Consider the finite part integral

$$\mathrm{I}f = \int_0^2 \frac{1 + x + x^2}{x^2}\,dx.$$

Since $\alpha = 1$ and the Taylor polynomial of degree $k = \lceil \alpha \rceil = 1$ at $y = 0$ consists of the terms of order $\leq k$ in the power representation of f, one obtains

$$\mathrm{I}f = \int_0^2 1\,dx + \int_0^2 \frac{1}{x^2}\,dx + \int_0^2 \frac{1}{x}\,dx = 2 - \frac{1}{2} + \log 2 = \frac{3}{2} + \log 2.$$

Note that definition (1.21) is still unsatisfactory in that it assumes that the singular factor in the integrand has the particular form $1/(x - y)^{\alpha+1}$. A definition of the finite part integral which is valid for more general singular factors is given in Section 1.4.3.

1.4.2. Properties of Univariate Hypersingular Integrals. The definition (1.21) implies that the finite part integral is a linear functional in the argument function,

$$(1.22)\quad \fint_a^b \frac{cf(x)+dg(x)}{(x-y)^{\alpha+1}}\,dx = c\fint_a^b \frac{f(x)}{(x-y)^{\alpha+1}}\,dx + d\fint_a^b \frac{g(x)}{(x-y)^{\alpha+1}}\,dx,$$

whenever the two right-hand side finite part integrals exist. From (1.22) it follows that the finite part integral is additive with respect to any partition of the integration interval $[a,b]$ into disjoint subintervals.[8]

An interesting property of finite part integrals is that they are the derivatives of other finite part integrals as well as of the Cauchy principal value integral.

THEOREM 1.4.1 (DERIVATIVES OF FINITE PART INTEGRALS). *For $a < y < b$ and $\alpha \in \mathbb{N}$ as well as for $y = a$ and $\alpha \notin \mathbb{N}$, the equality*

$$(1.23)\qquad \frac{d}{dy}\fint_a^b \frac{f(x)}{(x-y)^{\alpha}}\,dx = \alpha\fint_a^b \frac{f(x)}{(x-y)^{\alpha+1}}\,dx$$

holds. Furthermore, the finite part integrals for $a < y < b$ and $\alpha \in \mathbb{N}$ can be derived from the corresponding Cauchy principal value integral as follows:

$$(1.24)\qquad \fint_a^b \frac{f(x)}{(x-y)^{\alpha+1}}\,dx = \frac{1}{\alpha!}\frac{d^{\alpha}}{dy^{\alpha}}\fint_a^b \frac{f(x)}{x-y}\,dx.$$

Additionally, the following special form of the integration-by-parts rule is applicable to finite part integrals.

THEOREM 1.4.2 (INTEGRATION BY PARTS FOR FINITE PART INTEGRALS). *For $a < y < b$, it holds that*

$$(1.25)\quad \fint_a^b \frac{f(x)}{(x-y)^{\alpha+1}}\,dx = -\frac{1}{\alpha}\left(\frac{f(b)}{(b-y)^{\alpha}} - \frac{f(a)}{(y-a)^{\alpha}}\right) + \frac{1}{\alpha}\fint_a^b \frac{f'(x)}{(x-y)^{\alpha}}\,dx.$$

The transformation rule, however, is valid only for *linear* functions ψ.

THEOREM 1.4.3 (LINEAR TRANSFORMATION RULE FOR FINITE PART INTEGRALS). *Let ψ be the linear transformation mapping the interval $[\overline{a},\overline{b}] \subset \mathbb{R}$ onto $[a,b] \subset \mathbb{R}$. Then, for $a < y < b$ and $\alpha \in \mathbb{N}$ as well as for $y = a$ and $\alpha \notin \mathbb{N}$, the relationship*

$$(1.26)\qquad \fint_a^b \frac{f(x)}{(x-y)^{\alpha+1}}\,dx = \left(\frac{\overline{b}-\overline{a}}{b-a}\right)^{\alpha}\fint_{\overline{a}}^{\overline{b}} \frac{f(\psi(\overline{x}))}{(\overline{x}-\overline{y})^{\alpha+1}}\,d\overline{x}$$

holds with $y = \psi(\overline{y})$.

[8] Here and in the following, the finite part integral on a subinterval which does not contain the singularity y is simply defined as the corresponding Riemann integral.

1.4.3. Multivariate Hypersingular Integrals. Consider the integral

$$\text{(1.27)} \qquad \mathrm{I}_{\mathbf{y}} f := \int_B k(\mathbf{x}, \mathbf{y}) f(\mathbf{x})\, d\mathbf{x}, \qquad \mathbf{y} \in B,$$

where the function $\mathbf{x} \to k(\mathbf{x}, \mathbf{y})$ is smooth on B with the exception of a *singularity of order* $\alpha + N$ at $\mathbf{x} = \mathbf{y}$. That is, the function

$$\mathbf{x} \to \|\mathbf{x} - \mathbf{y}\|_2^{\alpha+N} k(\mathbf{x}, \mathbf{y})$$

is bounded in a neighborhood of $\mathbf{x} = \mathbf{y}$, but does *not* converge to $\mathbf{0}$ as $\mathbf{x} \to \mathbf{y}$.

If α is negative, then $\mathrm{I}_{\mathbf{y}} f$ exists as a (possibly improper) Riemann integral provided f is properly or improperly Riemann integrable on B and bounded in a neighborhood of $\mathbf{x} = \mathbf{y}$. For $\alpha \geq 0$, on the other hand, $\mathrm{I}_{\mathbf{y}} f$ does not generally exist in the Riemann sense. This section introduces an extension of the Riemann integral which assigns a meaningful and useful value to (1.27) for nonnegative values of α. As in Section 1.4.1, this generalized integral is obtained by omitting the divergent terms of the asymptotic expansion of $\mathrm{I}_{\mathbf{y}}(f; B \setminus S_N^{\|\cdot\|_2}(\mathbf{y}; \delta))$ for $\delta \to 0+$.

Suppose f is $k := \lceil \alpha \rceil$-times continuously differentiable in a neighborhood of $\mathbf{y}$. If α is an integer, assume additionally that the kth-order partial derivatives of f are Hölder continuous at $\mathbf{y}$. Let $T_k(f; \mathbf{y})$ denote the Taylor polynomial of f of degree k at y:

$$\begin{aligned}
\text{(1.28)}\ T_k(f; \mathbf{y})(\mathbf{x}) &= f(\mathbf{y}) + \sum_{i_1=1}^{N} \frac{\partial f}{\partial x_{i_1}}(\mathbf{y})(x_{i_1} - y_{i_1}) \\
&+ \sum_{i_1, i_2=1}^{N} \frac{1}{2!} \frac{\partial^2 f}{\partial x_{i_1} \partial x_{i_2}}(\mathbf{y})(x_{i_1} - y_{i_1})(x_{i_2} - y_{i_2}) \\
+ \cdots &+ \sum_{i_1, \ldots, i_k=1}^{N} \frac{1}{k!} \frac{\partial^k f}{\partial x_{i_1} \cdots \partial x_{i_k}}(\mathbf{y})(x_{i_1} - y_{i_1}) \cdots (x_{i_k} - y_{i_k}).
\end{aligned}$$

Since the finite part integral is supposed to be linear, the relationship

$$\begin{aligned}
⨍_B k(\mathbf{x}, \mathbf{y}) f(\mathbf{x})\, d\mathbf{x} &= ⨍_B k(\mathbf{x}, \mathbf{y})(f(\mathbf{x}) - T_k(f; \mathbf{y})(\mathbf{x}))\, d\mathbf{x} \\
&\quad + ⨍_B k(\mathbf{x}, \mathbf{y}) T_k(f; \mathbf{y})(\mathbf{x})\, d\mathbf{x}
\end{aligned}$$

must be satisfied for any useful definition of the finite part integral. The first integral on the right-hand side,

$$⨍_B k(\mathbf{x}, \mathbf{y})(f(\mathbf{x}) - T_k(f; \mathbf{y})(\mathbf{x}))\, d\mathbf{x},$$

can easily be seen to exist in the Riemann sense, whereas for the second integral

$$\begin{aligned} \mathrm{I}_{\mathbf{y};\delta} T_k(f;\mathbf{y}) &:= \mathrm{I}_{\mathbf{y}}(T_k(f;\mathbf{y}); B \setminus S_N^{\|\cdot\|_2}(\mathbf{y};\delta)) \\ &= \int_{B\setminus S_N^{\|\cdot\|_2}(\mathbf{y};\delta)} k(\mathbf{x},\mathbf{y}) T_k(f;\mathbf{y})(\mathbf{x})\, d\mathbf{x} \end{aligned}$$

exists for any $\delta \in \mathbb{R}_+$, but does not generally converge as $\delta \to 0+$. The behavior of $\mathrm{I}_{\mathbf{y};\delta}T_k(f;\mathbf{y})$ as $\delta \to 0+$ can be analyzed by using the N-dimensional spherical coordinates defined in (1.8) to parameterize $\mathbf{x}-\mathbf{y}$. $T_k(f;\mathbf{y})$ can then be expressed more concisely in the form

$$\text{(1.29)} \qquad \begin{aligned} T_k(f;\mathbf{y})(r,\varphi_1,\ldots,\varphi_{N-1}) &= p_0(\varphi_1,\ldots,\varphi_{N-1}) \\ &\quad + p_1(\varphi_1,\ldots,\varphi_{N-1})r \\ &\quad + p_2(\varphi_1,\ldots,\varphi_{N-1})r^2 \\ &\quad + \cdots + p_k(\varphi_1,\ldots,\varphi_{N-1})r^k, \end{aligned}$$

where the $p_j(\varphi_1,\ldots,\varphi_{N-1})$, $j = 0,1,\ldots,k$, are trigonometric polynomials in $\varphi_1,\ldots,\varphi_{N-1}$. Furthermore, for the sake of simplicity, assume that the boundary of B has an explicit representation in the angular coordinates $\varphi_1,\ldots,\varphi_{N-1}$, i.e., that there is a function

$$R : \Phi \subseteq [0,\pi] \times \cdots \times [0,\pi] \times [0,2\pi] \to \mathbb{R}_0^+$$

such that

$$\partial B = \{(r,\varphi_1,\ldots,\varphi_{N-1}) : (\varphi_1,\ldots,\varphi_{N-1})^{\mathrm{T}} \in \Phi,\ r = R(\varphi_1,\ldots,\varphi_{N-1})\}.$$

Case I: $k(\mathbf{x},\mathbf{y}) = r^{-\alpha-N} g(\varphi_1,\ldots,\varphi_{N-2},\varphi_{N-1})$. The function g is assumed to be smooth, π-periodic in $\varphi_1,\ldots,\varphi_{N-2}$, and 2π-periodic in φ_{N-1}. Using the representation (1.29) of the Taylor polynomial $T_k(f;\mathbf{y})$ as well as the relationships (1.9) and (1.10), the formula

$$\mathrm{I}_{\mathbf{y};\delta}T_k(f;\mathbf{y}) = \sum_{j=0}^{k} \int_\Phi (gqp_j)(\varphi_1,\ldots,\varphi_{N-1}) h_j(\delta,\varphi_1,\ldots,\varphi_{N-1}) d\varphi_1 \cdots d\varphi_{N-1}$$

with

$$h_j(\delta,\varphi_1,\ldots,\varphi_{N-1}) = \int_\delta^{\max\{\delta, R(\varphi_1,\ldots,\varphi_{N-1})\}} \frac{1}{r^{\alpha-j+1}}\, dr$$

can be derived. Clearly, $h_j(\delta,\varphi_1,\ldots,\varphi_{N-1}) = 0$ if $\delta \geq R(\varphi_1,\ldots,\varphi_{N-1})$, while otherwise

$$h_j(\delta,\varphi_1,\ldots,\varphi_{N-1}) = \begin{cases} \log R(\varphi_1,\ldots,\varphi_{N-1}) - \log\delta & \text{for } j = \alpha, \\ (j-\alpha)^{-1}(R(\varphi_1,\ldots,\varphi_{N-1})^{j-\alpha} - \delta^{j-\alpha}) & \text{for } j \neq \alpha \end{cases}$$

holds. As a result, the following asymptotic expansion can be derived:

$$\text{(1.30)}\quad \mathrm{I}_{\mathbf{y};\delta}T_k(f;\mathbf{y}) = a_0 + a_1 \log\delta + a_2\delta^{k-1-\alpha} + \cdots + a_{k+1}\delta^{-\alpha} + o(1) \quad \text{as} \quad \delta \to 0+,$$

where $a_0, a_1, \ldots a_{k+1} \in \mathbb{R}$ are constants and $o(1)$ indicates a term which converges to zero as $\delta \to 0+$. Now, the finite part integral of $T_k(f;\mathbf{y})$ is obtained by omitting the divergent terms in (1.30) and taking the limit of the remaining finite part as $\delta \to 0+$:

$$\text{(1.31)}\qquad ⨎_B k(\mathbf{x},\mathbf{y})T_k(f;\mathbf{y})(\mathbf{x})\,d\mathbf{x} := a_0.$$

Case II: General k. The asymptotic expansion (1.30) motivates the following definition.

DEFINITION 1.4.1 (FINITE PART OF A REAL FUNCTION). *Let the function $I : \delta \mapsto I(\delta)$ be defined for all $\delta \in (0, \delta_0)$, $\delta_0 > 0$. Assume that the following asymptotic expansion holds:*

$$\text{(1.32)}\quad I(\delta) = a_0 + a_1 \log\delta + a_2\delta^{-\alpha_1} + \cdots + a_{k+1}\delta^{-\alpha_k} + o(1) \quad \textit{as} \quad \delta \to 0+,$$

where $a_0, a_1, \ldots, a_{k+1}$ and $0 < \alpha_1 < \alpha_2 < \cdots < \alpha_k$ are real constants. Then, the finite part, p.f. $I(\delta)$, *of I is defined by*

$$\text{(1.33)}\qquad \text{p.f.}\, I(\delta) := a_0.$$

Now, if the presuppositions of Definition 1.4.1 hold for the function

$$\text{(1.34)}\qquad I : \delta \mapsto \int_{B\setminus S_N^{\|\cdot\|_2}(\mathbf{y};\delta)} k(\mathbf{x},\mathbf{y})f(\mathbf{x})\,d\mathbf{x},$$

then the following general definition of the finite part integral can be given:

$$\text{(1.35)}\qquad ⨎_B k(\mathbf{x},\mathbf{y})f(\mathbf{x})\,d\mathbf{x} := \text{p.f.}\, I(\delta).$$

For instance, it can easily be seen that (1.35) applies to any singular kernel k which can be expressed as a linear combination of kernels dealt with in Case I (provided, of course, f is sufficiently smooth).

It is important to note that definition (1.35) is consistent with the definitions of univariate, finite part integrals given in Section 1.4.1; i.e., the definitions in Section 1.4.1 are just particular instances of the general definition (1.35). Note also that if $\mathbf{y}$ is an interior point of B and if the coefficients $a_1, \ldots, a_{k+1}$ in the asymptotic expansion of the function I in (1.34) vanish,

$$a_1 = a_2 = \cdots = a_{k+1} = 0,$$

then the finite part integral (1.35) coincides with the Cauchy principal value integral (1.12).

1.5. Curve and Surface Integrals

Some applications, for instance, boundary integral methods (see Section 2.4), require integrals to be evaluated over *N-dimensional manifolds* $B \subseteq \mathbb{R}^M$, $N < M$. Using appropriate parameterizations, these integrals can be transformed into integrals over N-dimensional regions $\overline{B} \subseteq \mathbb{R}^N$ as defined in one of the foregoing sections. This section provides the basic concepts of and the specific transformation rules applicable to real *surface integrals*. For $N = 1$, *curve integrals* are obtained as particular instances of surface integrals. While surface integrals can be regarded as integrals of special *differential forms*, integrals of general differential forms are *not* considered in this section (see, for instance, Edwards [121]).

1.5.1. Riemann Integrals on Surfaces

Smooth Surfaces. A subset $\overline{B} \subseteq \mathbb{R}^N$ is said to be an *N-dimensional parameter set* if the *open* set $\overline{B} \setminus \partial\overline{B}$ is *connected*. A subset $B \subseteq \mathbb{R}^M$ is said to be an *N-dimensional surface* in $\mathbb{R}^M$ if there is an N-dimensional parameter set $\overline{B}$ and a continuous mapping $\psi : \overline{B} \to \mathbb{R}^M$ such that $\psi(\overline{B}) = B$. In this context, the mapping ψ is referred to as a *parameterization* of B. Note that ψ is not required to be a one-to-one mapping in this definition.

An N-dimensional surface B is *smooth* if there exists a continuously differentiable parameterization ψ such that the associated Jacobian matrix $J(\overline{\mathbf{x}})$ has the (maximum possible) rank N for all $\overline{\mathbf{x}} \in \overline{B}$. If an N-dimensional surface B in $\mathbb{R}^N$ can be represented as the union of a finite number of smooth surfaces, then B is said to be *piecewise smooth*. While the following definitions assume that the underlying surface B is smooth, their range of applicability can easily be extended to piecewise smooth surfaces B (by applying them to the smooth patches contained in B).

Area of Smooth Surfaces. Let the parameterization $\psi : \overline{B} \to \mathbb{R}^M$ of an N-dimensional surface B in $\mathbb{R}^M$ be one-to-one and smooth.[9] Furthermore, let $G(\overline{\mathbf{x}})$ denote the *Gram determinant* associated with ψ at a point $\overline{\mathbf{x}} \in \overline{B}$:

$$G(\overline{\mathbf{x}}) := \det(J(\overline{\mathbf{x}})\, J(\overline{\mathbf{x}})^{\top}). \tag{1.36}$$

Then, for any partition $\overline{\mathcal{P}} = \{\overline{B}_1^{(n)}, \ldots, \overline{B}_n^{(n)}\}$ of $\overline{B}$ into n disjoint subsets and for any choice of n points $\overline{\mathbf{x}}_i \in \overline{B}_i^{(n)}$, $i = 1, \ldots, n$, a Riemann sum is defined by

$$R_n(\psi) := \sum_{i=1}^{n} \mathrm{vol}(\overline{B}_i^{(n)}) \sqrt{G(\overline{\mathbf{x}}_i)}. \tag{1.37}$$

[9] Any parameter set $\overline{B}$ (and any subset of $\overline{B}$) is tacitly assumed to be (Jordan) measurable, i.e., to have a well-defined N-dimensional volume, in the remainder of Section 1.5.1.

Transformation Rules for Strongly Singular Integrals. An observation of crucial practical importance is that the transformation rule (1.39) does *not* generally hold for strongly singular surface integrals defined by (1.40) or (1.41). However, strongly singular surface integrals arise mainly in boundary integral methods for solving elliptic partial differential equations (see Section 2.4). For this particular problem class, one can show (see Kieser [232] and Schwab and Wendland [386]) the following.

(i) The transformation rule (1.39) is valid for surface integrals defined by (1.40) or (1.41) if $\mathbf{x}_0$ is an interior point of B and if the given integration problem results from a strongly elliptic boundary value problem.

(ii) If the conditions in (i) are violated or if B is not smooth at $\mathbf{x}_0$, then transformation rules valid for surface integrals defined by (1.40) or (1.41) can be obtained by adding suitable correction terms to the right-hand side in (1.39). Explicit formulas for these correction terms can be found in Schwab and Wendland [386].

Chapter 2

Computational Integration in Practice

In this chapter several prototypical areas of scientific computing in which computational integration plays a prominent role are dealt with. In practice, it often happens that integration, as a fundamental computational technique, is part of or preceded by reformulations of certain mathematical problems. For instance, integral transforms are often used to convert mathematical problems posed as ordinary or partial differential equations into algebraic equations. In this context computational integration is often required to (approximately) calculate the integral transforms involved (see Section 2.2).

Finite element methods are based on the fact that certain classes of partial differential equations are equivalent to variational equations. Computational integration comes into play as a tool for computing the coefficients of the stiffness matrices and right-hand sides of discretized variational equations (see Section 2.3). Boundary element methods rely on the transformation of partial differential equations into integral equations whose discretization also necessitates computational integration techniques (see Section 2.4).[12]

While computational integration techniques are an important part of various methods for solving differential equations, their usefulness is by no means restricted to this specific purpose. Computational integration plays, for instance, an important role in computational statistics (see Section 2.1). Moreover, in theoretical physics, computational integration is used to calculate Fermi–Dirac integrals (Pichon [332], Sagar [381],[382]) as well as Bose–Einstein integrals (Gautschi [158]). In computer graphics, computational integration methods are occasionally used for luminance computations (see, for instance, Buming, Wittels, and Fussell [46]). Computational integration is an indispensable tool in many other sectors of science and engineering as well.

The purpose of this chapter is not only to point out the relevance of com-

[12]It is not the authors' intention to give an overview of numerical methods for solving partial differential equations. Likewise, this section is not intended to comprehensively cover all of the aforementioned computational techniques. Instead, references to relevant publications are provided for readers who wish to extend their knowledge of these fields.

putational integration techniques on the basis of representative applications, but also to emphasize various requirements that computational integration software must meet. The most important of these requirements are:

useability for different types of integration regions and weight functions, for varying numbers of integrands, etc.,

reliability in attaining approximate results of a specific accuracy, and

a user-friendly interface which anticipates the kind and degree of user knowledge of relevant characteristics of a given integration problem.

Software which does not meet these requirements is only of restricted usefulness to the user. Some of the computational integration programs developed in the past are deficient in this respect. Only by carefully studying application areas and designing software accordingly can this pitfall be avoided.

Example (Statistical Computation). Statistical computation frequently requires the approximate integration of multiple integrands over the same integration region. Integrating multiple integrands simultaneously is often far more efficient than integrating them separately. Nevertheless, many numerical integration programs are only able to integrate *one* scalar function at a time, which, for reasons explained in Section 2.1, may significantly increase the computational effort needed to evaluate the integrands.

2.1. Computational Statistics

2.1.1. Fundamentals. *Random variables* are used for modeling phenomena which have, by their very nature, an indeterministic character. They are fundamental entities in all manner of statistical analyses (see, for instance, Berger [21]).

From a statistical point of view, a random variable X is fully characterized by its *probability distribution* P_X. P_X determines the probability $P_X(B)$ that an observation x of the phenomenon modeled by X—a *realization* of the random variable X—is contained in the particular subset B of the space of all possible results.[13] Usually, the set of all possible outcomes is the whole space $\mathbb{R}^N$ or a special subset of it. If $P_{\mathbf{X}}$ is *continuous*, i.e., has a *density* w, then

$$P_{\mathbf{X}}(B) = \int_B w(\mathbf{x})\, d\mathbf{x}.$$

2.1.2. The Computation of Stochastic Quantities. Not only probability distributions, but virtually *all* fundamental stochastic quantities are defined by, or can be represented by, appropriate integrals. The *cumulative distribution*

---

[13]Random variables are represented by uppercase letters; corresponding lowercase letters are used for their realizations.

function $F_{\mathbf{X}}$, for instance, is defined by

$$F_{\mathbf{X}}(b_1, \ldots, b_N) := P_{\mathbf{X}}((-\infty, b_1] \times \cdots \times (-\infty, b_N]) = \int_{-\infty}^{b_1} \cdots \int_{-\infty}^{b_N} w(\mathbf{x})\, d\mathbf{x},$$

the *mathematical expectation* $\mu(\mathbf{X})$ of $\mathbf{X}$ by

$$\mu(\mathbf{X}) := \int_{\mathbb{R}^N} w(\mathbf{x})\mathbf{x}\, d\mathbf{x}, \tag{2.1}$$

and the *variance-covariance matrix* $V(\mathbf{X})$ of $\mathbf{X}$ by

$$V(\mathbf{X}) := \int_{\mathbb{R}^N} w(\mathbf{x})(\mathbf{x} - \mu(\mathbf{X}))(\mathbf{x} - \mu(\mathbf{X}))^\top\, d\mathbf{x}. \tag{2.2}$$

Note that the integrands in (2.1) and (2.2) are vector- and matrix-valued functions, respectively. The integral of such vector or matrix integrands is defined by componentwise integration. For instance, the kth component of the vector $\mu(\mathbf{X})$ is defined by

$$(\mu(\mathbf{X}))_k := \int_{\mathbb{R}^N} w(\mathbf{x})x_k\, d\mathbf{x},$$

where x_k is the kth component of the vector $\mathbf{x} = (x_1, \ldots, x_N)^\top$.

From these examples it can be seen that the computation of probabilities and *moments* (means, variances, covariances, etc.) consists of the evaluation of integrals

$$\mathrm{I}f_j := \int_B w(\mathbf{x}) f_j(\mathbf{x})\, d\mathbf{x}, \qquad j = 1, 2, \ldots, m,$$

where $w \colon \mathbb{R}^N \to \mathbb{R}_0^+$ is the underlying density function. The functions

$$f_1, \ldots, f_m : \mathbb{R}^N \to \mathbb{R}$$

depend on the statistical entity to be computed. Note that in most cases $f_1, \ldots, f_m$ are fairly simple functions, such as constant, linear, or quadratic polynomials. In contrast, the density function w is usually much more complex and often contains transcendental functions, such as the exponential function in the multivariate, normal distribution.

Many statistical analyses require the computation of some or all of the aforementioned stochastic entities. If the associated integrals cannot be evaluated analytically then numerical techniques must be applied. This does *not* necessarily mean that numerical *integration* methods are used. For instance, the NAG library [311] as well as the IMSL STAT/LIBRARY [442] contain subroutines for numerically evaluating the cumulative distribution functions of many commonly used one-dimensional statistical distributions such as the *normal (Gaussian)* or the *Gamma distribution*. These subroutines usually apply tailor-made approximation techniques based on *Taylor series*, *asymptotic*, and/or *continued fraction expansions* of the integrals to be computed.

Example (Multivariate Normal Distribution). For a number of important statistical integration problems, specialized and highly efficient approximation techniques are *not* available (Genz [169]). A typical example for this is the computation of the cumulative N-dimensional normal distribution function

$$F(b_1, \ldots, b_N) = \frac{1}{\sqrt{(2\pi)^N \det V}} \int_{-\infty}^{b_1} \cdots \int_{-\infty}^{b_N} \exp\left(-\frac{1}{2}\mathbf{x}^\top V^{-1}\mathbf{x}\right) d\mathbf{x},$$

where $V \in \mathbb{R}^{N \times N}$ denotes the symmetric, positive definite variance-covariance matrix. Special approximation techniques for calculating F are available for dimensions $N = 1$ and $N = 2$ only (see, for instance, the NAG library [311] and the IMSL STAT/LIBRARY [442]). Software for calculating F for $N \geq 3$, for example, `NAG/g01hbf`, `TOMS/725` (Drezner [118]), or `MVNPACK` (Genz [168], [170]), is predominantly based on numerical integration schemes.

2.1.3. Bayesian Statistics. A particularly important source of integration problems for which no tailor-made computational techniques are available is *Bayesian statistics.* The goal of Bayesian statistics is to identify (quantities dependent on) an unknown state of a "system" by means of statistical investigations which take into account a priori information. The process of Bayesian statistics can be outlined briefly as follows.

The results of a statistical investigation are commonly modeled by a random variable $\mathbf{X} = (X_1, \ldots, X_M)$, where $X_1, \ldots, X_M$ are independent scalar random variables with a typical distribution P_X. This probability distribution depends on the unknown state of the system being investigated.

Example (Defective Shipments). The relative frequencies $x_1, \ldots, x_M$ of defective items received in M shipments from a factory can often be modeled by independent, identically distributed random variables $X_1, \ldots, X_M$. Their common distribution P_X depends on the state of the whole production system, for instance, on the qualification of the employees, the technical state of the production facilities, etc.

The space Θ of possible system states is usually the whole of or a particular subset of $\mathbb{R}^N$. The density w of P_X depends on the specific system state $\theta \in \Theta$, i.e., $w(x) = w(x|\theta)$. Frequently, there is some *prior* information concerning the specific value of θ. This information often comes from experience gained in comparable situations.

Example (Defective Shipments). Prior information about a supplier of a specific item may come from the quality of the supplier's previous shipments.

Prior information about the system state θ is generally not very precise. As a result, such information is often stated in terms of the probability of various possible values of θ occurring (Berger [21]). In this sense, prior information can be quantified in terms of a probability density $\pi(\theta)$ on Θ.

Bayesian analysis combines the prior information $\pi(\theta)$ with the specific sample information $\mathbf{x} = (x_1, \ldots, x_M)$ in order to derive the *posterior* distribution $\pi(\theta|\mathbf{x})$ of the required system state θ given $\mathbf{x}$. $\pi(\theta|\mathbf{x})$ reflects prior information concerning θ as well as information gained from the observations made. Subsequent statistical inferences are based on this posterior distribution $\pi(\theta|\mathbf{x})$.

The first step in determining the posterior density $\pi(\theta|\mathbf{x})$ is the evaluation of the *unconditional density* $m(\mathbf{x})$ of $\mathbf{X}$,

$$m(\mathbf{x}) = \int_\Theta w(\mathbf{x}|\theta)\pi(\theta)\,d\theta. \tag{2.3}$$

The density $g(\mathbf{x}, \theta)$ of the joint distribution of $\mathbf{X}$ and θ can be represented by

$$g(\mathbf{x}, \theta) = w(\mathbf{x}|\theta)\pi(\theta) \tag{2.4}$$

as well as by

$$g(\mathbf{x}, \theta) = \pi(\theta|\mathbf{x})m(\mathbf{x}). \tag{2.5}$$

Equating (2.4) and (2.5) yields

$$\pi(\theta|\mathbf{x}) = \frac{w(\mathbf{x}|\theta)\pi(\theta)}{m(\mathbf{x})}.$$

In practice, the predominant portion of the computational effort is in the evaluation of the integral (2.3). The function $w(\mathbf{x}|\theta)$ in (2.3) is the product of the M density functions $w(x_1|\theta), \ldots, w(x_M|\theta)$ and usually has a fairly complex structure. The dependence of the integrand in (2.3) on the observations $\mathbf{x} = (x_1, \ldots, x_M)$ generally impedes the application of tailor-made approximation techniques and the development of special numerical methods, as they exist for standard cumulative distribution functions. In most practical cases, (2.3) has to be evaluated using numerical integration. For recent work in this area, see Genz [169], Genz and Kass [171], and the references therein.

2.2. Integral Transforms

Integral transforms play an important role in pure and applied mathematics. They are used in practice for solving ordinary and partial differential equations as well as representing signals and systems.

2.2.1. Fundamentals. Integral transforms are linear integral operators (see Section 1.1.3) defined on vector spaces $\mathcal{F}$ of real- or complex-valued functions on domains $B \subseteq \mathbb{R}^N$. An integral transform $\mathcal{T}$ maps functions $f \in \mathcal{F}$ according to

$$F(\mathbf{y}) := (\mathcal{T}f)(\mathbf{y}) = \int_B w(\mathbf{x})\overline{k(\mathbf{x}, \mathbf{y})}f(\mathbf{x})\,d\mathbf{x}, \quad \mathbf{y} \in D, \tag{2.6}$$

onto other real- or complex-valued functions F defined on a domain $D \subseteq \mathbb{R}^N$. The function $k : B \times D \to \mathbb{R}$ ($\mathbb{C}$) is called the *kernel* (or *nucleus*) of the integral transform, and $w(\mathbf{x})$ is the corresponding *weight function*. The various integral transforms differ from each other in their underlying kernels and weight functions and/or in their integration regions.

In this section only one-dimensional integral transforms

$$(2.7)\quad F(y) := (\mathcal{T}f)(y) = \int_B w(x)\overline{k(x,y)}f(x)\,dx = \int_a^b w(x)\overline{k(x,y)}f(x)\,dx$$

are considered. All practically relevant multidimensional transforms (such as multivariate Fourier transforms) can be represented as repeated applications of one-dimensional transforms (see, for instance, Jerri [219]).

The integration region B in (2.7) is generally the infinite interval $B = (-\infty, \infty)$ or a semi-infinite interval such as $B = [0, \infty)$. Important examples of one-dimensional integral transforms are the *Laplace transform*

$$(2.8)\qquad F(y) = (\mathcal{T}_L f)(y) := \int_0^\infty \exp(-xy) f(x)\,dx,$$

the *(exponential) Fourier transform*

$$F(y) = (\mathcal{T}_F f)(y) := \int_{-\infty}^\infty \exp(-\mathrm{i}xy) f(x)\,dx,$$

and the *Hankel (Bessel) transform*

$$F_n(y) = (\mathcal{T}_{H_n} f)(y) := \int_0^\infty x J_n(xy) f(x)\,dx.$$

In this formula J_n is the *Bessel function* of the first kind of order n (see, for instance, Jerri [219]).

2.2.2. Application to Differential Equations. Integral transforms $\mathcal{T}$ are useful tools for solving *ordinary* differential equations because some important ordinary differential equations can be *algebraized*, i.e., transformed into *algebraic equations*, by subjecting the unknown solution function to an appropriate integral transform. Usually, the resulting algebraic equation is easier to solve than the original ordinary differential equation.

The application of appropriate integral transforms to *partial* differential equations results in differential equations with a reduced number of independent variables. Since the computational complexity of partial differential equations decreases rapidly when the number of independent variables is reduced, the transformed equations are usually far easier to solve than the original ones.

As soon as the transformed equations have been solved, the desired solution in the original function space $\mathcal{F}$ is obtained by subjecting the solution of the transformed equations to the appropriate *inverse* transformation $\mathcal{T}^{-1}$.

Inverse integral transforms are also integral operators. For instance, the *inverse Fourier transform* is given by

$$f(x) = (\mathcal{T}_F^{-1}F)(x) = \frac{1}{2\pi}\int_{-\infty}^{+\infty} \exp(\mathrm{i}xy)F(y)\,dy,$$

and the *inverse Hankel transform* by

$$f(x) = (\mathcal{T}_{H_n}^{-1}F_n)(x) = \int_0^{\infty} yF_n(y)J_n(xy)\,dy.$$

The *inverse Laplace transform* is given by the contour integral, usually referred to as *Bromwich's integral:*

$$f(x) = (\mathcal{T}_L^{-1}F)(x) = \frac{1}{2\pi\mathrm{i}}\int_{\gamma-\mathrm{i}\infty}^{\gamma+\mathrm{i}\infty} \exp(xs)F(s)ds, \tag{2.9}$$

provided that F is analytic in an open region which contains the right half-plane $\{\, z \;:\; \mathrm{Re}(z) \geq \gamma \,\}$. The minimum possible value γ_0 of γ is called the *abscissa of convergence* of F.

2.2.3. Finite Transforms and Fourier Series. The *finite transform* of $f \in \mathcal{F}$ with respect to a set of functions $\{k(.,\mathbf{y}_n)\}$ is defined by

$$F(\mathbf{y}_n) := (\mathcal{T}f)(\mathbf{y}_n) = \int_B w(\mathbf{x})\overline{k(\mathbf{x},\mathbf{y}_n)}f(\mathbf{x})\,d\mathbf{x}.$$

This definition of finite transforms is identical to the definition (2.6) of integral transforms with two exceptions. (i) The finite transform F is only defined for a *discrete* set of points $\{\mathbf{y}_n\}$. (ii) The functions $k(.,\mathbf{y}_n)$ are required to be *orthogonal* with respect to the weight function w on the domain $B \subseteq \mathbb{R}^N$; i.e.,

$$\int_B w(\mathbf{x})\overline{k(\mathbf{x},\mathbf{y}_n)}k(\mathbf{x},\mathbf{y}_m)\,d\mathbf{x} = \delta_{m,n}\|k(.,\mathbf{y}_n)\|_w^2,$$

where $\delta_{m,n}$ is the *Kronecker delta,*

$$\delta_{m,n} := \begin{cases} 0 & \text{if} \quad m \neq n, \\ 1 & \text{if} \quad m = n, \end{cases}$$

and the norm $\|k(.,\mathbf{y}_n)\|_w$ is given by

$$\|k(.,\mathbf{y}_n)\| := \sqrt{\int_B w(\mathbf{x})|k(\mathbf{x},\mathbf{y}_n)|^2\,d\mathbf{x}}.$$

The close relationship between finite and integral transforms is reflected in the fact that some finite transforms can be immediately obtained from integral transforms. Consider, for instance, the kernel $k(x, y) = \exp(ixy)$ of the Fourier transform. It can easily be seen that for all bounded intervals $[a, b]$ the functions

$$k(x, y_n) := \exp(ixy_n), \qquad y_n := \frac{2\pi}{b-a} n, \qquad n \in \mathbb{Z},$$

are orthogonal (with respect to $w(x) \equiv 1$) and that

$$\|k(x, y_n)\|_w^2 = b - a, \qquad n \in \mathbb{Z}.$$

As with integral transforms, finite transforms are used for algebraizing ordinary differential equations or for reducing the number of independent variables in partial differential equations. A special feature of finite transforms is that their *inverse* transforms can easily be represented by series, the *generalized Fourier series* with respect to $\{k(., \mathbf{y}_n)\}$:

$$f(\mathbf{x}) = \sum_n \frac{F(\mathbf{y}_n)}{\|k(\mathbf{x}, \mathbf{y}_n)\|_w^2} k(\mathbf{x}, \mathbf{y}_n). \tag{2.10}$$

By contrast, the inversion of integral transforms generally involves integral operators. Since, using (2.10), the original function f can easily be recovered from its *generalized Fourier coefficients* $\{F(\mathbf{y}_n)\}$, finite transforms are often used to efficiently represent functions (which are, in this context, referred to as signals).

2.2.4. Computational Aspects. There are several fundamentally different ways of performing integral or finite transforms and their corresponding inverse transforms. These methods are only briefly outlined in the remainder of this section. For a more thorough discussion, see Section 3.3.

If the function f (or F, in the case of an inverse transform) is available in closed form, one can try to compute the given integral transform manually. Manual integration is facilitated if the function f (or F) can be found in one of the numerous tables of integral transforms (such as those of Ditkin and Prudnikov [111] and Erdelyi et al. [132]). Such tables incorporate extensive lists of pairs (f, F) of functions and their transformed counterparts for various kinds of integral transforms. In some cases, functions in closed form can also be automatically transformed using *symbolic computation software* (see Chapter 4).

Still, these analytical approaches fail in a number of cases, for instance, when the function f (or F) is not available in closed ("analytical") form. This may happen, for example, when the argument F of an inverse transform is the solution of an algebraic or differential equation which cannot be solved explicitly. In such cases, numerical techniques must be used for computing integral transforms. As with the numerical computation of statistical integrals, this does *not* necessarily mean that numerical *integration* methods are used: as the number of practically relevant transforms is rather small, tailor-made numerical techniques have been developed to calculate them approximately. The most widely known among these special computational techniques is the fast Fourier transform (see, for instance, Van Loan [437]). Even so, such special techniques may in some way rely on numerical integration. This is true, for instance, of Weeks's method for inverting the Laplace transform (see Section 7.2.6).

There is a huge number of publications dealing with the numerical computation of integral transforms. Quadrature methods for integral transforms, i.e., computational techniques based on approximately evaluating integrals, are discussed in Chapter 7 in conjunction with methods for special integrals. Chapter 7 also provides references to computational methods which do *not* use numerical integration.

2.3. Finite Element Methods

In finite element methods for solving partial differential equations, the original partial differential equation is transformed into its *weak* formulation as a *variational equation* (see Section 2.3.1).

Terminology. The term "weak" in this context refers to the fact that the solution of a variational equation generally satisfies the original partial differential equation only in a *generalized sense* (see Section 2.3.1). Only under additional smoothness conditions can it be shown that the solution of a variational equation satisfies the original partial differential equation in the classical sense (see Großmann and Roos [184]). From this point of view, the variational equation is less stringent, i.e., weaker, than the original partial differential equation.

Since the space of possible solution functions is of infinite dimension, this variational equation, in general, cannot be solved exactly. Instead, by solving the variational equation over appropriately chosen finite-dimensional subspaces, the exact solution can be approximated with arbitrary accuracy (see Section 2.3.2). Solving such a *discretized* variational equation means solving a system of linear equations. The coefficients and the right-hand sides of these *Galerkin equations* are given in terms of integrals. In order to guarantee an efficient approximation of the exact solution, these integrals must be computed with a specific accuracy which depends on the finite-dimensional subspaces used for discretizing the variational equation (see Section 2.3.3). This particular accuracy requirement is one reason why symbolic integration methods are not usually used for evaluating the integrals associated with the Galerkin equations. It is simply not necessary to compute these integrals very accurately. The other reason is that often the integrals associated with the Galerkin equations cannot be evaluated analytically, which necessitates the use of numerical integration methods for their computation.

2.3.1. Variational Equations

The Poisson Equation. The specific variational equations solved in finite element methods depend on the partial differential equations being studied. The *Poisson equation*

$$\Delta u(\mathbf{x}) = g(\mathbf{x}) \quad \text{for all} \quad \mathbf{x} \in \Omega \tag{2.11}$$

is chosen as a *model problem* in the remainder of this section. However, most results hold for arbitrary *elliptic* partial differential equations. In (2.11) $\Delta := \partial^2/\partial x_1^2 + \cdots + \partial^2/\partial x_N^2$ denotes the *Laplacean operator* (in Cartesian coordinates); the domain $\Omega \subset \mathbb{R}^N$ is an open, bounded region. Furthermore, g is required to be continuous in $\operatorname{cl}\Omega$, i.e., $g \in C(\operatorname{cl}\Omega)$. The *Laplace equation* is obtained from the Poisson equation by setting $g(\mathbf{x}) \equiv 0$. Electrostatic fields, for example, can be determined by solving Poisson or Laplace equations.

Restrictions—most importantly, *boundary conditions*—have to be imposed on equation (2.11) in order to specify the solution u completely. The following discussion covers only the simplest boundary condition, the *homogeneous Dirichlet* condition

$$u(\gamma) = 0 \quad \text{for all} \quad \gamma \in \Gamma. \tag{2.12}$$ [14]

However, the results of this section can easily be extended to other types of boundary conditions.

Under the previously stated assumption concerning the smoothness of the right-hand side g and under additional assumptions concerning the smoothness of the boundary Γ, it can be shown that problem (2.11) has a unique solution

$$u^* \in C^2(\Omega) \cap C(\operatorname{cl}\Omega).$$

However, in most practical cases it is not possible to find u^* analytically. Instead, one tries to construct approximate solutions by *discretizing* the equation (2.11), i.e., by replacing the given infinite-dimensional problem with a properly chosen finite-dimensional problem. *Difference methods* are based on a spatial discretization of the region Ω and the approximation of differential operators using finite differences. In *finite element methods*, the original differential equation problem is first reformulated as a *variational* problem, which is then solved using a discretization method.

Weak Formulation of the Poisson Equation. The conversion of elliptic partial differential equations into their weak forms is achieved by applying *the multidimensional rule of integration by parts*. The conversion process itself is

[14] $\Gamma := \partial\Omega$ denotes the boundary of Ω.

irrelevant to the following discussion and is therefore not considered in detail. See Ciarlet [59] or Großmann and Roos [184] for thorough discussions.

Conversion of the model problem shows that the equality[15]

$$\int_\Omega \nabla u(\mathbf{x})^\top \cdot \nabla v(\mathbf{x})\, d\mathbf{x} = -\int_\Omega g(\mathbf{x})v(\mathbf{x})\, d\mathbf{x} \tag{2.13}$$

holds for all

$$v \in C_0^1(\operatorname{cl}\Omega),$$

where

$$C_0^1(\operatorname{cl}\Omega) := \{\, w \in C^1(\operatorname{cl}\Omega) \ :\ w(\gamma) = 0 \ \text{ for all } \ \gamma \in \Gamma\}.$$

Each solution u of the Poisson equation (2.11) satisfying the boundary condition (2.12) also satisfies (2.13) for all $v \in C_0^1(\operatorname{cl}\Omega)$. The reverse is also true; by choosing appropriate functions v, it can easily be seen that each function $u \in C^2(\Omega)$ satisfying (2.13) also satisfies (2.11).

Note that for equation (2.13) to be well defined, it is not necessary for the functions u, v, and g to meet the stringent smoothness conditions stated previously. For instance, (2.13) is also well defined if the function g, representing the problem data, is only *piecewise continuous* and if the possible solutions u and the functions v are assumed to be only *piecewise continuously differentiable* (Ciarlet [59]). Indeed, it can be proved that equation (2.13) has a unique solution for *square integrable* functions g, i.e., for functions g whose square g^2 is integrable over Ω. Let V denote the space of possible solution functions of (2.13) for square integrable functions g. V contains, for instance, all piecewise continuously differentiable functions. Note that since the functions in V are not necessarily contained in $C^2(\Omega)$, the solutions of equation (2.13) do not generally solve the original equation (2.11). This issue is not discussed here any further (for details, see Axelsson and Barker [11], Ciarlet [59], or Hackbusch [193]).

More abstractly, equation (2.13) can be written as a *variational equation*[16]

$$a(u,v) = (g,v) \quad \text{for all} \quad v \in V, \tag{2.14}$$

where

$$a(u,v) := \int_\Omega \nabla u(\mathbf{x})^\top \cdot \nabla v(\mathbf{x})\, d\mathbf{x}$$

is a *bilinear form* on $V \times V$, and

$$(g,v) := -\int_\Omega g(\mathbf{x})v(\mathbf{x})\, d\mathbf{x}$$

is linear in v.

---

[15] $\nabla := (\partial/\partial x_1, \ldots, \partial/\partial x_N)^\top$ denotes the *gradient operator*.

[16] The term "variational equation" refers to the fact that the solution of this equation can be shown to be equivalent to the solution of a properly defined *variational problem*, i.e., to the minimization of a properly defined linear functional on V. For details see Ciarlet [59].

2.3.2. The Galerkin Method. The variational equation (2.14) cannot generally be solved directly because the function space V which contains the solution u is of infinite dimension. The main idea of the *Galerkin method* is to approximate the exact solution by solving equation (2.14) over appropriately chosen finite-dimensional subspaces. More precisely, with an M_h-dimensional subspace V_h of V, the *discrete* problem

$$a(u_h, v_h) = (g, v_h) \quad \text{for all} \quad v_h \in V_h \tag{2.15}$$

has the same form as (2.14), but is solved in the finite-dimensional (discrete) space V_h.[17] Let $b_{h,1}, \ldots, b_{h,M_h}$ denote a basis for V_h. Due to the linearity of $a(u, .)$ and $(g, .)$, equation (2.15) is equivalent to

$$a(u_h, b_{h,j}) = (g, b_{h,j}), \qquad j = 1, 2, \ldots, M_h.$$

Representing the *discrete solution* u_h as a linear combination of the basis functions,

$$u_h = \sum_{i=1}^{M_h} s_i b_{h,i},$$

yields the so-called *Galerkin equations*

$$\sum_{i=1}^{M_h} s_i a(b_{h,i}, b_{h,j}) = (g, b_{h,j}), \qquad j = 1, 2, \ldots, M_h.$$

This is a system of linear equations for the unknown coefficients $s_1, \ldots, s_{M_h}$. Its matrix

$$A_h := (a(b_{h,i}, b_{h,j})), \tag{2.16}$$

the *stiffness matrix*[18], can be shown to be regular for all elliptic partial differential equations (Ciarlet [59]). Therefore, the discrete problem (2.15) always has a unique solution.

2.3.3. Discretization Using Finite Elements

---

[17] The parameter $h > 0$ characterizes the fineness of the discretization scheme on which V_h is based. Generally, it is assumed that $h_1 < h_2$ implies that functions $v \in V$ can be better approximated using functions $v_{h_1} \in V_{h_1}$ than by using functions $v_{h_2} \in V_{h_2}$. An appropriate definition of h in the context of finite element methods is given in Section 2.3.3.

[18] The term "stiffness matrix" stems from the early applications of finite element methods in structural mechanics.

Basic Principles. The *finite element method* (FEM) is a special technique for constructing the subspaces $V_h \subset V$ which occur in the Galerkin method. The subspaces V_h are called *finite element spaces.*

The finite element method is distinguished by the following three fundamental characteristics.

FEM 1. The domain Ω is decomposed into a finite number of nonempty, disjoint regions $\Omega_1, \ldots, \Omega_K$, the *finite elements,*

$$\operatorname{cl}\Omega = \operatorname{cl}\Omega_1 \cup \operatorname{cl}\Omega_2 \cup \cdots \cup \operatorname{cl}\Omega_K.$$

In practice, the decomposition $\{\Omega_1, \ldots, \Omega_K\}$ of Ω is constructed as follows. (i) The domain Ω is approximated using a *polyhedron* $\Omega^p \subseteq \Omega$. (ii) The polyhedron Ω^p is partitioned into a number of geometrically simple regions, usually N-dimensional *simplices* (resulting in *simplicial* finite elements) and/or N-dimensional *rectangles* (resulting in *rectangular* finite elements). Note that for the practically most important dimensions $N = 2$ and $N = 3$ the simplices are *triangles* and *tetrahedra,* respectively.

The remaining part of the domain, i.e., $\Omega \setminus \Omega^p$, may be treated in different ways. A common technique is, for instance, to approximate $\Omega \setminus \Omega^p$ using *isoparametric finite elements* (see Ciarlet [59] and Großmann and Roos [184]). Other methods can be found, for instance, in Hackbusch [193]. To keep things as simple as possible, Ω is in this section assumed to be a polyhedron, $\Omega^p = \Omega$, so that the problem of treating the remainder $\Omega \setminus \Omega^p$ does not arise.

FEM 2. The functions $v_h \in V_h$ are required to be at least continuous, $V_h \subset C(\operatorname{cl}\Omega)$. In addition, the functions $v_h \in V_h$ are assumed to be *piecewise polynomials* with respect to the decomposition $\{\Omega_1, \ldots, \Omega_K\}$; i.e., the restriction $v_h|_{\Omega_k}$ is a polynomial for any finite element Ω_k and any function $v_h \in V_h$.

From FEM 1 and FEM 2 it can easily be seen that equation (2.13) is meaningful for all functions $v_h \in V_h$. Hence, V_h is contained in V, $V_h \subset V$.

FEM 3. There exists a basis $\{b_{h,1}, \ldots, b_{h,M_h}\}$ for V_h in which the *support* of each basis function $b_{h,i}$, $\operatorname{supp} b_{h,i}$, is contained in a small number of finite elements. In addition, the basis $\{b_{h,1}, \ldots, b_{h,M_h}\}$ must be easily computable.

Note that if two basis functions $b_{h,i}$ and $b_{h,j}$ have disjoint support, then the corresponding coefficient $a_{ij} = a(b_{h,i}, b_{h,j})$ of the stiffness matrix vanishes. Therefore, FEM 3 ensures that most coefficients of the stiffness matrix are zero, i.e., that the stiffness matrix is *sparse.* This is of the utmost importance because solving the Galerkin equations with a dense stiffness matrix would require $O(M_h^3)$ floating-point operations. In contrast, sparse linear systems can generally be solved with considerably less computational effort using only $O(M_h^\alpha)$, $\alpha < 3$, floating-point operations (see, for instance, Hackbusch [192]).

Certain classes of linear systems with sparse matrices can even be solved using *multigrid methods* with *linear complexity* $O(M_h)$ (Hackbusch [190]).

Convergence Properties. The question arises as to whether or not convergence

$$\|u - u_{h_n}\| \to 0 \qquad \text{for} \qquad n \to \infty$$

can be guaranteed for a sequence of finite element spaces $V_{h_1}, V_{h_2}, \dots$. The following discussion is restricted to the simplest version of the finite element method, which is characterized by

$$\{ v_h|_{\Omega_k} \ : \ v_h \in V_h \} = \mathbb{P}_d^N|_{\Omega_k}, \quad k = 1, 2, \dots, K,$$

where $\mathbb{P}_d^N$ denotes the set of all N-variate polynomials of maximum degree d. In finite element methods, the discretization parameter h is defined as the *maximum diameter*[19] of the decomposition $\Omega_1, \dots, \Omega_K$,

$$h := \max\{ \operatorname{diam} \Omega_k \ : \ k = 1, 2, \dots, K \}.$$

Clearly, it cannot be expected that $u_{h_n} \to u$ unless $h_n \to 0$. However, this condition is *not* sufficient. In addition to $h_n \to 0$, the existence of a constant $\varphi_0 > 0$ such that all interior angles between the faces of all finite elements in V_{h_n}, $n = 1, 2, \dots$, are larger than φ_0, is required. This prevents the finite elements from degenerating, i.e., from becoming increasingly *flat*, as $n \to \infty$. With this additional condition, it can be proved that

$$\|u - u_h\| = O(h^d) \quad \text{for} \quad h \to 0, \tag{2.17}$$

provided that u is sufficiently smooth, i.e., $u \in C^{d+1}(\operatorname{cl} \Omega)$ (Ciarlet [59]).

The Calculation of the Galerkin Equations. In order to understand the role of computational integration in finite element methods, the origin of the coefficients of the stiffness matrix and the right-hand sides of the Galerkin equations must be taken into consideration. The computation is usually organized as follows: let K_j denote the set of the subscripts of those finite elements Ω_k which intersect the support of the basis function $b_{h,j}$,

$$K_j := \{ k \ : \ \operatorname{supp} b_{h,j} \cap \Omega_k \neq \emptyset \}.$$

Furthermore, let $a_{ij}^{(k)}$ and $b_j^{(k)}$ be the contributions to the coefficients $a_{ij} = a(b_{h,i}, b_{h,j})$ of the stiffness matrix (2.16) and to the right-hand sides $b_j = (g, b_{h,j})$ associated with the finite element Ω_k:

$$a_{ij}^{(k)} := \int_{\Omega_k} \nabla b_{h,i}(\mathbf{x})^\top \cdot \nabla b_{h,j}(\mathbf{x})\, d\mathbf{x} \tag{2.18}$$

[19]The *diameter* of a set S in a normed space is defined as the supremum of the distances of any two elements of S, $\operatorname{diam} S := \sup\{ \|x - y\| \ : \ x, y \in S \}$.

and

$$b_j^{(k)} := -\int_{\Omega_k} g(\mathbf{x}) b_{h,j}(\mathbf{x})\, d\mathbf{x}. \tag{2.19}$$

That is,

$$a_{ij} = \sum_{k \in K_i \cap K_j} a_{ij}^{(k)} \tag{2.20}$$

and

$$b_j = \sum_{k \in K_j} b_j^{(k)}.$$

In order to compute the values $a_{ij}^{(k)}$ and $b_j^{(k)}$, integrations over simplicial or rectangular regions Ω_k must be performed. The computation of the stiffness matrix (a_{ij}) according to (2.18) and (2.20) only involves polynomials because the Poisson equation is an elliptic partial differential equation with *constant coefficients*. Hence, the integrals occurring in (2.18) can easily be computed analytically. If an elliptic partial differential equation contains *non-constant* coefficients, then these coefficients enter the bilinear form $a(.,.)$ of the corresponding variational equation. Therefore, in many cases the computation of the stiffness matrix for general elliptic partial differential equations requires the use of numerical integration.

The computation of the quantities $b_j^{(k)}$ involves the representation of the problem data, the function g. Accordingly, the integrals (2.19) are *not* always analytically computable. For instance, the function g may not be available in closed form (as a formula). In such cases, numerical integration is necessary for the approximate evaluation of (2.19).

Requirements for Numerical Integration. Only approximate values $\hat{a}_{ij}$ and $\hat{b}_j$ can be computed for the entities a_{ij} and b_j, respectively, using numerical integration methods. As a result, the solution of the Galerkin equations with the approximate coefficients $\hat{a}_{ij}$ and right-hand sides $\hat{b}_j$ yields only an approximation $\hat{u}_h$ of the discrete solution u_h. The question arises as to whether the convergence result (2.17) also holds for the sequence of approximate discrete solutions $\{\hat{u}_h\}$, i.e., whether

$$\|u - \hat{u}_h\| = O(h^d) \quad \text{for} \quad h \to 0. \tag{2.21}$$

It is an important result in the theory of finite element methods that convergence (2.21) can be guaranteed, provided that the numerical integration scheme used meets the following requirement. The numerical integration scheme used for computing the entities $a_{ij}^{(k)}$ and $b_j^{(k)}$ must integrate all polynomials of maximum degree $2(d-1)$ over the integration region $B = \Omega_k$ exactly (see, for instance, Ciarlet [59] or Großmann and Roos [184]). For instance, if the approximation functions v_h are piecewise linear, the numerical integration scheme must be exact for all polynomials of degree $2(1-1) = 0$, i.e., for all constant polynomials. For piecewise quadratic approximation functions, the scheme must be exact for all polynomials of degree $2(2-1) = 2$; for piecewise cubic approximation functions, the scheme must be exact for all polynomials of degree $2(3-1) = 4$; etc. The construction of appropriate numerical integration schemes is discussed in Section 6.2.

Similar results can also be obtained when the domain Ω is *not* a polyhedron and isoparametric finite elements are used to approximate the boundary of Ω. For details, the reader is referred, for instance, to Ciarlet [59].

The elliptic *eigenvalue problem* is related to the elliptic boundary value problem outlined in this section (for details, see Andreev, Kacieva, and Vanmaele [6], Banerjee [14], and Banerjee and Osborn [15]).

2.4. Boundary Integral Methods

This section introduces the boundary integral method for solving partial differential equations. Laplace's equation (see Section 2.3.1) is chosen as a model problem to demonstrate the salient features of this method. However, most results hold for the more general case of homogeneous elliptic partial differential equations with constant coefficients.

In the boundary integral method, the original partial differential equation is transformed into an equivalent *boundary integral equation* (see Section 2.4.1). Generally, such a boundary integral equation cannot be solved exactly because the corresponding space of possible solution functions is infinite-dimensional. However, by solving the boundary integral equation over appropriately chosen finite-dimensional subspaces, an approximation to the exact solution can be found. Solving such a *discretized* boundary integral equation is equivalent to solving a linear system of equations. The coefficients of these equations are given by certain integrals, which usually cannot be evaluated analytically.

2.4.1. Boundary Integral Equations. In this section Laplace's equation

$$\Delta u(\mathbf{x}) = 0 \quad \text{for all} \quad \mathbf{x} \in \Omega \tag{2.22}$$

is reformulated as an integral equation. The conversion process itself is irrelevant to the following discussion and is therefore not considered in detail. See Atkinson [8] or Hackbusch [191] for thorough discussions.

The integration region in the resulting integral equation extends over the *boundary* Γ of the domain Ω. This is why the resulting integral equation is called a *boundary integral equation*.

The computational complexity of partial differential equations and integral equations grows rapidly as the problem dimension increases. Therefore, the most attractive feature of the resulting boundary integral equation (compared with the original Laplace equation) is the reduced problem dimension. Laplace's equation on a three-dimensional domain, for instance, is transformed into an integral equation over a two-dimensional region.

It should be noted that the boundary integral formulation of Laplace's equation given in this section is not unique. For extensive listings of boundary integral formulations for Laplace's equation, see, for instance, Atkinson [8] or Hackbusch [191].

Green's Functions. Let the function $v_{\mathbf{y}}\colon \mathbb{R}^N \setminus \{\mathbf{y}\} \to \mathbb{R}$ be defined by

$$v_{\mathbf{y}}(\mathbf{x}) := -\frac{1}{2\pi} \log \|\mathbf{x} - \mathbf{y}\|_2$$

for $N = 2$ and by

$$v_{\mathbf{y}}(\mathbf{x}) := \frac{1}{\mathrm{vol}(S_{N-1})} \|\mathbf{x} - \mathbf{y}\|_2^{-N+2} \tag{2.23}$$

for $N \geq 3$. In (2.23), $\mathrm{vol}(S_{N-1})$ denotes the volume of the $(N-1)$-dimensional unit sphere

$$S_{N-1} := \{\, \mathbf{x} \in \mathbb{R}^{N-1} \ : \ \|\mathbf{x}\|_2 \leq 1 \,\}.$$

It can easily be verified that the functions $v_{\mathbf{y}}$ satisfy Laplace's equation. More importantly, each solution of Laplace's equation can be represented in terms of the functions $v_{\mathbf{y}}$, which are hence called *fundamental solutions* or *Green's functions* (Hackbusch [191]).

Inhomogeneous Dirichlet Boundary Conditions. Consider Laplace's equation (2.22) subject to an *inhomogeneous* Dirichlet boundary condition

$$u(\gamma) = g(\gamma) \quad \text{for all} \quad \gamma \in \Gamma, \tag{2.24}$$

where $g \in C(\Gamma)$. By using the representation of the solutions of Laplace's equation in terms of Green's functions, equations (2.22) and (2.24) for u can be transformed into the following equation for $\rho := \partial u / \partial \nu$:[20]

$$(\mathcal{T}\rho)(\mathbf{y}) := \int_\Gamma v_{\mathbf{y}}(\gamma)\rho(\gamma)\, d\gamma = 2\, g(\mathbf{y}) + \int_\Gamma g(\gamma) \frac{\partial v_{\mathbf{y}}}{\partial \nu}(\gamma)\, d\gamma, \quad \mathbf{y} \in \Gamma. \tag{2.25}$$

The two integrals occurring in this equation are to be understood as surface integrals as defined in Section 1.5. (2.25) constitutes an integral equation for ρ. More precisely, since the integral operator $\mathcal{T}$ is linear, (2.25) is a linear integral equation, and since the integration region Γ does not depend on $\mathbf{y}$, (2.25) is a *Fredholm integral equation.*

It is important to observe that the *kernel* $v_{\mathbf{y}}(\gamma)$ is singular at $\gamma = \mathbf{y}$. Since the integral operator $\mathcal{T}$ in (2.25) can be defined as an *improper Riemann integral* (see Section 1.2), $v_{\mathbf{y}}$ is called a *weakly singular* kernel.

[20] $\partial u / \partial \nu$ denotes the partial derivative of u with respect to the outer unit normal vector ν of the boundary Γ.

2.4.2. Collocation Methods. In the remainder of this section, equation (2.25) has been chosen as a model problem in order to demonstrate the salient features of the numerical solution of boundary integral equations. Many of these results also hold for other kinds of boundary integral equations.

Generally, the boundary integral equation (2.25) cannot be solved directly because the space of possible solutions $C(\Gamma)$ is of infinite dimension. Methods for the numerical solution of the boundary integral equation (2.25) are based on the solution of appropriately modified equations in subspaces $V_h \subset C(\Gamma)$ of finite dimension M_h. Furthermore, most methods are based either on a *Galerkin approach* (see Section 2.3.2) or on a *collocation approach.* In the Galerkin approach, the boundary integral equation (2.25) is first transformed into a variational equation (see Section 2.3.1), whose discrete version is subsequently solved over V_h. Since the Galerkin approach has already been dealt with in Sections 2.3.1 and 2.3.2, this topic is not discussed here any further (for details, see Atkinson [8] or Hackbusch [191]).

In the collocation approach, the approximation to the solution is required to satisfy (2.25) for a finite set of carefully selected points $\mathbf{y}_1, \ldots, \mathbf{y}_{M_h} \in \Gamma$, rather than for *all* $\mathbf{y} \in \Gamma$. More precisely, using the notation

$$G(\mathbf{y}) := 2\,g(\mathbf{y}) + \int_\Gamma g(\gamma)\frac{\partial v_{\mathbf{y}}}{\partial \nu}(\gamma)\,d\gamma, \quad \mathbf{y} \in \Gamma, \tag{2.26}$$

equation (2.25) becomes

$$(\mathcal{T}\rho)(\mathbf{y}) = G(\mathbf{y}), \quad \mathbf{y} \in \Gamma. \tag{2.27}$$

For $\mathbf{y}_1, \ldots, \mathbf{y}_{M_h} \in \Gamma$, the *discrete problem*

$$(\mathcal{T}\rho_h)(\mathbf{y}_j) = G(\mathbf{y}_j), \qquad j = 1, \ldots, M_h, \quad \rho_h \in V_h, \tag{2.28}$$

has the same form as the continuous problem (2.27), but is solved only for the finite set of points $\mathbf{y}_1, \ldots, \mathbf{y}_{M_h} \in \Gamma$, not for the infinite set of points $\{\, \mathbf{y} \in \Gamma \,\}$. The *collocation points* $\mathbf{y}_1, \ldots, \mathbf{y}_{M_h} \in \Gamma$ should be chosen in such a way that, if (2.28) holds, then

$$(\mathcal{T}\rho)(\mathbf{y}) \approx G(\mathbf{y}) \quad \text{for all} \quad \mathbf{y} \in \Gamma.$$

For general finite-dimensional subspaces V_h, the construction of appropriate point sets may pose insurmountable difficulties (see, for instance, Atkinson [8]). However, for special classes of subspaces V_h, for instance, for the so-called *boundary element spaces* (which are basically finite element spaces on the boundary Γ), appropriate point sets $\mathbf{y}_1, \ldots, \mathbf{y}_{M_h}$ are known (see, for instance, Großmann and Roos [184]).

Let $b_{h,1}, \ldots, b_{h,M_h}$ denote a basis for V_h, and let the *discrete solution* ρ_h be represented as a linear combination of the basis functions,

$$\rho_h = \sum_{i=1}^{M_h} s_i b_{h,i}.$$

Since the integral operator $\mathcal{T}$ is linear, inserting this representation into (2.28) yields a system of M_h linear equations:

$$\sum_{i=1}^{M_h} s_i(\mathcal{T}b_{h,i})(\mathbf{y}_j) = G(\mathbf{y}_j), \qquad j = 1, 2, \ldots, M_h. \tag{2.29}$$

The Role of Computational Integration. In order to understand the role of computational integration in collocation-based boundary integral methods, the computation of the coefficients and the right-hand sides of the linear system (2.29) for the model problem (2.25) must be considered. By substituting the definition (2.25) of the integral operator $\mathcal{T}$ into (2.29), the coefficient matrix (a_{ij}) of the system of linear equations (2.29) can be obtained:

$$a_{ij} = (\mathcal{T}b_{h,i})(\mathbf{y}_j) = \int_\Gamma v_{\mathbf{y}_j}(\gamma) b_{h,i}(\gamma)\, d\gamma.$$

Similarly, by substituting the definition (2.26) of the function G into (2.29), the right-hand sides b_j of (2.29) can be derived:

$$b_j = G(\mathbf{y}_j) = 2\,g(\mathbf{y}_j) + \int_\Gamma g(\gamma) \frac{\partial v_{\mathbf{y}_j}}{\partial \nu}(\gamma)\, d\gamma.$$

The integrals defining the quantities a_{ij} and b_j are significantly more complex than the corresponding integrals (2.18) and (2.19) defining the coefficients and right-hand sides of Galerkin equations in finite element methods.[21] This is true for two reasons. (i) The integrals defining the coefficients a_{ij} involve Green's functions $v_{\mathbf{y}_j}$. Since the Green's function $v_{\mathbf{y}_j}$ is singular at $\gamma = \mathbf{y}_j$, the computation of a_{ij} requires the evaluation of a *singular* integral if $\mathbf{y}_j \in \operatorname{supp} b_{h,i}$. (ii) In boundary integral methods, all integrals are defined as *surface integrals* (see Section 1.5). When such a surface integral is transformed into a Riemann integral, the *Gram determinant* of the parameterization ψ enters the Riemann integral. This phenomenon may lead to a significantly more complex integrand in the Riemann integral than is apparent from the original surface integral formulation. For these reasons, appropriate numerical integration methods are required for the computation of the coefficients and right-hand sides of the discretized equations arising in boundary integral methods.

In contrast to the discussion concerning finite elements, no convergence results for boundary integral methods have been put forward in this section. This absence is due to the fact that the theory of boundary integral methods is significantly less developed than the theory of finite element methods. For instance, according to Atkinson [8], collocation methods have not been

[21] It should be noted that the coefficient matrix (a_{ij}) of (2.29) is generally *not* sparse. This is due to the fact that Green's functions $v_{\mathbf{y}_j}$ have *global* support, $v_{\mathbf{y}_j}|_{\Omega_k} \not\equiv 0, \quad k = 1, \ldots, M_h$.

analyzed for boundary integral equations of the first kind derived from the three-dimensional Laplace equation.

The study of numerical integration techniques for (boundary) integral equations is a very active field of research. Recent references are Atkinson and Venturino [9], Baker [13], Bertram [31], Eggermont and Lubich [122], Kieser, Schwab, and Wendland [233], Rosen and Cormack [376], Calio and Marchetti [49], Elliott [127], and Palamara Orsi [328].

Chapter 3

Fundamentals of Computational Integration

3.1. Integration Problems

The representative examples of integration[22] problems in Chapter 2 demonstrate that many scientific problems require the calculation of integrals

$$\mathrm{I}f := \int_B f(\mathbf{x})\, d\mathbf{x}$$

for given integrands $f : B \subseteq \mathbb{R}^N \to \mathbb{R}$ defined on given regions B.[23] Most of these integrals exist in the Riemann sense as either proper or improper integrals (see Sections 1.1 and 1.2, respectively).

However, for some applications (such as the boundary integral methods described in Section 2.4) it is necessary to compute integrals which exist only in a generalized sense as Cauchy principal value integrals (see Section 1.3) or as finite part integrals (see Section 1.4). Additionally, for p-dimensional manifolds B in $\mathbb{R}^N$, $p < N$, integrals on B are to be evaluated as surface integrals (see Section 1.5).

In many applications, *weight functions* $w : B \subseteq \mathbb{R}^N \to \mathbb{R}$ are included in the integrals:

$$\mathrm{I}_w f := \mathrm{I}(wf) = \int_B w(\mathbf{x}) f(\mathbf{x})\, d\mathbf{x}.$$

These weight functions are often nonnegative (for instance, in the statistical integration problems described in Section 2.1),

$$w(\mathbf{x}) \geq 0 \quad \text{for all} \quad \mathbf{x} \in B,$$

[22]The term "quadrature" is often used instead of "integration" for one-dimensional (univariate) problems. For multidimensional (multivariate) integration problems, the term "cubature" is common. This terminology enables a distinction between the problem of calculating integrals and the integration of differential equations.

[23]In this book, it is—unless explicitly stated otherwise—tacitly assumed that neither the integrand f nor the integration region B is *indeterminate*, i.e., defined in terms of unknown parameters.

or oscillatory (for instance, in the Fourier and Hankel integral transforms; see Section 2.2). In either case, a characteristic feature of weight functions is that, in contrast to integrands f, they are generally known analytically. This makes it possible to devise tailor-made computational methods for the various weight functions w used in practice.

Many numerical integration methods are based on approximating the integrand f using polynomial model functions p. This approach makes sense only if the integral $\mathrm{I}_w p$ exists for any polynomial $p \in \mathbb{P}^N$. *Nonnegative* weight functions satisfying this condition as well as $\mathrm{I}w > 0$ are said to be *admissible*.

The term *kernel function* and the symbol k are used for weight functions which depend on an additional parameter $\mathbf{y}$. Such kernel functions occur, for instance, in integral transforms and integral equations (see Section 2.2). Commonly used weight and kernel functions are described in Section 3.1.2.

Integral transforms, as well as several other applications, require a (possibly large) number m of integrals to be calculated as part of an algorithmic step:

$$\begin{aligned} \mathrm{I}_1 f_1 &= \int_{B_1} k(\mathbf{x}, \mathbf{y}_1) f_1(\mathbf{x})\, d\mathbf{x}, \\ &\vdots \\ \mathrm{I}_m f_m &= \int_{B_m} k(\mathbf{x}, \mathbf{y}_m) f_m(\mathbf{x})\, d\mathbf{x}. \end{aligned}$$

In the most general case, the m integrals may involve m different integration regions $B_1, \ldots, B_m$, weight functions $k(., \mathbf{y}_1), \ldots, k(., \mathbf{y}_m)$, and integrands $f_1, \ldots, f_m$. However, the underlying integration regions are often, for instance, in integral transforms, independent of the parameter m,

$$B_1 = B_2 = \cdots = B_m =: B.$$

The same independence property frequently holds for the weight functions,

$$k(., \mathbf{y}_1) = k(., \mathbf{y}_2) = \cdots = k(., \mathbf{y}_m) =: w.$$

The efficiency of computational integration methods for multiple integrals can be increased by exploiting such common problem components.

3.1.1. Integration Regions. Most computational (in particular, numerical) schemes used for (approximately) calculating integrals are formulated for certain "standardized" integration regions, which are prototypical examples of the integration regions that occur most frequently in practice. If a given integration region B is different from these standard regions, then the corresponding integration problem must be subjected to appropriate transformations in order to make these computational schemes applicable (see Section 3.6). The following list contains the most common standard integration regions along with their definitions. For a more extensive listing, see Stroud [423].

Region	**Notation**	**Definition**
entire space	E_N ($\mathbb{R}^N$)	$\{\mathbf{x} \in \mathbb{R}^N\}$
unit cube	C_N	$\{\mathbf{x} \in \mathbb{R}^N : \|\mathbf{x}\|_\infty \leq 1\}$
positive unit cube	C_N^+	$C_N \cap \left(\mathbb{R}_0^+\right)^N$
cubic shell	$C_N^{shell}(K_1, K_2)$	$\{\mathbf{x} \in \mathbb{R}^N : K_1 \leq \|\mathbf{x}\|_\infty \leq K_2\}$
unit sphere	S_N	$\{\mathbf{x} \in \mathbb{R}^N : \|\mathbf{x}\|_2 \leq 1\}$
spherical shell	$S_N^{shell}(K_1, K_2)$	$\{\mathbf{x} \in \mathbb{R}^N : K_1 \leq \|\mathbf{x}\|_2 \leq K_2\}$
surface of S_N	U_N	$\{\mathbf{x} \in \mathbb{R}^N : \|\mathbf{x}\|_2 = 1\}$
unit octahedron	G_N	$\{\mathbf{x} \in \mathbb{R}^N : \|\mathbf{x}\|_1 \leq 1\}$
unit simplex	T_N	$G_N \cap \left(\mathbb{R}_0^+\right)^N$

3.1.2. Weight and Kernel Functions

Univariate Weight and Kernel Functions. The following list contains the most frequently occurring weight functions together with the underlying integration region for one-dimensional integration (quadrature) problems. A more extensive listing can be found in Piessens [333].

Interval	**Weight Function** $w(x)$	**Name**
$[-1, 1]$	1	Legendre
$[-1, 1]$	$(1 - x^2)^{-1/2}$	Chebyshev, first kind
$[-1, 1]$	$(1 - x^2)^{1/2}$	Chebyshev, second kind
$[-1, 1]$	$(1 - x)^\alpha (1 + x)^\beta,\ \alpha, \beta > -1$	Jacobi
$[0, \infty)$	$\exp(-x)$	Laguerre
$[0, \infty)$	$x^\alpha \exp(-x),\ \alpha > -1$	Generalized Laguerre
$(-\infty, \infty)$	$\exp(-x^2)$	Hermite
$(-\infty, \infty)$	$1/\cosh x$	Hyperbolic Cosine

Similarly, the most common univariate kernel functions are:

Interval	**Kernel Function** $k(x, y)$	**Name**
$[-1, 1]$	$\|x - y\|^\alpha,\ \alpha > -1$	Algebraic
$[-1, 1]$	$\log \|x - y\|$	Logarithmic
$(-\infty, \infty)$	$\cos yx,\ \sin yx,\ y \in \mathbb{R}$	Trigonometric, Fourier
$[0, 2\pi]$	$\cos lx,\ \sin lx,\ l \in \mathbb{N}$	Finite Trigonometric, Fourier

In connection with Cauchy principal value integrals and finite part hypersingular integrals, the following kernel functions are of greatest importance:

Interval	Kernel Function $k(x, y)$	Name
$[-1, 1]$	$\|x-y\|^{-1}$	Hilbert
$[-1, 1]$	$\|x-y\|^{-n-1},\ n \in \mathbb{N}$	

Other weight functions are obtained by forming products of the preceding weight and kernel functions. An *algebraic-logarithmic* kernel, for instance, is (as the name implies) a product of algebraic and logarithmic kernel functions.

Multivariate Weight and Kernel Functions. Standard weight and kernel functions for multidimensional integration problems are obtained as follows. The space $\mathbb{R}^N$ is parameterized by a one-to-one function $\psi : \mathbb{R}^N \to \mathbb{R}^N$ such that the parameter set $\overline{B}$ corresponding to the original integration region B, $B = \psi(\overline{B})$, becomes an N-dimensional rectangle,

$$\overline{B} = [a_1, b_1] \times [a_2, b_2] \times \cdots \times [a_N, b_N].$$

By assigning weight functions $w_1, \ldots, w_N$ to the intervals $[a_1, b_1], \ldots, [a_N, b_N]$, a weight function $w : B \subseteq \mathbb{R}^N \to \mathbb{R}$ can be defined by

$$w(\mathbf{x}) := w_1((\psi^{-1}(\mathbf{x}))_1) w_2((\psi^{-1}(\mathbf{x}))_2) \cdots w_N((\psi^{-1}(\mathbf{x}))_N).$$

Example (Spherical Coordinates). The two-dimensional space $B = E_2$ can be parameterized by

$$x_1 = r\cos\varphi \qquad \text{and} \qquad x_2 = r\sin\varphi,$$

where $(r, \varphi) \in \overline{B} = [0, \infty) \times [0, 2\pi)$. Assigning the Laguerre weight function $w_1(r) := \exp(-r)$ to the interval $[0, \infty)$ and the Legendre weight function $w_2(\varphi) \equiv 1$ to $[0, 2\pi)$ yields the weight function

$$w(x_1, x_2) = \exp\left(-\sqrt{x_1^2 + x_2^2}\right)$$

for E_2.

The following common multidimensional weight functions are obtained using a parameterization in spherical coordinates:

Region	Weight Function $w(\mathbf{x})$	Name
E_N^r	$\exp(-\\|\mathbf{x}\\|_2)$	N-dimensional Laguerre
$E_N^{r^2}$	$\exp(-\\|\mathbf{x}\\|_2^2)$	N-dimensional Hermite

3.2. Problem Settings

The choice of an appropriate computational method for solving a given integration problem depends largely on the specific *problem setting*, i.e., on the form and contents of

- the information available regarding the input data, i.e., the integrand f and the integration region B;
- the requirements concerning the integration result, i.e., accuracy, reliability, etc.; and
- other properties expected from and restrictions imposed on admissible integration algorithms.

This section describes the various problem settings encountered in computational practice, while the various classes of computational integration methods and their suitability for particular problem settings are discussed in Section 3.3.

3.2.1. Input Data

The Integrand. The form and the content of available information about the integrand f has a significant impact on the choice as to an appropriate method for solving a computational integration problem. In most cases the problem belongs to one of the following three categories.

(i) Values $f(\mathbf{x}_1), \ldots, f(\mathbf{x}_n)$ of the integrand $f : B \subseteq \mathbb{R}^N \to \mathbb{R}$ are available only for a given, fixed set of points $\{\mathbf{x}_i \in B,\ i = 1, \ldots, n\}$. These points may be distributed either systematically or irregularly over the region B.

In this setting, the values $\{f(\mathbf{x}_i)\}$ are often results of measurements (representing, for instance, the result of experiments). A substantial amount of stochastic noise is often superimposed on such data.

(ii) The function f is defined for and can be evaluated at any point $\mathbf{x} \in B$. This situation is taken as the basis of most numerical integration programs. In this case, the integrand is usually described by a function procedure (subprogram) which supplies the value $f(\mathbf{x})$ of the integrand for any specified input argument $\mathbf{x} \in B$.

Example (Fortran). The integrand $f(x) = (1 + x^4)^{-1}$ can be described by the following Fortran 90 function subprogram:

```
FUNCTION f(x) RESULT (f_integrand)
   REAL, INTENT (IN) :: x
   REAL              :: f_integrand

   f_integrand = 1./(1. + x**4)

END FUNCTION f
```

The name `f` of this function subprogram is passed on to an integration subroutine which chooses suitable sample points $\mathbf{x}_1, \ldots, \mathbf{x}_n$ and evaluates $f(\mathbf{x}_1), \ldots, f(\mathbf{x}_n)$ by calling `f`.

(iii) The integrand is given by an explicit representation (a mathematical formula) which lends itself to symbolic manipulation.

Input data providing additional information about the integrand f can significantly facilitate the computational solution of integration problems. The most important types of auxiliary information concern the derivatives of the integrand.

(i) It may be known that the integrand f is k-times ($k \in \mathbb{N}$) or infinitely many times (continuously) differentiable.

(ii) Bounds M_k on the kth (partial) derivative(s) of f may be available.

(iii) The kth (partial) derivative(s) of f may be known, either as a discrete set of function values, as a subprogram, or as a mathematical formula.

Furthermore, information regarding the abscissas $\mathbf{x}_0$ at which the integrand f exhibits exceptional behavior is highly valuable.

(iv) The exact or approximate location of points $\mathbf{x}_0$ at which the integrand f is singular, discontinuous, or of reduced differentiability may be known.

(v) Information concerning the asymptotic behavior of $f(\mathbf{x})$ in the vicinity of such points, for instance, $f(\mathbf{x}) \sim \|\mathbf{x} - \mathbf{x}_0\|_2^\alpha$ as $\mathbf{x} \to \mathbf{x}_0$, may be available.

Unfortunately, in many practical integration problems this information is only partially available or not available at all.

Weight and Kernel Functions. Tailor-made computational integration methods exist for many *parameterized families* of weight and kernel functions (see Section 3.1.2). The particular integration problem under consideration has to be specified by providing parameter values which identify the kernel function.

The Integration Region. The integration regions of univariate integration problems are (collections of) intervals. These intervals are usually defined by their left and right endpoints a and b. Multivariate integration regions, on the other hand, can be represented in fundamentally different ways.

(i) *Standard Representation*: If the integration region B is the affine image of one of the standard regions described in Section 3.1.1, B can usually be specified in terms of a small number of geometric parameters. For instance, if B is a sphere, then it can be defined by specifying its center and its radius.

(ii) *Iterated Representation*: If a multivariate integral can be transformed into an iterated integral over intervals,

$$\mathrm{I}f = \int_B f(\mathbf{x})\,d\mathbf{x} = \int_{a_1}^{b_1}\int_{a_2(x_1)}^{b_2(x_1)} \cdots \int_{a_N(x_1,\ldots,x_{N-1})}^{b_N(x_1,\ldots,x_{N-1})} f(x_1,\ldots,x_N)\,dx_N \cdots dx_1,$$

then the integration region B can be specified by the N interval functions

$$a_k, b_k \colon \mathbb{R}^{k-1} \to \mathbb{R}, \qquad k = 1, 2, \ldots, N.$$

(iii) *Boundary Representation*: If the integration region $B \subset \mathbb{R}^N$ is bounded, B can be specified by providing an explicit, implicit, or parametric representation of the boundary $\Gamma = \partial B$. Implicit boundary representations are the most general form of specifying bounded integration regions. They are, however, generally unwieldy for the purposes of computational integration.

3.2.2. Integration Results. A crucial criterion for the choice of a computational method for solving a given integration problem is the *accuracy* requirement imposed on the results.

(i) *Mathematical Accuracy*: If an integration problem, in particular, the integrand f, is given in a closed, analytical form, it may be possible to provide an *exact mathematical formula* for the integral. The provision of such an exact mathematical formula is often necessary when the integration result is to be used in connection with mathematical proofs or derivations.

(ii) *Numerical Accuracy*: For many practical integration problems, it is not possible or advisable to determine the solution in terms of an exact mathematical formula (see Section 3.3). In these cases, one often tries to compute a solution $\mathrm{Q}f$ of the *numerical problem*

$$\begin{array}{ll} \textit{input data:} & f,\ B,\ \varepsilon \text{ (possibly also } w), \\ \textit{output data:} & \mathrm{Q}f \quad \text{with} \quad |\mathrm{Q}f - \mathrm{I}f| \le \varepsilon \end{array} \tag{3.1}$$

for a specific error tolerance ε.

(iii) *Miscellaneous Accuracy Requirements*: If a computational integration method is used within a more complex computational procedure, it may have to meet requirements specific to the overall procedure. For instance, integration methods used in finite element methods must be able to accurately integrate certain classes of low-order polynomials (see Section 2.3).

The *reliability* expected from the result, i.e., the probability that the result meets the given accuracy requirement, also has a crucial impact on the choice of an appropriate computational integration method.

The maximum accuracy and reliability which can realistically be expected from the results of computational integration depend crucially on the quality of the input data. If, for instance, the integrand f is not provided in closed form, a mathematically exact result cannot be obtained; if f is given in tabular form (at a discrete set of points) only, then in general neither the mathematical nor the numerical problem of integration can be solved.

3.2.3. Algorithmic Objectives and Restrictions. In order to be suitable for a given integration problem, a computational method must be able to handle the specific form of problem data and produce results which meet given accuracy and reliability requirements. A computational method may, however, have to fulfill other objectives and restrictions as well. For instance, since integrals are linear operators on the underlying space of integrands, it is often desirable that computational methods should also be *linear algorithms* (see Section 3.3.2).

In many applications, the computational effort needed to calculate (approximate) integral values provides a crucial indication of the quality of an algorithm. For instance, applications which require the computation of a large number of high-dimensional integrals may need an unacceptable amount of computation time if the integration algorithm used is inefficient. A thorough discussion of the performance aspects of computational integration algorithms can be found in Section 10.1.2.

3.3. Integration Methods

This section outlines the main computational methods available for solving integration problems. It highlights the salient features of each method and discusses its applicability in conjunction with the various problem settings described in Section 3.2.

3.3.1. Analytical Integration. Analytical integration methods aim at providing an exact mathematical formula, usually in terms of elementary functions, for an integral under consideration. For quadrature problems this task usually requires finding an elementary antiderivative for the given univariate integrand f. For cubature problems the multivariate integral usually has to be expressed, possibly after an appropriate change of variable, as an iterated integral, and elementary antiderivatives have to be found for all iterated univariate integrands.

Analytical integration must be used if an exact mathematical formula is the only acceptable result for the given integration problem (see Section 3.2). There are, however, inherent limitations on analytical integration:

(i) Analytical integration is not applicable if the integrand is not known "exactly" (as a formula), but is given only in tabular form or via a function subprogram (which evaluates, for instance, the integrand at given abscissas $\mathbf{x} \in B$ by approximately solving an algebraic or differential equation which cannot be solved explicitly).

(ii) The indefinite integrals of many univariate functions occurring in practice (even of many elementary functions) cannot be expressed as elementary formulas.

Example (Nonelementary Integral). The integration of $f(x) = \exp(-x^2)$ leads to a function (the *error function*) that *cannot* be written in terms of a finite number of algebraic, logarithmic, or exponential expressions (Rosenlicht [377]).

(iii) Multidimensional integrals can only be dealt with analytically if they can be transformed into iterated integrals whose integration boundaries are expressed explicitly in analytical form. If, for instance, the integration region B is given by an implicit representation (see, for instance, the example on page 205), an analytical conversion into an iterated integral may not be possible.

Additionally, even if an exact mathematical formula exists and can be found, there are always different, mathematically equivalent ways of expressing the result. The particular form obtained by applying an analytical integration method may prove to be unsuitable for practical purposes. For instance, evaluating the formula may be numerically unstable (due to cancellation, for instance) or even impossible (due to division by zero, for instance; see the examples in the remainder of this section).

Manual Integration. Using the transformation rule and integration by parts (see Section 1.1.5) as well as other special techniques, it is sometimes possible to transform a univariate integration problem in such a way that the antiderivative F of the transformed integrand f is known and can be represented in terms of elementary functions. According to the fundamental theorem of integral calculus,

$$\mathrm{I}f = \int_a^b f(x)\,dx = F(b) - F(a)$$

can be obtained immediately. For multivariate integrals, iterated integration must also be applied in order to transform the multivariate integration problem into an embedded sequence of univariate problems.

Example (Partial Fractions Decomposition). The integral

$$\int_0^1 f(x)\,dx = \int_0^1 \frac{dx}{1+x^4} \tag{3.2}$$

can be calculated analytically. By decomposing the integrand into partial fractions (see Section 4.3), F can be written in terms of elementary indefinite integrals:

$$\int \frac{dx}{1+x^4} = \frac{1}{4\sqrt{2}} \log \frac{x^2 + x\sqrt{2} + 1}{x^2 - x\sqrt{2} + 1} + \frac{1}{2\sqrt{2}} \arctan \frac{x\sqrt{2}}{1-x^2} + \text{const.}$$

However, this formula cannot be used directly to calculate a numerical value for (3.2) because at $x = 1$ a "division by zero" occurs. Suitable transformations lead to the alternative formula

$$\int_0^1 \frac{dx}{1+x^4} = \frac{1}{4\sqrt{2}} \log \frac{2+\sqrt{2}}{2-\sqrt{2}} + \frac{1}{2\sqrt{2}} \left[\arctan \frac{1}{\sqrt{2}-1} + \arctan \frac{1}{\sqrt{2}+1} \right],$$

which can be evaluated numerically.

Manual integration requires a significant amount of time and energy and is also a highly error-prone procedure. These problems are particularly severe for multidimensional integrals.

Tables of Integrals. Before computers became customary tools for symbolic manipulations, *integral tables* were used to avoid the problems associated with manual integration (i.e., the large amount of time required and the high probability of making mistakes).

Example (Table of Integrals). In the reference book by Bronstein and Semendjajew [42], the indefinite integral

$$\int \frac{dx}{a^4 + x^4} = \frac{1}{4a^3\sqrt{2}} \log \frac{x^2 + ax\sqrt{2} + a^2}{x^2 - ax\sqrt{2} + a^2} + \frac{1}{2a^3\sqrt{2}} \arctan \frac{ax\sqrt{2}}{a^2 - x^2}$$

can be found. As a result, the time-consuming decomposition of the integrand into partial fractions can be avoided; yet transformations are still required in order to evaluate the antiderivative at the terminals $x = 0$ and $x = 1$.

However, the probability of obtaining an incorrect result is still not negligible, even when using the relevant literature. In a survey of eight widespread (common) tables of integrals, Klerer and Grossman [234] found that the frequency of errors was larger than 5%.

Symbolic Integration. With the advent of symbolic integration methods, ways of integrating functions given in closed form (as formulas) using symbolic software systems have emerged.

Example (Mathematica). The software package MATHEMATICA (Wolfram [447]) can be used to calculate the indefinite integral

$$\int \frac{dx}{1+x^4}.$$

The interactive command `Integrate[(1+x^4)^(-1),x]` yields

$$\begin{aligned} \int \frac{dx}{1+x^4} &= \frac{\arctan\left(\frac{-\sqrt{2}+2x}{\sqrt{2}}\right)}{2^{3/2}} + \frac{\arctan\left(\frac{\sqrt{2}+2x}{\sqrt{2}}\right)}{2^{3/2}} \\ &\quad - \frac{\log(1-\sqrt{2}x+x^2)}{2^{5/2}} + \frac{\log(1+\sqrt{2}x+x^2)}{2^{5/2}}. \end{aligned}$$

This indefinite integral can easily be evaluated at the boundaries of the interval [0, 1]:

$$\int_0^1 \frac{dx}{1+x^4} = \frac{\arctan\left(\frac{2-\sqrt{2}}{\sqrt{2}}\right)}{2^{3/2}} + \frac{\arctan\left(\frac{2+\sqrt{2}}{\sqrt{2}}\right)}{2^{3/2}} - \frac{\log(2-\sqrt{2})}{2^{5/2}} + \frac{\log(2+\sqrt{2})}{2^{5/2}}.$$

If a user is sufficiently familiar with a particular symbolic computation system, then integrating a function symbolically is often less time-consuming and less error-prone than looking up the integrand in an integral table or performing the integration manually. Even more importantly, symbolic integration algorithms can determine whether or not an elementary integral exists for a given function; if so, they can determine the indefinite integral in a finite number of computational steps.

Example (Mathematica). The indefinite integral

$$\int \sin(\sin x)\, dx$$

cannot be expressed as an elementary function. MATHEMATICA, for instance, recognizes this fact automatically and informs the user.

By contrast, if one fails to calculate an elementary indefinite integral manually, the indefinite integral may still be an elementary function—one may just have applied inappropriate transformations. Similarly, if an integrand cannot be found in an integral table, it simply may be that the integral table is not extensive enough.

The usefulness of symbolic integration is limited by the principal restrictions on analytical integration described at the beginning of this section. Furthermore, the computation of the antiderivative F and the subsequent evaluation of F at the endpoints of the integration interval are often far more time-consuming than the numerical computation of the definite integral.

Example (Mathematica). The turnaround time for the symbolic determination of the indefinite integral (3.3) on an HP 9000/835 workstation is around 6 seconds. The definite integral

$$\int_0^1 \frac{dx}{1+x^4} \approx 0.8669\,72987\,33992$$

can be evaluated numerically with a relative error of $6.2 \cdot 10^{-15}$ using the 15-point Gauss–Kronrod formula implemented in the subroutine `QUADPACK/dqk15`. The computing time required is several orders of magnitude smaller than the time needed for the symbolic manipulations. In fact, the computing time is significantly smaller than 10 ms, the resolution of the available clock routines.

While symbolic computation software is based on mathematically proven algorithms which solve well-defined problems, the enormous extent and complexity of symbolic computation systems makes software implementation errors practically inevitable. Such errors can occur even with the most trivial mathematical tasks.

Example (Implementation Error). The MAPLE command `int(x*(1-x), x=0..1);` for calculating the simple integral

$$\int_0^1 x(1-x)\, dx = \frac{1}{6}$$

yields the *wrong*—negative!—result: $-1/6$.

A concise survey of the most important symbolic computation systems is given in Section 3.5.6, whereas a more detailed discussion of symbolic integration methods can be found in Chapter 4.

3.3.2. Numerical Integration. The analytical integration methods described in Section 3.3.1 are usually applied when an exact mathematical formula is the desired result of a computational integration procedure. However, as pointed out in Section 3.2.2, it is often not necessary to calculate such mathematically exact results, and it may simply not be possible to obtain them (see Section 3.3.1). For these reasons, it is often advisable to calculate an approximate solution Qf using *numerical* rather than analytical integration techniques.

Numerical Integration Using a Fixed Set of Data Points. If only a fixed, finite set of values

$$\{f(\mathbf{x}_1), \ldots, f(\mathbf{x}_n) : \mathbf{x}_i \in B \subseteq \mathbb{R}^N, \, i = 1, 2, \ldots, n\}$$

of the integrand is available, Qf is obtained by interpolating or approximating the data points $(\mathbf{x}_i, f(\mathbf{x}_i))$, $i = 1, \ldots, n$, using a function g chosen from a suitable linear space $\mathcal{M}$ of analytically integrable model functions. The exact integral value $\mathrm{I}g$ is then used as an approximation for $\mathrm{I}f$, $Qf := \mathrm{I}g$.

The quality of this type of integral approximation is determined by the quality of the approximation of f by g, which in turn depends, among other things, on the number and location of abscissas $\mathbf{x}_1, \ldots, \mathbf{x}_n$ in B. One of the main goals in numerical integration is therefore to find good, or even optimal sets of abscissas. Since this cannot be done when a fixed set of abscissas is given a priori, most of the numerical integration methods described in the remainder of this book are *not* applicable to integrands specified by a fixed set of function values.

The calculation of $\mathrm{I}g$ for a given model function g is relatively simple compared to the selection of suitable spaces $\mathcal{M}$ of model functions and the computation of good interpolation or approximation functions g. For instance, if the data points $\{(\mathbf{x}_i, f(\mathbf{x}_i))\}$ are results of measurements (describing the result of experiments, for instance), the abscissas $\{\mathbf{x}_i\}$ may be very irregularly distributed and a substantial amount of stochastic noise may be superimposed on the function values $\{f(\mathbf{x}_i)\}$. For both reasons, a model function g is likely to deviate significantly from the actual function f unless special care is taken when choosing the approximation method. Hence, integration problems of this type are to be regarded primarily as approximation problems, a thorough discussion of which is beyond the scope of this book. The interested reader should refer, for instance, to Ueberhuber [435].

The following software section provides references to subprograms which numerically evaluate definite integrals of common model functions, in particular, piecewise polynomial functions and polynomial splines. Software references for the computation of interpolation and approximation functions can be found in Ueberhuber [435].

Software (Univariate Integration Using a Fixed Set of Data Points). The subroutines `IMSL/MATH-LIBRARY/dbsitg` ≈ `CMLIB/dbsqad` ≈ `SLATEC/dbsqad` as well as `IMSL/MATH-LIBRARY/dcsitg` and `IMSL/MATH-LIBRARY/dppitg` compute definite integrals of B-splines, cubic spline functions, and other piecewise polynomials over bounded intervals $[a, b]$. The programs `CMLIB/dppqad` ≈ `SLATEC/dppqad` also compute definite integrals of B-splines. They differ from routines like `IMSL/MATH-LIBRARY/dbsitg` only in the way the B-splines are represented.

The subroutines `NAG/e02ajf` and `NAG/e02bdf` calculate definite integrals of piecewise Hermite interpolation polynomials and of cubic splines, respectively.

The subroutine `NAG/d01gaf` calculates an approximation to the integral of f on $[x_1, x_n]$ for given data points

$$\{(x_i, f(x_i)),\ i = 1, 2, \dots, n\}, \qquad x_1 < x_2 < \cdots < x_n.$$

On each subinterval $[x_i, x_{i+1}]$, $i = 2, 3, \dots, n-2$, the integrand f is approximated by a cubic polynomial interpolating the four data points

$$(x_{i-1}, f(x_{i-1})), \quad (x_i, f(x_i)), \quad (x_{i+1}, f(x_{i+1})), \quad (x_{i+2}, f(x_{i+2})).$$

In approximating f on the boundary intervals $[x_1, x_2]$ and $[x_{n-1}, x_n]$, the sets

$$\{(x_j, f(x_j)),\ j = 1, 2, 3, 4\} \qquad \text{and} \qquad \{(x_j, f(x_j)),\ j = n-3, n-2, n-1, n\}$$

are used. A similar method is used in the subroutine `SLATEC/davint`.

Software (Multivariate Integration Using a Fixed Set of Data Points). The subroutines `IMSL/MATH-LIBRARY/dbs2ig` and `IMSL/MATH-LIBRARY/dbs3ig` calculate definite integrals of tensor-product splines over two- and three-dimensional rectangles, respectively.

Numerical Integration Using Samples. The remaining chapters dealing with numerical integration consider only the solution of integration problems in which the integrand f can be evaluated at any point $\mathbf{x} \in B$. The approximate integration result $\mathrm{Q}f$ is in this case supposed to either solve the *numerical problem* (3.1) or meet other, application-specific accuracy requirements.

In most numerical integration methods, a finite number of values of the integrand, the *sampling information*

$$S(f) := (f(\mathbf{x}_1), f(\mathbf{x}_2), \dots, f(\mathbf{x}_n)),$$

is used for calculating the approximate value $\mathrm{Q}f$ for $\mathrm{I}f$. The sampling points $\mathbf{x}_1, \dots, \mathbf{x}_n$ are chosen by the numerical integration algorithm.

Among all possible ways of defining $\mathrm{Q}f$, *linear algorithms* of the form

$$\mathrm{Q}f = \mathrm{Q}_n f := \sum_{i=1}^{n} c_i f(\mathbf{x}_i) \tag{3.3}$$

are predominantly used.[24] Methods which do not belong to this category (with the notable exception of algorithms based on nonlinear extrapolation) are of minor importance in practice.

The n sampling points $\mathbf{x}_1, \ldots, \mathbf{x}_n$ in (3.3) are called *integration abscissas* (or *integration nodes*[25]); the values $c_1, \ldots, c_n$ are called *integration weights* (or *integration coefficients*). The weighted sum of function values (3.3) is called a numerical *n-point integration formula* if the abscissas and weights are chosen independently of the observed behavior of the integrand f.

Numerical integration techniques are, beginning with Chapter 5, dealt with thoroughly in this book.

3.4. The Conditioning of Integration Problems

In practice, the input data of integration problems, i.e., the integration region B and the integrand f, are often known only *approximately*. Such data errors limit the maximum accuracy which can reasonably be expected from the result of any computational procedure.

Numerical integration methods provide only approximate integral values for the problem data. The sensitivity of numerically calculated results to changes in the data may differ from the sensitivity of the exact integration results. If the numerically calculated results prove to be significantly more sensitive to data errors than the exact results, the numerical integration method used is numerically *unstable*.

It is therefore important to study the conditioning of both exact and numerical integration. This topic is discussed in detail in the remainder of this section.

3.4.1. Changes in the Integrand

Riemann Integrals. The *absolute* conditioning (sensitivity) of Riemann integrals $\mathrm{I}_w f$ with respect to changes in the integrand f is characterized by the smallest positive number $K_{f,w}$ such that

$$|\mathrm{I}_w \tilde{f} - \mathrm{I}_w f| \leq K_{f,w} \, \|\tilde{f} - f\|_\infty \tag{3.4}$$

holds for all functions $\tilde{f}$ which are integrable on B with respect to the weight function w.[26] If the absolute value function $|w|$ is integrable on B, i.e., if w

[24] The summation index usually starts with $i = 1$ and ends with $i = n$. However, it is sometimes preferable to change the upper and the lower bound of the summation index while retaining the meaning of n as the overall number of sampling points.

[25] In order to avoid confusion with the nodes of a parallel computer, the sampling points of a numerical integration procedure are consistently referred to in this book as integration *abscissas*.

[26] As the weight function w is usually known analytically, it is assumed there is no data error associated with w.

is *absolutely integrable*, the (absolute) *condition number* $K_{f,w}$ is given by the L_1-norm $\| \ \|_1$ of w. The estimate

$$\begin{aligned} |\, \mathrm{I}_w \tilde{f} - \mathrm{I}_w f \,| &= \left| \int_B w(\mathbf{x}) \tilde{f}(\mathbf{x})\, d\mathbf{x} - \int_B w(\mathbf{x}) f(\mathbf{x})\, d\mathbf{x} \right| \\ &\le \int_B |w(\mathbf{x})| |\tilde{f}(\mathbf{x}) - f(\mathbf{x})|\, d\mathbf{x} \\ &\le \|w\|_1 \, \|\tilde{f} - f\|_\infty \end{aligned} \tag{3.5}$$

shows that $\|w\|_1$ is an upper bound for $K_{f,w}$, i.e., $K_{f,w} \le \|w\|_1$. On the other hand, equality is attained in (3.5) for the perturbed functions

$$\tilde{f}(\mathbf{x}) := \begin{cases} f(\mathbf{x}) + c & \text{for} \quad w(\mathbf{x}) \ge 0, \\ f(\mathbf{x}) - c & \text{for} \quad w(\mathbf{x}) < 0, \end{cases}$$

which proves that $\|w\|_1 \le K_{f,w}$.

The absolute condition number $K_{f,w}$ depends only on the integral operator I_w, but *not* on the integrand f. The condition estimate (3.5) indicates that, for weight functions w with a *small* L_1-norm, the *mathematical* integration problem is *well* conditioned with respect to variations in the integrand.

For weight functions w with a *large* L_1-norm, the error $|\mathrm{I}_w \tilde{f} - \mathrm{I}_w f|$ can become large even when the data perturbation $\|\tilde{f} - f\|_\infty$ is small. Nevertheless, the *relative* conditioning in these cases, i.e., the smallest positive number $K_{f,w}^{\mathrm{rel}}$ such that

$$\frac{|\mathrm{I}_w \tilde{f} - \mathrm{I}_w f|}{|\mathrm{I}_w f|} \le K_{f,w}^{\mathrm{rel}} \frac{\|\tilde{f} - f\|_\infty}{\|f\|_\infty},$$

can still be small, indicating that the mathematical integration problem is well conditioned.

From the estimate (3.5) and the definition of $K_{f,w}^{\mathrm{rel}}$, the relationship

$$K_{f,w}^{\mathrm{rel}} = \frac{\|w\|_1 \|f\|_\infty}{|\mathrm{I}_w f|} \ge 1 \tag{3.6}$$

can easily be derived. In contrast to the absolute condition number $K_{f,w}$, the relative condition number $K_{f,w}^{\mathrm{rel}}$ depends on the integrand f under consideration. For nonnegative weight functions w, the constant functions $f \equiv c$ attain the optimal relative condition number $K_{f,w}^{\mathrm{rel}} = 1$. On the other hand, $K_{f,w}^{\mathrm{rel}}$ can become arbitrarily large if $|\mathrm{I}_w f|$ is small in comparison to $\|f\|_\infty$. In these cases small changes in the integrand (caused, for instance, by rounding errors) may entail large relative errors in the result.

Example (Integration Problem with Large Relative Condition Number). The integral

$$\mathrm{I} f := \int_0^1 \cos(a\pi x)\, dx, \quad a \in \mathbb{R}^+,$$

becomes zero for $a = 2, 4, 6, \ldots$. Since $\|w\|_1 = 1$ for the given weight function $w \equiv 1$ and $\|\cos(a\pi x)\|_\infty = 1$, the relative condition number

$$K_{f,w}^{\text{rel}} = 1/|\mathrm{I}_w f| = \left|\frac{a\pi}{\sin(a\pi)}\right|$$

is obtained from (3.6). This condition number becomes large when a is close to a zero of the denominator $\sin(a\pi)$, i.e., when $a \approx 2k$, $k \in \mathbb{N}$. For $a \approx 20$, for instance, the following values are obtained:

a	19.90	19.92	19.94	19.96	19.98	20
$K_{f,w}^{\text{rel}}$	202.3	251.6	333.3	500.3	999.7	∞

If w is not absolutely integrable, i.e., $\|w\|_1 = \infty$, $|\mathrm{I}_w\tilde{f} - \mathrm{I}_w f|$ cannot be bounded in terms of $\|\tilde{f} - f\|_\infty$, and consequently the mathematical problem of integration is, both absolutely and relatively, ill conditioned. Important examples of such weight functions w are the trigonometric (Fourier) weight functions $w(x) = \sin(yx)$ and $w(x) = \cos(yx)$, $y \in \mathbb{R}$, for the infinite interval $(-\infty, \infty)$. These weight functions occur in the direct and inverse Fourier transforms and in the inverse Laplace transform (see Section 2.2).

Generalized Integrals. Cauchy principal value integrals and Hadamard finite part integrals usually involve nonintegrable weight functions w, for instance, the Hilbert weight function $w(x) = 1/(x - y)$. Integrals of functions f with respect to w exist only in a generalized sense (see Sections 1.3 and 1.4). These integrals are *ill conditioned* with respect to changes in the integrand f. This fact can be demonstrated by considering the Hilbert transform

$$\mathrm{I}f = \fint_{-1}^{1} \frac{f(x)}{x}\, dx$$

and defining the functions $\tilde{f}_{\alpha,\varepsilon}$, $\alpha, \varepsilon > 0$, by

$$\tilde{f}_{\alpha,\varepsilon}(x) := \begin{cases} f(x) + \varepsilon|x|^\alpha & \text{for} \quad x \geq 0, \\ f(x) - \varepsilon|x|^\alpha & \text{for} \quad x < 0. \end{cases}$$

Clearly, $\|\tilde{f}_{\alpha,\varepsilon} - f\|_\infty = \varepsilon$ and

$$\mathrm{I}\tilde{f}_{\alpha,\varepsilon} - \mathrm{I}f = \mathrm{I}(\tilde{f}_{\alpha,\varepsilon} - f) = \frac{2\varepsilon}{\alpha}.$$

Hence, for a given level of error $\varepsilon > 0$, the difference $|\mathrm{I}\tilde{f}_{\alpha,\varepsilon} - \mathrm{I}f|$ can be made *arbitrarily large* by choosing the positive parameter α sufficiently close to 0.

Numerical Integration. The (absolute) condition number $K_{f,\mathrm{Q}}$ with respect to changes in the integrand f of the approximating linear functional Q in (3.3) is characterized by the smallest positive number $K_{f,\mathrm{Q}}$ such that

$$|\mathrm{Q}\tilde{f} - \mathrm{Q}f| \le K_{f,\mathrm{Q}} \, \|\tilde{f} - f\|_\infty.$$

It can easily be seen that $K_{f,\mathrm{Q}}$ is given by

$$K_{f,\mathrm{Q}} = \|\mathrm{Q}\|_1 := \sum_{i=1}^{n} |c_i|.$$

The estimate

$$\begin{aligned} |\mathrm{Q}\tilde{f} - \mathrm{Q}f| &= |\mathrm{Q}(\tilde{f} - f)| \\ &\le \sum_{i=1}^{n} |c_i| \, |\tilde{f}(\mathbf{x}_i) - f(\mathbf{x}_i)| \\ &\le \left(\sum_{i=1}^{n} |c_i| \right) \|\tilde{f} - f\|_\infty \end{aligned} \tag{3.7}$$

establishes $K_{f,\mathrm{Q}} \le \|\mathrm{Q}\|_1$. On the other hand, equality is attained in (3.7) for any perturbed function $\tilde{f}$ with

$$\tilde{f}(\mathbf{x}_i) := \begin{cases} f(\mathbf{x}_i) + c & \text{for} \quad c_i \ge 0, \\ f(\mathbf{x}_i) - c & \text{for} \quad c_i < 0, \end{cases}$$

for $i = 1, \ldots, n$, and consequently $\|\mathrm{Q}\|_1 \le K_{f,\mathrm{Q}}$.

If the weight function w is nonnegative and if Q yields the exact value $\mathrm{I}_w f$ for all constant functions $f(\mathbf{x}) \equiv c$, then the condition number $K_{f,\mathrm{Q}}$ of the approximating functional Q cannot be smaller than the condition number $K_{f,w}$ of the exact functional I_w:

$$K_{f,\mathrm{Q}} = \sum_{i=1}^{n} |c_i| \ge \sum_{i=1}^{n} c_i = \mathrm{I}w = \|w\|_1 = K_{f,w}.$$

The equality $K_{f,\mathrm{Q}} = K_{f,w}$ is attained if and only if all weights $c_1, \ldots, c_n$ are *nonnegative*. All linear integration algorithms Q for nonnegative weight functions w in practical use either have positive weights or their condition numbers $K_{f,\mathrm{Q}}$ are only insignificantly larger than the optimal condition number $K_{f,w}$.

The conditioning of *nonlinear* integration algorithms based on extrapolation methods (see Section 3.7) is a complex phenomenon. However, most extrapolation algorithms are potentially *ill conditioned*. Great care has to be taken to avoid the occurrence of numerical instabilities in such algorithms (see Brezinski and Redivo Zaglia [41]).

3.4.2. Changes in the Region. The conditioning of the exact integral $\mathrm{I}_w f = \mathrm{I}_w(f;B)$ with respect to changes in the integration region B can be considerably worse than the conditioning with respect to changes in the integrand f. Consider, for example, the one-dimensional integration region $B = [a,b]$ and the weight function $w \equiv 1$. Changes in the interval endpoints can be measured by appropriate norms on $\mathbb{R}^2$ such as the L_1-norm. The fundamental theorem of integral calculus shows that the (absolute) condition number $K_{f,[a,b]}$ of the integral with respect to small changes in the boundaries,

$$|\mathrm{I}(f;[\tilde{a},\tilde{b}]) - \mathrm{I}(f;[a,b])| \leq K_{f,[a,b]} \|(\tilde{a},\tilde{b})^\top, (a,b)^\top\|_1,$$

is given by $K_{f,[a,b]} = \max\{|f(a)|, |f(b)|\}$. The condition number $K_{f,[a,b]}$ is large if f is singular at or near the boundary of the integration region.

Example (Ill-Conditioned Quadrature Problem). The integral

$$\mathrm{I}(f;[0,b]) := \int_0^b (1-x)^{-0.9}\,dx = -10(1-b)^{0.1} + 10$$

is extremely sensitive to variations in the upper boundary b if b is close to 1:

b	$1-10^{-5}$	$1-10^{-10}$	$1-10^{-15}$	$1-10^{-20}$
$\mathrm{I}(f;[0,b])$	6.8377	9.0000	9.6838	9.9000
$K_{f,[0,b]} = f(b)$	$3.16 \cdot 10^4$	$1.00 \cdot 10^9$	$3.16 \cdot 10^{13}$	$1.00 \cdot 10^{18}$

On a computer implementing IEC/IEEE floating-point arithmetic, the distance between the floating-point numbers slightly smaller than 1 is given by

$$\Delta x = 2^{-24} \approx 5.96 \cdot 10^{-8} \qquad \textit{(single precision)}$$

or

$$\Delta x = 2^{-53} \approx 1.11 \cdot 10^{-16} \qquad \textit{(double precision)}.$$

Rounding b to the nearest floating-point number may, in the worst case, produce a rounding error of $2.98 \cdot 10^{-8}$ or $5.55 \cdot 10^{-17}$, respectively. Consider the integration problem with the exact right interval endpoint $b = 1 - 2.98 \cdot 10^{-8}$ or $b = 1 - 5.55 \cdot 10^{-17}$. Converting b into a single or double precision floating-point number yields $\Box b = 1$ in either case. While the exact integral for the exact problem data is given by $\mathrm{I}(f;[0,b]) = 8.23$ or $\mathrm{I}(f;[0,b]) = 9.76$, the exact integral for the rounded problem data is $\mathrm{I}(f;[0,\Box b]) = 10$. Thus, due to the poor conditioning of the integration problem for right interval endpoints $b \approx 1$, the exact solutions for the exact and rounded problem data may have *none* or only *one* significant decimal digit.

3.5. Software for Computational Integration

Using existing high-quality software for solving computational integration problems is generally preferable to developing new programs from scratch. Not only does the production of correct, bug-free software take a lot of time and energy, but the implementation of certain computational techniques (for instance, symbolic integration methods or reliable error estimation procedures for numerical integration) requires specialized knowledge available to only a small number of experts in the given field. Even if ready-made software proves to be unsatisfactory for a given problem, it is generally still easier to adapt available software to one's needs than to design it from scratch.

This section provides a survey of the various kinds of software available for solving computational integration problems and of the sources from which such software can be obtained. The software products mentioned are referred to throughout this book in conjunction with the specific computational methods they implement.

3.5.1. Software and the Internet. Computer software is commonly distributed using permanent storage devices such as magnetic tapes, floppy disks, or CD-ROMs, while information concerning available software is disseminated using print media such as journals, marketing brochures, etc. However, global electronic communication network services, the *Internet* in particular, have recently become increasingly important both to the dissemination of information and to the distribution of software. These services offer far quicker access to programs, documents, data, etc. than traditional communication methods, and they are constantly available (except when hardware or software faults occur or when maintenance work is carried out). Many software products described in this book can be obtained using Internet services.

It is beyond the scope of this book to discuss the various kinds of general Internet services, such as *e-mail*, *USENET news*, *FTP*, *telnet*, *WWW*, etc. It is generally assumed that the reader is familiar with these services as well as with the special terminology associated with them, like *Internet (e-mail) address*, *URL*, etc. Information about these topics can be found in any of the numerous Internet books, for instance, Krol [241] or Quarterman and Carl-Mitchell [345].

There are, however, special services dedicated to the distribution of (information about) mathematical software which are relevant to computational integration, but less well known. These services are briefly described in the remainder of this section.

NETLIB. The Internet service NETLIB (Network library; Dongarra and Grosse [115]) is of foremost importance to the distribution of *public domain* numerical software. NETLIB is a software repository containing not only numerical programs, subroutine packages, and program libraries, but also software development tools, program documentation, etc. The software repository is mirrored at a number of geographically dispersed sites all over the world. It is strongly recommended that one should access the nearest mirror site in order to keep access and transfer times low and to reduce the overall network traffic.

NETLIB can be accessed using the basic Internet services FTP-by-mail, anonymous FTP, and WWW. The easiest way to get started and to learn more about NETLIB is to send the e-mail message `send index` to the e-mail address `netlib@ornl.gov`. The automatically generated e-mail answer contains (i) information about how to access NETLIB via the aforementioned Internet services, (ii) alternative Internet addresses of a number of mirror sites, and (iii) a top-level index of NETLIB's contents.

Since NETLIB supports WWW accesses (the URL of NETLIB's home page is `www.netlib.org`) the various WWW clients (for instance, MOSAIC or NETSCAPE) available on various platforms provide graphical user interfaces to NETLIB. Alternatively, the tailor-made XNETLIB service provides, independently of the WWW, a powerful graphical user interface to NETLIB using the X-WINDOW system. XNETLIB clients can be obtained from NETLIB itself.

GAMS. The *Guide to Available Mathematical Software* (GAMS) database was created and is maintained by the National Institute of Standards (NIST) (Boisvert, Howe, and Kahaner [37]). The database stores information regarding a large number of subroutines contained in major program libraries and collections. The subroutines are ordered according to a hierarchical, problem-oriented classification scheme, the GAMS *index*. The GAMS database enables one not only to search for subroutines and retrieve information about them, but also to fetch the subroutines themselves, provided they are in the public domain.

The GAMS database can be accessed via a menu-oriented user interface by using the Internet service *telnet* to log in to `gams.nist.gov` with the username `gams`. By logging in as user `xgams`, an X-WINDOW graphical user interface is obtained instead. Client programs for both the `gams` and `xgams` services can be obtained by anonymous FTP from `enh.nist.gov:gams`. The GAMS database can also be accessed via the WWW system by using the URL `http://gams.nist.gov`.

3.5.2. Single Programs. In the 1960s two journals, *Numerische Mathematik* and *Communications of the ACM*, began to print single programs, written mostly in Algol 60. Although *Numerische Mathematik* stopped printing programs some time ago, many of the published programs are, in updated form, still used in current program packages and libraries. The ACM (Association of Computing Machinery), on the other hand, still runs the "Collected Algorithms of the ACM" series. Programs have been published in the journal *Transactions on Mathematical Software (TOMS)* since 1975. A new journal dedicated to the publication of numerical algorithms and programs has recently been founded under the name *Numerical Algorithms*. *TOMS* and *Numerical Algorithms* journal programs are stored in the directories `toms` and `numeralgo`, respectively, of NETLIB (see Section 3.5.1).

Most of the programs published in journals are written using Fortran 77. More recently, the languages Fortran 90, Fortran 95, and C have been used by an increasing number of authors. Since the programs are peer reviewed, software quality standards, i.e., portability, reliability, efficiency, etc., are generally high. Although the quality of program documentation is sometimes unsatisfactory, the most severe deficiency is that they lack the infrastructure and the services, i.e., maintenance and support, that usually come with commercial products: the author of a published program cannot generally be expected to help users with their problems or to fix program bugs promptly.

Programs and algorithms are published occasionally in other journals as well, for instance, in the *Journal of Computational and Applied Mathematics*. The corresponding, unrefereed software can often be obtained by writing directly to the author(s) of the respective article. There is, however, no guarantee as to the quality of the software.

An alternative source of single programs is the *Numerical Recipes* book series by Press et al. [342], [343], which contains a large number of well-structured sample programs illustrating numerical methods. These programs are, however, *not* meant to be used as production codes; no particular effort has been made to make them highly reliable, efficient, etc. These sample programs are thus not considered in this book.

3.5.3. Software Packages. A software package is a systematic compilation of programs for a specific set of problems. The subroutine package QUADPACK, developed by Piessens et al. [339], provides a number of programs for the numerical solution of univariate integration problems. Since its publication in 1983, QUADPACK has become a de facto standard for one-dimensional numerical integration. Most major program libraries (see Section 3.5.4) have adopted and incorporated QUADPACK (with only minor modifications). The QUADPACK subroutine suite can also be obtained directly from the directory `quadpack` on NETLIB.

Recently, a C++ package for bivariate integration, CUBPACK++, has been developed and published as algorithm 764 of *TOMS*. Updated versions of this package can be obtained via anonymous FTP from `ftp.cs.kuleuven.ac.be:/pub/NumAnal-ApplMath/Cubpack`. It is beyond the scope of this book to adequately describe the novel (from the viewpoint of numerical integration) object-oriented concepts and techniques used in this package, and therefore CUBPACK++ will not be referenced in this book any further. Detailed mathematical and algorithmic information can be found in Cools, Laurie, and Pluym [62]; a user manual is contained in Cools, Laurie, and Pluym [63].

3.5.4. Program Libraries. Most of the numerical integration programs referenced in this book can be found in one of the major numerical libraries. This section provides a concise overview of libraries, relevant to numerical integration, which are available as either portable source codes or binary codes for a broad spectrum of hardware/system software platforms. The reader may, however, also find numerical integration programs in local, platform-dependent libraries, such as ESSL, SCILIB, etc.

NAG. The Numerical Algorithms Group Ltd. (NAG) provides a comprehensive Fortran 77 numerical library, containing in its most recent release, Mark 17, as many as 1152 documented subroutines. The software pointers given in this book refer to this library. NAG also distributes Fortran 90 and C libraries, which provide more convenient subroutine interfaces, but are not as extensive as the Fortran 77 library.

Moreover, NAG has recently released a new library of numerical routines, called the NAG Parallel Library, specifically designed for parallel and distributed computer systems (shared and distributed memory parallel computers, networks and clusters of workstations, etc.). This library is aimed at users with computational problems large enough to make efficient use of the increased processing power and memory capacity of multiple processors. NAG's homepage address on the WWW is `http://www.nag.co.uk`.

IMSL. Visual Numerics Inc., formerly known as IMSL (International Mathematical and Statistical Libraries), distributes (among other things) Fortran 77, Fortran 90, and C/C++ libraries for numerical and statistical computing. The software pointers given in this book refer generally to the Fortran 77 library, which is the most extensive of the IMSL libraries. Visual Numerics Inc.'s homepage address on the WWW is `http://www.vni.com`.

CMLIB. NIST's Core Math Library (CMLIB) is a public domain collection of high-quality, easily portable Fortran 77 packages which solve standard problems in many areas of mathematics and statistics. It contains approximately 750 subroutines. More detailed information about CMLIB as well as single CMLIB programs can be obtained using the GAMS system.

SLATEC. The SLATEC Common Mathematical Library is a collection of Fortran 77 subprograms for a wide variety of mathematical problems. The primary impetus for the development of this library was to provide portable, nonproprietary mathematical software for supercomputers at American research laboratories.

SLATEC is distributed by the Energy Science and Technology Software Center via `http://www.doe.gov/html/osti/estsc/estsc.html` on the WWW. The SLATEC library can be accessed directly from NETLIB (see the directory `slatec`) or via GAMS.

Harwell Subroutine Library. The Harwell Subroutine Library is a suite of Fortran 77 subroutines for scientific computation developed by the Numerical Analysis Group at Rutherford Appleton Laboratory and distributed by AEA Technology. For more detailed information see `http://www.rl.ac.uk/departments/ccd/numerical/hsl/hsl.html`.

3.5.5. Software References in this Book. Numerical programs designed to solve computational integration problems are referred to throughout this book in conjunction with the specific computational methods that they implement. Each program is briefly characterized by describing its basic algorithmic structure and/or the most important input/output parameters, and any particular merits or shortcomings known to the authors are mentioned. Whenever both single and double precision versions of a subprogram are available, only the user-callable name of the double precision routine is provided because it is, for reasons of reliability and accuracy, generally recommended to avoid using single precision routines.

The notational system used to identify the various programs is similar to the syntax used for UNIX filenames. For instance,

`NAG/d01gaf`

denotes the numerical integration program `d01gaf` in the NAG Fortran 77 library [311], while

`TOMS/698`

refers to Algorithm 698 in *TOMS*. The following list contains the various path names used in this book and the software repositories they refer to:

`CMLIB/`	CMLIB library
`HARWELL/`	Harwell library
`IMSL/MATH-LIBRARY/`	IMSL Fortran 77 MATH/LIBRARY
`IMSL/STAT-LIBRARY/`	IMSL Fortran 77 STAT/LIBRARY
`NAG/`	NAG Fortran 77 library
`NAG-PAR/`	NAG Parallel Library
`NUMERALGO/`	`Numerical Algorithms` collection
`QUADPACK/`	QUADPACK package
`SLATEC/`	SLATEC library
`TOMS/`	*TOMS* collection

The notation

$$\texttt{IMSL/MATH-LIBRARY/dqdag} \approx \texttt{QUADPACK/dqag}$$

indicates that two programs are identical in most major algorithmic aspects, although minor differences (for instance, in the argument lists) are possible.

3.5.6. Interactive Programming Systems. Standard numerical software has traditionally been provided in the form of single programs, subroutine packages, and libraries. In order to accomplish a specific task, the user has to write a main program which sets up all necessary data structures and calls the required subprograms, passing the appropriate data as arguments.

Recently, however, interactive programming systems have become increasingly popular for both numerical and symbolic computing. These systems provide the user with an interactive interface, generally in the form of a command-based interpreter. In order to perform a given task, the user starts the interpreter, formulates the required mathematical operations using the syntax of the command language, and passes the command string to the interpreter. The interpreter then parses the string, sets up the necessary data structures, and calls the required subroutines.

The main advantage of interactive programming systems is that they allow for an *explorative programming style*, in which the user does not plan all computational steps in advance, but rather determines each computational step depending on previously obtained results. Moreover, by hiding most internal software details (subroutine parameters, data structures, calling sequences, etc.) from the user, interactive programming systems facilitate the *rapid prototyping* of programs. Interactive programming systems are therefore most suitable for solving small- or medium-sized nonrepetitive (one-of-a-kind) computational problems. They are less suitable for solving repetitive large problems (with varying data), especially when performance is a critical issue.

Numerical Interactive Programming Systems. There are several interactive programming systems for numerical computation, the most widely known of which is MATLAB (Hill and Moler [206], Coleman and Van Loan [60]). Compared to MATLAB, all other systems are of relatively minor significance.

MATLAB. MATLAB (Matrix Laboratory) was initially designed to perform matrix calculations and numerical linear algebra. However, MATLAB's functionality can be extended significantly by incorporating special-purpose, domain-specific *toolboxes*. In conjunction with numerical integration, the *NAG Foundation Toolbox*, which provides MATLAB access to the large number of numerical routines contained in the NAG Foundation Library, is of foremost importance. The NAG Foundation Toolbox consists of about 240 subroutines whose function names and calling syntax are derived from the NAG Foundation Library, a subset of the NAG Fortran 77 library. As a result, users familiar with the NAG Fortran 77 library can easily access the NAG Foundation Toolbox functions. More information concerning MATLAB can be found at the WWW address `http://www.mathworks.com/`.

Symbolic Interactive Programming Systems. A large number of software systems for symbolic computation (sometimes called computer algebra systems) have emerged over the last 30 years, a considerable number of them within the last decade. This development is partly due to the progress made in symbolic computation research, which has enabled such systems to solve a growing number of mathematical problems using algorithms of increasing efficiency. This has led to a continuing and growing interest amongst researchers, instructors, and students in such systems.

Advances in other areas of computer science have also contributed to the current popularity of symbolic computation systems. When dealing with complex real-world problems, symbolic computations can be fairly demanding in terms of both the number of operations and the amount of storage space required. Therefore, only large mainframe computers were suitable platforms for symbolic computation in the early 1980s. The access to such mainframe computers was, however, often cumbersome and by no means conducive to the highly interactive way in which one usually wants to use symbolic computation systems. Since then, workstations and personal computer systems have become available. Nowadays, due to improvements in computer hardware and architecture, these systems are so powerful that they can perform most algebraic calculation tasks. The development of graphical user interfaces for both workstations and personal computer systems has made symbolic computation systems far easier to use.

Interactive programming systems attempt to keep the inner structure of their software modules transparent (invisible) to the user. This approach proves to be particularly well suited for symbolic computation: symbolic software is based on mathematically proven algorithms which solve well-defined problems and can therefore be expected to be, with the notable exception of implementation errors, reliable. Hence, there is no need for the user of symbolic software to modify modules in order to improve their robustness and/or accuracy, as the user of numerical software sometimes must do—symbolic software modules can be used as black boxes. Moreover, data structures and module dependencies tend to be rather complicated in symbolic computation, which makes setting up data structures and calling subroutines a cumbersome and error-prone process.

This section provides short descriptions of the most commonly used symbolic computation systems (including the Internet addresses of the main distributors). A more extensive compilation can be obtained via the WWW document `http://symbolicnet.mcs.kent.edu/www-sites.html`. Except for DERIVE, all of the following software packages are available on a wide range of platforms, including personal computers, workstations, and supercomputer systems.

REDUCE. Of all symbolic computation systems still in use, REDUCE [286] has the longest tradition; its first version dates back to 1967. Since then it has undergone several fundamental revisions, and the most recent version, REDUCE 3.6, is dated July 1995. In contrast to more recent symbolic software packages, REDUCE is implemented using LISP, a language primarily used in artificial intelligence. Over the years a large number of independent third-party developers have contributed modules to REDUCE, providing special functions for specific application domains, thereby greatly extending REDUCE's basic capabilities. Detailed information about REDUCE, including references to other sources of information, can be found in the WWW document `http://www.rrz.uni-koeln.de/REDUCE/`.

MACSYMA. MACSYMA [88] was developed at MIT and first became available in 1971. MACSYMA incorporated the most advanced symbolic computation software in the 1970s and early 1980s. However, due to financial difficulties, the development of MACSYMA could not keep up with the rapid pace of progress in symbolic mathematics and software engineering in the 1980s. Recently, a change in ownership has brought a new influx of funds and has led to a revitalization of the MACSYMA system. For detailed information on MACSYMA, see `http://www.macsyma.com/`.

MAPLE. MAPLE [54] was designed at the University of Waterloo. It first appeared in 1981. MAPLE consists of a small kernel implemented in the programming language C and a large mathematical library written in the MAPLE command language which is interpreted by the kernel. The kernel incorporates only the most fundamental functions, i.e., integer and rational arithmetic, simple polynomial manipulation, efficient memory management, and the command interpreter. The small size of the kernel facilitates the porting of MAPLE to different platforms. For detailed information about MAPLE, see `http://www.maplesoft.com/`.

An improved graphical user interface to MAPLE is, among other things, provided by MATHCAD, which is distributed by Mathsoft, Inc.; see `http://www.mathsoft.com/`.

MATHEMATICA. MATHEMATICA [447], which was first released in 1988, was developed by Stephen Wolfram. The rapid commercial success of MATHEMATICA can be attributed mainly to its graphical capabilities which, at the time of its first appearance, significantly surpassed the graphics of all other existing systems. MATHEMATICA is implemented using the programming language C. More information can be found in `http://www.wri.com/`.

DERIVE. Among the major symbolic software systems, DERIVE [368] is an exception in two respects. First, it supports a menu-driven, instead of a command-based, interactive user interface and is therefore not designed to be used as a programming system. Second, rather than being available on a broad range of computer platforms, DERIVE runs only on MS DOS systems. Indeed, it was the first symbolic computation system to run efficiently on PC systems, which is why DERIVE is still popular among PC users. Information about DERIVE can be obtained from `http://www.mathware.com/derive.html`.

AXIOM. AXIOM [218] (formerly known as SCRATCHPAD) was developed at IBM's Thomas J. Watson Research Center and was initially available on IBM systems only. AXIOM is now distributed and developed by NAG and runs on a number of workstation systems and PCs. AXIOM is commonly acknowledged to be, from a mathematical point of view, the most coherent and sound symbolic software system (see, for instance, Davenport, Siret, and Tournier [88]) on the market. AXIOM subroutines can be used via the command-based interpreter, but they can also be called directly from user programs written in Fortran or C. Further information can be found in `http://www.nag.co.uk/symbolic/AX.html`.

3.6. The Preprocessing of Integrals

Numerical schemes for approximately calculating integrals cannot always, or at least not always *effectively*, be used for integration problems encountered in practice. For instance, the given integration region may be different from the standard integration regions described in Section 3.1.1. In such cases, the integration problem has to undergo a *preprocessing phase*; i.e., it has to be transformed manually or automatically into an equivalent integration problem which is better suited for numerical treatment. The most important preprocessing techniques, which are based on the mathematical properties of integrals described in Chapter 1, are discussed in this section.

Preprocessing techniques are frequently used for the construction of integration formulas (see Sections 5.1.4, 5.3.5, 6.1.3, 6.1.4, and 6.1.5). However, preprocessing techniques can be used for other purposes as well. For instance, *adaptive subdivision algorithms* (see Section 8.2.2) are normally applicable only if the integration region is a bounded N-dimensional cube. In order to make an adaptive subdivision algorithm applicable to integrals over unbounded and/or noncubic regions, the respective integral has to undergo an appropriate preprocessing phase.

3.6.1. The Transformation of Integrals. According to the transformation rule, given in Theorem 1.1.4, the equality

$$\mathrm{I}f = \int_B f(\mathbf{x})\, d\mathbf{x} = \int_{\overline{B}} \overline{f}(\overline{\mathbf{x}})\, d\overline{\mathbf{x}} \tag{3.8}$$

holds with the transformed integrand

$$\overline{f}(\overline{\mathbf{x}}) = |\det J(\overline{\mathbf{x}})|\, f(\psi(\overline{\mathbf{x}})), \quad \overline{\mathbf{x}} \in \overline{B}, \tag{3.9}$$

provided the transformation function $\psi : \overline{B} \subseteq \mathbb{R}^N \to B$ is continuously differentiable, and the Jacobian matrix $J(\overline{\mathbf{x}})$ is regular for all $\mathbf{x} \in \overline{B}$.

Preprocessing integrals using appropriate transformations functions ψ is often done when constructing integration formulas for regions B which differ from the standard regions $\overline{B}$ described in Section 3.1.1. Using an appropriate transformation ψ, the formula

$$\overline{Q}_n\overline{f} = \sum_{i=1}^{n} \overline{c}_i \overline{f}(\overline{\mathbf{x}}_i)$$

for the approximate integration of functions $\overline{f} : \overline{B} \to \mathbb{R}$ is turned into an integration formula

$$Q_n f = \sum_{i=1}^{n} c_i f(\mathbf{x}_i)$$

for functions $f : B \to \mathbb{R}$, and the transformed abscissas and weights are given by

$$\mathbf{x}_i := \psi(\overline{\mathbf{x}}_i), \quad c_i := |\det J(\overline{\mathbf{x}}_i)|\, \overline{c}_i, \quad i = 1, 2, \dots, n$$

(see Section 6.1.3). In other words, the abscissas of Q_n are simply images of the abscissas of $\overline{Q}_n$ with respect to the transformation ψ. The weights of Q_n are obtained by multiplying the corresponding weights of $\overline{Q}_n$ by the absolute values of the determinant of the Jacobian matrix at the respective abscissas.

The application of the transformed integration formula Q_n to a function $f : B \to \mathbb{R}$ can also be interpreted as the application of the original integration formula $\overline{Q}_n$ to the transformed integrand $\overline{f}$ in (3.9). This consideration shows that the error $Q_n f = \overline{Q}_n \overline{f}$ is small if $\overline{f}$ is an integrand for which the original integration formula $\overline{Q}_n$ is accurate. Whether or not this is the case depends on both the original integrand f *and* on the transformation ψ, specifically, its Jacobian J. For instance, $\overline{Q}_n$ can often be expected to be accurate if $\overline{f} \in C^k(\overline{B})$, $k \in \mathbb{N}$. Transformations ψ which map bounded regions $\overline{B}$ onto unbounded regions B (see Section 7.2.1) have singularities at the boundary of $\overline{B}$. In such cases, the condition $\overline{f} \in C^k(\overline{B})$ is not generally satisfied.

Two classes of *general purpose* transformations are described in Section 5.3.4 and 7.2.1. A thorough knowledge and understanding of the integration problem at hand may, however, enable the development of *tailor-made* transformations. Special transformations for statistical integration problems, for instance, are described by Genz [169].

If the integration problem is given in analytical form, the transformation of integrals can sometimes be facilitated by using symbolic computation software (see Davenport [87]).

Example (General-Purpose and Tailor-Made Transformations). Consider the integral

$$\mathrm{I}f = \int_{-\infty}^{\infty} \exp(-x^2) f(x)\, dx, \tag{3.10}$$

i.e., the integral of f with respect to the Hermite weight function $w(x) = \exp(-x^2)$. The infinite interval of integration $(-\infty, \infty)$ can be transformed into a finite interval by using the general-purpose transformation

$$\psi(\overline{x}) = \frac{\overline{x}}{1 - |\overline{x}|}, \qquad -1 < \overline{x} < 1.$$

Because

$$\psi'(\overline{x}) = \frac{1}{(1 - |\overline{x}|)^2},$$

the integral (3.10) is transformed into

$$\mathrm{I}f = \int_{-1}^{1} \frac{1}{(1 - |\overline{x}|)^2} \exp\left(-\left(\frac{\overline{x}}{1 - |\overline{x}|}\right)^2\right) f\left(\frac{\overline{x}}{1 - |\overline{x}|}\right) d\overline{x}. \tag{3.11}$$

On the other hand, the substitution $x = t/\sqrt{2}$ transforms (3.10) into

$$\mathrm{I}f = \frac{1}{\sqrt{2}} \int_{-\infty}^{\infty} \exp(-t^2/2) f(t/\sqrt{2})\, dt.$$

The infinite interval of integration $(-\infty, \infty)$ can now be transformed into a finite interval by using the tailor-made transformation

$$t = \psi(\overline{x}) := \Phi^{-1}(\overline{x}),$$

where

$$\Phi(t) = \frac{1}{\sqrt{2\pi}} \int_{-\infty}^{t} \exp(-u^2/2)\, du$$

is the distribution function for the standard normal distribution. Because

$$\psi'(\overline{x}) = \frac{1}{\Phi'(t)} = \sqrt{2\pi} \exp(t^2/2),$$

integral (3.10) is transformed into

$$\mathrm{I}f = \sqrt{\pi} \int_{0}^{1} f\left(\Phi^{-1}(\overline{x})/\sqrt{2}\right) d\overline{x}. \tag{3.12}$$

A comparison of (3.11) and (3.12) shows that the integrand in (3.12) has a much simpler structure. On the other hand, the integrand in (3.11) contains only elementary functions, whereas the integrand in (3.12) contains Φ^{-1}, which is not an elementary function. Φ^{-1} can be evaluated numerically, for instance, using the subroutine `IMSL/STAT-LIBRARY/danorin` or `NAG/g01faf`. As a result, it is not clear which of the integrals (3.11) and (3.12) is easier to compute numerically.

3.6.2. The Decomposition of Integrals. Integrals are *additive* with respect to the partitioning of B (the region of integration) into pairwise disjoint subregions $B_1, \ldots, B_L$,

$$\int_B f(\mathbf{x})\, d\mathbf{x} = \int_{B_1} f(\mathbf{x})\, d\mathbf{x} + \int_{B_2} f(\mathbf{x})\, d\mathbf{x} + \cdots + \int_{B_L} f(\mathbf{x})\, d\mathbf{x}$$

(see Section 1.1.3). Such a decomposition is usually part of the preprocessing when a complicated integration region B does not allow for the straightforward application of numerical integration schemes. The subregions $B_1, \ldots, B_L$ are chosen to be of a simpler form than B. For instance, an integral over a polyhedron B is usually computed by partitioning B into a finite number L of simplices and/or rectangles $B_1, \ldots, B_L$, for which numerical integration schemes exist.

It is important to know which error tolerances to impose on the numerical computations of the integrals over the subregions $B_1, \ldots, B_L$ in order to meet a given accuracy requirement for the overall integral over B. This problem is addressed in the context of adaptive numerical integration algorithms (see Section 8.2.2).

3.6.3. The Iteration of Integrals. The decomposition of an integration region B results in subregions $B_1, \ldots, B_L$ which are supposed to be geometrically simpler than B, but are of the same dimension N as B. In contrast, the iteration of an integral according to Theorem 1.1.5 is meant to decompose an integration problem into integrals of reduced dimension. The rationale behind this approach is that the construction of efficient numerical integration schemes becomes more and more difficult as the problem dimension N increases. The iteration of high-dimensional integrals makes it possible to employ numerical schemes for low-dimensional integrals.

The order of iteration may have a profound impact on the computational complexity of the subsequent numerical integration procedure.

Example (Iteration Order). Consider the two-dimensional integral

$$\mathrm{I}f := \int_B 1\, dx,$$

where B is defined by

$$B := \{\, (x_1, x_2) \in [0,1]^2 \;:\; 0 \le x_1 \le \sqrt{1 - x_2}\,\}.$$

Iterating $\mathrm{I}f$ yields either

$$\mathrm{I}f = \int_0^1 \int_0^{\sqrt{1-x_2}} dx_1 dx_2 = \int_0^1 \sqrt{1 - x_2}\, dx_2 \tag{3.13}$$

or

$$\mathrm{I}f = \int_0^1 \int_0^{1-x_1^2} dx_2 dx_1 = \int_0^1 (1 - x_1^2)\, dx_1. \tag{3.14}$$

The integrand $\sqrt{1-x_2}$ in (3.13) has a cusp at $x_2 = 1$, while $1 - x_1^2$ in (3.14) is an easily integrable second-degree polynomial. Hence, the numerical calculation of (3.13) is significantly more difficult than that of (3.14).

It is important to know which error tolerances to impose on the inner integrals of an iterated integral in order to meet a given accuracy requirement for the overall integral. This issue is addressed, for instance, in Fritsch, Kahaner, and Lyness [153].

Software (Iterated Integrals). Both `NAG/d01daf` and `IMSL/MATH-LIBRARY/dtwodq` compute two-dimensional integrals represented in iterated form (see Section 3.2.1).

3.7. The Postprocessing of Integrals

Preprocessing techniques are designed to transform integration problems in such a way that the transformed problem is more amenable to numerical techniques than the original one. By contrast, postprocessing techniques are designed to improve the accuracy of numerically obtained results by appropriately transforming a sequence $\{s_m := \mathrm{Q}_{n_m} f\}$ of approximate integral values. Such transformations are called *sequence transformations* because they map sequences $\{s_m\}$ onto sequences $\{\hat{s}_m\}$. If the original sequence $\{s_m\}$ converges, $s_m \to s$ as $m \to \infty$, and if the transformed sequence $\{\hat{s}_m\}$ is supposed to converge faster than the original sequence $\{s_m\}$,

$$\text{(3.15)} \qquad \lim_{m\to\infty} \frac{\hat{s}_m - s}{s_m - s} = 0, \quad \text{i.e.,} \quad \hat{s}_m - s = o(s_m - s),$$

then sequence transformations are referred to as *acceleration algorithms*.

3.7.1. Acceleration Techniques.

There is *no universal* acceleration algorithm which can speed up the convergence of all convergent sequences $\{s_m\}$. The nonexistence of a universal acceleration scheme still holds for restricted sets of convergent series, for instance, for the set of all (strictly) monotonic sequences and for the set of all partial sums of alternating series (see Brezinski and Redivo Zaglia [41]).

Instead, acceleration algorithms can be constructed based on specific *asymptotic expansions* of the error sequence $\{s_m - s\}$,

$$\text{(3.16)} \quad s_m - s = g(m; a_1, \dots, a_J) + o(g(m; a_1, \dots, a_J)) \quad \text{as} \quad m \to \infty.$$

The unknown parameters $a_1, \dots, a_J \in \mathbb{R}$ are independent of m and

$$\lim_{m\to\infty} g(m; a_1, \dots, a_J) = 0$$

holds for any admissible parameter vector $(a_1, \dots, a_J)^\top$.

If the term $o(g(m; a_1, \dots, a_J))$ in (3.16) were actually zero, then the $J+1$ unknowns $s; a_1, \dots, a_J$ could be determined by simultaneously solving the $J+1$

equations

$$(3.17) \qquad s_{\overline{m}} = s + g(\overline{m}; a_1, \ldots, a_J), \quad \overline{m} = m, m+1, \ldots, m+J,$$ [27]

for any arbitrary $m \in \mathbb{N}$. The solution $s; a_1, \ldots, a_J$ of (3.17) would be independent of the choice $m \in \mathbb{N}$, and thus the unknown limit s could be determined from any finite section $s_m, s_{m+1}, \ldots, s_{m+J}$, $m \in \mathbb{N}$, of the sequence $\{s_m\}$.

In general, however, the term $o(g(m; a_1, \ldots, a_J))$ in (3.16) does *not* vanish. Hence, the solution $s^{(m)}; a_1^{(m)}, \ldots, a_J^{(m)}$ of (3.17),

$$(3.18) \quad s_{\overline{m}} = s^{(m)} + g(\overline{m}; a_1^{(m)}, \ldots, a_J^{(m)}), \quad \overline{m} = m, m+1, \ldots, m+J,$$

depends on the index $m \in \mathbb{N}$ chosen. From (3.16) and (3.18), the following representation of the error in $s^{(m)}$, $m = 1, 2, 3, \ldots$, is obtained:

$$s^{(m)} - s = g(m; a_1, \ldots, a_J) - g(m; a_1^{(m)}, \ldots, a_J^{(m)}) + o(g(m; a_1, \ldots, a_J)).$$

If the function $g(m; a_1^{(m)}, \ldots, a_J^{(m)})$ exhibits the same asymptotic behavior as $g(m; a_1, \ldots, a_J)$,

$$g(m; a_1^{(m)}, \ldots, a_J^{(m)}) \sim g(m; a_1, \ldots, a_J) \qquad \text{as} \quad m \to \infty,$$

then $s^{(m)} - s = o(g(m; a_1, \ldots, a_J))$; i.e., the sequence $\{s^{(m)}\}$ converges faster to s than the original sequence $\{s_m\}$. The transformed sequence $\{\hat{s}_m\}$ is defined by $\hat{s}_m := s^{(m)}$, $m = 1, 2, 3, \ldots$. In acceleration methods, therefore, since

$$\hat{s}_m = s_m - g(m; a_1^{(m)}, \ldots, a_J^{(m)}),$$

it is often said that $\hat{s}_m$ is obtained by *eliminating* $g(m; a_1, \ldots, a_J)$. Whether or not the acceleration property $\hat{s}_m - s = o(s_m - s)$ actually holds for the transformed sequence $\{\hat{s}_m\}$ needs to be verified mathematically. Such analysis is not carried out in this book. For more information on the subject, see Brezinski and Redivo Zaglia [41] and the numerous references therein.

The aforementioned definition of the accelerated sequence $\{\hat{s}_m\}$ can also be interpreted as follows: the equations (3.17) can be regarded as interpolation conditions for the unknown function

$$\overline{m} \mapsto s + g(\overline{m}; a_1, \ldots, a_J),$$

and the calculated function

$$(3.19) \qquad \overline{m} \mapsto s^{(m)} + g(\overline{m}; a_1^{(m)}, \ldots, a_J^{(m)})$$

[27] In the remainder of this section, it is tacitly assumed that this system of equations can be solved analytically.

interpolates the data points

$$(m, s_m),\ (m+1, s_{m+1}), \ldots, (m+J, s_{m+J}).$$

Because $s_{\overline{m}} \to s$ for $\overline{m} \to \infty$, it is to be expected that good approximations for s can be obtained by evaluating the limit $\overline{m} \to \infty$ for the interpolation function (3.19). The convergence $g(\overline{m}; a_1^{(m)}, \ldots, a_J^{(m)}) \to 0$ as $\overline{m} \to \infty$ shows that this limit is equal to $s^{(m)}$; i.e., $s^{(m)}$ is obtained by *extrapolating* the value of the interpolation function (3.19) at $\overline{m} = \infty$. Accordingly, acceleration algorithms are often referred to as *extrapolation algorithms*.

A number of numerical integration schemes give rise to asymptotic error expansions which make suitable acceleration algorithms applicable to sequences $\{s_m = Q_{n_m} f\}$ of numerical integration results. The specific error expansions are discussed in conjunction with the corresponding integration methods in the following chapters.[28] Since the same, or similar, types of error expansions are valid for a number of numerical integration schemes, an acceleration algorithm may be applicable to more than one numerical integration method. Therefore, a short overview of the most important acceleration algorithms, their underlying asymptotic expansions, and the available software is given in this section. A comprehensive and thorough discussion of the properties and algorithmic details of extrapolation algorithms can be found in Brezinski and Redivo Zaglia [41].

3.7.2. Aitken's Δ^2 Process. Let the function $g(m; a_1, \ldots, a_J)$ in (3.16) be given by $J = 2$ and

$$g(m; a_1, a_2) = a_1 a_2^m, \qquad a_1 \neq 0,\ |a_2| < 1. \tag{3.20}$$

The system of equations (3.18) then becomes

$$\begin{aligned} s_m &= s^{(m)} + a_1^{(m)} \left(a_2^{(m)}\right)^m, \\ s_{m+1} &= s^{(m)} + a_1^{(m)} \left(a_2^{(m)}\right)^{m+1}, \\ s_{m+2} &= s^{(m)} + a_1^{(m)} \left(a_2^{(m)}\right)^{m+2}. \end{aligned} \tag{3.21}$$

Clearly, the parameter $a_2^{(m)}$ can be obtained by

$$a_2^{(m)} = \frac{s_{m+1} - s^{(m)}}{s_m - s^{(m)}}$$

[28] Examples of the application of extrapolation algorithms are also given in connection with particular integration methods.

as well as

$$a_2^{(m)} = \frac{s_{m+2} - s^{(m)}}{s_{m+1} - s^{(m)}}.$$

Equating the two right-hand sides gives rise to a quadratic equation for $s^{(m)}$, from which the formula

$$\text{(3.22)} \qquad s^{(m)} = \frac{s_m s_{m+2} - s_{m+1}^2}{s_{m+2} - 2s_{m+1} + s_m} = s_{m+2} - \frac{(\Delta s_{m+1})^2}{\Delta^2 s_m}$$

can be derived. In (3.22) the recursive definition

$$\text{(3.23)} \qquad \left.\begin{array}{rcl} \Delta s_m & := & s_{m+1} - s_m \\ \Delta^{k+1} s_m & := & \Delta^k s_{m+1} - \Delta^k s_m \end{array}\right\} \quad \text{for} \quad m, k \in \mathbb{N}$$

of the (forward) *difference operator* Δ^k is used.

The sequence transformation defined by (3.22) is called *Aitken's* Δ^2 *process*. It can be shown that the transformed sequence $\{\hat{s}_m := s^{(m)}\}$ indeed converges faster than the original sequence $\{s_m\}$, provided that the asymptotic expansion (3.16) with (3.20) holds.

3.7.3. Richardson's Extrapolation. Let the function $g(m; a_1, \dots, a_J)$ in (3.16) be given by

$$\text{(3.24)} \qquad g(m; a_1, \dots, a_J) = a_1 t_m + a_2 t_m^2 + \cdots + a_J t_m^J,$$

where the auxiliary sequence $\{t_m\}$ converges to zero. It is assumed that any element t_m of the auxiliary sequence is either defined a priori (independently of $\{s_m\}$) or can be calculated from a finite number of sequence elements s_m.

Since (3.24) is a homogeneous polynomial in t_m of degree J, the quantities $s^{(m)}; a_1^{(m)}, \dots, a_J^{(m)}$ satisfying the equations (3.18) are the coefficients of the unique polynomial of degree J which interpolates the data points

$$(t_m, s_m), (t_{m+1}, s_{m+1}), \dots, (t_{m+J}, s_{m+J}).$$

Since $s^{(m)}$ is the constant term in the interpolation polynomial, determining $s^{(m)}$ amounts to evaluating the interpolation polynomial at the abscissa 0. Hence, the quantities $s^{(m)} =: s^{(m;J)}$ can be obtained recursively using the *Neville–Aitken scheme* (see, for instance, Ueberhuber [435]) as follows:

$$\text{(3.25)} \qquad \left.\begin{array}{rcl} s^{(m;0)} & := & s_m \\ s^{(m;j)} & := & \dfrac{t_{m+j} s^{(m;j-1)} - t_m s^{(m+1;j-1)}}{t_{m+j} - t_m} \end{array}\right\} \quad \text{for } m \in \mathbb{N},\ j = 1, 2, \dots, J.$$

The instance (3.25) of the Neville–Aitken scheme is generally referred to as *Richardson's extrapolation*. For each $j = 0, \dots, J$, $s^{(m;j)}$ is the value of the unique polynomial of degree j interpolating the data points

$$(t_m, s_m), (t_{m+1}, s_{m+1}), \dots, (t_{m+j}, s_{m+j}).$$

Hence, the transformed sequence $\{ s^{(m;j)} : m = 1, 2, 3, \dots \}$ can be regarded as the result of eliminating the first j terms in the asymptotic error form (3.24). Since $t^{k+1} = o(t^k)$ as $t \to 0$, these first j terms are, from an asymptotic point of view, the most significant contributions to the error $s_m - s$. Performing Richardson's extrapolation (3.25) is thus not only computationally efficient and numerically stable, it also enables one to increase the number of most significant error terms eliminated *incrementally*. This property is of particular importance because the maximum number J of terms for which the error expansion (3.24) holds is often not initially known in practice.

Similar to Richardson's extrapolation, most practical extrapolation algorithms are based on recursive implementations of sequence transformations: the asymptotic error is represented in the form

$$\begin{aligned}(3.26)\quad s_m - s \;=\;& g_1(m; a_{j_0+1}, \dots, a_{j_1}) + g_2(m; a_{j_1+1}, \dots, a_{j_2}) + \cdots \\ &+ g_K(m; a_{j_{K-1}+1}, \dots, a_{j_K}) + o(g_K(m; a_{j_{K-1}+1}, \dots, a_{j_K})),\end{aligned}$$

where

$$0 = j_0 < j_1 < \cdots < j_K = J$$

and, for $k = 1, 2, \dots, K-1$,

$$g_{k+1}(m; a_{j_k+1}, \dots, a_{j_{k+1}}) = o(g_k(m; a_{j_{k-1}+1}, \dots, a_{j_k})) \qquad \text{as} \quad m \to \infty.$$

The transformed sequences $\{ s^{(m;k)} : m = 1, 2, 3, \dots \}$ are recursively obtained for $k = 1, 2, \dots, K$ by eliminating the first k most significant terms in (3.26). The corresponding recurrence formula makes it possible to efficiently calculate $s^{(m;k)}$ from sequence elements $s^{(m;k-1)}$, in the same way as (3.25) does. This approach is feasible only if a suitable recurrence formula can be found for a given asymptotic expansion (3.26), which in many cases is difficult.

3.7.4. Shank's Transformation. Let the functions g_k in (3.26) be given by $j_k := 2k$ and

$$(3.27)\qquad g_k(m; a_{2k-1}, a_{2k}) = a_{2k-1} a_{2k}^m, \qquad a_{2k-1} \neq 0, \ |a_{2k-2}| < |a_{2k}|$$

for $k = 1, 2, \dots, K$ ($|a_2| < 1$). Note that for $K = 1$ the asymptotic expansion (3.20) for Aitken's Δ^2 process is obtained.

The process of calculating the transformed sequence $\{\hat{s}_m := s^{(m;K)}\}$ by eliminating the error terms $g_1, \dots, g_K$ is referred to as *Shank's transformation*.

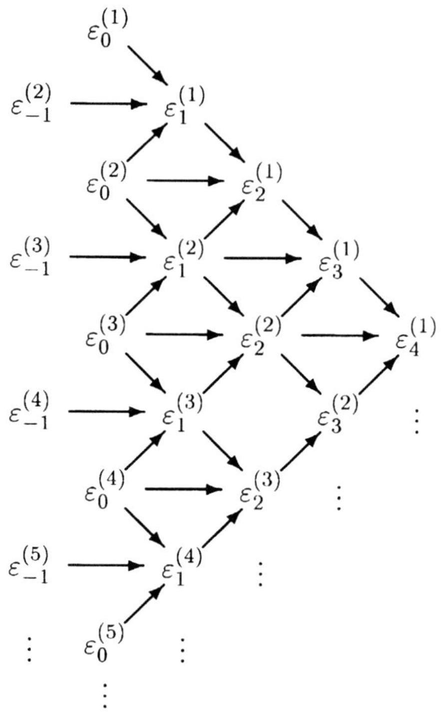

FIG. 3.1. *Dependency graph of the ε-algorithm.*

Shank's transformation is usually performed recursively using the ε-*algorithm* (Wynn [449]):

$$\begin{aligned} \varepsilon_{-1}^{(m)} &:= 0, \\ \varepsilon_{0}^{(m)} &:= s_m, \quad m = 1,2,3,\ldots, \\ \varepsilon_{k+1}^{(m)} &:= \varepsilon_{k-1}^{(m+1)} + \frac{1}{\varepsilon_k^{(m+1)} - \varepsilon_k^{(m)}}, \qquad m = 1,2,3,\ldots, \qquad k = 0,1,\ldots,2K-1. \end{aligned}$$

In the transformed sequences $\{s^{(m;k)} := \varepsilon_{2k}^{(m)}\}$, the first k most significant terms $g_1, \ldots, g_k$ in (3.27) are eliminated. A graphical representation of the data dependencies in the ε-algorithm is given in Fig. 3.1.

3.7.5. Levin's Transformations. Let the functions g_k in (3.26) be given by $j_k := k$, and

$$g_k(m; a_k) = t_m a_k (m + m_0)^{-k+1}, \quad k = 1, 2, \ldots, K, \tag{3.28}$$

where the known real constant m_0 is not a negative integer and the auxiliary sequence $\{t_m\}$ has the properties stated in Section 3.7.3. Transformations for sequences with asymptotic error expansions of type (3.28) are called *Levin's transformations*. More specifically, the particular choice

$$t_m := (m + m_0)\Delta s_{m-1}$$

gives rise to Levin's *u-transform*,

$$t_m := \Delta s_{m-1}$$

leads to Levin's *t-transform*, while

$$t_m := -\Delta s_{m-1}\Delta s_m/\Delta^2 s_{m-1}$$

results in Levin's *v-transform* (Levin [258]).[29] Other choices of t_m are possible (see, for instance, Smith and Ford [408], [409]).

Recurrence formulas for Levin's transformations as well as a detailed discussion of their convergence properties can be found in Brezinski and Redivo Zaglia [41].

3.7.6. *W*-Transform. Let the functions g_k in (3.26) be given by $j_k := k$, and

$$g_k(m; a_k) = g(t_m)a_k t_m^{k-1}, \quad k = 1, 2, \ldots, K, \tag{3.29}$$

where g is continuous and the auxiliary sequence $\{t_m\}$ converges to zero. Note that for $g(t) \equiv 1$ and $a_0 = 0$ the asymptotic expansion (3.29) reduces to (3.24), the asymptotic expansion from Richardson's extrapolation. For $k = 1, \ldots, K$, the transformed sequences $\{s^{(m;k)}\}$, in which the error terms $g_1, \ldots, g_k$ are eliminated, can be calculated recursively using the following *W-algorithm* (Sidi [391]):

$$\left.\begin{aligned} M_0^{(m)} &:= \frac{s_m}{g(t_m)} \\ N_0^{(m)} &:= \frac{1}{g(t_m)} \\ M_k^{(m)} &:= \frac{M_{k-1}^{(m+1)} - M_{k-1}^{(m)}}{t_{m+k} - t_m} \\ N_k^{(m)} &:= \frac{N_{k-1}^{(m+1)} - N_{k-1}^{(m)}}{t_{m+k} - t_m} \\ s^{(m;k)} &:= \frac{M_k^{(m)}}{N_k^{(m)}} \end{aligned}\right\} \text{ for } \quad m \in \mathbb{N}, \quad k = 1, 2, \ldots, K. \tag{3.30}$$

[29] The definition $s_0 := 0$ is used for Δs_0.

3.7.7. Summation Process. Let the functions g_k in (3.26) be given by $j_k := k$, and

$$g_k(m; a_k) = a_k b_k^m, \quad |b_{k-1}| < |b_k|, \quad k = 1, 2, \ldots, K, \quad |b_1| < 1. \tag{3.31}$$

The asymptotic expansion (3.31) is identical to (3.27) except that the bases $b_1, \ldots, b_K$ in (3.31) are, in contrast to the bases $a_2, \ldots a_{2K}$ in (3.27), assumed to be known. The corresponding acceleration method is known as a *summation process* and can be implemented using the following simple recursive algorithm:

$$\left.\begin{array}{lll} s^{(m;0)} & := & s_m \\ s^{(m;k)} & := & \dfrac{s^{(m+1;k-1)} - b_k s^{(m;k-1)}}{1 - b_k} \end{array}\right\} \text{ for } \quad m \in \mathbb{N}, \quad k = 1, 2, \ldots, K. \tag{3.32}$$

The limit case of the recursion (3.32) for $b_k \to -1$, $k = 1, \ldots, K$, is the *Euler transform*:

$$\left.\begin{array}{lll} s^{(m;0)} & := & s_m \\ s^{(m;k)} & := & \dfrac{s^{(m+1;k-1)} + s^{(m;k-1)}}{2} \end{array}\right\} \text{ for } \quad m \in \mathbb{N}, \quad k = 1, 2, \ldots, K. \tag{3.33}$$

It can easily be seen that these transformations accelerate the convergence of sequences $\{s_m\}$ for which the relationship

$$\lim_{m \to \infty} \frac{s_{m+1} - s}{s_m - s} = b, \quad b \in [-1, 1), \tag{3.34}$$

or, equivalently,

$$\lim_{m \to \infty} \frac{\Delta s_{m+1}}{\Delta s_m} = b, \quad b \in [-1, 1), \tag{3.35}$$

holds. Such sequences are said to be *linearly* convergent for $b \in (-1, 1)$ and *logarithmically* convergent for $b = -1$. Specifically, if $b_1 = b$ in (3.32) or in (3.33) for $b \in (-1, 1)$ or $b = -1$, respectively, then the sequence $\{s^{(m;1)}\}$ converges to s faster than the original sequence $\{s_m\}$.

3.7.8. The *E*-Algorithm. The range of applicability of the different acceleration algorithms introduced so far is by no means disjoint. For instance, Shank's transformation is a generalization of Aitken's Δ^2 process, and the transformed sequences $\{s^{(m)}\}$ in (3.22) and $\{\varepsilon_{2k}^{(m)}\}$, $k = 1$, in the ε-algorithm coincide. The most general extrapolation algorithm known to date, the *E-algorithm*, applies to sequences for which the functions g_k in the asymptotic expansion (3.26) are given by $j_k := k$, $J = K$, and

$$g_k(m; a_k) = a_k t_m^{(k)}, \quad k = 1, 2, \ldots, K. \tag{3.36}$$

The auxiliary sequences $\{t_m^{(j)}\}$, $j = 1, \ldots, J$, are assumed to have the properties stated in Section 3.7.3.

The asymptotic error expansions underlying *all* sequence transformations introduced so far can be expressed in the general form (3.36). For instance, the equality

$$a_1 a_2^m = \frac{s_{m+1} - s_m}{a_2 - 1}$$

can easily be obtained from (3.20). This leads to the following alternative way to express the error expansion (3.20) underlying Aitken's Δ^2 process:

$$g(m; a_1, a_2) = (a_2 - 1)^{-1} \Delta s_m,$$

which is clearly of the general form (3.36).

Since

$$g(m; a_1, \ldots, a_K) = g_1(m; a_1) + \cdots + g_K(m; a_K) = a_1 t_m^{(1)} + \cdots + a_K t_m^{(K)}$$

is linear in the unknown coefficients $a_1, \ldots, a_J$, the equations (3.18) for determining the quantities $s^{(m)}; a_1^{(m)}, \ldots, a_J^{(m)}$ are linear. The unknown $s^{(m)}$, which is of primary interest, can be expressed using Cramer's rule,

$$(3.37) \qquad s^{(m)} = \frac{\begin{vmatrix} s_m & s_{m+1} & \cdots & s_{m+J} \\ t_m^{(1)} & t_{m+1}^{(1)} & \cdots & t_{m+J}^{(1)} \\ \vdots & \vdots & & \vdots \\ t_m^{(J)} & t_{m+1}^{(J)} & \cdots & t_{m+J}^{(J)} \end{vmatrix}}{\begin{vmatrix} 1 & 1 & \cdots & 1 \\ t_m^{(1)} & t_{m+1}^{(1)} & \cdots & t_{m+J}^{(1)} \\ \vdots & \vdots & & \vdots \\ t_m^{(J)} & t_{m+1}^{(J)} & \cdots & t_{m+J}^{(J)} \end{vmatrix}}.$$

Formula (3.37) clearly shows that $s^{(m)}$ exists and can be computed as long as the determinant in the denominator of (3.37) is nonzero. However, the determinants in (3.37) should not be used in practice to calculate $s^{(m)}$. $s^{(m)} =: s^{(m;J)}$ can be calculated recursively instead by using the *Mühlbach scheme*, which extends the Neville–Aitken scheme for interpolation polynomials to general *Chebyshev systems* (see Brezinski [40] and Mühlbach [308]).

While the E-algorithm encompasses the range of applicability of most other sequence transformations, the E-algorithm should *not* be used when a special algorithm, particularly adapted to the given asymptotic error expansion, is available. The E-algorithm generally proves to be less efficient, in terms of the number of required arithmetic operations, than tailor-made acceleration algorithms.

Software (Extrapolation Algorithms). The subroutines CMLIB/dea and NAG/c06baf both implement Wynn's ε-algorithm.

TOMS/602 provides an implementation of Levin's u-transform. Additionally, the algorithm incorporates a stopping criterion which terminates the sequence transformation at an index m when the (estimated) accumulated rounding error in $\hat{s}_m$ becomes comparable in size to the (estimated) truncation error $\hat{s}_m - s$ (for details, see Fessler, Ford, and Smith [148]).

The general E-algorithm is described in Brezinski [40] and implemented in TOMS/585.

The companion floppy disk to Brezinski and Redivo Zaglia [41] contains a number of Fortran 77 subroutines implementing most of the extrapolation algorithms described in the book. A driver program is provided for each subroutine, demonstrating the subroutine's proper use. Most acceleration algorithms useful for numerical integration purposes, such as Aitken's Δ^2 method, Richardson's extrapolation, the ε-algorithm, Levin's transformations, and the G-transform, are provided.

Part II

Symbolic Integration

Chapter 4

Symbolic Integration

The problem of *integration in finite terms* or *integration in closed form* has been the subject of mathematical research for more than 150 years and has attracted the interest of a number of outstanding investigators. Indeed, it was the French mathematician Joseph Liouville who first extensively studied the problem of integration in closed form. *Liouville's principle*, presented in 1833, describes the structure of all elementary functions[30] whose antiderivatives can again be expressed as elementary functions. Used by Liouville to prove that the indefinite integral $\int \exp(x^2)\,dx$ is *not* elementary, Liouville's principle forms the basis of all modern, constructive, algebraic approaches to integration in finite terms.

Today's computational methods also rely heavily on the work of another 19th century French mathematician, Charles Hermite. His method of integrating rational functions (Hermite [204]) has proved to be more generally applicable to rational expressions of elementary functions and is, therefore, a main building block in all algorithms for integrating elementary functions.

After Liouville, work on the constructive approach to integrating transcendental, elementary functions was largely stalled for about a century. It was not until the 1940s that Ritt [371] achieved a major breakthrough by developing *differential algebra*, which made it possible to study the problem of integrating elementary functions from a purely algebraic point of view, i.e., without referring to calculus in any way.

In the 1960s, the widespread use of computers and the development of computer languages for algebraic computation gave rise to renewed interest in symbolic integration. The research efforts culminated in Risch's fundamental algorithm (Risch [369]). Risch's algorithm computes an elementary indefinite

[30]Informally speaking, *elementary functions* can be expressed using, in addition to rational and algebraic functions, a finite number of exponential, trigonometric, and/or hyperbolic functions and their corresponding inverse mappings. *Transcendental* elementary functions do not involve any algebraic expressions. Stringent formal definitions of (transcendental) elementary functions are given in Section 4.1.5.

integral F for *any* given transcendental, elementary function f in a finite number of steps, or proves that such an antiderivative does not exist. All contemporary work in symbolic integration is influenced by and based on Risch's fundamental results.

4.1. Representations and Operations in Algebraic Computation

Users of symbolic computation systems (see Section 3.5.6) often exhibit a rather uncritical attitude towards the capabilities of these systems and the results they produce. To begin with, they underestimate the inherent limitations of algebraic computation, especially when it comes to performing analytical tasks: even simple integration problems or differential equations, expressed in terms of a small number of elementary functions, may have solutions which *cannot* be represented in closed form. And then, users are often not aware of the way mathematical objects are represented in symbolic computation systems and of the set of mathematical operations which can be performed accurately. For instance, in order to obtain numerical results, nonalgebraic (usually multiple precision) computations are often used to evaluate algebraic expressions. Such computations are prone to the types of errors generally encountered in numerical computing, for instance, the loss of accuracy due to the cancellation of the most significant digits.

This section deals with the ability or lack of ability of symbolic computation systems to represent mathematical objects and perform operations on them. The reader is assumed to have a basic understanding of fundamental algebraic concepts (see, for instance, van der Waerden [436]). Emphasis is given to fundamental concepts and ideas rather than to technical details. More detailed technical discussions of this subject can be found in Davenport, Siret, and Tournier [88] and Geddes, Czapor, and Labahn [161].

4.1.1. Integer Calculations. The most basic class of objects dealt with in algebraic computation is the set of integers $\mathbb{Z}$. In contrast to most programming languages, symbolic computation systems do *not* simply represent integers i by one of the integer data types implemented in the computer's processor. Rather, integers i are represented as variable-length arrays $(d_{J(i)}, \ldots, d_2, d_1, d_0)$ of values of a machine-dependent integer data type. Each array element corresponds to a digit in the digit expansion (6.34) of i in base b:

$$i = \sum_{j=0}^{J(i)} d_j(i) b^j.$$

The maximum size of such an array—and thus the maximum representable integer—is *not* determined in advance and is only limited by the available resources, i.e., the storage space.

The set of basic calculations which can be performed on integers comprises rational operations, addition, multiplication, and division with remainder, as well as relational operations: $=, >, <, \geq, \leq$. Using these basic operations, algorithms for more complex operations can be implemented. Most importantly, the Euclidean algorithm can be used to compute the greatest common divisor of two integers.

4.1.2. Rational Calculations. All other mathematical objects in algebraic computation are derived, directly or indirectly, from the set of integers $\mathbb{Z}$ using one of the *extension mechanisms* in algebra. Such mechanisms extend the range of applicability of certain operators from an initial set A to a more comprehensive set $\overline{A}$, $A \subset \overline{A}$. From an abstract point of view, all extension techniques follow the same procedure. To begin with, the elements $\overline{a} \in \overline{A}$ are defined in terms of elements $a \in A$. Next, the equality of and operations on elements of $\overline{A}$ are defined in terms of operations on the constituent elements of A. Finally, the term *extension* for $\overline{A}$ is justified by showing that the initial algebraic structure A is *isomorphic* (i.e., algebraically equivalent) to a properly chosen subset of $\overline{A}$.

An important example of this procedure is the construction of *quotient fields*. Given an *integral domain*[31] D, addition and multiplication operations as well as the equality of elements can be defined on the set

$$Q(D) := \{(p, q) \ : \ p, q \in D, q \neq 0\}$$

by

$$(p, q) + (r, s) := (ps + qr, qs) \qquad \text{and} \qquad (p, q)\,(r, s) := (pr, qs)$$

as well as

$$(p, q) = (r, s) \quad \Leftrightarrow \quad ps = qr.$$

$Q(D)$ can be shown to be a *field*.[32] The elements of $Q(D)$ are commonly written as fractions $p/q := (p, q)$. Applying this extension mechanism to the integral domain $D = \mathbb{Z}$ yields the field of all rational numbers, $\mathbb{Q} = Q(\mathbb{Z})$.

4.1.3. Polynomial Calculations. A characteristic feature of symbolic computation systems is their ability to perform *polynomial computations*. Given a *commutative ring* R, the set

$$R[x] := \left\{ p(x) = \sum_{k=0}^{K} a_k x^k \ : \ a_0, a_1, \ldots, a_K \in R, \ K \in \mathbb{N}_0 \right\} \tag{4.1}$$

[31] In an integral domain, $a \cdot b = 0$ implies $a = 0$ or $b = 0$.

[32] In a field, there is an inverse multiplicative element a^{-1} for each element $a \neq 0$: $aa^{-1} = 1$. Hence, division by a is well defined as long as $a \neq 0$.

consists of all polynomials p in x with coefficients in R.[33]

$$\deg p := \max\{k : a_k \neq 0\}$$

is referred to as the degree of a polynomial $p \in R[x]$. The equality of two polynomials

$$p(x) = \sum_{k=0}^{K} a_k x^k \qquad \text{and} \qquad q(x) = \sum_{l=0}^{L} b_l x^l$$

is defined by

$$(4.2) \quad p = q \quad :\Leftrightarrow \quad (\deg p = \deg q \quad \text{and} \quad a_k = b_k \text{ for } k = 1, 2, \ldots, \deg p).$$

Their sum and product are defined by

$$\begin{aligned}(p+q)(x) &:= \sum_{m=0}^{\max(K,L)} (a_m + b_m)x^m, \\ (4.3) \qquad (p \cdot q)(x) &:= \sum_{n=0}^{K+L} \left(\sum_{m=0}^{n} a_m b_{n-m} \right) x^n.\end{aligned}$$

It can be shown that $R[x]$ is, like R, a commutative ring. Moreover, if R is a field, then $R[x]$ forms an integral domain, indeed a *Euclidean domain*.[34] Hence, quotient fields $Q(R[x])$ can be derived from polynomial rings $R[x]$, provided the elements of R – the *constants* – form a field. For notational convenience $Q(R[x])$ is usually abbreviated to $R(x)$.

Applying polynomial extension to $\mathbb{Z}$ ($\mathbb{Q}$) yields the set $\mathbb{Z}[x]$ ($\mathbb{Q}[x]$) of all polynomials with integer (rational) coefficients. The quotient fields $\mathbb{Z}(x)$ and $\mathbb{Q}(x)$ are equal and form the set of all rational expressions – or functions, if x is interpreted as an independent variable – in x over $\mathbb{Q}$.

Note that the equality of and the operations on elements of $R[x]$ are defined in terms of operations on the coefficients of the polynomials involved and are completely independent of the indeterminate x. x can thus be replaced by any other symbol without changing the definitions or the properties of polynomial calculations.

4.1.4. Algebraic Calculations. So far, the range of applicability of symbolic computation has been extended to polynomial and rational calculations involving whole numbers and indeterminate symbols. A further extension can be achieved by introducing *algebraic numbers* and *functions*.

[33]If the symbol x is the name of an independent variable, the polynomials $p(x)$ can be interpreted as polynomial functions.

[34]In a *Euclidean domain* the Euclidean algorithm can be used to compute the greatest common divisor of two elements.

Field Extensions Based on Irreducible Polynomials. Let F be a field and $m(x)$ an *irreducible* polynomial[35] over F. Then, the elements of the algebraic structure $F[x]/m(x)$ are defined to be the elements of the polynomial ring $F[x]$; analogously, addition and multiplication in $F[x]/m(x)$ are defined by (4.3) in the same way as addition and multiplication in $F[x]$. What distinguishes $F[x]/m(x)$ from $F[x]$ is a different concept of equality: two polynomials p and q are equal in $F[x]/m(x)$ if and only if the difference $p - q \in F[x]$ is divisible by m,

$$p = q \quad :\Leftrightarrow \quad m \mid (p-q). \tag{4.4}$$

It can easily be seen that $F[x]/m(x)$ forms a field and that, because

$$F \subset F[x]/m(x),$$

$F[x]/m(x)$ is an extension field of F. Since $m \mid m$, the equality $m(x) = 0$ holds in $F[x]/m(x)$. Moreover, if a polynomial p over F is of degree

$$\deg p =: K \geq L := \deg m$$

and a_K and b_L denote the leading coefficients of p and m, respectively, then p is equal to the polynomial

$$p - (a_K/b_L)m =: p_1$$

in $F[x]/m(x)$, and $\deg p_1 < \deg p$. If $\deg p_1 \geq \deg m$, then this procedure can be repeated with the polynomial p_1 in place of p. This shows that each element of $F[x]/m(x)$ is equal to a (unique) polynomial of degree $< \deg(m)$.

Let y denote an arbitrary indeterminate other than x. Since the coefficients of the irreducible polynomial $m(x) \in F[x]$ are elements of F and

$$F \subset F[x]/m(x),$$

$m(y)$ is a polynomial in y over the field $F[x]/m(x)$; i.e.,

$$m(y) \in (F[x]/m(x))[y].$$

Note that the original indeterminate x is a linear polynomial in x over F and, as such, an element of $F[x]/m(x)$. Indeed, as the equality $m(x) = 0$ holds in $F[x]/m(x)$, the polynomial $m(y)$ in y over $F[x]/m(x)$ has the zero x in $F[x]/m(x)$. In other words, the extension field $F[x]/m(x)$ of F has the property that the algebraic equation $m = 0$, while unsolvable in F, has a (not necessarily unique) solution in $F[x]/m(x)$.

[35] A polynomial $m(x) \in F(x)$ is *irreducible* over a field F if for any factorization $m(x) = m_1(x)m_2(x)$, $m_1(x), m_2(x) \in F(x)$, at least one of the polynomials $m_1(x)$ or $m_2(x)$ is constant, i.e., has degree zero.

Algebraic Elements. More generally, an element α of an extension field $\overline{F}$ of F is said to be *algebraic* over F if there is a polynomial $m(x) \in F[x]$ such that $m(\alpha) = 0$. Clearly, m can be chosen as an irreducible polynomial over F. In this case, $F[x]/m(x)$ is *isomorphic* to the *smallest subfield* of $\overline{F}$, denoted $F(\alpha)$, containing both F and α; i.e.,

$$F[x]/m(x) \equiv F(\alpha)$$

can be regarded as the field obtained by adjoining the algebraic element α to the field F. Since the equality of and the operations on elements of $F[x]/m(x)$ are constructively defined in terms of operations on elements of F, computer algebra systems can operate in algebraic extension fields $F[x]/m(x) \equiv F(\alpha)$ provided they can operate in the underlying fields F.

By successively adjoining *algebraic numbers*, i.e., *real* numbers which are algebraic over $\mathbb{Q}$, $\alpha_1, \alpha_2, \ldots, \alpha_n$ to the field of rational numbers $\mathbb{Q}$, *algebraic number fields*

$$\begin{aligned}
&\mathbb{Q}(\alpha_1), \\
\mathbb{Q}(\alpha_1, \alpha_2) \quad &:= \quad \mathbb{Q}(\alpha_1)(\alpha_2), \\
&\vdots \\
\mathbb{Q}(\alpha_1, \ldots, \alpha_n) \quad &:= \quad \mathbb{Q}(\alpha_1, \ldots, \alpha_{n-1})(\alpha_n)
\end{aligned}$$

can be obtained and dealt with constructively.

Elements of and calculations in algebraic number fields are abstractly defined in terms of polynomials and their operations. However, for practical purposes, *numerical* representations of algebraic numbers are frequently required. Calculating the numerical value of an algebraic number α means approximately solving the polynomial equation $m(x) = 0$ for the associated (irreducible) polynomial m with rational coefficients, which is often not a trivial task (in particular, when the degree of m is high). Nevertheless, it is possible to determine the numerical value of *any* algebraic number α with *arbitrary precision* (Davenport, Siret, and Tournier [88]).

This method of extending fields F by adjoining elements algebraic over F is applicable to any field F (not just to number fields). Extension fields obtained by adjoining algebraic elements to the field $F = \mathbb{Q}(x)$ of all rational functions are said to be fields of *algebraic functions*.

4.1.5. Transcendental Calculations

Transcendental Elements. An element α of an extension field $\overline{F}$ of F is said to be *transcendental* over F if $p(\alpha) \neq 0$ holds for all $p(x) \in F[x]$, i.e., if α is not algebraic. A transcendental element α of the number (function) field $\mathbb{Q}$ ($\mathbb{Q}(x)$) is referred to as a *transcendental number* (*transcendental function*). As a first step, algebraic computation systems deal with transcendental elements on a purely symbolic level by treating them as indeterminates: transcendental constants such as π and transcendental functions such as e^x are treated in essentially the same way as simple indeterminates such as x are. By using the mechanisms described in the previous sections, i.e., by forming polynomial rings and quotient fields, the range of applicability of rational computations can be extended to include a finite number of transcendental expressions.

Example (Manipulation of Transcendental Expressions). A suitable algebraic structure for manipulating expressions containing, in addition to rational numbers and functions, the transcendentals π and e^x is given by the field $\mathbb{Q}(\pi)(x, e^x)$, which can be constructed as follows. First, the polynomial ring in π with coefficients in $\mathbb{Q}$ and its quotient field $\mathbb{Q}(\pi)$ are constructed. $\mathbb{Q}(\pi)$ is the smallest subfield of $\mathbb{R}$ containing $\mathbb{Q}$ and π. Next, $\mathbb{Q}(\pi)(x)$, the set of rational functions with coefficients in $\mathbb{Q}(\pi)$, is formed. Finally, the polynomial ring in e^x with coefficients in $\mathbb{Q}(\pi)(x)$ and its quotient field $\mathbb{Q}(\pi)(x)(e^x) =: \mathbb{Q}(\pi)(x, e^x)$ are constructed.

Transformation Rules. The main shortcoming in dealing with transcendental entities as meaningless symbols is that identities between transcendental expressions, for instance, between $e^x \cdot e^{-x}$ and 1, are not recognized. This problem is overcome by incorporating knowledge regarding *transformation rules* such as

$$\begin{aligned} \exp(x+y) &= \exp(x)\exp(y), \\ \log(xy) &= \log(x) + \log(y), \text{ and} \\ \exp(\log(x)) &= \log(\exp(x)) = x \end{aligned} \tag{4.5}$$

into symbolic computation systems. While applying such transformation rules to transcendental expressions always yields equivalent expressions, there remains the question of whether or not the reverse is also true, i.e., whether or not, for a given set of transformation rules, *any* pair of equivalent expressions can be transformed into one another using the given rules. A set of transformation rules for which this property holds is said to be *complete*.

Elementary Functions. The completeness issue has been resolved for a special class of transcendental functions, transcendental *elementary* functions. Informally speaking, elementary functions are obtained from an appropriate field of algebraic functions by adjoining a finite number of exponential, trigonometric, and/or hyperbolic expressions as well as their inverse expressions. Note that all transcendental elementary functions can be expressed in terms of the *imaginary unit* i as well as the *exponential* and *logarithmic* functions using *Euler's formula*

$$\exp(\mathrm{i}x) = \sin(x) + \mathrm{i}\cos(x).$$

Therefore, a formal definition of elementary functions can be given as follows. Let F be a field of (real or complex) univariate functions (in the variable x), and $\overline{F} \supseteq F$, an extension field (containing functions in x) of F.

(i) An element $f \in \overline{F}$ is said to be *logarithmic* over F if there exists a function $g \in F$ such that $f' = g'/g$ or (equivalently) $f = \log(g)$ holds.

(ii) An element $f \in \overline{F}$ is said to be *exponential* over F if there is a function $g \in F$ such that $f'/f = g'$ or (equivalently) $f = \exp(g)$ holds.

Let $F(f)$ denote the smallest subfield of $\overline{F}$ containing F and f.

(i) $F(f)$ is said to be an *elementary* extension of F if f is either algebraic, logarithmic, or exponential over F.

(ii) $F(f)$ is said to be a *transcendental* elementary extension of F if f is transcendental and either logarithmic or exponential over F.

Let $F(f_1, f_2, \ldots, f_n)$ denote the function field obtained by successively adjoining the elements

$$f_1, f_2, \ldots, f_n \in \overline{F}$$

to F, i.e.,

$$F(f_1, f_2, \ldots, f_i) := F(f_1, f_2, \ldots, f_{i-1})(f_i), \qquad i = 1, 2, \ldots, n.$$

(iii) An extension field G of F is said to be *(transcendental) elementary* if there exists a representation $G = F(f_1, f_2, \ldots, f_n)$ such that each

$$F(f_1, f_2, \ldots, f_i), \qquad i = 1, 2, \ldots, n,$$

is a (transcendental) elementary extension of $F(f_1, f_2, \ldots, f_{i-1})$.

Let K be a subfield of the field of complex numbers $\mathbb{C}$. A (transcendental) elementary extension F of $K(x)$, the field of rational functions over K, is said to be a *field of (transcendental) elementary functions*. Thus, a function f is formally defined as *(transcendental) elementary* if there exists a field of (transcendental) elementary functions F such that $f \in F$.

Risch's Structure Theorem. It can be shown that the transformation rules given in (4.5) are indeed complete for the set of elementary functions; i.e., if two different elementary functions f_1 and f_2 yield equal values $f_1(x) = f_2(x)$ for any eligible abscissa x, then they can be transformed into one another using the rules stated in (4.5). This result is an immediate consequence of *Risch's structure theorem* (Risch [370]). This theorem provides a method for testing whether or not a function f is exponential or logarithmic over a field of elementary functions F, is transcendental over F, and whether or not the sets of constants in F and in the extension field $F(f)$ are the same.

Unfortunately, Risch's structure theorem has been established so far only for transcendental *functions*. For transcendental *numbers*, Risch's structure theorem, although conjectured to hold (Ax [10]), has not yet been proven. It is not even known whether or not the two fundamental transcendental constants e and $\pi = -i\log(-1)$ are independent, i.e., whether or not one of them can be expressed in terms of the other using an algebraic expression.

If two transcendental constants a and b (for instance, $a = e$ and $b = \pi$) cannot be proved or disproved to be algebraically independent of one another, then adjoining a and b to a rational or algebraic number field F gives rise to the following problem. The elements of $F(a, b)$ are polynomials $p(a, b)$ in a and b with coefficients in F. Since it is not known whether or not a and b are algebraically independent, the possibility that some nontrivial elements $r(a, b) \in F(a, b)$, i.e., polynomials in a and b with nonzero coefficients, actually vanish cannot be excluded. When such a hypothetical element r is added to an arbitrary element $s(a, b) \in F(a, b)$, the equality $s(a, b) + r(a, b) = s(a, b)$ is obtained. Since it is not known whether such hypothetical elements r exist, it is not clear whether every problem of equality in $F(a, b)$ can be decided. As explained in Section 4.2, this phenomenon has an impact on the solvability of symbolic integration problems.

4.1.6. Multiple Precision Calculations.

Systems for symbolic computation generally use *multiple precision software* to perform numerical calculations. Multiple precision software implements floating-point numbers and floating-point operations using suitable data structures and algorithms.[36] The range of the floating-point system and the accuracy of floating-point operations can easily be changed by modifying certain algorithmic parameters and are thus completely independent of the particular properties of a given processor. The default level of accuracy used by symbolic computation systems in numerical calculations can be changed by the user at any time, and the maximum level of accuracy is only limited by the available resources (the storage space and the processor time).

However, the flexibility and the potentially high accuracy of multiple precision arithmetic should not obscure the fact that the implemented floating-point operations are prone to the same kinds of errors that occur in hardware-coded floating-point operations. Even the numerical evaluation of a symbolically derived, mathematically exact result may yield a completely wrong number (see, for instance, the example on page 109). It is therefore generally advisable to examine the numerical results produced by symbolic computation systems critically and to at least check their plausibility.

4.2. The Problem of Symbolic Integration

Within the context of symbolic computation, the general problem of constructive indefinite integration is defined as follows. For two given classes of (real or

[36] See Ueberhuber [435] for a discussion of multiple precision software.

complex) functions $\mathcal{F}$ and $\mathcal{G}$, the integration problem associated with $\mathcal{F}$ and $\mathcal{G}$ is finding an algorithm which, for every member $f \in \mathcal{F}$, either computes an antiderivative $g \in \mathcal{G}$ of f, $f = g'$, or proves that there is no such element $g \in \mathcal{G}$. The most relevant classes $\mathcal{F}$ of eligible integrands are the set of all rational and the set of all elementary functions, while the class $\mathcal{G}$ of admissible antiderivatives is usually chosen to be the set of all elementary functions.

Undecidable Equalities. The definition of the class of admissible integrands $\mathcal{F}$ may involve the adjunction of two or more transcendental constants. If it is impossible to decide whether these transcendental constants are algebraically dependent, then the problem of equality of two elements in $\mathcal{F}$ generally cannot be resolved (see Section 4.1.5). This implies that the integration problem from $\mathcal{F}$ to $\mathcal{G}$ is unsolvable if there is a function $f \in \mathcal{F}$ for which no antiderivative $g \in \mathcal{G}$ exists: for a constant $c \in F$, the function cf has an antiderivative in $\mathcal{G}$ if and only if $c = 0$. Therefore, an algorithm can solve the integration problem from $\mathcal{F}$ to $\mathcal{G}$ only if every problem of equality between a constant c and 0 can be resolved.

When symbolic computation systems cannot establish the algebraic dependence or independence of transcendental constants, they assume these constants to be independent. This approach is justified by the fact that all antiderivatives calculated remain correct even if the independence assumption proves to be incorrect (Davenport [85]). The only effect of a possible dependence is that some functions may be incorrectly diagnosed as being nonintegrable.

Field Extensions in Symbolic Integration. Symbolic computation systems can constructively deal with algebraic structures only if these structures are obtained from the fundamental set $\mathbb{Z}$ using a finite number of appropriate extension operations (see Section 4.1). This primarily excludes number fields $\mathbb{R}$ of all algebraic structures derived from it, such as the field $\mathbb{R}(x)$ of rational functions with real coefficients. This implies that the relevant classes $\mathcal{F}$ and $\mathcal{G}$ of admissible integrands and antiderivatives, respectively, are not amenable to the computational capabilities of symbolic computation systems. Computer algebra systems therefore have to adapt the algebraic structures used for integrating functions $f \in \mathcal{F}$ to the given problem. This is done by extending the initial field $\mathbb{Q}(x)$ incrementally during the course of the computational procedure by adjoining at each computational step those algebraic or transcendental elements which are required to perform it.

Suitable algorithms for symbolic integration are often required to be performable with as few algebraic field extensions as possible (see Section 4.1.4). The main reason for this requirement lies in the high computational cost incurred in using algebraic field extensions. Some algebraic field extensions may, however, be inevitable simply because they are needed to express the algorithm's final solution (see Section 4.3). Algorithms which are performable with just this minimum number of unavoidable extensions are said to be *rational*.

4.3. Integration of Rational Functions

Partial Fraction Decompositions. The indefinite integration of rational functions is generally regarded as a rather trivial subject. Every student in a beginners' calculus course is taught the following basic procedure for integrating rational functions. Let $r = p/q$, with p and q polynomials over $\mathbb{R}$ or $\mathbb{C}$, be a rational function. Then, based on the factorization (over $\mathbb{C}$) of the denominator q into linear factors,

$$q = (x - a_1)^{n_1}(x - a_2)^{n_2} \cdots (x - a_k)^{n_k}, \quad a_1, a_2, \ldots, a_k \in \mathbb{C}, \tag{4.6}$$

the rational function r can be decomposed into *partial fractions*:

$$r = \sum_{i=1}^{k} \frac{p_i(x)}{(x - a_i)^{n_i}},$$

where each p_i, $i = 1, \ldots, k$, is a polynomial of degree $< n_i$. Representing $p_i(x)$ in the monomial basis $\{1, x - a_i, \ldots, (x - a_i)^{n_i - 1}\}$,

$$p_i = \sum_{j=0}^{n_i - 1} b_{i,j}(x - a_i)^j,$$

yields

$$r = \sum_{i=1}^{k} \sum_{j=1}^{n_i - 1} \frac{b_{i,j}}{(x - a_i)^{n_i - j}}.$$

This decomposition of r can be integrated easily, yielding

$$\int r(x)\,dx = \sum_{i=1}^{k} \sum_{j=1}^{n_i - 2} \frac{b_{i,j}}{(j + 1 - n_i)(x - a_i)^{n_i - j - 1}} + \sum_{i=1}^{k} b_{i,n_i - 1} \log(x - a_i). \tag{4.7}$$

Note that in this representation the antiderivative of r is quite naturally split into rational and logarithmic parts.

However, symbolic computation systems do *not* integrate rational functions by applying partial fraction decompositions. The main reason is that the algebraic extensions required to perform the partial fraction decomposition of r may not be necessary in order to express the final result, the antiderivative of r; i.e., integrating rational functions using partial fraction decompositions does *not* generally constitute a rational algorithm.

Example (Decomposition into Partial Fractions). Because

$$\int r(x)\,dx := \int \frac{1}{x^3+x}\,dx = \frac{1}{2}\log\left(\frac{x^2}{x^2+1}\right),$$

the only extension of $\mathbb{Q}(x)$ required to integrate the rational function r is the adjunction of the transcendental function $\log(x^2/(x^2+1))$. By contrast, decomposing r into partial fractions,

$$r(x) = \frac{1}{x} - \frac{1/2}{x+\mathrm{i}} - \frac{1/2}{x-\mathrm{i}},$$

and subsequently integrating

$$\int r(x)\,dx = \log(x) - \frac{1}{2}\log(x+\mathrm{i}) - \frac{1}{2}\log(x-\mathrm{i})$$

requires the extension of $\mathbb{Q}(x)$ by $\mathrm{i}; x, \log(x), \log(x+\mathrm{i})$, and $\log(x-\mathrm{i})$. The imaginary unit i, in particular, need not appear in the final result at all.

Additionally, even if no algebraic extensions are required to perform a partial fraction decomposition, the complete factorization (4.6) of the denominator q into linear factors is computationally expensive for polynomials of high degree.

Rational Algorithms. Methods used in practice for integrating rational functions avoid introducing superfluous field extensions. Hermite [204] was the first to develop a rational algorithm for calculating the *rational part* in (4.7). An algorithmically simpler method was devised by Horowitz [211]. As for the *logarithmic part* in (4.7), the first rational algorithm was developed by Rothstein [379] and Trager [427]. A recent improvement of the Rothstein–Trager method can be found in Lazard and Rioboo [256]. For detailed descriptions and proofs of these algorithms, see Geddes, Czapor, and Labahn [161].

While rational algorithms avoid introducing superfluous algebraic extensions, the occurrence of algebraic numbers in the final result, the indefinite integral, generally cannot be avoided. Since algebraic numbers are represented by irreducible polynomials in symbolic computation systems (see Section 4.1.4), certain parameters in the output of integration routines may be specified in terms of the zeros of irreducible polynomials.

Example (Integration of Rational Functions). The command

```
F := int(1/(x^3+x+1),x);
```

in MAPLE yields

$$F := \sum_{\_R = RootOf(31\_Z^3 - 3\_Z - 1)} \_R \log\left(x - \frac{62\_R^2}{9} + \frac{31\_R}{9} + 4/9\right). \tag{4.8}$$

The roots of quartic polynomials can always be expressed as radicals, which gives rise to more familiar expressions. But even the zeros of cubic and quartic polynomials are not necessarily, at least not by default, represented in terms of radicals by symbolic computation systems because the resulting expressions tend to be fairly complicated.

Example (Integration of Rational Functions). The exact roots of the irreducible polynomial $31\_Z^3 - 3\_Z - 1$ in (4.8) can be obtained using the command

```
x := allvalues(_Z^3-3*_Z-1);
```

which, after simplification, gives

$$x_1 = \frac{\left(3844 + 372\sqrt{93}\right)^{2/3} + 124}{62\sqrt[3]{3844 + 372\sqrt{93}}},$$

$$x_{2,3} = -\frac{\left(3844 + 372\sqrt{3}\sqrt{31}\right)^{2/3} + 124 \pm \sqrt{-1}\left(3844 + 372\sqrt{3}\sqrt{31}\right)^{2/3}\sqrt{3} - \pm 124\sqrt{-1}\sqrt{3}}{124\sqrt[3]{3844 + 372\sqrt{3}\sqrt{31}}}.$$

These radicals can be used in (4.8), yielding a rather awkward, lengthy expression for the indefinite integral.

Alternatively, since the numerical values of algebraic numbers can be evaluated with arbitrary precision, relatively compact representations of antiderivatives can be obtained by replacing algebraic numbers with approximate numerical values. Note, however, that the resulting function is only an approximation for the exact indefinite integral.

If a rational integrand f contains unknown parameters, i.e., is indeterminate, the occurrence of roots of irreducible polynomials in the antiderivative of f may significantly reduce the usefulness of the symbolic integration result for practical purposes. In particular, the set of operations which can be performed on the antiderivative is limited if the unknown parameters enter the irreducible polynomial(s) contained in the antiderivative. For instance, it may not be possible to represent the antiderivative using numerical constants or to differentiate the antiderivative with respect to the unknown parameters.

4.4. Integration of Elementary Functions

Students in a beginners' calculus course are taught how to reduce complicated integrals involving elementary functions to simple ones using integration by parts and the transformation rule (see Section 1.1.5). Unfortunately, finding the appropriate transformations is a highly heuristic, unsystematic procedure, requiring a great deal of experience and intuition. This is why, until very recently, it was not generally believed that systematic, constructive methods for integrating elementary functions existed. Nevertheless, such methods have been devised and have subsequently been incorporated into existing symbolic computation systems.

However, these systematic algorithms for integrating elementary functions can be computationally very expensive. This is why symbolic computation systems initially attempt to integrate elementary functions by applying the same, computationally cheap, heuristic techniques (including look-up in integral tables) a (good) student would use. Only if this approach fails are systematic symbolic integration algorithms applied.

4.4.1. Liouville's Principle. One of the most puzzling phenomena in the indefinite integration of elementary (and rational) functions is the apparent lack of any systematic relationship between the form of an integrand f and the form of its antiderivative F: a rational integrand may give rise to an inverse trigonometric integral,

$$\int \frac{1}{1+x^2}\,dx = \arctan(x) + c,$$

while an almost identical integrand has an inverse hyperbolic antiderivative,

$$\int \frac{1}{1-x^2}\,dx = \operatorname{artanh}(x) + c.$$

An algebraic integrand may result in an inverse trigonometric integral,

$$\int \frac{1}{\sqrt{1-x^2}}\,dx = \arcsin(x) + c,$$

while the antiderivative of an inverse trigonometric integrand may contain inverse trigonometric and algebraic functions,

$$\int \arcsin(x)\,dx = x\arcsin(x) + \sqrt{1-x^2} + c.$$

On the other hand, despite this apparent irregularity, formula (4.7) shows that the indefinite integral of any *rational* function can be expressed in terms of a rational function as well as a linear combination of logarithms of rational functions. By expressing all transcendental functions in terms of exponential and logarithmic functions as well as the imaginary unit i, a similar relationship can be shown to hold for the integrands in the previous examples:

$$\begin{aligned}
\int \frac{1}{1+x^2}\,dx &= -\frac{\mathrm{i}}{2}\log\frac{x+i}{x-\mathrm{i}}, \\
\int \frac{1}{1-x^2}\,dx &= \frac{1}{2}\log\frac{1+x}{1-x}, \\
\int \frac{1}{\sqrt{1-x^2}}\,dx &= -\mathrm{i}\log(\mathrm{i}x + \sqrt{1-x^2}), \\
\int \arcsin(x)\,dx &= -\mathrm{i}\int \log(\mathrm{i}x + \sqrt{1-x^2})\,dx \\
&= -\mathrm{i}x\log(\mathrm{i}x + \sqrt{1-x^2}) + \sqrt{1-x^2}.
\end{aligned}$$

The common pattern found in these examples can informally be described as follows. If an integrand contains certain (rational, algebraic, etc.) types of functions, the corresponding indefinite integral can be expressed in terms of these functions and linear combinations of logarithms of these functions. This relationship between integrands and their indefinite integrals is described formally in the following theorem.

THEOREM 4.4.1 (LIOUVILLE'S PRINCIPLE). *Let F be a field of functions, and let K denote the field of constants contained in F. Assume that the function $f \in F$ has an antiderivative $g \in G$, $f = g'$, where G is an elementary extension field of F with the same field of constants K. Then, the function f is of the form*

$$f = v_0' + \sum_{i=1}^{n} c_i \frac{v_i'}{v_i}, \tag{4.9}$$

where $c_1, \ldots, c_n \in K$ and $v_0, v_1, \ldots, v_n \in F$.

Proof. See Geddes, Czapor, and Labahn [161].

By virtue of formula (4.9), Liouville's principle describes the structure of all those functions f of a field F whose antiderivatives are contained in elementary extensions of F. Note that the indefinite integral of any such function f can be expressed as

$$\int f(x)\,dx = v_0 + \sum_{i=1}^{n} c_i \log v_i,$$

i.e., in terms of a function $v_0 \in F$ and a linear combination of logarithms of functions $v_i \in F$. This general theoretical result confirms the previously stated informal empirical rule.

4.4.2. Algorithms for the Integration of Elementary Functions. A function f is formally defined as elementary if f is contained in an elementary extension $K(x, f_1(x), \ldots, f_n(x))$ of the field of rational functions $K(x)$ with coefficients in $K \subseteq \mathbb{C}$ (see Section 4.1.5). It can be assumed that the field of constants K is obtained from $\mathbb{Q}$ by adjoining a finite number of algebraic and/or transcendental numbers.[37] Since operations in $K(x, f_1(x), \ldots, f_n(x))$ are defined constructively, $K(x, f_1(x), \ldots, f_n(x))$ provides an appropriate computational framework for performing symbolic manipulations on f.

The first computational step when integrating a given elementary function f is thus the determination of a constant field K and elementary extension functions $f_1(x), \ldots, f_n(x)$ such that $f \in K(x, f_1(x), \ldots, f_n(x))$. As explained in Section 4.1.5, Risch's structure theorem is used at this stage to determine whether the elementary extension functions $f_i(x)$, $i = 1, \ldots, n$, are transcendental or algebraic over $K(x, f_1(x), \ldots, f_{i-1}(x))$.

[37]For any element f of a field of elementary functions $K(x, f_1(x), \ldots, f_n(x))$, there is a subfield of constants $K' \subseteq K$ obtained from $\mathbb{Q}$ by adjoining a finite number of algebraic and/or transcendental numbers such that $f \in K'(x, f_1(x), \ldots, f_n(x))$.

The recursive nature of the definition of $K(x, f_1(x), f_2(x), \ldots, f_n(x))$,

$$K(x, f_1(x), \ldots, f_i(x)) := K(x, f_1(x), \ldots, f_{i-1}(x))(f_i(x)), \quad i = 1, 2, \ldots, n,$$

suggests that a general method for integrating elementary functions can be obtained by recursively applying a basic module which performs the following task. For any field F of elementary functions, any function g which is elementary over F, and any function f_g contained in the elementary extension field $F(g)$, the module either

(i) directly solves the symbolic integration problem for f_g, i.e., calculates an elementary integral of f_g;

(ii) determines that the indefinite integral of f_g is not elementary; or

(iii) reduces the problem of integrating f_g to a finite number of symbolic integration problems involving functions in F.

A program which recursively applies such a basic module to

$$f \in K(x, f_1(x), f_2(x), \ldots, f_n(x))$$

terminates after a finite number of computational steps. Upon termination, the given integration problem has either been resolved or has been reduced to a finite number of integration problems involving rational integrands. In the latter case the constituent rational integration problems can be solved using the techniques described in Section 4.3, and the resulting indefinite integrals can be used to resolve the original elementary integration problem.

The basic module must implement a *polyalgorithm* because the actual algorithm used to perform the aforementioned task depends to a large extent on the problem data, i.e., the class F of elementary functions and the particular type of the extension function g. Risch's [369] fundamental achievement was to devise an algorithm capable of dealing with fields of transcendental elementary functions F and elementary functions g which are transcendental over F. Hence, Risch's algorithm effectively solves the problem of integrating transcendental elementary functions.

Subsequently, research efforts have concentrated on developing methods which cope with algebraic field extensions. Trager [428] devised an algorithm for fields of transcendental elementary functions F and nth root functions g over F, i.e., $g^n \in F$. The algorithm developed by Davenport [86] works for fields of transcendental elementary functions F and general algebraic functions g over F. Trager [429] presented an algorithm for the same problem setting which has the advantage of being rational. Finally, Bronstein [43] solved the problem for general fields F of elementary functions and elementary extensions g which are algebraic over F. However, further algorithmic improvements are still possible since Bronstein's algorithm is not completely rational. The state-of-the-art in symbolic integration methods for elementary functions is described in Bronstein [44].

4.5. Integration of Nonelementary Functions

Many elementary functions do not have elementary indefinite integrals. Some of these indefinite integrals are of such importance to mathematical theory that they define *special functions* such as the *error function*

$$\text{erf}(x) := \frac{2}{\sqrt{\pi}} \int_0^x \exp(-t^2)\, dt, \tag{4.10}$$

the *exponential integral function*

$$\text{Ei}(x) := \int_x^\infty \frac{\exp(-t)}{t}\, dt, \quad x > 0, \tag{4.11}$$

or the *logarithmic integral function*

$$\text{Li}(x) := \text{Ei}(\log x), \quad x > 1. \tag{4.12}$$

Optimized algorithms and programs have been developed to evaluate special functions such as these (see, for instance, the NAG and IMSL libraries). Therefore, it is desirable, if possible, to express indefinite integrals of elementary functions—provided that no elementary integral exists—in terms of special functions. It is also important to be able to symbolically integrate expressions involving special functions.

Expressing integrals in terms of special functions as well as symbolically integrating special functions are problems actively studied by a number of researchers. Singer, Saunders, and Caviness [400] generalize Liouville's principle by allowing special functions such as the error function. the logarithmic integral, and *Fresnel integrals* to appear in the indefinite integral of an elementary function. Cherry [57] describes an algorithm which computes an antiderivative of a given transcendental elementary function in terms of elementary functions and the logarithmic integral or proves that no such antiderivative exists. Knowles [235], [236] devises an algorithm that can handle certain *transcendental Liouvillian* functions.[38]

A completely different class of problems is dealt with in Piquette [340]. Piquette describes a method for integrating functions which are products of general functions f and special functions, required to satisfy certain recurrence relations. The set of special functions which can be treated in this approach comprises most special functions occurring in theoretical physics, such as Bessel, Legendre, Hermite, and Laguerre functions.

[38] Liouvillian extensions are obtained by adjoining antiderivatives of elements $f \in F$ to fields of functions F.

4.6. Definite Integration

Once an antiderivative F of a function f has been obtained, the definite integral of f over any interval $[a, b]$ can be computed by evaluating the expression $F(b) - F(a)$. While calculating $F(b) - F(a)$ for given interval endpoints a and b is usually straightforward, the following pitfalls should be avoided. Most importantly, even if the antiderivative F is well defined at the interval endpoints a and b, F may be unbounded on $[a, b]$, i.e., may have singularities in the interior of $[a, b]$. In this case, f is not integrable over $[a, b]$ although $F(b) - F(a)$ yields a finite value.

This problem can be avoided to a certain degree by using the definite integration operation provided by most symbolic computation systems. When this operation is used, the integrand and the endpoints a and b of the integration interval have to be supplied as parameters, which enables symbolic computation systems to check for the presence of (nonintegrable) singularities in $[a, b]$.[39] However, symbolic computation systems *cannot* always be relied upon to detect singularities within the integration interval and to inform the user of the nonintegrability of the given integrand, even when the definite integration operation is used.

Example (Nonintegrable Integrand). The polynomial $q(x) := 1 + x + x^3$ has the simple real root

$$x_0 = -\frac{\left(108 + 12\sqrt{93}\right)^{2/3} - 12}{6\sqrt[3]{108 + 12\sqrt{93}}} \approx -0.6823.$$

Hence, the rational function $r(x) := 1/q(x)$ is not integrable on any interval which contains x_0. Nevertheless, the MAPLE command `int(1/(x^3+x+1),x=-1..1);` for integrating r over $[-1, 1]$ yields

$$\sum_{\_R=RootOf(31\_Z^3-3\_Z-1)} \_R \log\left(\frac{13}{9} - \frac{62\_R^2}{9} + \frac{31\_R}{9}\right)$$
$$- \sum_{\_R=RootOf(31\_Z^3-3\_Z-1)} \_R \log\left(-5/9 - \frac{62\_R^2}{9} + \frac{31\_R}{9}\right).$$

This is just the result obtained by evaluating $F(1) - F(-1)$, where F is the antiderivative of r defined in (4.8). However, the numerical value of $F(1) - F(-1)$ calculated by MAPLE is *complex* (!): $\mathrm{I}(r; [-1, 1]) \approx 1.32 - 1.31\mathrm{i}$.

This incorrect result can be avoided: since the zeros of q can be computed with arbitrary precision, it is possible to detect whether the pole x_0 of r is contained in $[-1, 1]$. The situation is different when it comes to transcendental functions, for instance, when the denominator of a rational expression involves transcendental functions. In these cases, the poles generally cannot be located accurately, and for this

[39] It is assumed that neither a nor b are indeterminate.

reason they may remain undetected. For instance, the denominator of the function

$$f(x) := \frac{1 - \cos(x)}{(\sin(x) - x + 1/10)^2}$$

has a double zero at $x_0 \approx 0.854$. The MAPLE command `int(1/f,x=0..1);` yields

$$-\frac{-100 + 100\sin(1)}{10\sin(1) - 9},$$

although f is not integrable over $[0, 1]$.

Even if f is integrable over $[a, b]$, the functional expression given for the definite integral $\mathrm{I}(f; [a, b])$ may involve indefinite forms, making it impossible to evaluate $\mathrm{I}(f; [a, b])$ numerically.

Example (Indefinite Form Resulting from Definite Integration). The values of the polynomial

$$q(x) := x^2 + 2cx + 2$$

for abscissas $x \in [0, 1]$ are located in the interval $[\min(2, 3+2c), \max(2, 3+2c)]$. Hence, the rational function $f := 1/q$ is analytic in $[0, 1]$ if $c > -3/2$. The MAPLE command `int(1/(x^2+2*c*x*2),x=0..1);` yields

$$-\operatorname{arctanh}\left(\frac{1+c}{\sqrt{-2+c^2}}\right)\frac{1}{\sqrt{-2+c^2}} + \operatorname{arctanh}\left(\frac{c}{\sqrt{-2+c^2}}\right)\frac{1}{\sqrt{-2+c^2}}, \tag{4.13}$$

which is an indefinite form for $c = \sqrt{2}$.

Another problem to note is that evaluating the symbolic expression for a definite integral may be numerically unstable.

Example (Numerical Instability). Using the MAPLE command `convert(%,an);` (4.13) can be transformed into a logarithmic form:

$$\begin{aligned} &-\left(\log\left(\frac{1+c}{\sqrt{-2+c^2}}+1\right)1/2 - \log\left(1 - \frac{1+c}{\sqrt{-2+c^2}}\right)1/2\right)\frac{1}{\sqrt{-2+c^2}} \\ &+\left(\log\left(\frac{c}{\sqrt{-2+c^2}}+1\right)1/2 - \log\left(1 - \frac{c}{\sqrt{-2+c^2}}\right)1/2\right)\frac{1}{\sqrt{-2+c^2}}. \end{aligned} \tag{4.14}$$

(4.14) is still not defined for $c = \sqrt{2}$. Additionally, evaluating (4.14) is numerically unstable for values of c close to $\sqrt{2}$. While the exact value of (4.14) for $c = \sqrt{2}$ is $(2 + \sqrt{2})^{-1} \approx 0.2929$, the following numerical results are obtained for $c = \sqrt{2 + 10^{-m}}$, $m = 2, 4, \ldots, 18$, using a multiple precision arithmetic with 10 digits accuracy:[40]

m	2	4	6	8	10	12	14	16	18
(4.14)	0.2927	0.2929	0.2929	0.2929	0.2935	0.2950	0.3000	0.5000	0.0000

[40]This example is due to Stetter [418].

4.7. Symbolic Methods for Preprocessing Integration Problems

Frequently, the role of symbolic computation is to complement and facilitate rather than to replace numerical calculations. The efficiency and reliability of numerical integration algorithms can often be increased significantly by transforming the given integration problem into an equivalent form better suited for numerical treatment (see Section 3.6). Unfortunately, performing the mathematical operations required to implement suitable transformations (change of variable, integration by parts, etc.) is usually a tiresome and error-prone procedure. In these situations, symbolic computation systems can help to accelerate and increase the reliability of the transformation procedure. The task of finding appropriate transformations, i.e., transformations which convert the given integration problem into a numerically more tractable one, is, however, still up to the user.

Example (Transformation of Integration Problems). The integral

$$\mathrm{I}f := \int_0^1 \frac{\sin(x^2)}{x^{5/2}}\,dx$$

cannot be calculated symbolically. The numerical computation of $\mathrm{I}f$ proves difficult because the integrand has a singularity at $x = 0$. The MAPLE command

```
intparts(int(sin(x^2)/x^(5/2),x=0..1),sin(x^2));
```

which applies integration by parts to $\mathrm{I}f$, yields

$$\mathrm{I}f = -\frac{2\sin(1)}{3} - \int_0^1 -\frac{4\cos(x^2)}{3\sqrt{x}}\,dx.$$

A similar second step gives

$$\mathrm{I}f = -\frac{2\sin(1)}{3} + \frac{8\cos(1)}{3} + \int_0^1 \frac{16\sin(x^2)x^{3/2}}{3}\,dx.$$

The remaining right-hand side integral does not contain singularities and is thus better suited to numerical calculation than the original integration problem. The smoothness of the integrand can be further increased by repeated applications of integration by parts.

Part III

Numerical Integration

Chapter 5

Univariate Integration Formulas

Any numerical method for discretizing and approximating integrals is based, in one way or another, on the use of appropriate integration formulas (see Section 3.3.2). This chapter deals with integration formulas

$$(5.1) \qquad \mathrm{Q}f = \mathrm{Q}_n f := \sum_{i=1}^{n} c_i f(x_i)$$

for univariate, i.e., quadrature, problems. Section 5.1 discusses a variety of approaches to and methods of constructing quadrature formulas. Many quadrature formulas used in practice are obtained by approximating the integrand f in the integration interval $[a, b]$ using a *single* polynomial. The resulting *simple* formulas are discussed in Section 5.2. Alternatively, quadrature formulas can be constructed by approximating f using a *piecewise polynomial*. The *compound* formulas obtained in this way are the topic of Section 5.3.

Characteristics of Quadrature Formulas. Many important formulas are *symmetric*; i.e., if the abscissas are numbered in such a way that

$$x_1 < x_2 < \cdots < x_n,$$

then the following relationships hold:

$$x_i = (a + b) - x_{n-i+1} \quad \text{and} \quad c_i = c_{n-i+1}, \qquad i = 1, 2, \ldots, \lfloor (n+1)/2 \rfloor.$$

Another interesting class of formulas is the class of *equal-weight formulas*, which, as the name implies, have n equal weights, $c_1 = c_2 = \cdots = c_n$.

The abscissas $x_1, \ldots, x_n$ of all practical quadrature formulas Q_n are located in the interval of integration $[a, b]$. If the endpoints of the integration interval $[a, b]$ are the outermost abscissas of Q_n, then Q_n is said to be *closed*. By contrast, if all abscissas of Q_n are located in the interior of $[a, b]$, Q_n is said to be *open*.

5.1. Construction of Quadrature Formulas

The construction of quadrature formulas involves two main tasks. First, in order to obtain efficient integration methods, the abscissas and weights of formulas Q_n should be chosen in such a way that the accuracy requirement (see Section 3.2.2) is met using a minimum number n of abscissas. Then, if integration formulas are to be used for solving, for a class of admissible integrands $\mathcal{F}$, the numerical problem (3.1) for any given error tolerance ε, one must be able to construct sequences of integration formulas $\{\mathrm{Q}_n\}$ which converge for any $f \in \mathcal{F}$: $\mathrm{Q}_n f \to \mathrm{I}f$ as $n \to \infty$.

5.1.1. Formula Construction Using Riemann Sums.

Using Riemann sums (1.1) as quadrature formulas, it is possible to approximate *any* Riemann integral (1.2) with arbitrary accuracy. Therefore, the numerical problem (3.1) can, in principle, be solved for *any* Riemann integrable function f and for *any* accuracy requirement ε.[41]

The easiest way to obtain a convergent sequence of Riemann sums is to divide the interval $[a, b]$ into n subintervals of equal length:

$$x_i := a + ih, \quad i = 0, 1, \ldots, n; \qquad h := \frac{b-a}{n}.$$

An obvious choice of the sampling points $\xi_i \in [x_{i-1}, x_i]$ is either the right-hand or the left-hand endpoint of the subinterval. If the abscissas ξ_i are chosen as the right-hand endpoints, $\xi_i := x_i$,

$$R_n^r f := \sum_{i=1}^{n} (x_i - x_{i-1}) f(x_i) = h \sum_{i=1}^{n} f(x_i);$$

otherwise, when $\xi_i := x_{i-1}$,

$$R_n^l f := \sum_{i=1}^{n} (x_i - x_{i-1}) f(x_{i-1}) = h \sum_{i=1}^{n} f(x_{i-1}). \tag{5.2}$$

Hence, two equal-weight quadrature formulas with

$$c_1 = c_2 = \cdots = c_n = h$$

have been derived. Convergence

$$\begin{aligned} R_n^r f &\to \mathrm{I}f, \quad \text{as} \quad n \to \infty, \quad \text{and} \\ R_n^l f &\to \mathrm{I}f, \quad \text{as} \quad n \to \infty, \end{aligned}$$

---

[41] In practice, however, data errors, the use of floating-point numbers, etc. limit the maximum attainable accuracy.

for any Riemann integrable function f follows directly from the fact that R_n^r and R_n^l are Riemann sums and $h \to 0$ as $n \to \infty$.

In order to assess the efficiency of R_n^r and R_n^l, measured by the number n of f-values required to attain a specific accuracy, bounds on the absolute error $Q_n f - If$ are needed. If f is *continuous*, the discretization error $R_n^r f - If$ can be estimated in terms of the *modulus of continuity*

$$\omega(f;\delta) := \max\{|f(x_1) - f(x_2)| \ : \ x_1, x_2 \in [a,b], \ |x_1 - x_2| \le \delta\}.$$

THEOREM 5.1.1. *If $f \in C[a,b]$, then*

$$|R_n^r f - If| \le (b-a)\,\omega\left(f; \frac{b-a}{n}\right). \tag{5.3}$$

Proof.

$$\begin{aligned} |R_n^r f - If| &= \left| h\sum_{i=1}^n f(x_i) - \int_a^b f(x)\,dx \right| \\ &= \left| \sum_{i=1}^n \int_{x_{i-1}}^{x_i} [f(x_i) - f(x)]\,dx \right| \\ &\le \sum_{i=1}^n \int_{x_{i-1}}^{x_i} |f(x_i) - f(x)|\,dx \\ &\le \sum_{i=1}^n \int_{x_{i-1}}^{x_i} \omega(f;h)\,dx \\ &= nh\,\omega(f;h) = (b-a)\,\omega\left(f; \frac{b-a}{n}\right). \end{aligned}$$

□

The same error bound is valid for $R_n^l f - If$. Of course, this bound characterizes the convergence of $\{R_n^r f\}$ and $\{R_n^l f\}$ only in a *worst*-case setting. In general, if additional restrictions, such as the existence and continuity of certain derivatives, are imposed on the set $\mathcal{F}$ of integrands, then the discretization error can be characterized more accurately.

Example (Riemann Sums). The n-point Riemann sums $R_n^l f_i$ were calculated for the three integrands

$$f_1(x) = x, \quad f_2(x) = \sqrt{x}, \quad \text{and} \quad f_3(x) = \sin \pi x$$

over the interval $[0,1]$ (see Table 5.1 and Figure 5.1).

The exact discretization error for f_1 is $R_n^l f_1 - If_1 = -(2n)^{-1}$, while (5.3) gives the error bound $\omega(f;n^{-1}) = n^{-1}$. Hence, the error bound is, independently of the number n of abscissas, larger than the actual error only by a factor of two.

The modulus of continuity for the integrand f_2 in $[0,1]$ is $\omega(f_2;\delta) = \sqrt{\delta}$; thus, the error bound for $|R_n^l f_2 - If_2|$ is $1/\sqrt{n}$. Using $n = 4$ quadrature abscissas, the error bound is 0.5, whereas the actual error is only -0.1484. Using $n = 4096$ sampling

TABLE 5.1. *Values of Riemann sums R_n^l.*

N	$R_n^l f_1$	$R_n^l f_2$	$R_n^l f_3$
4	0.3750 000	0.5182 831	0.6035 534
16	0.4687 500	0.6323 312	0.6345 732
64	0.4921 875	0.6584 584	0.6364 920
256	0.4980 469	0.6646 632	0.6366 118
1024	0.4995 117	0.6661 723	0.6366 193
4096	0.4998 779	0.6665 431	0.6366 200
exact	*0.5000 000*	*0.6666 666*	*0.6366 198*

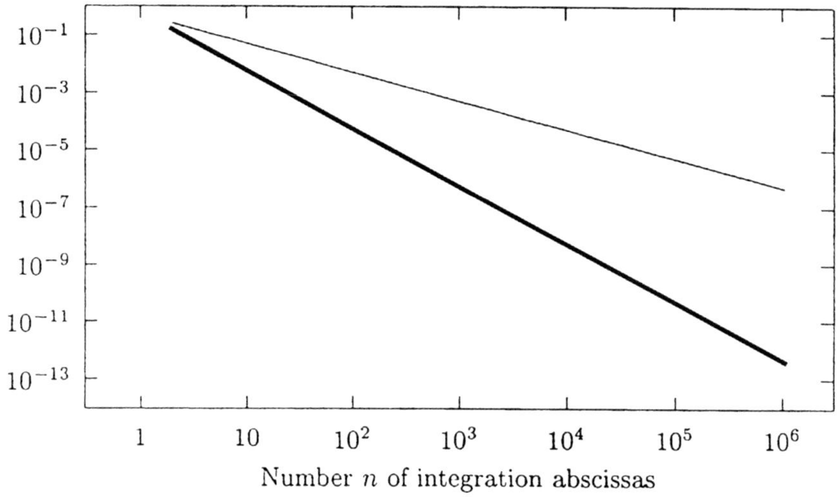

FIG. 5.1. *Errors in the Riemann sums R_n^l for the integrands f_2 (—) and f_3 (**—**) over* $[0, 1]$.

points, the error bound $1.563 \cdot 10^{-2}$ is considerably larger than the actual error $-1.235 \cdot 10^{-4}$. Hence, the convergence rate of R_n^l for f_2 is significantly higher than can be expected from the error bound (5.3). The reason for this discrepancy is that (5.3) is based on the modulus of continuity with respect to the *whole* integration interval $[a, b]$. However, the low degree of smoothness of the integrand f_2 at $x = 0$ has only a local influence near this endpoint, not a global influence on $[0, 1]$.

The Riemann sum $R_n^l f_3$ provides the most accurate numerical results. An error of $-5 \cdot 10^{-7}$ is attained for as few as $n = 1024$ sampling points, whereas (5.3) yields the considerably larger error bound $3.068 \cdot 10^{-3}$.

The convergence of the simple Riemann sums R_n^r and R_n^l is unacceptably slow for most practical integration problems. There are, however, families of quadrature formulas $\{\mathrm{Q}_n f\}$ which, in many cases, converge to $\mathrm{I}f$ much faster than the primitive rules $\{R_n^r f\}$ and $\{R_n^l f\}$ and which can be demonstrated to be families of Riemann sums. The family of Gauss formulas is such an example (see Section 5.2.7). Unfortunately, it is not possible to devise such formulas simply by looking at the definition of Riemann sums.

5.1.2. Formula Construction Using Approximation. Most efficient quadrature formulas are obtained by approximating the integrand f using a model function $g \in \mathcal{M}$ and taking the definite integral $\mathrm{I}g$ as an approximation to $\mathrm{I}f$. This approach is practical only if the indefinite integral of the model function g can be explicitly expressed as a simple formula which can be evaluated using a small number of arithmetic operations. From this point of view, polynomials are especially attractive model functions because they are easily integrated in closed form.

If the approximating function g satisfies

$$\|g - f\|_\infty \le \frac{\varepsilon}{b-a},$$

then the error bound

$$\begin{aligned} |\mathrm{I}g - \mathrm{I}f| &= \left| \int_a^b g(x)\,dx - \int_a^b f(x)\,dx \right| \\ &\le \int_a^b |g(x) - f(x)|\,dx \\ &\le (b-a)\|g - f\|_\infty \\ &\le \varepsilon \end{aligned}$$

can be derived. Therefore, approximating functions which are sufficiently close to f with respect to the L_∞ norm are suitable candidates for solving the numerical integration problem (3.1).

Example (Integration of Model Functions). Suppose the integral

$$\mathrm{I}f = \int_0^1 \sqrt{x}\,\Gamma(x+1)\,dx$$

is to be calculated with the maximum (absolute) error $\varepsilon = 10^{-3}$. The following approximation can be found in Hart et al. [200]:

$$\Gamma(x+1) = a_0 + a_1 x + a_2 x^2 + a_3 x^3 + e(x),$$

$$a_0 = 0.9991\,0836,\ a_1 = 0.4497\,361,\ a_2 = 0.2855\,737,\ a_3 = 0.2646\,888,$$

where

$$|e(x)| \le 9 \cdot 10^{-4} \quad \text{for all} \quad x \in [0,1].$$

Hence,

$$\mathrm{I}f = \frac{2}{3}a_0 + \frac{2}{5}a_1 + \frac{2}{7}a_2 + \frac{2}{9}a_3 + \int_0^1 \sqrt{x}e(x)\,dx = 0.9863\,789 + \eta$$

is obtained, and the absolute error η is bounded by

$$|\eta| \leq \frac{2}{3} 9 \cdot 10^{-4},$$

i.e., $\mathrm{I}f \in [0.9857, 0.9870]$.

In order to construct an algorithm which is able to solve the numerical problem (3.1) for any given error tolerance ε, a *sequence* of model functions $\{g_m\}$ is needed for which the convergence

$$g_m \to f \qquad \text{as} \qquad m \to \infty$$

(with respect to the L_∞ norm) can be guaranteed.

Example (Bernstein Polynomials). Bernstein polynomials play an important role in approximation theory because they asymptotically approximate continuous functions on compact intervals with arbitrary accuracy. The Bernstein polynomial for $f \in C[0,1]$ of degree d is given by

$$B_d(f)(x) := \sum_{i=0}^{d} \binom{d}{i} f\left(\frac{i}{d}\right) x^i (1-x)^{d-i}.$$

The sequence $\{B_d(f)\}$ converges to f *uniformly* on $[0,1]$. The relationship

$$\int_0^1 B_d(f)(x)\,dx = \sum_{i=0}^{d} \binom{d}{i} f\left(\frac{i}{d}\right) \int_0^1 x^i (1-x)^{d-i}\,dx = \frac{1}{d+1} \sum_{i=0}^{d} f\left(\frac{i}{d}\right)$$

leads to a $(d+1)$-point quadrature formula with the abscissas and weights

$$x_i = \frac{i-1}{d}, \quad c_i = \frac{1}{d+1}, \qquad i = 1, \ldots, d+1.$$

This formula corresponds to the mean value of the integrand under an equidistant subdivision of the interval $[0,1]$. Due to the slow convergence $B_d(f) \to f$, the formulas obtained by integrating Bernstein polynomials do not show any substantial improvement compared with the simple rules R_n^r and R_n^l.

Example (Taylor Polynomials). The truncated expansion of a function into a Taylor series, i.e., the Taylor polynomial, can be integrated, and an estimate of the truncation error can be obtained from the remainder term of the expansion. If, for example, a function is expanded into a Taylor series around 0,

$$f(x) = f(0) + f'(0)x + \frac{f''(0)}{2!}x^2 + \cdots + \frac{f^{(d)}(0)}{d!}x^d + \cdots,$$

and the radius of convergence is greater than 1, then the following approximation is obtained:

$$\int_{-1}^{1} f(x)\,dx \;\approx\; \mathrm{I}(T_d(f;0)) \;=\; 2\left(f(0) + \frac{f''(0)}{3!} + \frac{f^{(4)}(0)}{5!} + \cdots + \frac{f^{(2\lfloor d/2 \rfloor)}(0)}{(2\lfloor d/2 \rfloor + 1)!}\right).$$

For $f \in C^{(d+1)}[-1,1]$, the remainder term can be bounded by

$$|f(x) - T_d(f;0)(x)| \le \frac{M_{d+1}}{(d+1)!}|x|^{d+1}, \quad x \in [-1,1],$$

from which

$$|\mathrm{I}f - \mathrm{I}(T_d(f;0))| \le \frac{2M_{d+1}}{(d+2)!}$$

can be derived.

Product Integration. In practical applications it is often necessary to integrate functions of the product form wf, where f is the actual integrand and w is an additional weight function (see Section 3.1). Frequently, the weight function w is significantly more difficult to approximate using model functions than the integrand f. This is true, for instance, for singular and/or rapidly oscillatory weight functions (see Section 3.1.2). If the weight function is difficult to approximate, quadrature formulas obtained by approximating the product wf cannot be expected to yield accurate results.

This problem can be overcome by approximating only the integrand f using suitable model functions $g \in \mathcal{M}$ and by taking the integral $\mathrm{I}_w g = \mathrm{I}(wg)$ of g with respect to the weight function w as an approximation for $\mathrm{I}_w f = \mathrm{I}(wf)$. This modified approximation approach to constructing quadrature formulas is referred to as *product integration*, and the resulting formulas are said to be *product formulas*. Product integration is practical only if the integral of g with respect to w can be expressed explicitly in closed form. This requires the weight function w to be known analytically, regardless of the space of model functions $\mathcal{M}$ used. Polynomials are suitable model functions, as they can be integrated in closed form with respect to most common weight functions (see Section 3.1.2).[42]

The validity of product integration stems from the error bound

$$|\mathrm{I}_w g - \mathrm{I}_w f| \le \|w\|_1 \|g - f\|_\infty \tag{5.4}$$

derived from the condition estimate (3.5). If the weight function is absolutely integrable on $[a,b]$, then (5.4) shows that the error becomes arbitrarily small when g approximates f (with respect to the L_∞ norm) with sufficient accuracy. It is assumed to hold throughout the remainder of this chapter that w is absolutely integrable.

Product integration is particularly suitable for integrating a single integrand f with respect to several different weight functions $w_1, \dots, w_m$. In this case, a single model function g is sufficient for approximating f in *all* m integrals, while m model functions would be required for approximating the

[42]Here and in the remainder of this chapter, it is tacitly assumed that the product $w \cdot P$ is Riemann integrable on $[a,b]$ for any polynomial $P \in \mathbb{P}$.

products $w_1 f, \ldots, w_m f$ in the conventional approximation approach to formula construction. In general, product integration significantly reduces the computational effort required to calculate m integrals. Additionally, the estimate (5.4) shows that the resulting integration errors

$$\mathrm{I}_{w_1} g - \mathrm{I}_{w_1} f, \ldots, \mathrm{I}_{w_m} g - \mathrm{I}_{w_m} f$$

can be expected to be of similar magnitude if the condition numbers $\|w_1\|_1, \ldots, \|w_m\|_1$ are of comparable size.

Approximation Using Polynomial Interpolation. The approximation of the integrand by means of *polynomial interpolation* is of the greatest importance to the construction of quadrature formulas. The unique polynomial $P_{n-1} \in \mathbb{P}_{n-1}$ interpolating the integrand f at the n distinct points $x_1, \ldots, x_n$,

$$P_{n-1}(x_i) = f(x_i), \quad i = 1, \ldots, n,$$

can be represented as a linear combination of *fundamental Lagrange polynomials* $\varphi_{n-1,i}$:

$$\begin{aligned} P_{n-1}(x) &= \sum_{i=1}^{n} f(x_i)\varphi_{n-1,i}(x), \\ \varphi_{n-1,i}(x) &:= \prod_{\substack{j=1 \\ j\neq i}}^{n} \frac{x - x_j}{x_i - x_j}, \quad i = 1, 2, \ldots, n. \end{aligned} \tag{5.5}$$

According to

$$\mathrm{I}_w P_{n-1} = \int_a^b w(x) P_{n-1}(x)\, dx = \sum_{i=1}^{n} f(x_i) \int_a^b w(x)\varphi_{n-1,i}(x)\, dx,$$

the weights of the quadrature formula $\mathrm{Q}_n f := \mathrm{I}_w P_{n-1}$ are given by

$$c_i := \int_a^b w(x)\varphi_{n-1,i}(x)\, dx, \qquad i = 1, 2, \ldots, n. \tag{5.6}$$

This formula can also be derived using a different approach: given the n abscissas $x_1, \ldots, x_n$, the weights $c_1, \ldots, c_n$ are chosen in such a way that $\mathrm{Q}_n f$ *exactly* integrates the elements of a basis $\{b_0, b_1, \ldots, b_{n-1}\}$ for $\mathbb{P}_{n-1}$. For the *monomial* basis $\{1, x, x^2, \ldots, x^{n-1}\}$, for example, the quadrature weights are obtained by solving the following system of linear equations:

$$\begin{aligned} c_1 + c_2 + \cdots + c_n &= \int_a^b w(x)\, dx, \\ c_1 x_1 + c_2 x_2 + \cdots + c_n x_n &= \int_a^b w(x) x\, dx, \end{aligned}$$

$$\vdots$$

(5.7)
$$c_1 x_1^{n-1} + c_2 x_2^{n-1} + \cdots + c_n x_n^{n-1} = \int_a^b w(x) x^{n-1}\, dx.$$

The coefficient matrix for this system of linear equations is the *Vandermonde matrix* $V(x_1, x_2, \ldots, x_n)$, whose determinant is a product of differences of the abscissas (Davis [90]):

(5.8)
$$\det V(x_1, \ldots, x_n) = \prod_{i=2}^{n} \prod_{j=1}^{i-1} (x_i - x_j).$$

If the abscissas $x_1, \ldots, x_n$ are distinct, which is part of the presupposition, then (5.8) implies that $V(x_1, \ldots, x_n)$ is nonsingular and thus there is a uniquely determined set of integration weights $c_1, \ldots, c_n$ satisfying the *moment equations* (5.7).[43] On the other hand, the quadrature formula defined by (5.6), which is derived using fundamental Lagrange polynomials, integrates the monomials $1, x, \ldots, x^{n-1}$ exactly. Hence, the unique solution of (5.7) is given by the weights (5.6).

DEFINITION 5.1.1 (INTERPOLATORY QUADRATURE FORMULA). *A quadrature formula* Q_n *with the abscissas* $x_1, \ldots, x_n$ *is said to be interpolatory if its weights* $c_1, \ldots, c_n$ *are given by (5.6) or, equivalently, if its weights satisfy the moment equations (5.7). An interpolatory quadrature formula can be interpreted as the exact integral of the polynomial which interpolates* f *at the quadrature abscissas.*

Given the general error bound (5.4) for quadrature formulas based on integrating approximation functions g, the absolute discretization error of an interpolatory quadrature formula can easily be estimated in terms of the approximation error $e_{n-1}(f) := \|P_{n-1} - f\|_\infty$:

$$|Q_n f - I_w f| = |I_w P_{n-1} - I_w f| \le \|w\|_1\, e_{n-1}(f).$$

For sufficiently smooth integrands f, the approximation error $e_{n-1}(f)$ can be bounded as follows.

THEOREM 5.1.2 (ERROR IN INTERPOLATION POLYNOMIALS). *For* $f \in C^{(n)}[a,b]$ *the approximation error* $e_{n-1}(f)$ *satisfies*

(5.9)
$$e_{n-1}(f) \le \frac{M_n}{n!} \|\omega_n\|_\infty, \quad M_n := \|f^{(n)}\|_\infty,$$

where the polynomial $\omega_n \in \mathbb{P}_n$ *is defined by*

$$\omega_n(x) = (x - x_1)(x - x_2) \cdots (x - x_n).$$

[43]For $w \not\equiv 1$ the equations (5.7) are often referred to as the *modified* moment equations.

Proof. See Ueberhuber [435].

Quadrature formulas such as interpolatory quadrature formulas, which are derived by approximating the integrand f in the integration interval $[a, b]$ using a *single* polynomial P_d, are said to be *simple*. By contrast, *compound formulas* are obtained by approximating the integrand f using a *piecewise polynomial*, i.e., by subdividing the integration interval $[a, b]$ into a number of subintervals and approximating f using different polynomials (of the same degree) in different subintervals. These two types of quadrature formulas differ fundamentally in the way their integration errors can be reduced. The error in a simple quadrature formula is generally decreased by increasing the degree d of the approximation polynomial P_d; by contrast, the error of a compound formula is decreased by refining the subdivision of the integration interval $[a, b]$, while fixing the degree of the polynomials used in each subinterval for approximating f.

A number of different choices of the interpolation points $x_1, \ldots, x_n$ are possible, leading to a number of different numerical quadrature formulas. In order to establish which formulas should be used for a given problem, the properties of the most important simple and compound formulas are discussed in Sections 5.2 and 5.3.

5.1.3. Formula Construction Using Acceleration. If a suitable asymptotic expansion for the truncation errors $\mathrm{Q}_{n_m} f - \mathrm{I}f$ in a sequence of quadrature formulas $\{\mathrm{Q}_{n_m} f =: s_m\}$ is available, then acceleration algorithms can be used to obtain a sequence $\{\hat{s}_m\}$ which converges faster to $\mathrm{I}f$ than the original sequence s_m (see Section 3.7).

Example (Acceleration of Simple Riemann Sums). If the sequence of the Riemann sums $\{R_n^l\}$, with $n = 2^1, 2^2, 2^3, \ldots$ quadrature abscissas, obtained for the integral

$$\mathrm{I}f = \int_0^1 \sqrt{x}\, dx = \frac{2}{3}$$

is examined closely, it becomes clear that the error is approximately halved each time the number of sampling points is doubled:

n	$R_n^l f - \mathrm{I}f$
2	$-3.131 \cdot 10^{-1}$
4	$-1.484 \cdot 10^{-1}$
8	$-7.104 \cdot 10^{-2}$
16	$-3.434 \cdot 10^{-2}$
32	$-1.673 \cdot 10^{-2}$
64	$-8.208 \cdot 10^{-3}$

n	$R_n^l f - \mathrm{I}f$
128	$-4.047 \cdot 10^{-3}$
256	$-2.003 \cdot 10^{-3}$
512	$-9.944 \cdot 10^{-4}$
1024	$-4.944 \cdot 10^{-4}$
2048	$-2.462 \cdot 10^{-4}$
4096	$-1.236 \cdot 10^{-4}$

i.e.,

$$R_n^l f - \mathrm{I}f \approx \frac{A}{n}; \tag{5.10}$$

$$R_{2n}^l f - \mathrm{I}f \approx \frac{A}{2n}, \tag{5.11}$$

where A is a constant independent of n. Multiplying (5.11) by 2 and subtracting (5.10) leads to

$$(2R^l_{2n}f - R^l_n f) - \mathrm{I}f \approx 0;$$

i.e., it is to be expected that $(2R^l_{2n}f - R^l_n f)$ is a better approximation to $\mathrm{I}f$ than $R^l_{2n}f$. This is indeed the case; for instance,

$$2R^l_{128}f - R^l_{64}f = 2 \cdot 0.6626\,195 - 0.6584\,584 = 0.6667\,806$$

has an error of $1.139 \cdot 10^{-4}$, which is comparable to the error in $R^l_{4096}f$. Thus, by utilizing the knowledge of the error structure as given by (5.10) and (5.11), it is possible to construct a new quadrature formula

$$\mathrm{Q}_{2n}f := \{2R^l_{2n}f - R^l_n f\}$$

which is significantly more accurate than $R^l_{2n}f$.

If the transformed sequence elements $\hat{s}_1, \hat{s}_2, \ldots$ correspond to linear combinations of the original quadrature formulas, then they themselves can be interpreted as quadrature formulas:

$$\hat{s}_m := \sum_{j=0}^{J} d_j^{(m)} \mathrm{Q}_{n_{m+j}} f = \sum_{j=0}^{J} d_j^{(m)} \sum_{i=1}^{n_{m+j}} c_i^{(j)} f(x_i^{(j)}) =: \sum_{i=1}^{\hat{n}_m} \hat{c}_i f(\hat{x}_i).$$

As a result, *linear* acceleration algorithms can be regarded as methods of constructing new quadrature formulas. However, the acceleration algorithms described in Section 3.7 are mostly *nonlinear*, and their results *cannot* be interpreted as quadrature formulas. Exceptions are Richardson's extrapolation (see Section 3.7.3), provided the auxiliary sequence $\{t_m\}$ is independent of the sequence $\{s_m\}$, and, more generally, particular instances of the E-algorithm (see Section 3.7.8), provided the auxiliary sequences $\{t_m^{(1)}\}, \ldots, \{t_m^{(J)}\}$ are independent of the sequence $\{s_m\}$.

Suitable asymptotic expansions for the integration error can be obtained for the *compound* quadrature rules $k \times \mathrm{Q}$ described in Section 5.3. More specifically, error expansions of the form

$$(k \times \mathrm{Q})f - \mathrm{I}f = Ak^{-\alpha} + Bk^{-\beta} + Ck^{-\gamma} + \cdots \tag{5.12}$$

can be obtained for particular classes $\mathcal{F}$ of integrands. For instance, if the integrand is sufficiently smooth, the Euler–Maclaurin sum formula (see Theorem 5.3.3) shows that the exponents in (5.12) for the compound trapezoidal rule are given by

$$\alpha = 2, \quad \beta = 4, \quad \gamma = 6, \quad \ldots.$$

In this case Richardson's extrapolation can be applied, which gives rise to the family of *Romberg formulas* (see Section 5.3.6). For integrands with low-order continuity or differentiability (with algebraic or logarithmic singularities, for instance), the exponents in (5.12) depend on certain analytic properties of the integrand (Lyness and Ninham [279]). If the exponents are known *explicitly*, special linear instances of the E-algorithm can be applied to speed up convergence. If the exponents are *not* known then only nonlinear extrapolation can be used. Nonlinear extrapolation methods are discussed mainly in conjunction with integration techniques for special kinds of integrals (see Chapter 7).

5.1.4. Formula Construction Using Transformations. Many quadrature formulas are derived for particular intervals of integration. Gauss formulas, for instance, are often derived for the interval $[-1, 1]$. In order to make such formulas applicable to other intervals, the formulas have to undergo appropriate transformations. More precisely, since only *affine* transformations map general polynomials onto polynomials of the *same degree*, only such affine transformations are generally appropriate for transforming quadrature formulas based on polynomial approximation. Nonlinear transformations are only relevant in the context of IMT formulas (see Section 5.3.5).

Fortunately, any two compact intervals $[a, b]$ and $[\overline{a}, \overline{b}]$ can be mapped onto each other using affine transformations. $\overline{x} \in [\overline{a}, \overline{b}]$ is mapped onto $x \in [a, b]$ by

$$x = \gamma\overline{x} + \beta, \qquad \overline{x} = \frac{1}{\gamma}\big(x - \beta\big), \tag{5.13}$$

with the constants

$$\gamma := \frac{b - a}{\overline{b} - \overline{a}}, \qquad \beta := \frac{a\overline{b} - \overline{a}b}{\overline{b} - \overline{a}}.$$

Hence,

$$\int_a^b w(x) f(x)\, dx = \gamma \int_{\overline{a}}^{\overline{b}} \overline{w}(\overline{x}) \overline{f}(\overline{x})\, d\overline{x},$$

where

$$\overline{w}(\overline{x}) := w(\gamma\overline{x} + \beta), \qquad \overline{f}(\overline{x}) := f(\gamma\overline{x} + \beta).$$

The affine transformation (5.13) converts the quadrature formula

$$\int_{\overline{a}}^{\overline{b}} \overline{w}(\overline{x}) \overline{f}(\overline{x})\, d\overline{x} = \sum_{i=1}^{n} \overline{c}_i \overline{f}(\overline{x}_i) + \overline{E}(\overline{f})$$

into

$$\int_a^b w(x) f(x)\, dx = \sum_{i=1}^{n} c_i f(x_i) + E(f),$$

with abscissas and weights given by

$$x_i := \gamma\overline{x}_i + \beta, \quad c_i := \gamma\overline{c}_i, \quad i = 1, 2, \ldots, n,$$

and $E(f) = \gamma\overline{E}(\overline{f})$ (see, for instance, Stroud [424]).

5.2. Simple Interpolatory Quadrature Formulas

Interpolatory quadrature formulas are uniquely characterized by the choice of their abscissas $x_1, \ldots, x_n$. The most important sets of quadrature abscissas (which are discussed later in more detail) are:

Abscissas	**Class of Formulas**
equidistant subdivision	*Newton–Cotes* formulas
including endpoints	*closed* Newton–Cotes formulas
without endpoints	*open* Newton–Cotes formulas
zeros of orthogonal polynomials	*Gauss* formulas
Legendre polynomials P_n	Gauss–*Legendre* formulas
Laguerre polynomials L_n	Gauss–*Laguerre* formulas
Hermite polynomials H_n	Gauss–*Hermite* formulas
Jacobi polynomials $P_n^{\alpha,\beta}$	Gauss–*Jacobi* formulas
zeros of	
$[P_{n-1}(x) + P_n(x)]$	*Radau* formulas
$(x^2 - 1)\, P'_{n-1}$	*Lobatto* formulas
Chebyshev extrema	(practical) *Clenshaw–Curtis* formulas
Chebyshev zeros	(classical) *Clenshaw–Curtis* formulas

Software (Interpolatory Quadrature Formulas). `TOMS/655` contains subroutines which compute, for given abscissas $x_1, \ldots, x_n$, the weights $c_1, \ldots, c_n$ for the corresponding interpolatory quadrature formula Q_n. In addition, `TOMS/655` includes programs which, instead of returning the weights $c_1, \ldots, c_n$, evaluate the computed quadrature formula for a user-supplied integrand f.

Any admissible weight function w can be handled if the user provides the moment $\mathrm{I}w$ as well as the *Jacobi matrix* (5.25) of w. Furthermore, `TOMS/655` contains routines for evaluating $\mathrm{I}w$ and setting up the Jacobi matrix for a number of common positive weight functions w, among them most of the admissible classical weight functions described in Section 3.1.2. In order to obtain interpolatory formulas with respect to any of these predefined weight functions, the user only has to assign a certain value to a subroutine parameter instead of having to provide $\mathrm{I}w$ and the Jacobi matrix.

`TOMS/655` also allows for *multiple* abscissas by applying *Hermite–Birkhoff interpolation*: if the multiplicity of a sampling point x_i is equal to k_i, $k_i \in \{2, 3, 4, \ldots\}$, then the corresponding Hermite–Birkhoff interpolation polynomial interpolates not only f, but also the derivatives $f', f'', \ldots, f^{(k_i-1)}$ at x_i.

5.2.1. Degree of Accuracy.

An important characteristic of every quadrature formula is its *degree of accuracy* (or *degree of exactness*).

DEFINITION 5.2.1 (DEGREE OF ACCURACY). *The degree of accuracy of a quadrature formula* Q_n *is* D *if*

$$\mathrm{Q}_n x^k = \mathrm{I}_w x^k, \qquad k = 0, 1, \ldots, D,$$

$$Q_n x^{D+1} \neq I_w x^{D+1},$$

i.e., if Q_n *is exact for all polynomials of degree* $d \leq D$ *and if there is at least one polynomial of degree* $d = D+1$ *for which* Q_n *is not exact.*

Clearly, the degree of accuracy D of a simple interpolatory quadrature formula Q_n satisfies $D \geq n-1$. The following theorem shows that the converse is also true.

THEOREM 5.2.1. *Any* n*-point quadrature formula* Q_n *with a degree of accuracy* $D \geq n-1$ *is interpolatory.*

Proof. Since the degree of accuracy of Q_n is $D \geq n-1$, the weights $c_1, \ldots, c_n$ of Q_n satisfy the moment equations (5.7) for the given abscissas $x_1, \ldots, x_n$ of Q_n. However, the unique solution of the moment equations (5.7) for given abscissas $x_1, \ldots, x_n$ are the weights of the corresponding interpolatory quadrature formula. Hence, Q_n must be interpolatory. □

The significance of the degree of accuracy stems from the following theorem, which relates the error in a quadrature formula with degree of accuracy D to the error with respect to the L_∞ norm in the best approximation polynomial P_D^*.

THEOREM 5.2.2. *Any quadrature formula* Q_n *with a degree of accuracy* $D \geq 0$ *satisfies the inequality*

$$(5.14) \qquad |Q_n f - I_w f| \leq \left(\|w\|_1 + \sum_{i=1}^{n} |c_i| \right) e_D^*(f),$$

where

$$e_D^*(f) := \inf\{\|P_D - f\|_\infty : P_D \in \mathbb{P}_D\}.$$

If all quadrature weights of Q_n *are nonnegative, then*

$$(5.15) \qquad |Q_n f - I_w f| \leq (\|w\|_1 + Iw)\, e_D^*(f).$$

Proof. $P_D \in \mathbb{P}_D$ implies $Q_n P_D = I_w P_D$ and thus

$$\begin{aligned} |Q_n f - I_w f| &= |Q_n f - Q_n P_D + Q_n P_D - I_w P_D + I_w P_D - I_w f| \\ &\leq |Q_n f - Q_n P_D| + |I_w P_D - I_w f| \\ &= |Q_n (P_D - f)| + |I_w (P_D - f)|. \end{aligned}$$

The two right-hand side terms can be bounded by

$$\begin{aligned} |Q_n (P_D - f)| &= \left| \sum_{i=1}^{n} c_i (P_D(x_i) - f(x_i)) \right| \\ &\leq \sum_{i=1}^{n} |c_i| |P_D(x_i) - f(x_i)| \end{aligned}$$

$$\begin{aligned}
&\leq \|P_D - f\|_\infty \sum_{i=1}^{n} |c_i|, \\
|\mathrm{I}_w(P_D - f)| &= \left| \int_a^b w(x)(P_D(x) - f(x))\, dx \right| \\
&\leq \int_a^b |w(x)|\, \|P_D - f\|_\infty\, dx \\
&= \|w\|_1\, \|P_D - f\|_\infty .
\end{aligned}$$

Consequently, for any $P_D \in \mathbb{P}_D$, the error estimate

$$|\mathrm{Q}_n f - \mathrm{I}_w f| \leq \left(\|w\|_1 + \sum_{i=1}^{n} |c_i| \right) \|P_D - f\|_\infty$$

holds. The sharpest error estimate is obtained with the best approximation polynomial $P_D^* \in \mathbb{P}_D$ with respect to the L_∞ norm: $\|P_D^* - f\|_\infty =: e_D^*(f)$. If all the weights of the quadrature formula Q_n are nonnegative, then

$$\sum_{i=1}^{n} |c_i| = \sum_{i=1}^{n} c_i = \int_a^b w(x)\, dx = \mathrm{I}w$$

because Q_n is assumed to be exact for the constant polynomial $P_0 \equiv 1$. □

5.2.2. Convergence. If the coefficients of a sequence $\{\mathrm{Q}_n\}$ of interpolatory quadrature formulas are *nonnegative* (for all n), then the error bound (5.15) implies the convergence

$$\mathrm{Q}_n f \to \mathrm{I}_w f \qquad \text{as} \qquad n \to \infty$$

for all $f \in C[a, b]$, provided the interval of integration $[a, b]$ is bounded. The latter condition on $[a, b]$ is assumed throughout the remainder of this chapter, unless explicitly stated otherwise. The aforementioned convergence result is due to the *Weierstrass approximation theorem* (Davis [90]), which guarantees that

$$e_D^*(f) \to 0 \qquad \text{as} \qquad D \to \infty$$

for all continuous functions f. It can even be shown that any such sequence of interpolatory quadrature formulas converges even for all $f \in \mathcal{R}[a, b]$.

THEOREM 5.2.3. *Let $\{\mathrm{Q}_n\}$ denote a sequence of quadrature formulas with respect to an admissible weight function w on the bounded interval $[a, b]$, and let all coefficients of $\{\mathrm{Q}_n\}$ be nonnegative. Then the convergence of $\{\mathrm{Q}_n\}$ for all $f \in C[a, b]$ implies the convergence of $\{\mathrm{Q}_n\}$ for all Riemann integrable functions $f \in \mathcal{R}[a, b]$.*

Proof. See Davis and Rabinowitz [92].

The relationship

$$e_D^*(f) \leq e_{D-1}^*(f) \qquad \text{for all} \quad D \in \mathbb{N},$$

along with (5.15), implies that the error in interpolatory formulas (with nonnegative weights) can be expected to decrease as the degree of accuracy increases. Hence, among all n-point interpolatory quadrature formulas, those with nonnegative weights and maximum degree of accuracy are of particular importance. These are the Gauss formulas, i.e., formulas whose abscissas are the zeros of orthogonal polynomials (see Section 5.2.7).

Interestingly, interpolatory formulas with a particularly high degree of accuracy, $D \geq 2n-2$, can generally be shown to have positive weights.

THEOREM 5.2.4. *The coefficients $c_1, \dots, c_n$ of any n-point quadrature formula Q_n with respect to an admissible weight function w with a degree of accuracy $D \geq 2n-2$ are all positive: $c_i > 0$ for $i = 1, \dots, n$.*

Proof. As the quadrature formula Q_n is exact for all polynomials in $\mathbb{P}_{2n-2}$, it also is exact for the polynomials

$$\varphi_{n-1,j}^2 \in \mathbb{P}_{2n-2}, \quad j = 1, 2, \dots, n,$$

where $\varphi_{n-1,j}$ denotes the jth fundamental Lagrange polynomial (5.5) of degree $d = n-1$ for the abscissas of Q_n. Thus

$$0 < \int_a^b w(x)\varphi_{n-1,j}^2(x)\,dx = \sum_{i=1}^{n} c_i \varphi_{n-1,j}^2(x_i) = c_j, \quad j = 1, 2, \dots, n. \qquad \square$$

If all weights of Q_n are nonnegative, bounds on the truncation error $\mathrm{Q}_n f - \mathrm{I}_w f$ can be derived directly from bounds on $e_D^*(f)$, which enables one to characterize the rate of convergence $\mathrm{Q}_n f \to \mathrm{I}_w f$. According to *Jackson's theorems* (Cheney [56]), for instance,

$$e_D^*(f) \leq C_1\, \omega\left(f; \frac{b-a}{2D}\right)$$

holds for any $f \in C[a, b]$, which implies

$$e_D^*(f) \leq C_2 \frac{b-a}{D}$$

for Lipschitz continuous functions f. Additionally, if $f \in C^{(k)}[a, b]$, $k \in \mathbb{N}$, then

$$e_D^*(f) \leq C_3 \frac{M_k (b-a)^k}{D^k},$$

where M_k is a bound for $|f^{(k)}|$ on $[a, b]$.[44]

[44] C_1, C_2, and C_3 are constants independent of f.

However, *without* additional information about f such as bounds on certain derivatives or on the modulus of continuity (as assumed in Jackson's theorems), nothing can be said about the *rate* of convergence of $\{\mathrm{Q}_n f\}$. The following theorem shows that the rate of convergence can be arbitrarily slow.

THEOREM 5.2.5. *If* $\{\mathrm{Q}_n f\}$ *converges to* $\mathrm{I}f$ *for all* $f \in C[a,b]$,

$$\mathrm{Q}_n f \to \mathrm{I}f \qquad \textit{as} \quad n \to \infty,$$

then for every convergent sequence $\{\lambda_n\}$ *(with any arbitrarily slow rate of convergence),*

$$\lambda_n \in \mathbb{R}, \qquad \lambda_n \geq 0, \qquad \lim_{n\to\infty} \lambda_n = 0,$$

there exists a function $f \in C[a,b]$ *for which the discretization error in* $\{\mathrm{Q}_n f\}$ *converges even more slowly:*

$$|\mathrm{Q}_n f - \mathrm{I}f| \geq \lambda_n \quad \textit{for all} \quad n \in \mathbb{N}.$$

Proof. See Lipow and Stenger [263].

5.2.3. Asymptotic Error Behavior. Many interpolatory quadrature formulas, such as the closed Newton–Cotes formulas (see Section 5.2.4), are used primarily as basic rules in compound quadrature formulas. Other interpolatory quadrature formulas, like the Gauss formulas (see Section 5.2.7), are often used in adaptive subdivision algorithms (see Section 8.2.2). In both cases, a fixed interpolatory quadrature formula Q_n is applied repeatedly (in a properly transformed form) to intervals $[a,b]$ of decreasing length $h := b - a$. The efficiency of compound formulas as well as adaptive subdivision algorithms is therefore influenced by the asymptotic behavior of the integration error $\mathrm{Q}_n(f;[a,b]) - \mathrm{I}(f;[a,b])$ as $h = b - a \to 0+$.

THEOREM 5.2.6. *Let* Q_n *be a quadrature formula with degree of accuracy* $D \geq 0$, *and assume that the integrand* f *is* $(D+1)$*-times continuously differentiable on* $[a,b]$, $f \in C^{(D+1)}[a,b]$. *Then there exists a constant* C *such that the error bound*

$$|\mathrm{Q}_n(f;[\alpha,\beta]) - \mathrm{I}(f;[\alpha,\beta])| \leq C\,(\beta - \alpha)^{D+2} \tag{5.16}$$

holds for any subinterval $[\alpha,\beta] \subseteq [a,b]$.

Proof. The theorem can easily be proved by expanding the integrand f into its Taylor series around the interval center $(a+b)/2$. □

The error bound (5.16) implies the rate of convergence

$$|\mathrm{Q}_n(f;[a,b]) - \mathrm{I}(f;[a,b])| = O(h^{D+2}) \qquad \text{as} \quad h := b - a \to 0+.$$

This result confirms the importance of the degree of accuracy as a key property of interpolatory quadrature formulas.

5.2.4. Closed Newton–Cotes Formulas. Closed Newton–Cotes formulas Q_n, $n = 2, 3, 4, \ldots$, are interpolatory quadrature formulas for the weight function $w(x) \equiv 1$ based on the equidistant quadrature abscissas

$$x_i = a + (i-1)\frac{b-a}{n-1}, \qquad i = 1, 2, \ldots, n.$$

The weights $c_1, c_2, \ldots, c_n$ can be obtained by integrating the corresponding fundamental Lagrange polynomials.

Example (Trapezoidal Rule). Perhaps the most important closed Newton–Cotes formula is the *trapezoidal rule*

$$Q_2(f;[a,b]) = \frac{b-a}{2}(f(a) + f(b)).$$

This formula is obtained by interpolating the data points $(a, f(a))$ and $(b, f(b))$ using a linear polynomial $P_1 \in \mathbb{P}_1$, which leads to a trapezoid bounded by the lines $x = a$, $x = b$, the x-axis, and the chord between $(a, f(a))$ and $(b, f(b))$. The area of this trapezoid is used as an estimate of the required integral.

The first four closed Newton–Cotes formulas are ($f_i := f(x_i)$)

$$\begin{aligned}
Q_2 f &= \frac{b-a}{2}(f_1 + f_2) && (\textit{trapezoidal rule}),\\
Q_3 f &= \frac{b-a}{6}(f_1 + 4f_2 + f_3) && (\textit{Simpson's rule}),\\
Q_4 f &= \frac{b-a}{8}(f_1 + 3f_2 + 3f_3 + f_4) && (\textit{Simpson's 3/8 rule}),\\
Q_5 f &= \frac{b-a}{90}(7f_1 + 32f_2 + 12f_3 + 32f_4 + 7f_5) && (\textit{Boole's rule}).
\end{aligned}$$

While the weights of the above low-order Newton–Cotes formulas are all positive, the closed Newton–Cotes formula Q_9,

$$Q_9 f = \frac{b-a}{28350}(989f_1 + 5888f_2 - 928f_3 + 10496f_4 - 4540f_5 + \cdots),$$

and *all* formulas with $n \geq 11$ abscissas contain some *negative* coefficients. Hence, the relationship

$$\sum_{i=1}^{n} |c_i| > b - a$$

holds for Newton–Cotes formulas of higher degree; for Q_9, for example,

$$\sum_{i=1}^{n} |c_i| = \frac{6857}{4725}(b-a) \approx 1.45(b-a).$$

Moreover, the following theorem holds.

THEOREM 5.2.7 (KUSMIN). *Let $c_1^{(n)}, \ldots, c_n^{(n)}$ be the coefficients of the closed Newton–Cotes formulas Q_n. Then*

$$\sum_{i=1}^{n} |c_i^{(n)}| \to \infty \quad \text{as} \quad n \to \infty. \tag{5.17}$$

Proof. See Engels [131].

The increase in the condition numbers $K_{f,Q_n} = \sum |c_i^{(n)}|$ of the Newton–Cotes formulas Q_n with growing n signifies an amplified impact of the errors (perturbations) in the function values $f(x_i)$ on the result $Q_n f$. Hence, for large values of n, Newton–Cotes formulas are *numerically unstable*.

Furthermore, when it comes to proving the convergence of Newton–Cotes formulas for all $f \in C[a,b]$, neither of the inequalities (5.14) or (5.15) applies because high-order Newton–Cotes formulas contain negative weights, and the sequence $\{K_{f,Q_n} = \sum |c_i^{(n)}|\}$ is unbounded. Indeed, it can be proved that the Newton–Cotes formulas do *not* converge for all integrands which are analytic in the interval of integration (Engels [131]). The convergence of closed Newton–Cotes formulas can only be guaranteed for functions analytic in a certain "oval" domain of the complex plane containing the real interval $[a,b]$ (Krylov [250]).

The practical significance of Newton–Cotes formulas stems from the fact that they are used in compound quadrature formulas. Since the degree of accuracy of the Newton–Cotes formula Q_n is at least $D \geq n-1$, the error estimate

$$|Q_n f - If| = O(h^{n+1}) \quad \text{as} \quad h := (b-a) \to 0+$$

can be obtained from (5.16) for $f \in C^{(n)}[a,b]$. More accurate characterizations of the integration error $Q_n f - If$ can be derived using Peano's theorem (see, for instance, Davis and Rabinowitz [92]).

Example (Trapezoidal Rule). The trapezoidal rule is obtained by integrating the linear polynomial $P_1 \in \mathbb{P}_1$ which interpolates f at the abscissas $x_1 = a$ and $x_2 = b$. The inequality for the approximation error

$$e_1(f) = \|P_1 - f\|_\infty \leq \frac{(b-a)^2}{8} M_2,$$

derived from (5.9), leads to the error estimate

$$|Q_2 f - If| \leq \frac{(b-a)^3}{8} M_2.$$

Using Peano's theorem, it is possible to improve this result:

$$Q_2 f - If = \frac{(b-a)^3}{12} f''(\xi), \quad \xi \in [a,b], \tag{5.18}$$

and consequently

$$|Q_2 f - If| \leq \frac{(b-a)^3}{12} M_2.$$

Accurate error terms are known for all Newton–Cotes formulas (see, for example, Engels [131]):

$$\begin{aligned}
\mathrm{Q}_2 f - \mathrm{I} f &= \frac{(b-a)^3}{12} f''(\xi_2), \\
\mathrm{Q}_3 f - \mathrm{I} f &= \frac{(b-a)^5}{2880} f^{(4)}(\xi_3), \\
\mathrm{Q}_4 f - \mathrm{I} f &= \frac{(b-a)^5}{6480} f^{(4)}(\xi_4), \\
\mathrm{Q}_5 f - \mathrm{I} f &= \frac{(b-a)^7}{1935360} f^{(6)}(\xi_5),
\end{aligned}$$

with $\xi_k \in [a, b]$, $k = 2, \ldots, 5$. These error expressions show a special feature of the Newton–Cotes formulas: for formulas with an *odd* number of abscissas, the exponent of $(b - a)$ is larger by one than can be expected from the number of abscissas. Correspondingly, the degree of accuracy of these formulas is $D = n$ rather than $D = n - 1$, as is the case with Newton–Cotes formulas with an even number of abscissas. This phenomenon results from the fact that any *symmetric* quadrature formula for the interval $[a, b]$ exactly integrates all polynomials $(x - (a + b)/2)^d$ of *odd* degree d.

5.2.5. Open Newton–Cotes Formulas. Open Newton–Cotes formulas $\mathrm{Q}_n, n = 1, 2, \ldots$, are interpolatory quadrature formulas for the weight function $w(x) \equiv 1$ based on the abscissas

$$x_i = a + i\,\frac{b-a}{n+1}, \qquad i = 1, 2, \ldots, n.$$

The first three open Newton–Cotes formulas are $(f_i := f(x_i))$

$$\begin{aligned}
\mathrm{Q}_1 f &= (b-a) f_1 \qquad\qquad (\textit{midpoint rule}), \\
\mathrm{Q}_2 f &= \frac{b-a}{2}(f_1 + f_2), \\
\mathrm{Q}_3 f &= \frac{b-a}{3}(2f_1 - f_2 + 2f_3).
\end{aligned}$$

The corresponding error terms are

$$\begin{aligned}
\mathrm{Q}_1 f - \mathrm{I} f &= -\frac{(b-a)^3}{24} f''(\xi_1), \\
\mathrm{Q}_2 f - \mathrm{I} f &= -\frac{(b-a)^3}{36} f''(\xi_2), \\
\mathrm{Q}_3 f - \mathrm{I} f &= -\frac{7(b-a)^5}{23040} f^{(4)}(\xi_3).
\end{aligned}$$

Like their closed counterparts, the open Newton–Cotes formulas with odd numbers n of abscissas have higher degrees of accuracy, $D = n$, than formulas with even numbers of points, where $D = n-1$.

5.2.6. Clenshaw–Curtis Formulas. Interpolation polynomials using the zeros

$$x_i = \xi_i := \cos\frac{(2i-1)\pi}{2n}, \qquad i = 1, 2, \ldots, n, \tag{5.19}$$

or the extrema

$$x_i = \eta_i := \cos\frac{(i-1)\pi}{n-1}, \qquad i = 1, 2, \ldots, n,$$

of Chebyshev polynomials as interpolation nodes generally approximate a given function on the interval $[-1, 1]$ more accurately than interpolation polynomials based on equidistant abscissas (Rivlin [372]). *Clenshaw–Curtis formulas*[45] (quadrature formulas based on Chebyshev abscissas transformed to the integration interval $[a, b]$) can thus be expected to be more suitable for numerical integration purposes than Newton–Cotes formulas. This expectation proves to be true. For instance, Clenshaw–Curtis formulas for the weight function $w(x) \equiv 1$ have positive weights (Fejér [147], Imhof [214]), which guarantees the convergence

$$\mathrm{Q}_n f \to \mathrm{I} f \quad \text{as} \quad n \to \infty \qquad \text{for all} \quad f \in C[a, b].$$

More generally, Clenshaw–Curtis formulas for weight functions

$$w \in L_p[a, b], \qquad p > 1,$$

can be shown to converge for all Riemann integrable functions f (Sloan and Smith [406]). The degree of accuracy of n-point Clenshaw–Curtis formulas, however, is in general no better than $D = n - 1$.

For computational purposes Chebyshev extrema are preferred to Chebyshev zeros: the transition from Q_n to Q_{2n-1} requires only $n - 1$ new values of f to be computed when Chebyshev extrema are used, whereas $2n$ new values must be calculated when making the transition from Q_n to Q_{2n} if Chebyshev zeros are used. This is why Chebyshev extrema are said to be *practical* Clenshaw–Curtis abscissas, and Chebyshev zeros are referred to as *classical* Clenshaw–Curtis abscissas. Classical Clenshaw–Curtis formulas are also called *Fejér formulas*.

[45] Clenshaw–Curtis formulas for weight functions $w(x) \not\equiv 1$ are often referred to as *modified* Clenshaw–Curtis formulas.

While explicit formulas for the weights in Clenshaw–Curtis rules Q_n can be given (see, for instance, Davis and Rabinowitz [92]), implementations of Clenshaw–Curtis formulas often avoid explicitly computing integration weights. Instead, computationally efficient implementations of Clenshaw–Curtis formulas are usually based on expanding the polynomial P_d, $d = n-1$, which interpolates f at the practical or classical abscissas in terms of Chebyshev polynomials. For practical Clenshaw–Curtis formulas, the expansion

$$(5.20)\quad P_d(x) = \sum_{j=0}^{d}{}'' a_j T_j(x) := \frac{a_0}{2} + a_1 T_1(x) + a_2 T_2(x) + \cdots + \frac{a_d}{2} T_d(x)$$

is used.[46] Because

$$(5.21)\qquad \mathrm{Q}_n f = \mathrm{I}_w P_d = \sum_{j=0}^{d}{}'' a_j \, \mathrm{I}_w T_j,$$

computing $\mathrm{Q}_n f$ requires calculating the coefficients $a_0, a_1, \ldots, a_d$ of the Chebyshev expansion as well as evaluating the *moments* $\mathrm{I}_w T_0, \mathrm{I}_w T_1, \ldots, \mathrm{I}_w T_d$.[47]

Explicit formulas for the coefficients $a_0, a_1, \ldots, a_d$ can easily be derived using the *discrete orthogonality* property of the Chebyshev polynomials (see, for instance, Ueberhuber [435]):

$$a_j = \frac{2}{d}\sum_{i=0}^{d}{}'' f(\eta_i) T_j(\eta_i) = \frac{2}{d}\sum_{i=0}^{d}{}'' f(\eta_i)\cos\left(\frac{ij\pi}{d}\right), \quad j = 0, 1, \ldots, d.$$

Similar expressions can be obtained for the coefficients in classical Clenshaw–Curtis formulas. A variant of the *fast Fourier transform* makes it possible to calculate all $n = d+1$ coefficients using only $O(n \log n)$ arithmetic operations (Gentleman [163]).

For a number of practically relevant weight functions w, the moments $\mathrm{I}_w T_0, \mathrm{I}_w T_1, \ldots, \mathrm{I}_w T_d$ can be shown to satisfy particular linear recurrence relations. These recurrence relations are usually used to efficiently calculate the moments $\mathrm{I}_w T_0, \mathrm{I}_w T_1, \ldots, \mathrm{I}_w T_d$. Note that recursion schemes based on linear recurrence relations can be numerically unstable. Special care may have to be taken to avoid such numerical instabilities. An extensive list of recurrence relations for various weight functions w as well as a discussion of numerically stable ways of employing these recurrence relations can be found in Piessens [333].

[46]The double prime on the summation symbol indicates that the terms with the first and last subscript are to be halved.

[47]For weight functions $w(x) \not\equiv 1$, $\mathrm{I}_w T_0, \mathrm{I}_w T_1, \ldots, \mathrm{I}_w T_d$ are often referred to as *modified* moments.

Software (Clenshaw–Curtis Formulas). The routines CMLIB/dqc25s ≈ QUADPACK/dqc25s ≈ SLATEC/dqc25s evaluate the modified 25-point Clenshaw–Curtis formula with respect to weight functions w with algebraic or algebraic-logarithmic endpoint singularities for any integrand f. Additionally, an embedded 13-point Clenshaw–Curtis formula is used to calculate an error estimate.

The subroutines CMLIB/dqc25f ≈ QUADPACK/dqc25f ≈ SLATEC/dqc25f and the subroutines CMLIB/dqc25c ≈ QUADPACK/dqc25c ≈ SLATEC/dqc25c perform the same function for trigonometric and Hilbert weight functions, respectively.

The subroutine IMSL/MATH-LIBRARY/dfqrul calculates, for any given number of abscissas, the abscissas and weights of the corresponding Fejér formula (the classical Clenshaw–Curtis formula). Moreover, the abscissas and weights of modified Fejér formulas can be calculated for weight functions with algebraic or algebraic-logarithmic endpoint singularities as well as for Hilbert weight functions.

5.2.7. Gauss Formulas. All interpolatory quadrature formulas Q_n have degrees of accuracy $D \geq n-1$. The degree of accuracy *cannot* generally be expected to be larger than $n-1$. For *given* n abscissas $x_1, \ldots, x_n$, the remaining n parameters (the quadrature coefficients $c_1, \ldots, c_n$) of an interpolatory quadrature formula Q_n are uniquely determined by requiring Q_n to be exact for all polynomials in the n-dimensional linear space $\mathbb{P}_d$, $d = n-1$. There is no reason why Q_n should be exact for polynomials of degree $d > n-1$. If, however, *all* $2n$ parameters of an n-point formula Q_n (the n coefficients *and* the n abscissas) are chosen appropriately, then formulas with maximal degrees of accuracy can be obtained. These formulas are exact for all polynomials in the $2n$-dimensional linear space $\mathbb{P}_d$, $d = 2n-1$.

Since a quadrature formula Q_n with degree of accuracy $D = 2n-1$ is, according to Theorem 5.2.1, necessarily interpolatory, the weights $c_1, \ldots, c_n$ of Q_n can be calculated (for instance, by solving the moment equations (5.7)) if the abscissas $x_1, \ldots, x_n$ are determined. The proper choice for the abscissas can be derived as follows. Let $P_{2n-1} \in \mathbb{P}_{2n-1}$ be an arbitrary polynomial of maximum degree $d = 2n-1$. Using the Euclidean algorithm, P_{2n-1} can be split,

$$P_{2n-1}(x) = r(x)\omega_n(x) + s(x),$$

where $r, s \in \mathbb{P}_{n-1}$ and

$$\omega_n(x) := (x - x_1)(x - x_2) \cdots (x - x_n)$$

with the unknown abscissas $x_1, \ldots, x_n$. Hence,

$$\begin{aligned} \int_a^b w(x) P_{2n-1}(x)\, dx &= \int_a^b w(x) r(x) \omega_n(x)\, dx + \int_a^b w(x) s(x)\, dx \\ &= \sum_{i=1}^n c_i r(x_i) \omega_n(x_i) + \sum_{i=1}^n c_i s(x_i) + E(P_{2n-1}) \end{aligned}$$

$$= \sum_{i=1}^{n} c_i s(x_i) + E(P_{2n-1}),$$

because

$$\omega_n(x_i) = 0 \quad \text{for} \quad i = 1, 2, \ldots, n.$$

If the abscissas could be selected in such a way that, for all $r \in \mathbb{P}_{n-1}$, the identity

$$\int_a^b w(x) r(x) \omega_n(x)\, dx = 0 \tag{5.22}$$

holds, then choosing the quadrature weights according to the principle of interpolation leads to

$$\int_a^b w(x) s(x)\, dx = \sum_{i=1}^{n} c_i s(x_i),$$

i.e., to the required result

$$E(P_{2n-1}) = 0 \quad \text{for all} \quad P_{2n-1} \in \mathbb{P}_{2n-1}.$$

The question of whether or not the abscissas $x_1, \ldots, x_n$ can be chosen in such a way that ω_n satisfies (5.22) is linked to the theory of *orthogonal polynomials*, the relevant aspects of which are outlined in the following section.

Orthogonal Polynomials. Let w denote an admissible weight function. Then an *inner product* is defined on the set of all univariate polynomials $\mathbb{P}$ by

$$\langle P, Q \rangle_w := \mathrm{I}_w(PQ) = \int_a^b w(x) P(x) Q(x)\, dx \tag{5.23}$$

for all $P, Q \in \mathbb{P}$. The inner product $\langle\, ,\, \rangle_w$ induces a norm on $\mathbb{P}$:

$$\|P\|_w := \sqrt{\langle P, P \rangle_w}.$$

Two polynomials $P \not\equiv 0$ and $Q \not\equiv 0$ are said to be *w-orthogonal* if

$$\mathrm{I}_w(PQ) = \int_a^b w(x) P(x) Q(x)\, dx = 0.$$

If, additionally, $\|P\|_w = \|Q\|_w = 1$, P and Q are said to be *w-orthonormal*.

Let

$$\mathcal{P}^w := \{P_d^w \;:\; d = 0, 1, 2, \ldots\}$$

be a set of pairwise w-orthogonal polynomials with

$$\deg P_d^w = d, \quad d = 0, 1, 2, \ldots.$$

The set $\mathcal{P}^w$ is said to be *orthogonal with respect to the weight function* w. Sets of orthogonal polynomials $\mathcal{P}^w$ can be obtained for any weight function by applying the *Gram–Schmidt orthogonalization* procedure to the set of monomials $\{1, x, x^2, \ldots\}$ (Isaacson and Keller [217]). The Gram–Schmidt orthogonalization theorem also shows that the members of two sets

$$\mathcal{P}^w = \{P_d^w\} \quad \text{and} \quad \tilde{\mathcal{P}}^w = \{\tilde{P}_d^w\}$$

of orthogonal polynomials differ only by multiplicative constants,

$$P_d^w = C_d \tilde{P}_d^w, \quad C_d \in \mathbb{R}, \quad C_d \neq 0, \quad d = 0, 1, 2, \ldots,$$

and therefore have the same zeros. In general, the Gram–Schmidt orthogonalization procedure requires the evaluation of a d-term recurrence formula in calculating the dth element in an orthogonal sequence. In the orthogonalization of polynomials with respect to inner products $\langle\, , \rangle_w$, however, the Gram–Schmidt procedure takes a particularly simple form.

THEOREM 5.2.8 (RECURRENCE FORMULA FOR ORTHOGONAL POLYNOMIALS). $\mathcal{P}^w = \{P_d^w\}$ *is a set of* w*-orthogonal polynomials if and only if the three-term recurrence formula*

$$P_{d+1}^w = (a_d x - b_d) P_d^w - c_d P_{d-1}^w, \quad d = 0, 1, 2, \ldots,$$

with

$$\begin{aligned} a_d &\neq 0, \qquad d = 0, 1, 2, \ldots, \\ b_d &= a_d \frac{\langle x P_d^w, P_d^w \rangle_w}{\|P_d^w\|_w^2}, \quad d = 0, 1, 2, \ldots, \\ c_d &= \frac{a_d}{a_{d-1}} \frac{\|P_d^w\|_w^2}{\|P_{d-1}^w\|_w^2}, \quad d = 1, 2, 3, \ldots, \end{aligned} \tag{5.24}$$

is satisfied.[48]

Proof. See Engels [131].

Note that the orthogonal polynomials P_d^w, $d = 0, 1, \ldots,$ are *monic*[49] if and only if $P_0^w = 1$ and $a_d = 1, d = 0, 1, \ldots,$ holds. In this case the coefficients c_d are all positive, and the tridiagonal symmetric matrix

$$J_n^w := \begin{pmatrix} b_0 & \sqrt{c_1} & 0 & & \mathbf{0} \\ \sqrt{c_1} & b_1 & \sqrt{c_2} & & \\ & \sqrt{c_2} & \ddots & \ddots & \\ & & \ddots & & \sqrt{c_{n-1}} \\ \mathbf{0} & & & \sqrt{c_{n-1}} & b_{n-1} \end{pmatrix} \tag{5.25}$$

[48] $P_{-1} = 0$ is assumed in this definition.

[49] A polynomial P_d of degree d is said to be *monic* if P_d is of the form $P_d(x) = x^d + a_{d-1}x^{d-1} + \cdots + a_1 x + a_0$.

is the nth-order *Jacobi matrix* for w.

Using the terminology introduced in this section, the polynomial $\omega_n \in \mathbb{P}_n$ in (5.22) can be said to be w-orthogonal to all polynomials in $\mathbb{P}_{n-1}$ and hence ω_n is equal, up to a multiplicative constant, to the w-orthogonal polynomial P_n^w, and the abscissas $x_1, \ldots, x_n$ are the zeros of P_n^w. In order to be usable as quadrature abscissas, the zeros $x_1, \ldots, x_n$ of P_n^w have to be simple and real and must be located within the interval of integration $[a, b]$. The following theorem shows that these requirements are indeed met.

THEOREM 5.2.9 (ZEROS OF ORTHOGONAL POLYNOMIALS). *Let* $\mathcal{P}^w = \{P_d^w\}$ *be a set of* w*-orthogonal polynomials. Then, each polynomial* P_d^w *has* d *simple, real zeros in the interior of* $[a, b]$.

Proof. See Ueberhuber [435].

It can easily be seen that the zeros of the orthogonal polynomial P_n^w are equal to the eigenvalues of the Jacobi matrix J_n^w. Additionally, the weights $c_1, \ldots, c_n$ of the interpolatory quadrature formula using the zeros of P_n^w as abscissas can be obtained as follows:

$$c_i = x_{i,1}^2 \mathrm{I}w, \qquad i = 1, 2, \ldots, n,$$

where $(x_{i,1}, \ldots, x_{i,n})^\top$, $i = 1, \ldots, n$, is the ith (with respect to the Euclidean norm) normalized eigenvector of J_d^w (Golub and Welsch [180]). Practical methods for calculating interpolatory quadrature formulas based on the zeros of P_d^w use suitable algorithms (for instance, special variants of the QR-algorithm) for computing eigenvalues and eigenvectors of tridiagonal, symmetric matrices.

Properties of Gauss Formulas. The *Gauss formula* G_n^w with respect to the admissible weight function w is defined as the unique n-point, interpolatory quadrature formula that uses the zeros $x_1, \ldots, x_n$ of the w-orthogonal polynomial P_n^w, of degree n, as quadrature abscissas. The names of the Gauss formulas G_n^w for the most important weight functions w (see Section 3.1.2), such as Gauss–*Legendre*, Gauss–*Jacobi*, Gauss–*Laguerre*, Gauss–*Hermite*, etc. formulas, indicate the underlying system of orthogonal polynomials. The Gauss formulas for $w \equiv 1$ are often denoted by G_n and referred to as *Gauss formulas* (rather than Gauss–Legendre formulas).

Since the abscissas of G_n^w are equal to the zeros of P_n^w, $G_n^w(P_n^w)^2 = 0$ holds, whereas the exact integral $\mathrm{I}_w(P_n^w)^2$ is nonzero. G_n^w is therefore *not* exact for the polynomial $(P_n^w)^2 \in \mathbb{P}_{2n}$. The results concerning Gauss quadrature obtained so far can be summarized as follows.

THEOREM 5.2.10. *The maximum degree of accuracy of an* n*-point quadrature formula*

$$\mathrm{Q}_n f = \sum_{i=1}^{n} c_i f(x_i)$$

with respect to an admissible weight function w is $D = 2n - 1$. This degree of accuracy is attained if and only if Q_n is chosen as the Gauss formula G_n^w with respect to w.

Gauss formulas G_n^w exist and attain the maximum degree of accuracy whenever w is an admissible weight function, irrespective of whether the integration interval $[a, b]$ is bounded or not. The following convergence results, however, hold only for bounded intervals $[a, b]$. According to Theorem 5.2.4 all Gauss formulas G_n^w have positive weights. Thus the convergence

$$G_n^w f \to \mathrm{I}_w f \qquad \text{as} \quad n \to \infty$$

for all $f \in C[a, b]$ can be derived from Theorem 5.2.2, which, as a result of Theorem 5.2.3, ensures the convergence of Gauss formulas for all Riemann integrable functions f.

Software (Gauss Formulas). The subroutine `IMSL/MATH-LIBRARY/dgqrul` calculates, for any given number of abscissas n, the abscissas and weights for the Gauss formulas G_n^w with respect to many classical weight functions w. More generally, the subroutine `IMSL/MATH-LIBRARY/dgqrcf` calculates the abscissas and weights of Gauss formulas G_n^w with respect to any admissible weight function w. The user supplies the weight function w *implicitly* by providing the moment $\mathrm{I}w$, as well as the first n recurrence coefficients in (5.24) for the special sequence of monic orthogonal polynomials $\{P_d^w\}$.

The subroutine `IMSL/MATH-LIBRARY/drecqr` reverses the function of the subroutine `IMSL/MATH-LIBRARY/dgqrcf` in the following sense: for n given abscissas and weights for a Gauss quadrature formula G_n^w, it computes the moment $\mathrm{I}w$ as well as the first n recurrence coefficients in (5.24) for the special sequence of monic orthogonal polynomials $\{P_d^w\}$. The moment and the recurrence coefficients for many classical, admissible weight functions w can be obtained more easily using `IMSL/MATH-LIBRARY/dreccf`.

The subroutine `NAG/d01bbf` calculates the weights and abscissas of Gauss quadrature formulas G_n^w with respect to many classical admissible weight functions w. The number n of abscissas can be chosen from a given set of admissible values. The subroutine `NAG/d01baf` directly evaluates $G_n^w f$ for any given integrand f instead of returning the abscissas and weights of G_n^w. The subroutine `NAG/d01bcf` can be used to compute, for *any* given number n of abscissas, the weights and abscissas of Gauss formulas G_n^w with respect to most classical admissible weight functions w.

`TOMS/655` contains subroutines that, for any given number n of abscissas, calculate the weights and abscissas of Gauss quadrature formulas G_n^w with respect to most classical admissible weight functions w or that evaluate $G_n^w f$ for any given integrand f. Any admissible weight function w can be handled if the user provides the moment $\mathrm{I}w$ as well as the Jacobi matrix (5.25) of w.

`TOMS/726` contains a subroutine for calculating the recurrence coefficients in (5.24) for the special sequence of monic orthogonal polynomials $\{P_d^w\}$ with respect to a number of classical weight functions w. Several other subroutines help calculate these recurrence coefficients for nonclassical weight functions w. Additionally, a subroutine is provided which computes the abscissas and weights of the Gauss formulas G_n^w for given recurrence coefficients.

Gauss-Type Formulas. The definition of Gauss formulas G_n^w presupposes that the weight function w is nonnegative on the interval of integration $[a, b]$: if the weight function w is of mixed sign, (5.23) does *not* define an inner product on $\mathbb{P}$. Hence, for weight functions w of mixed sign, quadrature formulas Q_n with maximum degree of accuracy $D = 2n - 1$ do not generally exist.

Nevertheless, quadrature formulas Q_n with degrees of accuracy D equal to or *greater than* $2n - 1$ exist for *particular* weight functions w of mixed sign. Such formulas are referred to as *Gauss-type formulas*. Gauss-type formulas exist, for instance, for all weight functions w which are *odd*[50],

$$w(x) = -w(-x), \quad x \in [-1, 1],$$

and nonnegative on the interval $[0, 1]$. Let $G_n^{\overline{w}}$ denote the n-point Gauss formula with respect to the weight function $\overline{w}(x) := w(\sqrt{x})$ for the interval $[0, 1]$, and let $x_1, \ldots, x_n$ and $c_1, \ldots, c_n$ be the abscissas and weights of $G_n^{\overline{w}}$. The $2n$-point formula $\overline{G}_{2n}^w$ is defined by

$$\overline{G}_{2n}^w f := \sum_{i=1}^{n} \frac{c_i}{2\sqrt{x_i}} \Big(f(\sqrt{x_i}) - f(-\sqrt{x_i}) \Big).$$

Let the function h be defined by

$$h \colon x \mapsto x f(x^2).$$

These definitions immediately entail the relationship

$$\overline{G}_{2n}^w h = G_n^{\overline{w}} f,$$

whereas the relationship $I_w h = I_{\overline{w}} f$ can be obtained using the transformation $x^2 = \overline{x}$ in $I_w h$. Since $G_n^{\overline{w}}$ is exact for the monomials x^d, $d = 0, 1, \ldots, 2n - 1$, $\overline{G}_{2n}^w$ is exact for the monomials x^{2d+1}, $d = 0, 1, \ldots, 2n - 1$:

$$\overline{G}_{2n}^w x^{2d+1} = G_n^{\overline{w}} x^d = I_{\overline{w}} x^d = I_w x^{2d+1}.$$

In other words, $\overline{G}_{2n}^w$ is exact for all monomials of odd degree $d \leq 4n - 1$. Furthermore,

$$\overline{G}_{2n}^w x^{2d} = 0 = I_w x^{2d} \quad \text{for all} \quad d \in \mathbb{N};$$

i.e., $\overline{G}_{2n}^w$ is exact for all monomials of even degree. Consequently, $\overline{G}_{2n}^w$ is exact for all polynomials $P \in \mathbb{P}_{4n}$.

[50] For the sake of notational simplicity, the interval of integration is assumed to be $[-1, 1]$.

5.2.8. Radau and Lobatto Formulas. While Gauss formulas use all the $2n$ available parameters (the n weights and n abscissas) of an n-point quadrature formula to attain the maximum degree of accuracy $D = 2n - 1$, some sampling points are *preassigned* in Radau and Lobatto formulas. Only the remaining abscissas and all the weights are determined in such a way that the resulting formula attains the highest possible degree of accuracy. As with Gauss formulas, Radau and Lobatto formulas can be obtained for *any* admissible weight function w.

Radau formulas have just *one* preassigned abscissa: the left-hand endpoint $x_1 = a$ or the right-hand endpoint $x_n = b$. Using the remaining $2n - 1$ parameters, it is possible to construct formulas with degree of accuracy $D = 2n - 2$. *Lobatto* formulas have *two* preassigned abscissas: $x_1 = a$ *and* $x_n = b$. The resulting formulas have the degrees of accuracy $D = 2n - 3$. Both Radau and Lobatto formulas can be shown to have positive weights and are therefore, according to Theorem 5.2.3, guaranteed to converge for all Riemann integrable functions.

n-point Radau and Lobatto formulas can be obtained from *modified* Jacobi matrices in the same way that Gauss formulas G_n^w are obtained from the Jacobi matrices J_n^w (Golub [178]). Practical methods for calculating Radau and Lobatto formulas are therefore based on computing the eigenvalues and eigenvectors of tridiagonal, symmetric matrices.

Radau and Lobatto formulas are predominantly used in numerical methods for solving differential and integral equations.

Software (Radau and Lobatto Formulas). The subroutine `IMSL/MATH-LIBRARY/dgqrul` calculates, for any given number of abscissas, the abscissas and weights of the corresponding Radau and Lobatto formulas with respect to many classical admissible weight functions. More generally, the subroutine `IMSL/MATH-LIBRARY/dgqrcf` calculates the abscissas and weights of Radau and Lobatto formulas with respect to any admissible weight function w. The user supplies the weight function w implicitly by providing the moment Iw as well as the first n recurrence coefficients in (5.24) for the special sequence of monic, orthogonal polynomials $\{P_d^w\}$. Both programs can be used to calculate quadrature formulas of maximum possible degree with up to two, arbitrarily located, preassigned abscissas.

`TOMS/726` contains subroutines for calculating the abscissas and weights for Radau and Lobatto formulas with respect to any admissible weight function w (given by the corresponding sequence of recurrence coefficients).

5.2.9. Gauss–Kronrod Formulas. In practice, it is often necessary (for instance, for obtaining error estimates (see Section 8.1.1)) to compute two or more formulas with increasing degrees of accuracy. In order to keep the computational cost as low as possible, each formula should (at least partially) reuse the f-values calculated previously. In this respect, Gauss formulas have a severe disadvantage: the Gauss formulas G_n and G_m, $n \neq m$, do not have any common abscissas (except, possibly, the midpoint of the integration interval).

Gauss formulas do not, therefore, permit the reuse of previously calculated f-values.

This problem can be overcome by using a method devised by Kronrod [248]: taking the n abscissas of G_n as preassigned abscissas, a $(2n+1)$-point formula is constructed in such a way that the resulting degree of accuracy is as high as possible. Kronrod proved that the *Gauss–Kronrod* formulas obtained in this way attain degrees of accuracy $D = 3n+1$ (for even n) or $D = 3n+2$ (for odd n). The $n+1$ additional abscissas are located in the intervals

$$(a, x_1),\ (x_1, x_2),\ (x_2, x_3), \dots, (x_n, b),$$

with $x_1, \dots, x_n$ denoting the abscissas of G_n. A detailed characterization of Kronrod abscissas can be found in Piessens and Branders [334].

In contrast to Gauss, Radau, and Lobatto formulas, Gauss–Kronrod formulas do *not* exist for all admissible weight functions. An overview of results on the existence and nonexistence of Gauss–Kronrod formulas for various weight functions can be found in Notaris [325].

Software (Gauss–Kronrod Formulas). The routines `CMLIB/dqk`n $\approx$ `QUADPACK/dqk`n $\approx$ `SLATEC/dqk`n, $n = 15, 21, 31, 41, 51, 61$, evaluate the n-point Gauss–Kronrod formula for any given integrand f. Additionally, an error estimate is calculated based on the corresponding embedded Gauss formula.

The paper by Piessens and Branders [334] contains a Fortran program that calculates the abscissas and weights of Gauss formulas G_n as well as the corresponding Kronrod abscissas and weights for any given number of abscissas n.

5.2.10. Gauss-Type Formulas with Preassigned Abscissas.

Gauss, Radau, Lobatto, and Gauss–Kronrod formulas are examples of quadrature formulas obtained using the following approach. For fixed n, prescribed abscissas $x_1, \dots, x_n$, and a given number $m \in \mathbb{N}$, m additional abscissas $x_{n+1}, \dots, x_{n+m}$, and $n+m$ weights $c_1, \dots, c_{n+m}$ are chosen in such a way that the resulting formula,

$$Q_{n+m}f = \sum_{i=1}^{n+m} c_i f(x_i) \approx I_w f,$$

attains a degree of accuracy which corresponds, at least, to the number of free parameters: $D \geq n + 2m - 1$. In Gauss formulas none of the abscissas are fixed in advance, whereas one or both endpoints of the integration interval are preassigned in Radau and Lobatto formulas. In Gauss–Kronrod formulas the n abscissas of the Gauss formula G_n are prescribed.

In Section 5.2.7 interpolatory quadrature formulas using the zeros of orthogonal polynomials as abscissas were shown to attain the maximum possible degree of accuracy. Analogously, it can be shown that the condition

$$(5.26) \qquad \int_a^b w(x)v(x)\omega_m(x)P_{m-1}(x)\,dx = 0 \qquad \text{for all} \quad P_{m-1} \in \mathbb{P}_{m-1}$$

must be satisfied, where

$$v(x) := (x - x_1)\cdots(x - x_n) \quad \text{and} \quad \omega_m(x) := (x - x_{n+1})\cdots(x - x_{n+m}),$$

in order for the quadrature formula Q_{n+m} to be exact for all $P_{n+2m-1} \in \mathbb{P}_{n+2m-1}$. In other words, the abscissas $x_{n+1}, \ldots, x_{n+m}$ have to be chosen as the zeros of the "orthogonal" polynomial $\omega_m(x)$ of degree m with respect to the weight function wv.

The term orthogonal is placed in quotation marks because the weight function wv is *not* generally admissible, i.e., v is of mixed sign, if at least one preassigned abscissa is located in the interior of $[a, b]$. For weight functions wv of mixed sign, suitable orthogonal polynomials $\omega_m(x)$ (with m simple, real zeros in $[a, b]$) satisfying (5.26) generally do not exist. Such weight functions occur, for instance, when the abscissas of the Gauss formula G_n^w are prescribed. This is why Kronrod-type extensions of Gauss formulas G_n^w exist only for particular choices of w such as the Legendre weight function $w(x) \equiv 1$.

If, however, only endpoints of the integration interval are prescribed, then v does not change sign in $[a, b]$ and so wv (or $-wv$) is an admissible weight function. This is why Radau and Lobatto formulas can be shown to exist for any admissible weight function w.

Although the problem of constructing Gauss-type formulas with preassigned abscissas has been studied by a number of authors (see, for instance, Krylov [250], Monegato [304], Patterson [331], and the references therein), no comprehensive theory has been developed so far. Nevertheless, a number of useful results have been obtained. Most importantly, Patterson [330] generalizes Kronrod's idea of extending Gauss formulas by adding $2n + 2$ abscissas to the Kronrod extension K_{2n+1} of G_n. The resulting formula attains the degree of accuracy $D = 6n + 5$. For particular values of n, further Gauss-type extensions are possible. For example, a sequence of 1-, 3-, 7-, 15-, 31-, 63-, 127-, 255-point Gauss-type extension formulas is given in Patterson [330]. Such formulas are referred to as *Patterson formulas*.

Software (Gauss-Type Formulas with Preassigned Abscissas). `TOMS/672` contains a subroutine for calculating the abscissas and weights of Gauss-type formulas with preassigned abscissas. The weight function has to be specified by providing a subroutine which returns the recurrence coefficients (5.24). Note that this subroutine cannot be guaranteed to terminate successfully because the sought-after formula may not exist.

The nonadaptive integration program `QUADPACK/dqng` uses the sequence of 10-, 21-, 43-, and 87-point Patterson formulas, while `TOMS/699` is based on the sequence of 1-, 3-, 7-, 15-, 31-, 63-, 127-, and 255-point Patterson formulas.

5.3. Compound Quadrature Formulas

In Section 5.2 a number of *simple* interpolatory quadrature formulas, i.e., quadrature formulas obtained by integrating a *single* interpolation polynomial on the given interval of integration, are discussed. This section introduces formulas which are the result of integrating *piecewise* polynomials. The resulting *compound* (or *composite*) quadrature formulas can also be obtained by decomposing the interval $[a, b]$ into subintervals and by applying a simple interpolatory formula to each subinterval. Throughout this section it is assumed that $[a, b]$ is subdivided into k intervals by $k + 1$ *equidistant* partition points[51]

$$a = \bar{x}_0 < \bar{x}_1 < \cdots < \bar{x}_k = b, \qquad \bar{x}_j = a + j\,\frac{b-a}{k}, \quad j = 0, 1, \ldots, k.$$

An n-point quadrature formula Q_n (in a properly transformed form) is applied to each of the k subintervals $[\bar{x}_0, \bar{x}_1], \ldots, [\bar{x}_{k-1}, \bar{x}_k]$.[52] The resulting compound quadrature formula is denoted by $k \times Q_n$.

If Q_n is a *closed* formula, the number of abscissas in the compound formula $k \times Q_n$ is smaller than $k\,n$: the abscissas $\bar{x}_1, \ldots, \bar{x}_{k-1}$ at the endpoints of the subintervals are counted only once because f is to be evaluated only once at these abscissas. Thus, the number of abscissas in such formulas $k \times Q_n$ is $k(n-1)+1$.

Suppose the abscissas and weights of Q_n are given for the standard interval $[-1, 1]$:

$$Q_n f = \sum_{i=1}^{n} c_i f(x_i) \approx \int_{-1}^{1} f(x)\,dx, \qquad x_i \in [-1, 1],\ i = 1, 2, \ldots, n.$$

The transformed formula for $[\bar{x}_{j-1}, \bar{x}_j]$ is then given by

$$\int_{\bar{x}_{j-1}}^{\bar{x}_j} f(x)\,dx \approx \frac{b-a}{2k} \sum_{i=1}^{n} c_i f(x_{ij}), \qquad j = 1, 2, \ldots, k,$$

where

$$x_{ij} := \bar{x}_{j-1} + \frac{b-a}{2k}(1 + x_i), \quad i = 1, 2, \ldots, n, \quad j = 1, 2, \ldots, k.$$

The compound formula $k \times Q_n$ for $[a, b]$ can thus be explicitly written as

$$(5.27) \qquad (k \times Q_n) f \;=\; \sum_{j=1}^{k} \frac{b-a}{2k} \sum_{i=1}^{n} c_i f(x_{ij}) = \frac{b-a}{2k} \sum_{j=1}^{k} \sum_{i=1}^{n} c_i f(x_{ij}).$$

[51] *Non*equidistant subdivisions of $[a, b]$ are discussed in Section 8.2.2 in connection with adaptive integration algorithms.

[52] It is assumed that the weight function $w(x) \equiv 1$ on $[a, b]$ is used.

5.3.1. Convergence of Compound Formulas. The convergence of compound quadrature formulas can easily be guaranteed for all properly Riemann integrable functions f.

THEOREM 5.3.1. *Let* Q_n *be an* n*-point quadrature formula with degree of accuracy* $D \geq 0$. *Then for all properly Riemann integrable functions* f,

$$(k \times Q_n)f \to If \qquad as \qquad k \to \infty.$$

Proof. Changing the order of the summation in (5.27) leads to

$$(k \times Q_n)f \quad = \quad \frac{b-a}{2k} \sum_{j=1}^{k} \sum_{i=1}^{n} c_i f(x_{ij}) = \frac{1}{2} \sum_{i=1}^{n} c_i \frac{b-a}{k} \sum_{j=1}^{k} f(x_{ij}).$$

For any $k = 1, 2, \ldots$ and $i = 1, 2, \ldots, n$,

$$\frac{b-a}{k} \sum_{j=1}^{k} f(x_{ij}) \tag{5.28}$$

is a Riemann sum because $x_{ij} \in [\bar{x}_{j-1}, \bar{x}_j]$ and $\bar{x}_j - \bar{x}_{j-1} = (b-a)/k$ for $j = 1, \ldots, k$. As a result, the sum (5.28) converges to If as $k \to \infty$. On the other hand, according to the assumption that $f \equiv 1$ is integrated exactly, it follows that

$$\sum_{i=1}^{n} c_i = 2,$$

which leads to

$$\lim_{k \to \infty} (k \times Q_n)f = \frac{1}{2} \sum_{i=1}^{n} c_i \lim_{k \to \infty} \frac{b-a}{k} \sum_{j=1}^{k} f(x_{ij}) = \frac{1}{2} \sum_{i=1}^{n} c_i If = If. \qquad \square$$

By restricting the class of integrands, it is possible to characterize the *order* of convergence $(k \times Q_n)f \to If$.

THEOREM 5.3.2. *Let* Q_n *be an* n*-point quadrature formula for which the error for integrands* $f \in C^p[a, b]$ *is given by*

$$Q_n f - \int_a^b f(x)\, dx = C\, (b-a)^{p+1} f^{(p)}(\xi), \qquad \xi \in [a, b], \tag{5.29}$$

where C *is a constant independent of* a, b, *and* f. *Then the error in* $k \times Q_n$,

$$E_{k \times Q_n}(f) := (k \times Q_n)f - \int_a^b f(x)\, dx,$$

can be characterized by

$$\lim_{k \to \infty} k^p E_{k \times Q_n}(f) = C\, (b-a)^p \left[f^{(p-1)}(b) - f^{(p-1)}(a)\right]; \tag{5.30}$$

i.e., the convergence of $E_{k \times Q_n}(f) \to 0$ *as* $k \to \infty$ *is at least of order* p.

Proof. The error $E_{k\times Q_n}(f)$ is the sum of the integration errors in Q_n in the subintervals $[\bar{x}_0, \bar{x}_1], \ldots, [\bar{x}_{k-1}, \bar{x}_k]$:

$$\begin{aligned} E_{k\times Q_n}(f) &= \sum_{j=1}^{k} C\left(\frac{b-a}{k}\right)^{p+1} f^{(p)}(\xi_j) \\ &= C\left(\frac{b-a}{k}\right)^{p+1} \sum_{j=1}^{k} f^{(p)}(\xi_j), \qquad \xi_j \in [\bar{x}_{j-1}, \bar{x}_j]. \end{aligned}$$

Passing to the limit

$$\lim_{k\to\infty} \frac{b-a}{k} \sum_{j=1}^{k} f^{(p)}(\xi_j) = \int_a^b f^{(p)}(x)\, dx = f^{(p-1)}(b) - f^{(p-1)}(a)$$

establishes the assertion (5.30). □

Suitable representations (5.29) of the error in $Q_n f$ exist for the Newton–Cotes formulas (see Sections 5.2.4 and 5.2.5), but are also known for other types of formulas, such as Gauss, Radau, and Lobatto formulas (see Davis and Rabinowitz [92]). Since (5.29) means that Q_n has at least the degree of accuracy $D = p - 1$, (5.30) implies

$$E_{k\times Q_n}(f) = O(k^{-D-1}).$$

This convergence result can also be obtained directly from Theorem 5.2.6 without assuming the existence of the special form of error representation (5.29).

5.3.2. Compound Trapezoidal Rule. The compound trapezoidal rule $T_l := l \times T$,

$$T_l f = h\left[\frac{1}{2} f(a) + f(a+h) + \cdots + f(a+(l-1)h) + \frac{1}{2} f(b)\right],$$

with step length $h := (b-a)/l$ is of special importance. (5.18), along with Theorem 5.3.2, leads to the following asymptotic error estimate:

$$T_l f - \mathrm{I} f \approx \frac{h^2}{12}(f'(b) - f'(a)).$$

The error in the $(l+1)$-point formula T_l can be characterized asymptotically even more accurately by the following theorem (provided f is sufficiently smooth).

THEOREM 5.3.3 (EULER–MACLAURIN SUM FORMULA). *If $f \in C^{2k+1}[a, b]$, then*

$$T_l f - \mathrm{I} f = \frac{B_2}{2!} h^2 [f'(b) - f'(a)] + \frac{B_4}{4!} h^4 [f^{(3)}(b) - f^{(3)}(a)] + \cdots$$

$$(5.31) \qquad \cdots + \frac{B_{2k}}{(2k)!} h^{2k} [f^{(2k-1)}(b) - f^{(2k-1)}(a)] + \\ + h^{2k+1} \int_a^b \overline{P}_{2k+1}\left(l \frac{x-a}{b-a}\right) f^{(2k+1)}(x)\,dx,$$

where the constants

$$B_{2j} = (-1)^{j-1}(2j)! \sum_{i=1}^{\infty} 2(2i\pi)^{-2j}, \qquad j = 1, 2, 3, \ldots,$$

are the Bernoulli numbers

$$B_2 = 1/6, \quad B_4 = -1/30, \quad B_6 = 1/42, \quad B_8 = -1/30, \quad \ldots,$$

and $\overline{P}_{2k+1}$ *is the function*

$$\overline{P}_{2k+1}(x) := (-1)^{k-1} \sum_{i=1}^{\infty} 2(2i\pi)^{-2k-1} \sin(2\pi i x).$$

Proof. See Davis and Rabinowitz [92].

The error expansion (5.31) shows, among other things, that the compound trapezoidal rule gives outstandingly accurate results for smooth integrands whose odd derivatives assume equal values at the endpoints of the integration interval. This is true, for example, for all $(b-a)$-periodic integrands f. By applying the compound trapezoidal rule T_l to an integrand $f \in C^{2k+1}[a,b]$ with

$$f^{(j)}(a) = f^{(j)}(b), \qquad j = 1, 2, \ldots, 2k-1,$$

and

$$|f^{(2k+1)}(x)| \le M_{2k+1} \qquad \text{for all } x \in [a,b],$$

the following result concerning the rate of the convergence $T_l f \to \mathrm{I}f$ can be obtained from (5.31):

$$|T_l f - \mathrm{I}f| \le C h^{2k+1}, \quad h = (b-a)/l.$$

Theorems concerning the rate of convergence for analytic functions can be found, for instance, in Davis and Rabinowitz [92].

Example (Trapezoidal Rule for Periodic Integrands). The function

$$f(x) = \frac{1}{1 + r\sin(2j\pi x)}, \qquad |r| < 1, \quad j \in \mathbb{Z},$$

has a period of length j^{-1} and is differentiable infinitely many times. The rate of convergence of the compound trapezoidal rule T_l is therefore better than $O(l^{-k})$ for *any* $k \in \mathbb{N}$. The absolute error $|T_l f - \mathrm{I}f|$ (for the parameters $r = 0.5$, $j = 5$, $a = 0$, and $b = 1$) is shown in Figure 5.2.

The four different curves represent, from left to right, the absolute error $|T_l f - If|$ for the following choices of l: (i) l is an odd number, but not a multiple of 5; (ii) l is an even number, but not a multiple of 5; (iii) l is an odd multiple of 5; and (iv) l is an even multiple of 5. The special impact of the factor 5 on the behavior of the error is related to the location of the zeros of the function $\sin(10\pi x)$ in the denominator of f.

The function f was also integrated using Gauss formulas G_{l+1}. In general, the error in G_{l+1} is significantly larger than the error of the compound trapezoidal rule T_l, which contains the same number $l+1$ of abscissas.

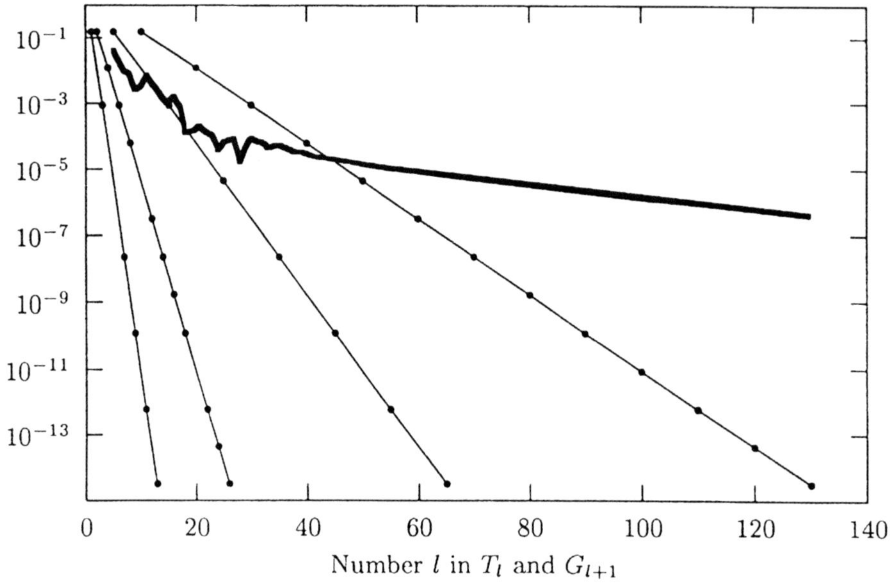

FIG. 5.2. *Absolute errors in the compound trapezoidal rules T_l (—) and the Gauss formulas G_{l+1} (—).*

5.3.3. Generalized Euler–Maclaurin Sum Formulas.

The Euler–Maclaurin sum formula (5.31) provides a very specific asymptotic characterization of the integration error of compound trapezoidal rules $T_l = l \times T$ for smooth integrands. It can be shown that the Euler–Maclaurin sum formula can be subsumed under a more general asymptotic error expansion valid for a large class of compound quadrature rules $k \times Q_n$.

THEOREM 5.3.4 (GENERALIZED EULER–MACLAURIN SUM FORMULA). *Let Q_n be a quadrature formula of precision D which assigns the same weight to the endpoint abscissas a and b.*[53] *Then, the error of the compound quadrature*

[53]Choosing open quadrature formulas Q_n, i.e., assigning a zero weight to the endpoints, is admissible in this theorem.

formulas $k \times \mathrm{Q}_n$ *for integrands* $f \in C^{p+1}[a,b]$ *can be expanded as*

$$(5.32) \qquad (k \times \mathrm{Q}_n)f - \mathrm{I}f = \sum_{q=1}^{p-D} a_q k^{-D-q} + O(k^{-p-1}).$$

The coefficients $a_1, \ldots, a_{p-D}$ *depend on the basic rule* Q_n *and the integrand* f*, but not on the partitioning factor* k*. Furthermore, the coefficients* a_q *vanish for odd* $D+q$ *if the basic rule* Q_n *is symmetric.*

Proof. See Lyness and Ninham [279].

Generalizations of the Euler–Maclaurin sum formula for *singular* integrands are discussed in Section 7.3.4.

5.3.4. Periodizing Transformations. Compound trapezoidal rules T_l prove to be much less effective when applied to nonperiodic integrands than when applied to periodic integrands (or integrands which have smooth, periodic extensions). In order to accelerate the convergence $T_l f \to \mathrm{I}f$, nonperiodic integrands can be subjected to *periodizing transformations.*

A number of such transformations, based on the transformation rule for univariate integrals (1.6), have been proposed. In these transformations the independent variable x is transformed according to $x = \psi(\overline{x})$, where the function ψ maps the standard interval $[0,1]$ monotonically onto itself. This gives rise to the relationship

$$\int_0^1 f(x)\,dx = \int_0^1 f(\psi(\overline{x}))\psi'(\overline{x})\,d\overline{x}.$$

DEFINITION 5.3.1 (POLYNOMIAL TRANSFORMATIONS). *The polynomial transformations (Hua and Wang [212]) are defined by*

$$(5.33) \qquad \psi(\overline{x}) = D_\alpha \int_0^{\overline{x}} (t(1-t))^{\alpha-1}\,dt, \quad \alpha = 2,3,4,\ldots,$$

with the scaling factor

$$D_\alpha := \left(\int_0^1 (t(1-t))^{\alpha-1}\,dt\right)^{-1}.$$

It is easy to verify that

$$\psi^{(k)}(0) = \psi^{(k)}(1) = 0 \quad \text{for} \quad k = 1,2,\ldots,\alpha-1.$$

As a result, if the integrand f is in $C^{\alpha-2}[0,1]$, then the transformed integrand $(f \circ \psi)\psi'$ is also in $C^{\alpha-2}[0,1]$ and all its derivatives up to the maximum order of $\alpha - 2$ vanish at the boundary. The transformed integrand $(f \circ \psi)\psi'$ thus has a smooth, periodic extension. The parameter α can be used to control the smoothness of the transformed integrand at the boundaries of the integration interval. Using *harmonic analysis*, the asymptotic error bound $T_l(f \circ \psi)\psi' - \mathrm{I}f = O(l^{-\alpha})$ can be shown to hold if f is sufficiently smooth (see Section 6.5.4).

The above properties of polynomial transformations also hold for the following type of transformations.

DEFINITION 5.3.2 (SINE TRANSFORMATIONS). *The sine transformations (Sidi [396]) are defined by*

$$\psi(\overline{x}) = D_\alpha \int_0^{\overline{x}} (\sin \pi t)^{\alpha-1}\, dt, \quad \alpha = 2, 3, 4, \ldots, \tag{5.34}$$

with the scaling factor

$$D_\alpha := \left(\int_0^1 (\sin \pi t)^{\alpha-1}\, dt \right)^{-1}.$$

Example (Periodizing Transformations). The integral

$$\mathrm{I}f = \int_0^1 \sin(10x)\, dx = \cos(10)/10 \approx -5.44 \cdot 10^{-2} \tag{5.35}$$

was approximately calculated using the trapezoidal rules T_l. The periodizing transformation ψ was chosen as the polynomial transformation with parameter $\alpha = 3$. For $l = 40$ the errors $T_{40}^l f - \mathrm{I}f = 2.84 \cdot 10^{-4}$ and $T_{40}^l (f \circ \psi)\psi' - \mathrm{I}f = -3.17 \cdot 10^{-8}$ were obtained. Hence, the integration error was reduced by a factor $\approx 10^4$ by subjecting the integrand to the polynomial periodizing transformation.

Iri, Moriguti, and Takasawa [216] have proposed the so-called *IMT transformations.*

DEFINITION 5.3.3 (IMT TRANSFORMATIONS). *The IMT transformations are defined by*

$$\psi(\overline{x}) = D_c \int_0^{\overline{x}} \exp\left(-\frac{c}{t(1-t)} \right) dt, \tag{5.36}$$

where the parameter c can be chosen from $\mathbb{R}_+$. The scaling factor D_c is given by

$$D_c := \left(\int_0^1 \exp\left(-\frac{c}{t(1-t)} \right) dt \right)^{-1}.$$

It can easily be verified that *all* derivatives of ψ vanish at the boundaries 0 and 1,

$$\psi^{(k)}(0) = \psi^{(k)}(1) = 0, \quad k = 1, 2, 3, \ldots.$$

As a result, if the integrand f is in $C^\infty[0,1]$, then the transformed integrand $(f \circ \psi)\psi'$ is also in $C^\infty[0,1]$ and all its derivatives vanish at the boundary, and so the transformed integrand $(f \circ \psi)\psi'$ has a smooth, periodic extension. The parameter c determines how fast the derivatives of ψ converge to zero at the boundaries 0 and 1.

DEFINITION 5.3.4 (TANH-TRANSFORMATIONS). *The tanh-transformations (Sag and Szekeres [380]) are defined by*

$$\psi(\overline{x}) = \frac{1}{2}\left(1 + \tanh\left(\frac{c}{2}\left(\frac{1}{1-\overline{x}} - \frac{1}{\overline{x}}\right)\right)\right), \quad c \in \mathbb{R}_{+}. \tag{5.37}$$

DEFINITION 5.3.5 (DOUBLE EXPONENTIAL TRANSFORMATIONS). *The double exponential transformations (Beckers and Haegemans [20]) are defined by*

$$\psi(\overline{x}) = \frac{1}{2}\left(1 + \tanh\left(c \sinh\left(d\left(\frac{1}{1-\overline{x}} - \frac{1}{\overline{x}}\right)\right)\right)\right), \quad c, d \in \mathbb{R}_{+}. \tag{5.38}$$

Both the tanh-transformations and the double exponential transformations have the same property as the IMT transformation: all derivatives of ψ vanish at the boundaries 0 and 1. The parameters c and d determine how fast the derivatives of ψ converge to zero at the boundaries. These transformations are hence also called IMT-type transformations.

5.3.5. Formulas Obtained by Periodization. Applying compound trapezoidal rules T_l to periodized integrands $(f \circ \psi)\psi'$ can also be regarded as applying the quadrature formulas Q_n, $n = l - 1$, with abscissas and weights defined by

$$x_i = \psi(i/l), \quad c_i = \psi'(i/(n+1))/(n+1), \quad i = 1, 2, \ldots, n,$$

to the original integrand f. These formulas Q_n, although they are derived from the compound trapezoidal rules T_l, $l = n + 1$, are *not* compound formulas. Nevertheless, the sequence $\{Q_1 f, Q_3 f, Q_5 f, \ldots\}$ of formulas inherits the advantageous property that all f-values computed in a specific formula are reused in all subsequent formulas from the sequence $\{T_3 f, T_5 f, T_7 f, \ldots\}$ of trapezoidal rules.

Different types of formulas are obtained for different periodizing transformations. In particular, formulas based on the IMT transformation (5.36) are referred to as *IMT formulas*.

IMT-type transformations have one main advantage: even if the original integrand f has a boundary singularity, the transformed integrand $(f \circ \psi)\psi'$ is often continuously differentiable at this point due to the rapid decline in ψ' near the boundaries. IMT (type) formulas are therefore especially suited to integrands with (integrable) singularities at the endpoints of the integration interval. However, the rapid decline in ψ' near the boundaries may give rise to numerical instabilities. For instance, the computation of $\psi'(x)$ for abscissas x close to the interval endpoints may cause floating-point underflows.

Example (Inaccuracy due to Transformation). Consider the computation of the integral

$$\mathrm{I}f := \int_0^1 (1-x)^{-0.9}\,dx$$

using a 26-point, compound trapezoidal rule T_{25}. It can easily be seen that applying the tanh-transformation removes the singularity of the integrand at the boundary $b = 1$.

For the choice $c = 2$ in (5.37), the value of ψ at $\overline{x}_{24} = 0.96$ is

$$x_{24} := \psi(\overline{x}_{24}) \approx 1 - 1.5 \cdot 10^{-21},$$

and $f(x_{24}) \approx 5.4 \cdot 10^{18}$ is obtained for the corresponding integrand value. Furthermore, the weight $c_{24} := \psi'(\overline{x}_{24})/25$ of the transformed trapezoidal integration formula is given by $c_{24} \approx 7.8 \cdot 10^{-20}$, and thus the abscissa $\overline{x}_{24}$ makes a contribution of

$$c_{24} f(x_{24}) \approx 4.2 \cdot 10^{-1}$$

to the overall numerical integration result.

The problem with evaluating $c_{24} f(x_{24})$ using double precision IEC/IEEE arithmetic is that the transformed abscissa x_{24} must be approximated using a floating-point number $\Box x_{24}$. The two nearest floating-point numbers are $\Box x_{24} = 1$ and $\Box x_{24} \approx 1 - 1.1 \cdot 10^{-16}$. The first choice is clearly unsuitable, since $f(1) = \infty$, whereas for the second choice $f(1 - 1.1 \cdot 10^{-16}) \approx 2.3 \cdot 10^{14}$ is obtained, which results in an error of

$$c_{24}(f(x_{24}) - f(\Box x_{24})) \approx 4.2 \cdot 10^{-1}$$

in the transformed trapezoidal rule. As the exact integral value is $\mathrm{I}f = 10$, this means a relative error of $4.2 \cdot 10^{-2}$ is encountered.

5.3.6. Romberg Formulas. The Euler–Maclaurin expansion (5.31) shows that the error $T_l f - \mathrm{I}f$ in the compound trapezoidal rule T_l has the following structure (provided that the integrand is sufficiently smooth):

$$T_l f - \mathrm{I}f = C_2 h^2 + C_4 h^4 + C_6 h^6 + \cdots. \tag{5.39}$$

The constants $C_2, C_4, \ldots$ depend only on the integrand f, but not on the step size $h = h_l = (b-a)/l$. Hence, the error expansion (3.24) holds for the sequence $\{s_m := T_{l_m} f\}$ of compound trapezoidal rules if the auxiliary sequence $\{t_m\}$ is defined by $t_m := l_m^{-2}$, $m = 1, 2, \ldots$. Consequently, Richardson's extrapolation (3.25) can be used to accelerate the convergence of the sequence $\{T_{l_m} f\}$. In particular, for the special sequence $l_m := 2^{m-1}$, $m = 1, 2, \ldots$, the following recurrence scheme, which was first described by Romberg [375], is obtained from (3.25):[54]

$$\left.\begin{aligned} T_m^0 &:= T_{2^{m-1}} f \\ T_m^j &:= \frac{4^j T_{m+1}^{j-1} - T_m^{j-1}}{4^j - 1} \end{aligned}\right\} \quad \text{for } m,\ j = 1, 2, 3, \ldots. \tag{5.40}$$

[54] The entity T_m^j in (5.40) corresponds to the entity $s^{(m;j)}$ in (3.25).

The recurrence (5.40) enables one to construct the following triangular array of approximate integral values, generally called the *T-table*:

$$\begin{array}{llllll} T_1 = T_1^0 & & & & & \\ T_2 = T_2^0 & T_1^1 & & & & \\ T_4 = T_3^0 & T_2^1 & T_1^2 & & & \\ T_8 = T_4^0 & T_3^1 & T_2^2 & T_1^3 & & \\ T_{16} = T_5^0 & T_4^1 & T_3^2 & T_2^3 & T_1^4 & \\ \vdots \quad \vdots \quad \vdots & \vdots & \vdots & \vdots & \vdots & \ddots \end{array}$$

Since each T_m^j, $m, j = 1, 2, \ldots$, is a linear combination of the compound trapezoidal formulas $T_{2^{m-1}}, \ldots, T_{2^{m+j-1}}$, the entity T_m^j can be interpreted as a quadrature formula which contains the same abscissas as $T_{2^{m+j-1}}$. These formulas T_m^j, $m, j = 1, 2, \ldots$, are called *Romberg formulas*.

The following theorems summarize important properties of Romberg formulas.

THEOREM 5.3.5. *Each Romberg formula $T_1^j, T_2^j, T_4^j, \ldots$ has a degree of accuracy $D = 2j + 1$.*

Proof. See Bauer, Rutishauser, and Stiefel [17].

THEOREM 5.3.6. *Every Romberg formula T_m^j is a Riemann sum.*

Proof. See Baker [12].

Because

$$D = 2j + 1 < 2^j \quad \text{for} \quad j \geq 3,$$

the Romberg formulas $T_m^3, T_m^4, \ldots$ are *not* interpolatory. Nevertheless, since Romberg formulas are Riemann sums, both the column sequences

$$\{T_1^j f, T_2^j f, T_3^j f, \ldots\}$$

and the diagonal sequences

$$\{T_m^0 f, T_m^1 f, T_m^2 f, \ldots\}$$

in the T-table converge to $\mathrm{I}f$ for any Riemann integrable function f.

When applied to smooth integrands f, the Romberg formulas T_m^j usually prove highly effective. For integrands f of reduced smoothness, on the other hand, convergence can be rather slow.

Example (Romberg Formulas). Figure 5.3 shows the errors in the Romberg formulas $T_1^0, T_1^1, \ldots, T_1^9$ applied to the integrals

$$\text{(5.41)} \qquad \mathrm{I}f_k := \int_0^1 (k + 3/2)x^{k+1/2}\, dx = 1, \qquad k = 0, 1, \ldots, 4.$$

The error curves demonstrate the relationship between the rate of convergence and the smoothness of the integrand. The error for the integrand $f_4 \in C^4[0, 1]$ is considerably smaller than the error for the continuous integrand f_0.

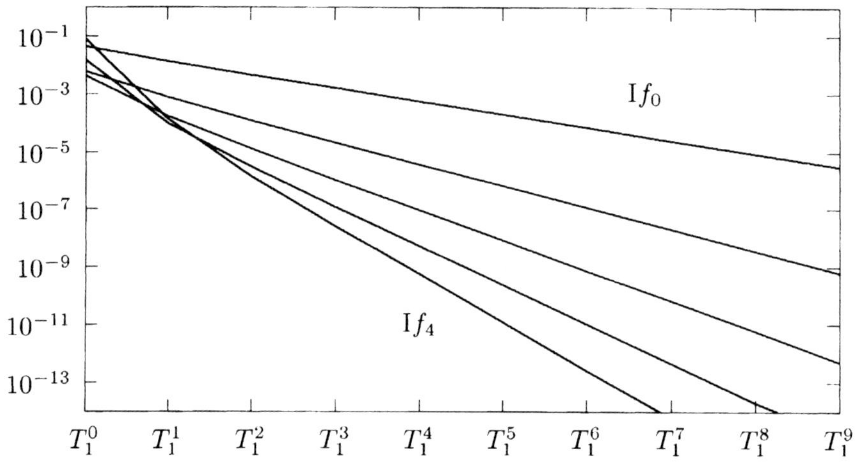

FIG. 5.3. *Errors in Romberg formulas.*

Chapter 6

Multivariate Integration Formulas

Most univariate integration formulas used in practice are based on (piecewise) polynomial interpolation (see Chapter 5). The construction of effective multivariate polynomial integration formulas proves, for the reasons explained in Section 6.2, to be significantly more difficult than the construction of univariate polynomial formulas. This is why methods which are *not* based on polynomial interpolation, such as *Monte Carlo methods* (see Section 6.4) and *number-theoretic methods* (see Sections 6.3 and 6.5), are often used as alternative methods for the construction of cubature techniques.

6.1. Construction of Cubature Formulas

This section deals with *general* principles employed in constructing cubature formulas. In particular, methods which can be used to make existing formulas for specific integration regions applicable to integration regions of different shapes and/or dimensions are described.

6.1.1. Formula Construction Using Riemann Sums. Multidimensional Riemann integrals (1.4) are defined as limits of appropriate sequences of Riemann sums (1.3). Riemann sums can thus be used, at least in principle, to approximate *every* multivariate Riemann integral with arbitrary accuracy.

If the integration region B is an N-dimensional cube and the integrand f is continuous on B, then the error of a Riemann sum can, analogously to (5.3), be bounded in terms of the *multidimensional modulus of continuity* of the integrand,

$$\omega(f;\delta) := \max\{|f(\mathbf{x}_1) - f(\mathbf{x}_2)| \ : \ \mathbf{x}_1, \mathbf{x}_2 \in B,\ \|x_1 - x_2\| \le \delta\},$$

where $\|\ \|$ is a norm on $\mathbb{R}^N$. However, if the integration region B is not an N-dimensional cube, the integral of f on B is defined as the integral of the modified integrand $c_B f$ on B', where B' is a cube containing the integration region B (see Section 1.1.2). The integrand $c_B f$ on B' is, in general, discontinuous on the boundary ∂B of B (unless, of course, f is continuous and vanishes on ∂B). It is therefore not possible to bound the discretization error in Riemann sums on noncubic domains simply in terms of the modulus of continuity of the integrand f. In fact, for noncubic regions B, the *regularity* of the boundary ∂B can also have a substantial impact on the discretization error of Riemann sums.

The convergence of multivariate integration rules based on simple Riemann sums is unacceptably slow for most practical problems. This is why Riemann sums are only of minor importance to the construction of multivariate integration formulas and algorithms.

6.1.2. Formula Construction Using Approximation. The approximation approach can be used to construct quadrature formulas ($N = 1$) as well as cubature formulas ($N \geq 2$).

The construction of integration formulas is usually based on approximating the integrand f using a model function $g \in \mathcal{M}$ and taking the definite integral $\mathrm{I}_w g$ as an approximation for $\mathrm{I}_w f$. The approximation approach is based on the error estimate (3.5),

$$|\,\mathrm{I}_w g - \mathrm{I}_w f\,| \leq \|w\|_1 \|g - f\|_\infty,$$

which is independent of the dimension N of the integration problem. If the approximation function g satisfies

$$\|g - f\|_\infty \leq \frac{\varepsilon}{\|w\|_1},$$

then the resulting error bound $|\mathrm{I}_w g - \mathrm{I}_w f| \leq \varepsilon$ shows that approximation functions which are sufficiently close to f with respect to the L_∞ norm are suitable candidates for solving the numerical integration problem (3.1). The approximation approach can therefore be used for constructing quadrature formulas ($N = 1$) as well as cubature formulas ($N \geq 2$).

The approximation approach is practical only if the definite integral of the model function over the given integration region can be expressed explicitly as a simple formula which can be evaluated using a small number of arithmetic operations. From this point of view, polynomials are especially attractive model functions because they are easily integrated over polyhedra and because their definite integrals can be expressed using only rational operations. Integrals of polynomials over N-dimensional spheres and their surfaces can be expressed in closed form in terms of the *Gamma function* (see, for instance, Davis and Rabinowitz [92]). Integrals of polynomials over other important integration regions can be found in, for example, Engels [131] and Stroud [423].

Example (Taylor Polynomials). The expansion of a function into a Taylor series can be integrated, and an estimate of the truncation error can be obtained from the remainder term. If, for example, a function can be expanded into a Taylor series around $\mathbf{0}$,

$$\begin{aligned} f(\mathbf{x}) &= f(\mathbf{0}) + \sum_{i_1=1}^{N} \frac{\partial f}{\partial x_{i_1}}(\mathbf{0}) x_{i_1} + \sum_{i_1,i_2=1}^{N} \frac{1}{2!} \frac{\partial^2 f}{\partial x_{i_1} \partial x_{i_2}}(\mathbf{0}) x_{i_1} x_{i_2} \\ &\quad \cdots + \sum_{i_1,\ldots,i_d=1}^{N} \frac{1}{d!} \frac{\partial^d f}{\partial x_{i_1} \cdots \partial x_{i_d}}(\mathbf{0}) x_{i_1} \cdots x_{i_d} + \cdots, \end{aligned}$$

and this Taylor series converges for all $\mathbf{x} \in C_N$, the following approximation can be obtained:

$$\begin{aligned} (6.1) \qquad \int_{C_N} f(\mathbf{x})\, d\mathbf{x} &\approx f(\mathbf{0}) + \sum_{i_1=1}^{N} \frac{\partial f}{\partial x_{i_1}}(\mathbf{0}) \int_{C_N} x_{i_1}\, d\mathbf{x} \\ &\quad + \frac{1}{2!} \sum_{i_1,i_2=1}^{N} \frac{\partial^2 f}{\partial x_{i_1} \partial x_{i_2}}(\mathbf{0}) \int_{C_N} x_{i_1} x_{i_2}\, d\mathbf{x} \\ &\quad \cdots + \frac{1}{d!} \sum_{i_1,\ldots,i_d=1}^{N} \frac{\partial^d f}{\partial x_{i_1} \cdots \partial x_{i_d}}(\mathbf{0}) \int_{C_N} x_{i_1} \cdots x_{i_d}\, d\mathbf{x}. \end{aligned}$$

The *monomials* $x_{i_1} \cdots x_{i_l}$ in (6.1) can easily be integrated analytically over C_N. For $f \in C^{(d+1)}(C_N)$, the remainder term for $\mathbf{x} \in C_N$ is given by

$$f(\mathbf{x}) - T_d(f;\mathbf{0})(\mathbf{x}) = \frac{1}{(d+1)!} \sum_{i_1,\ldots,i_{d+1}=1}^{N} \frac{\partial^{d+1} f}{\partial x_{i_1} \cdots \partial x_{i_{d+1}}}(\theta(\mathbf{x})\mathbf{x}) x_{i_1} \cdots x_{i_{d+1}},$$

where $\theta(\mathbf{x}) \in (0,1)$. By integrating (bounds on) this remainder formula, bounds on the integration error can be derived.

(Piecewise) polynomials are therefore the only class of approximation functions which are widely used in practice to construct cubature formulas. The resulting integration formulas are dealt with in Section 6.2.

6.1.3. Formula Construction Using Transformation. In Section 3.6.1 it is shown that the numerical computation of integrals can often be facilitated by transforming the given integration problem using the transformation rule (1.7). In this section it is demonstrated that the transformation rule can also be used for the construction of cubature formulas.

Many cubature formulas $\overline{\mathrm{Q}}_n$ are specific to particular integration regions $\overline{B}$, usually for one of the standard integration regions described in Section 3.1.1. In order to make such a formula $\overline{\mathrm{Q}}_n$ applicable to other integration regions B, $\overline{\mathrm{Q}}_n$ has to undergo appropriate transformations.

Let $\psi : \overline{B} \subset \mathbb{R}^N \to \mathbb{R}^N$ denote a continuously differentiable, one-to-one transformation from $\overline{B}$ onto B, and assume that the Jacobian matrix J of ψ is nonsingular on $\overline{B}$, i.e.,

$$\det J(\overline{\mathbf{x}}) \neq 0 \quad \text{for all} \quad \overline{\mathbf{x}} \in \overline{B}.$$

Then, the multidimensional transformation rule (1.7) yields

$$\int_B w(\mathbf{x}) f(\mathbf{x})\, d\mathbf{x} = \int_{\overline{B}} \overline{w}(\overline{\mathbf{x}}) |\det J(\overline{\mathbf{x}})| \overline{f}(\overline{\mathbf{x}})\, d\overline{\mathbf{x}},$$

where

$$\overline{w}(\overline{\mathbf{x}}) = w(\psi(\overline{\mathbf{x}})) \qquad \text{and} \qquad \overline{f}(\overline{\mathbf{x}}) = f(\psi(\overline{\mathbf{x}})).$$

The transformation ψ converts the cubature formula

$$\int_{\overline{B}} \overline{w}(\overline{\mathbf{x}}) |\det J(\overline{\mathbf{x}})| \overline{f}(\overline{\mathbf{x}})\, d\overline{\mathbf{x}} = \sum_{i=1}^{n} \overline{c}_i |\det J(\overline{\mathbf{x}}_i)| \overline{f}(\overline{\mathbf{x}}_i) + \overline{E}(|\det J| \overline{f})$$

into the cubature formula

$$\int_B w(x) f(\mathbf{x})\, d\mathbf{x} = \sum_{i=1}^{n} c_i f(\mathbf{x}_i) + E(f)$$

with abscissas and weights given by

$$\mathbf{x}_i := \psi(\overline{\mathbf{x}}_i), \quad c_i := \overline{c}_i |\det J(\overline{\mathbf{x}}_i)|, \quad i = 1, 2, \ldots, n,$$

and an error term $E(f) = \overline{E}(|\det J| \overline{f})$. For affine transformations

$$\psi(\overline{\mathbf{x}}) = A\overline{\mathbf{x}} + b, \quad A \in \mathbb{R}^{N \times N}, \quad b \in \mathbb{R}^N,$$

these relations are particularly simple:

$$\mathbf{x}_i := A\overline{\mathbf{x}}_i + b, \quad c_i := |\det A| \overline{c}_i, \quad i = 1, 2, \ldots, n; \qquad E(f) = |\det A| \overline{E}(\overline{f}).$$

6.1.4. Formula Construction Using Iteration. In Section 3.6.3 it is shown that integral iteration can be used to preprocess multivariate integrals in such a way that their numerical computation becomes easier. The iteration of integrals can also be used to facilitate the construction of multidimensional integration formulas.

DEFINITION 6.1.1 (CARTESIAN PRODUCT RULE). *Suppose the integration region B is the Cartesian product $B = B_1 \times B_2$ of the regions $B_1 \subseteq \mathbb{R}^{N_1}$ and $B_2 \subseteq \mathbb{R}^{N_2}$, and let*

$$\mathrm{Q}_{n_1}^{(1)} f_1 = \sum_{i_1=1}^{n_1} c_{i_1}^{(1)} f_1(\mathbf{x}_{i_1}^{(1)}) \quad \textit{and} \quad \mathrm{Q}_{n_2}^{(2)} f_2 = \sum_{i_2=1}^{n_2} c_{i_2}^{(2)} f_2(\mathbf{x}_{i_2}^{(2)})$$

denote integration formulas for the two integrals

$$\mathrm{I}_1 f_1 := \int_{B_1} f_1(\mathbf{x}^{(1)})\, d\mathbf{x}^{(1)} \quad \textit{and} \quad \mathrm{I}_2 f_2 := \int_{B_2} f_2(\mathbf{x}^{(2)})\, d\mathbf{x}^{(2)},$$

respectively. Then, the Cartesian product rule[55] $Q_{n_1}^{(1)} \times Q_{n_2}^{(2)}$ *for the integral over* B *is defined by*

$$(Q_{n_1}^{(1)} \times Q_{n_2}^{(2)})f := \sum_{i_1=1}^{n_1} \sum_{i_2=1}^{n_2} c_{i_1}^{(1)} c_{i_2}^{(2)} f(\mathbf{x}_{i_1}^{(1)}, \mathbf{x}_{i_2}^{(2)}).$$

THEOREM 6.1.1. *If* $Q_{n_1}^{(1)}$ *integrates* f_1 *exactly over* B_1 *and* $Q_{n_2}^{(2)}$ *integrates* f_2 *exactly over* B_2*, then the Cartesian product rule* $Q_{n_1}^{(1)} \times Q_{n_2}^{(2)}$ *integrates the product*

$$f(\mathbf{x}^{(1)}, \mathbf{x}^{(2)}) := f_1(\mathbf{x}^{(1)}) f_2(\mathbf{x}^{(2)}), \quad \mathbf{x}^{(1)} \in B_1, \ \mathbf{x}^{(2)} \in B_2,$$

exactly over the region $B := B_1 \times B_2$.

Proof.

$$\begin{aligned} If &= \int_{B_1 \times B_2} f(\mathbf{x}^{(1)}, \mathbf{x}^{(2)})\, d(\mathbf{x}^{(1)}, \mathbf{x}^{(2)}) = \int_{B_1} \int_{B_2} f_1(\mathbf{x}^{(1)}) f_2(\mathbf{x}^{(2)})\, d\mathbf{x}^{(2)}\, d\mathbf{x}^{(1)} \\ &= \int_{B_1} f_1(\mathbf{x}^{(1)})\, d\mathbf{x}^{(1)} \int_{B_2} f_2(\mathbf{x}^{(2)})\, d\mathbf{x}^{(2)} = \sum_{i_1=1}^{n_1} c_{i_1}^{(1)} f_1(\mathbf{x}_{i_1}^{(1)}) \sum_{i_2=1}^{n_2} c_{i_2}^{(2)} f_2(\mathbf{x}_{i_2}^{(2)}) \\ &= \sum_{i_1=1}^{n_1} \sum_{i_2=1}^{n_2} c_{i_1}^{(1)} c_{i_2}^{(2)} f_1(\mathbf{x}_{i_1}^{(1)}) f_2(\mathbf{x}_{i_2}^{(2)}) = (Q_{n_1}^{(1)} \times Q_{n_2}^{(2)})f. \end{aligned}$$ □

Example (Cartesian Product Rules). Consider the bivariate integral

$$If = \int_{C_2} f(x_1, x_2)\, dx_1 dx_2. \tag{6.2}$$

The two-dimensional unit cube is a Cartesian product, $C_2 = [-1,1] \times [-1,1]$. Thus, if the one-dimensional integration formulas $Q_{n_1}^{(1)}$ and $Q_{n_2}^{(2)}$ are both applicable to the unit interval $[-1,1]$, then the Cartesian product rule $Q_{n_1}^{(1)} \times Q_{n_2}^{(2)}$ is a formula for (6.2).

Let, for instance, $Q_{n_1}^{(1)}$ and $Q_{n_2}^{(2)}$ be chosen as the Gauss–Legendre formulas (see Section 5.2.7) G_{n_1} and G_{n_2}, defined for the unit interval. Since G_{n_1} and G_{n_2} are exact for all polynomials of maximum degree $d = 2n_1 - 1$ and $d = 2n_2 - 1$, respectively, the Cartesian product rule $G_{n_1} \times G_{n_2}$ is exact for all polynomials

$$P \in \operatorname{span}\{\, x_1^{k_1} x_2^{k_2} \ : \ k_1 \le 2n_1 - 1,\ k_2 \le 2n_2 - 1 \,\}.$$

In particular, $G_{n_1} \times G_{n_2}$ is exact for all polynomials $P \in \mathbb{P}^2$ of total degree

$$d \le 2 \min\{n_1, n_2\} - 1$$

(see Section 6.2.1).

[55] Cartesian product rules are often referred to as product rules. However, in order to avoid confusion with product integration methods (see Section 5.1.2), this simplified terminology is generally avoided in this book.

Cartesian product rules can be generalized as follows.

DEFINITION 6.1.2 (K-FOLD CARTESIAN PRODUCT RULES). *Suppose the integration region B is the Cartesian product of $K \geq 2$ regions $B_1, \ldots, B_K$, i.e., $B = B_1 \times \cdots \times B_K$. Let*

$$\mathrm{Q}_{n_k}^{(k)} f_k = \sum_{i_k=1}^{n_k} c_{i_k}^{(k)} f_k(\mathbf{x}_{i_k}^{(k)}), \quad k = 1, 2, \ldots, K,$$

denote integration formulas for the integrals

$$\mathrm{I}_k f_k := \int_{B_k} f_k(\mathbf{x}^{(k)})\, d\mathbf{x}^{(k)}, \quad k = 1, 2, \ldots, K,$$

respectively. Then, the K-fold Cartesian product rule $\mathrm{Q}_{n_1}^{(1)} \times \cdots \times \mathrm{Q}_{n_K}^{(K)}$ for the integral over B is defined by

$$(\mathrm{Q}_{n_1}^{(1)} \times \cdots \times \mathrm{Q}_{n_K}^{(K)}) f := \sum_{i_1=1}^{n_1} \cdots \sum_{i_K=1}^{n_K} c_{i_1}^{(1)} \cdots c_{i_K}^{(K)} f(\mathbf{x}_{i_1}^{(1)}, \ldots, \mathbf{x}_{i_K}^{(K)}).$$

Theorem 6.1.1 can easily be generalized for K-fold Cartesian product rules $\mathrm{Q}_{n_1}^{(1)} \times \cdots \times \mathrm{Q}_{n_K}^{(K)}$.

THEOREM 6.1.2. *If $\mathrm{Q}_{n_k}^{(k)}$ integrates f_k exactly over B_k for $k = 1, \ldots, K$, then the Cartesian product rule $\mathrm{Q}_{n_1}^{(1)} \times \cdots \times \mathrm{Q}_{n_K}^{(K)}$ integrates the product*

$$f(\mathbf{x}^{(1)}, \ldots, \mathbf{x}^{(K)}) := f_1(\mathbf{x}^{(1)}) \cdots f_K(\mathbf{x}^{(K)}), \quad \mathbf{x}^{(k)} \in B_k, \ k = 1, 2, \ldots, K,$$

exactly over $B := B_1 \times \cdots \times B_K$.

Proof. See the proof of Theorem 6.1.1. □

A characteristic property of Cartesian product rules is that the number of abscissas n in $\mathrm{Q}_{n_1}^{(1)} \times \cdots \times \mathrm{Q}_{n_K}^{(K)}$ is the product $n = n_1 \cdots n_K$ of the numbers of abscissas n_k in the constituent rules $\mathrm{Q}_{n_k}^{(k)}$, $k = 1, \ldots, K$. Since the constituent formulas $\mathrm{Q}_{n_k}^{(k)}$, $k = 1, \ldots, K$, are usually chosen as one-dimensional integration formulas, the number of abscissas in $\mathrm{Q}_{n_1}^{(1)} \times \cdots \times \mathrm{Q}_{n_K}^{(K)}$ increases rapidly when the dimension $N = K$ of the integral increases. This rapid increase in the computational effort makes Cartesian product rules useless for practical purposes except for moderate dimensions, say $N \leq 5$.

Example (Number of Abscissas in Cartesian Product Rules). $(\mathrm{Q}_n)^K$ denotes Cartesian product formulas $\mathrm{Q}_{n_1}^{(1)} \times \cdots \times \mathrm{Q}_{n_K}^{(K)}$ whose constituent one-dimensional integration formulas are identical,

$$\mathrm{Q}_{n_1}^{(1)} = \mathrm{Q}_{n_2}^{(2)} = \cdots = \mathrm{Q}_{n_K}^{(K)} = \mathrm{Q}_n.$$

Clearly, this formula has n^K abscissas. Increasing the dimension from N to $N+1$ augments the number of abscissas by the factor n. Even for the moderate value $n = 10$, the number of abscissas is increased by one order of magnitude with each additional dimension.

Most multidimensional weight functions w are products of several one-dimensional weight functions (see Section 3.1.2). The concept of Cartesian product rules can be extended to, and Theorem 6.1.2 can easily be shown to hold for, the corresponding weighted integrals $\mathrm{I}_w f$.

Generalized Cartesian Product Rules. Cartesian product rules can be generalized for integration regions B which are *not* Cartesian products. Assume that the integral $\mathrm{I}f$ can be iterated,

$$\mathrm{I}f = \int_B f(\mathbf{x})\,d\mathbf{x} = \int_{B_1} \int_{B_2(\mathbf{x}^{(1)})} f(\mathbf{x}^{(1)}, \mathbf{x}^{(2)})\,d\mathbf{x}^{(2)}\,d\mathbf{x}^{(1)} = \int_{B_1} f_1(\mathbf{x}^{(1)})\,d\mathbf{x}^{(1)},$$

where

$$f_1(\mathbf{x}^{(1)}) := \int_{B_2(\mathbf{x}^{(1)})} f(\mathbf{x}^{(1)}, \mathbf{x}^{(2)})\,d\mathbf{x}^{(2)}.$$

Furthermore, assume that, for each $\mathbf{x}^{(1)} \in B_1$, there is a continuously differentiable one-to-one transformation $\psi_{\mathbf{x}^{(1)}}$ from $\overline{B}_2$ onto $B_2(\mathbf{x}^{(1)})$ for which the Jacobian matrix $J_{\mathbf{x}^{(1)}}$ of $\psi_{\mathbf{x}^{(1)}}$ is regular on $\overline{B}_2$.

Let

$$\mathrm{Q}^{(1)}_{n_1} f_1 = \sum_{i_1=1}^{n_1} c^{(1)}_{i_1} f_1(\mathbf{x}^{(1)}_{i_1})$$

be an integration formula for the integral

$$\mathrm{I}_1 f_1 = \int_{B_1} f_1(\mathbf{x}^{(1)})\,d\mathbf{x}^{(1)}.$$

Then, $\mathrm{I}f$ can be approximated by

$$\text{(6.3)} \qquad \mathrm{I}f \approx \sum_{i_1=1}^{n_1} c^{(1)}_{i_1} f_1(\mathbf{x}^{(1)}_{i_1}) = \sum_{i_1=1}^{n_1} c^{(1)}_{i_1} \int_{B_2(\mathbf{x}^{(1)}_{i_1})} f(\mathbf{x}^{(1)}_{i_1}, \mathbf{x}^{(2)})\,d\mathbf{x}^{(2)}.$$

In addition, let

$$\mathrm{Q}^{(2)}_{n_2} f_2 = \sum_{i_2=1}^{n_2} c^{(2)}_{i_2} f_2(\mathbf{x}^{(2)}_{i_2})$$

be an integration formula for the integral

$$\mathrm{I}_2 f_2 = \int_{\overline{B}_2} f_2(\mathbf{x}^{(2)})\,d\mathbf{x}^{(2)}.$$

Each of the n_1 integrals

$$\int_{B_2(\mathbf{x}^{(1)}_{i_1})} f(\mathbf{x}^{(1)}_{i_1}, \mathbf{x}^{(2)})\,d\mathbf{x}^{(2)}, \quad i_1 = 1, 2, \ldots, n_1,$$

can be approximated using the n_2-point rule $Q^{(2)}_{n_2}$ transformed onto the region $B_2(\mathbf{x}^{(1)}_{i_1})$:

$$\int_{B_2(\mathbf{x}^{(1)}_{i_1})} f(\mathbf{x}^{(1)}_{i_1}, \mathbf{x}^{(2)})\, d\mathbf{x}^{(2)} \approx \sum_{i_2=1}^{n_2} c^{(2)}_{i_2} |\det J_{\psi_{\mathbf{x}^{(1)}_{i_1}}}(\mathbf{x}^{(2)}_{i_2})|\, f(\mathbf{x}^{(1)}_{i_1}, \psi_{\mathbf{x}^{(1)}_{i_1}}(\mathbf{x}^{(2)}_{i_2}))$$

for $i_1 = 1, \ldots, n_1$. Combining this result with (6.3) yields

$$\text{(6.4)} \qquad \mathrm{I}f \approx \sum_{i_1=1}^{n_1} c^{(1)}_{i_1} f_1(\mathbf{x}^{(1)}_{i_1})$$

$$\approx \sum_{i_1=1}^{n_1} \sum_{i_2=1}^{n_2} c^{(1)}_{i_1} c^{(2)}_{i_2} |\det J_{\psi_{\mathbf{x}^{(1)}_{i_1}}}(\mathbf{x}^{(2)}_{i_2})|\, f(\mathbf{x}^{(1)}_{i_1}, \psi^{i_1}(\mathbf{x}^{(2)}_{i_2})).$$

The integration formula on the right-hand side of (6.4) is said to be a *generalized Cartesian product formula* for $\mathrm{I}f$.

Example (Generalized Cartesian Product Rules). Consider the integral

$$\mathrm{I}f = \int_0^1 \int_{-x^{(1)}}^{x^{(1)}} f(x^{(1)}, x^{(2)})\, dx^{(2)} dx^{(1)}.$$

Note that the integration region is the triangle with the vertices

$$\mathbf{v}_1 = (0,0), \quad \mathbf{v}_2 = (1,1), \quad \text{and} \quad \mathbf{v}_3 = (1,-1).$$

Let the one-dimensional integration formulas $Q^{(1)}_n$ and $Q^{(2)}_n$ be chosen as the compound Gauss Legendre formula $n \times G_1$ (the compound midpoint formula) transformed onto the interval $[0,1]$:

$$(n \times G_1)f = \frac{1}{n} \sum_{i=1}^{n} f\left(\frac{i - \frac{1}{2}}{n}\right).$$

In order to be made applicable to the interval $[-x^{(1)}, x^{(1)}]$, $n \times G_1$ has to be subjected to the affine transformation

$$\psi_{x^{(1)}} : \overline{x}^{(2)} \mapsto x^{(1)}(2\overline{x}^{(2)} - 1).$$

Note that

$$\psi'_{x^{(1)}}(\overline{x}^{(2)}) = 2x^{(1)}$$

is independent of $\overline{x}^{(2)}$. After carrying out a small number of algebraic manipulations, the following generalized Cartesian product formula can be derived:

$$Q_{n^2} f = \frac{2}{n^3} \sum_{i_1=1}^{n} \left(i_1 - \frac{1}{2}\right) \sum_{i_2=1}^{n} f\left(\frac{i_1 - \frac{1}{2}}{n}, \frac{i_1 - \frac{1}{2}}{n} \frac{2i_2 - 1 - n}{n}\right).$$

Software (Cartesian Product Rules). The subroutine `NAG/d01fbf` evaluates, for any given integrand f, the Cartesian product rule

$$(\mathrm{Q}_{N_1}^{(1)} \times \mathrm{Q}_{N_2}^{(2)} \times \cdots \times \mathrm{Q}_{N_K}^{(K)})f$$

for the given univariate factor formulas $\mathrm{Q}_{N_k}^{(k)}$, $k = 1, \ldots, K$. The user has to provide the factor formulas by supplying their weights and abscissas. The number K of factor formulas (which equals the dimension N of the integral) can be chosen between 1 and 20.

Cartesian product rules using one-dimensional Gauss formulas are used in the adaptive integration subroutine `IMSL/MATH-LIBRARY/dqand` for the calculation of integrals over N-dimensional ($N \leq 20$) cubes.

6.1.5. Formula Construction Using Domain Decomposition.

In Section 3.6.2 the preprocessing of integration problems based on a subdivision of the integration region into pairwise disjoint subregions $B_1, \ldots, B_L$ is discussed. This kind of preprocessing is employed when the integration region is so complicated that numerical integration schemes cannot be applied directly. In this section, it is shown that this domain decomposition approach can also be used for the construction of cubature formulas.

DEFINITION 6.1.3 (MULTIVARIATE COMPOUND INTEGRATION FORMULAS). *Let*

$$\mathrm{Q}_{n_l}^{(l)} f = \sum_{i_l=1}^{n_l} c_{i_l}^{(l)} f(\mathbf{x}_{i_l}^{(l)}), \qquad l = 1, 2, \ldots, L,$$

denote integration formulas for the integrals

$$\int_{B_l} f(\mathbf{x})\, d\mathbf{x} = \mathrm{Q}_{n_l}^{(l)} f + E^{(l)}(f), \quad l = 1, 2, \ldots, L.$$

Then, the compound integration formula $\mathrm{Q}_{n_1}^{(1)} + \cdots + \mathrm{Q}_{n_L}^{(L)}$ *for the integral of* f *over* $B := B_1 \cup \cdots \cup B_L$ *is defined by*

$$(\mathrm{Q}_{n_1}^{(1)} + \cdots + \mathrm{Q}_{n_L}^{(L)})f = \sum_{l=1}^{L} \sum_{i_l=1}^{n_l} c_{i_l}^{(l)} f(\mathbf{x}_{i_l}^{(l)}).$$

Clearly, the error in $(\mathrm{Q}_{n_1}^{(1)} + \cdots + \mathrm{Q}_{n_L}^{(L)})f$ is given by

$$(6.5) \qquad (\mathrm{Q}_{n_1}^{(1)} + \cdots + \mathrm{Q}_{n_L}^{(L)})f - \mathrm{I}f = E^{(1)}(f) + \cdots + E^{(L)}(f).$$

k^N-Copy Rules. Without any additional assumptions concerning the integration region B, the subdivision $B_1, \ldots, B_L$, and the integration formulas $\mathrm{Q}_{n_1}^{(1)}, \ldots, \mathrm{Q}_{n_L}^{(L)}$, formula (6.5) constitutes the maximum information available about the error in the compound formula $\mathrm{Q}_{n_1}^{(1)} + \cdots + \mathrm{Q}_{n_L}^{(L)}$. By contrast, as shown, for instance, in Section 5.3.2, the error in univariate compound formulas can be characterized far more accurately. Most importantly, the Euler–Maclaurin sum formula (5.31) provides an asymptotic error expansion for the compound trapezoidal rule T_l, which makes it possible to accelerate the convergence of trapezoidal rules T_l.

Asymptotic error expansions can be obtained for univariate compound rules for the following reasons. (i) The integration interval $[a, b]$ is subdivided by *equidistant* partition points. (ii) The same (properly transformed) quadrature formula is applied to each of the resulting subintervals.

Similarly, asymptotic error expansions and corresponding convergence acceleration schemes can be obtained for multivariate compound formulas (i) if the subdivision of the integration region B into the subregions $B_1, \ldots, B_L$ is highly regular and (ii) if the same (properly transformed) cubature formula is applied to each of the subregions $B_1, \ldots, B_L$.

In the following discussion, the integration region is assumed to be the unit cube C_N. Most of the results obtained for C_N also hold when the integration region is chosen as the unit simplex T_N. Let the interval $[-1, 1]$ be subdivided into k intervals by $k+1$ equidistant partition points

$$-1 = \overline{x}_0 < \overline{x}_1 < \ldots < \overline{x}_{k-1} < \overline{x}_k = 1.$$

Then, the set of the k^N subcubes

$$[\overline{x}_{l_1}, \overline{x}_{l_1+1}] \times [\overline{x}_{l_2}, \overline{x}_{l_2+1}] \times \cdots \times [\overline{x}_{l_N}, \overline{x}_{l_N+1}],$$
$$l_j = 0, 1, \ldots, k-1, \quad j = 1, 2, \ldots, N,$$

is a partition $\mathcal{P}$ of C_N. The compound formula obtained by transforming and applying the n-point cubature formula Q_n for $[-1, 1]^N$ to each of the k^N subcubes is denoted by $k^N \times \mathrm{Q}_n$ and is said to be the k^N*-copy rule* of Q_n.

If it is assumed that the basic formula Q_n integrates at least all constant polynomials P_0 exactly, then the convergence

$$(k^N \times \mathrm{Q}_n)f \to \mathrm{I}f \qquad \text{as} \qquad k \to \infty$$

can be guaranteed for all properly Riemann integrable functions f.[56] Similarly, asymptotic expansions of the Euler–Maclaurin type (5.32) can be derived for compound formulas $k^N \times \mathrm{Q}_n$, provided that the basic formula Q_n exactly integrates all polynomials P_d of total degree $d \leq D$ (de Doncker [101]). In this case asymptotic expansions for the error $(k^N \times \mathrm{Q}_n)f - \mathrm{I}f$ can also be derived for integrands f with specific kinds of singularities (see Section 7.3.4). These results suggest that the basic formulas Q_n should be exact for certain classes of polynomials. Such formulas are discussed in Section 6.2.

[56]This statement can be proved in the same way as Theorem 5.3.1.

The Number of Abscissas in Copy Rules. Compound formulas $k^N \times Q_n$ can usually be expected to have $k^N n$ abscissas. However, if some of the abscissas of Q_n are located on the boundary of C_N, the number of abscissas in $k^N \times Q_n$ can sometimes be reduced significantly.

Example (Copy Rules with Reduced Numbers of Abscissas). If the abscissas of the integration formula Q_{2^N} for C_N are given by

$$\{\mathbf{x}_1, \ldots, \mathbf{x}_{2^N}\} = \{-1, 1\}^N,$$

the copy rule $k^N \times Q_{2^N}$ has only $(k+1)^N$ instead of $k^N n = (2k)^N$ abscissas.

Nevertheless, since the intersection of more than 2^N subcubes in $\mathcal{P}$ is always empty, the total number of abscissas cannot be reduced by more than a factor of 2^N. Hence, the number of abscissas in N-dimensional copy rules $k^N \times Q_n$ is at least $(k/2)^N n$.

For constant k, the number of abscissas in $k^N \times Q_n$ grows *exponentially* as the dimension N increases. For large values of N, this rapid growth limits the number of subdivisions k which can be performed with an acceptable computational effort. As a result, compound formulas $k^N \times Q_n$ are hardly ever considered for numerical integration problems with dimensions $N > 3$.

6.2. Polynomial Formulas

Since polynomials can easily be integrated in closed form over most standard regions, (piecewise) polynomials appear to be suitable model functions for use in the approximation approach to cubature formula construction. *Univariate* polynomial approximation functions $P \in \mathbb{P}_{n-1}$ are obtained by requiring P to interpolate the values of the integrand at a suitably chosen set of abscissas $x_1, \ldots, x_n \in B = [a, b]$. Section 6.2.1 describes the difficulties arising when this approach is extended to multivariate functions.

Univariate interpolatory integration formulas Q_n are obtained by (exactly) integrating interpolation polynomials or, equivalently, by requiring Q_n to be exact for all polynomials $P \in \mathbb{P}_{n-1}$. Section 6.2.2 shows that the second condition is more suitable for defining multivariate interpolatory formulas than the straightforward interpolation condition and discusses important properties of the resulting formulas.

Particular methods of constructing simple interpolatory cubature formulas, based on *invariant theory* and *ideal theory*, are described in Sections 6.2.3 and 6.2.4, respectively.

6.2.1. Multivariate Polynomial Approximation

Polynomial Spaces. The following are a number of notational conventions, definitions, and facts concerning multivariate polynomials used throughout the remainder of this section.

$\mathbb{P}^N$ denotes the set of *all* multivariate polynomials in N variables with coefficients in $\mathbb{R}$. $\mathbb{P}^N$ is a linear vector space of infinite dimension over the field of real numbers $\mathbb{R}$.

Multivariate *monomials* $x_1^{d_1} x_2^{d_2} \cdots x_N^{d_N}$ are concisely written as

$$\mathbf{x}^{\mathbf{d}}, \quad \mathbf{d} = (d_1, \ldots, d_N)^\top \in \mathbb{N}_0^N.$$

The set of all N-variate monomials $\{\mathbf{x}^{\mathbf{d}} \ : \ \mathbf{d} \in \mathbb{N}_0^N\}$ is a basis for $\mathbb{P}^N$; i.e., any polynomial $P \in \mathbb{P}^N$ can be uniquely represented as a linear combination of a finite number of monomials,

$$P = \sum_{j=1}^{J} a_j \mathbf{x}^{\mathbf{d}_j}, \qquad a_j \neq 0, \quad j = 1, 2, \ldots, J.$$

DEFINITION 6.2.1 (DEGREE OF MULTIVARIATE POLYNOMIALS). *The degree of a monomial* $\mathbf{x}^{\mathbf{d}}$ *is defined by*

$$\deg \mathbf{x}^{\mathbf{d}} := \|\mathbf{d}\|_1 = d_1 + d_2 + \cdots + d_N.$$

The degree of a general polynomial

$$P = \sum_{j=1}^{J} a_j \mathbf{x}^{\mathbf{d}_j}, \qquad a_j \neq 0, \quad j = 1, 2, \ldots, J,$$

is defined as

$$\deg P := \max\{\|\mathbf{d}_1\|_1, \|\mathbf{d}_2\|_1, \ldots, \|\mathbf{d}_J\|_1\}.$$

The set of all N-variate polynomials of maximum degree d, denoted by

$$\mathbb{P}_d^N := \{P \in \mathbb{P}^N \ : \ \deg P \leq d\},$$

is a linear subspace of $\mathbb{P}^N$. Because

$$\mathbb{P}_d^N = \operatorname{span}\{\mathbf{x}^{\mathbf{d}} \ : \mathbf{d} \in \mathbb{N}_0^N, \ |\mathbf{d}| \leq d\},$$

the space $\mathbb{P}_d^N$ is of finite dimension.

THEOREM 6.2.1. *The dimension of* $\mathbb{P}_d^N$ *is given by*

$$\dim(d, N) := \dim \mathbb{P}_d^N = |\{\mathbf{x}^{\mathbf{d}} \ : \ |\mathbf{d}| \leq d\}| = \binom{N+d}{N}.$$

Proof. See Stroud [423].

The following table demonstrates the rapid growth of $\dim(d, N)$, the number of coefficients of an N-variate polynomial of maximum degree d, as the dimension N increases.

N	$d = 1$	$d = 5$	$d = 10$	$d = 15$
1	2	6	11	16
5	6	252	3 003	15 504
10	11	3 003	184 756	3 268 760
15	16	15 504	3 268 760	155 117 520

Polynomial Model Functions. In this section the problem of choosing suitable n-dimensional spaces of polynomial model functions $\mathcal{M}_{\mathbb{P}}$ for given numbers $n \in \mathbb{N}$ is considered. Generally speaking, $\mathcal{M}_{\mathbb{P}}$ should be chosen such that all $f \in \mathcal{F}$ (where $\mathcal{F}$ is a relevant class of functions) can be approximated by polynomials $P \in \mathcal{M}_{\mathbb{P}}$ with an accuracy as high as possible.

For continuous *univariate* functions f, the Weierstrass approximation theorem (see Section 5.2.2) shows that f can be approximated with arbitrary accuracy by polynomials, provided the maximum polynomial degree is chosen sufficiently large. Jackson's theorems (Cheney [56]) suggest that a high approximation accuracy can be attained, even for small polynomial degrees, provided f is sufficiently smooth. A similar conclusion can also be drawn from the Taylor expansion of smooth functions (see Section 5.1.2). Hence, when looking for a suitable n-dimensional space of univariate polynomial model functions, the choice $\mathcal{M}_{\mathbb{P}} = \mathbb{P}_{n-1}$ is reasonable. Only supplementary information about the function f may make the choice of more appropriate spaces of model functions possible.

Example (Even Functions). If the approximation problem is restricted to even functions $f\colon [a,b] \to \mathbb{R}$,

$$f(-x + x_c) = f(x + x_c) \quad \text{for all} \quad x \in [a,b], \quad x_c := (a+b)/2,$$

then the space of model functions can be chosen as the space of even polynomials

$$\mathcal{M}_{\mathbb{P}} := \operatorname{span}\{1, (x-x_c)^2, (x-x_c)^4, \ldots, (x-x_c)^{2n-2}\}.$$

For dimensions $N \geq 2$, the problem of choosing suitable n-dimensional spaces of polynomial model functions $\mathcal{M}_{\mathbb{P}}$ is more intricate. The multidimensional generalization of the Weierstrass approximation theorem (see Section 6.2.2) shows that continuous multivariate functions f can be approximated with arbitrarily high accuracy by polynomials $P \in \mathbb{P}_d^N$, provided the maximum polynomial degree d is chosen sufficiently large. Again, the multivariate Taylor series expansion (see the example on page 156) suggests that a high approximation accuracy can be attained even for small degrees d, provided the integrand is sufficiently smooth. For both reasons, the choice of the polynomial spaces $\mathbb{P}_d^N$, $d = 0, 1, \ldots$, as spaces of model functions $\mathcal{M}_{\mathbb{P}}$ appears to be highly reasonable for multiple dimensions as well.

However, for $N \geq 2$, the dimensions $\dim(d, N)$ of the spaces

$$\mathbb{P}_d^N, \quad d = 0, 1, 2, \ldots,$$

do *not* form a contiguous sequence of natural numbers, i.e.,

$$\{\dim(d,N) : d \in \mathbb{N}_0\} \neq \mathbb{N}.$$

If $n \notin \{\dim(d,N) : d \in \mathbb{N}_0\}$, there is no obvious choice of the space of model functions $\mathcal{M}_{\mathbb{P}}$: without supplementary information about $\mathcal{F}$, there are no theoretical grounds for assuming that a certain choice of $\mathcal{M}_{\mathbb{P}}$, with $\dim \mathcal{M}_{\mathbb{P}} = n$, would give a better approximation accuracy for functions $f \in \mathcal{F}$ than other possible choices of $\mathcal{M}_{\mathbb{P}}$.

Polynomial Interpolation. This section discusses the approximation of functions $f \in \mathcal{F}$ by interpolating f at n distinct points $\mathbf{x}_1, \ldots, \mathbf{x}_n \in \mathbb{R}^N$ using polynomials P chosen from suitable n-dimensional spaces of model functions $\mathcal{M}_{\mathbb{P}}$. Generally speaking, this interpolation problem can be solved for arbitrarily prescribed function values $f(\mathbf{x}_1), \ldots, f(\mathbf{x}_n)$ if and only if the linear functionals

$$\mathbf{x}_1^*, \mathbf{x}_2^*, \ldots, \mathbf{x}_n^* \ : \ \mathcal{M}_{\mathbb{P}} \to \mathbb{R},$$

defined by

$$\mathbf{x}_i^*(P) := P(\mathbf{x}_i), \quad i = 1, 2, \ldots, n, \tag{6.6}$$

are linearly independent over $\mathcal{M}_{\mathbb{P}}$ (see, for instance, Davis [90]). If this independence requirement is met, the interpolation polynomial is unique if and only if $\dim \mathcal{M}_{\mathbb{P}} = n$.

For the dimension $N = 1$, there is a "natural" choice of the space of model functions, $\mathcal{M}_{\mathbb{P}} = \mathbb{P}_{n-1}$, for any $n \in \mathbb{N}$. Additionally, for any n data points $x_1, \ldots, x_n$, the interpolation problem

$$P(x_i) = f(x_i), \quad i = 1, 2, \ldots, n,$$

for $P \in \mathbb{P}_{n-1}$ has a unique solution (see Section 5.1.2). Furthermore, sets of abscissas $x_1, \ldots, x_n$ for which the approximation quality of the (integration formula based on the) interpolation polynomial P can be expected to be high or, in a properly defined sense, even optimal are known. Choosing $x_1, \ldots, x_n$ as the Chebyshev zeros (5.19), for instance, gives rise to interpolation polynomials whose worst-case error for functions $f \in C^{(n)}[a, b]$, $\|f^{(n)}\|_\infty \leq M_n$, is optimal.

For dimensions $N \geq 2$, on the other hand, the situation is quite different: "natural" choices of the space of model functions, $\mathcal{M}_{\mathbb{P}} = \mathbb{P}_d^N$, exist only for particular values of n,

$$n \in \{\dim(d, N) : d \in \mathbb{N}_0\}.$$

Additionally, even if the number n of interpolation points is chosen equal to the dimension $\dim \mathbb{P}_d^N$ of one of the eligible spaces $\mathbb{P}_d^N$ of model functions, it may happen that the linear functionals $\mathbf{x}_1^*, \ldots, \mathbf{x}_n^*$ are *not* linearly independent over $\mathbb{P}_d^N$.

Example (Nonexistence and Nonuniqueness of Interpolation Polynomials). Consider the interpolation of a function $f \colon \mathbb{R}^2 \to \mathbb{R}$ at the points

$$\mathbf{x}_1 = (0, 0)^\top, \ \mathbf{x}_2 = (1/2, 1/2)^\top, \ \text{and } \mathbf{x}_3 = (1, 1)^\top$$

using polynomials $P \in \mathbb{P}_1^2 = \operatorname{span}\{1, x_1, x_2\}$. This interpolation problem is generally unsolvable even though the number of interpolation points is equal to the dimension of $\mathbb{P}_1^2$.

Since the functions in $\mathbb{P}_1^2$ are affine and the interpolation points $\mathbf{x}_1, \mathbf{x}_2$, and $\mathbf{x}_3$ lie on the straight line $\{\mathbf{x} : x_1 = x_2\}$, the function f can be interpolated at $\mathbf{x}_1, \mathbf{x}_2$, and $\mathbf{x}_3$ if and only if the points $(\mathbf{x}_1, f(\mathbf{x}_1))$, $(\mathbf{x}_2, f(\mathbf{x}_2))$, $(\mathbf{x}_3, f(\mathbf{x}_3))$ lie on a straight line in $\mathbb{R}^3$. If this is the case and if $P \in \mathbb{P}_1^2$ is a corresponding interpolation polynomial, *all* polynomials of the form $P + \lambda(x_1 - x_2) \in \mathbb{P}_1^2$, $\lambda \in \mathbb{R}$, satisfy the interpolation conditions.

Recently, de Boor [94], [95] and de Boor and Ron [96], [97], [98], [99] developed a method for constructing n-dimensional polynomial spaces

$$\mathcal{M}_{\mathbb{P}} = \mathcal{M}_{\mathbb{P}}(\mathbf{x}_1, \mathbf{x}_2, \ldots, \mathbf{x}_n) \tag{6.7}$$

for any n data points $\mathbf{x}_1, \ldots, \mathbf{x}_n$ such that the linear functionals $\mathbf{x}_1^*, \ldots, \mathbf{x}_n^*$ are linearly independent over $\mathcal{M}_{\mathbb{P}}(\mathbf{x}_1, \ldots, \mathbf{x}_n)$. The polynomial space (6.7) and the corresponding multivariate interpolation polynomial can be shown to have a number of intriguing and, from a mathematical point of view, appealing properties. In particular,

$$\mathcal{M}_{\mathbb{P}} = \mathcal{M}_{\mathbb{P}}(\mathbf{x}_1, \mathbf{x}_2, \ldots, \mathbf{x}_n) = \mathbb{P}_d^N$$

holds if $n = \dim \mathbb{P}_d^N$ and $\mathbf{x}_1^*, \ldots, \mathbf{x}_n^*$ are linearly independent over $\mathbb{P}_d^N$. There are, however, no theoretical or experimental grounds for assuming that the resulting interpolation polynomials approximate functions $f \in \mathcal{F}$ better than interpolation polynomials obtained by using other eligible model spaces $\mathcal{M}_{\mathbb{P}}$. It is also not clear as to how the approximation error for interpolation polynomials in $\mathcal{M}_{\mathbb{P}}(\mathbf{x}_1, \ldots, \mathbf{x}_n)$ for $n \notin \{\dim(d, N) : d \in \mathbb{N}_0\}$ compares to the approximation error for interpolation polynomials in

$$\mathcal{M}_{\mathbb{P}}(\mathbf{x}_1, \mathbf{x}_2, \ldots, \mathbf{x}_n) = \mathbb{P}_d^N \quad \text{for} \quad n \in \{\dim(d, N) : d \in \mathbb{N}_0\}.$$

Another problem with multivariate polynomial interpolation is that there is no general characterization of sets of abscissas $\mathbf{x}_1, \ldots, \mathbf{x}_n$ for which the approximation quality of the (integration formula based on the) corresponding interpolation polynomial P can be expected to be high or even, in a properly defined sense, optimal.

The numerous problems and issues involved in multivariate polynomial interpolation described in this section provide convincing evidence that a straightforward application of the interpolation approach to cubature formula construction is *not* possible.

6.2.2. Simple Interpolatory Formulas. Univariate interpolatory integration formulas Q_n with prescribed abscissas $x_1, \ldots, x_n$ can be obtained alternatively by requiring Q_n to integrate all polynomials in $\mathbb{P}_{n-1}$ exactly, i.e., by requiring Q_n to have a degree of accuracy $D \geq n-1$. For interpolatory quadrature formulas Q_n, the error bound (see Theorem 5.2.2)

$$|Q_n f - I_w f| \leq \left(\|w\|_1 + \sum_{i=1}^{n} |c_i| \right) e_D^*(f), \tag{6.8}$$

with

$$e_D^*(f) := \inf\{\|P_D - f\|_\infty : P_D \in \mathbb{P}_D\}, \tag{6.9}$$

holds for $D \geq n-1$. However, the error bound (6.8) is true for any quadrature formula Q_n with a degree of accuracy $D \geq 0$, independently of whether or not Q_n is interpolatory. Since the straightforward interpolation approach to cubature formula construction is not feasible, it seems reasonable to construct cubature formulas Q_n by extending the aforementioned alternative construction method to multiple dimensions, i.e., by requiring that Q_n integrate all polynomials of maximum degree $d = D$ exactly.

DEFINITION 6.2.2 (DEGREE OF ACCURACY). *The degree of accuracy of an N-dimensional cubature formula Q_n is D if*

$$\begin{aligned} Q_n \mathbf{x}^{\mathbf{d}} &= I_w \mathbf{x}^{\mathbf{d}} \quad \text{for all } \mathbf{d} \in \mathbb{N}_0^N \text{ with } \|\mathbf{d}\|_1 \leq D, \\ Q_n \mathbf{x}^{\mathbf{d}} &\neq I_w \mathbf{x}^{\mathbf{d}} \quad \text{for at least one } \mathbf{d} \in \mathbb{N}_0^N \text{ with } \|\mathbf{d}\|_1 = D+1, \end{aligned}$$

i.e., if Q_n is exact for all polynomials in N variables of degree $d \leq D$ and if there is at least one polynomial of degree $d = D+1$ for which Q_n is not exact.

Interpolatory Formulas and Moment Equations. For n given distinct abscissas $\mathbf{x}_1, \ldots, \mathbf{x}_n$, the requirement that Q_n must integrate all polynomials of maximum degree $d = D$ exactly means that the weights $c_1, \ldots, c_n$ of the cubature formula

$$Q_n f = \sum_{i=1}^{n} c_i f(\mathbf{x}_i)$$

to be constructed are chosen in such a way that

$$Q_n P_d = I_w P_d \qquad \text{for all} \quad P_d \in \mathbb{P}_d^N. \tag{6.10}$$

Since $\mathbb{P}_d^N$ is a vector space of dimension $\dim(d, N)$ and both operators I_w and Q_n are linear functionals over $\mathbb{P}_d^N$, it is sufficient that the equality (6.10) hold for just one particular basis $\{b_1, \ldots, b_{\dim(d,N)}\}$ of $\mathbb{P}_d^N$:

$$Q_n b_j = I_w b_j, \quad j = 1, 2, \ldots, \dim(d, N). \tag{6.11}$$

The resulting equations (6.11) are called *moment equations.* For a fixed choice of the basis $\{b_1, \ldots, b_{\dim(d,N)}\}$ and of the abscissas $\mathbf{x}_1, \ldots, \mathbf{x}_n$, the requirement (6.11) defines the system of linear equations

$$\sum_{i=1}^{n} c_i b_j(\mathbf{x}_i) = I_w b_j, \quad j = 1, 2, \ldots, \dim(d, N). \tag{6.12}$$

DEFINITION 6.2.3 (INTERPOLATORY INTEGRATION FORMULA). *An integration formula* Q_n *with the abscissas* $\mathbf{x}_1, \ldots, \mathbf{x}_n$ *is said to be interpolatory if its weights* $c_1, \ldots, c_n$ *are the unique solution of the system of equations* (6.12).

Note that for $N = 1$, Definitions 5.1.1 and 6.2.3 of interpolatory integration formulas coincide.

Example (Interpolatory Cubature Formula). Consider the integral operator

$$\mathrm{I}f = \int_{T_2} f(x_1, x_2)\, dx_1 dx_2,$$

where T_2 denotes the two-dimensional unit triangle with the vertices

$$\mathbf{v}_1 = (0,0)^\top, \quad \mathbf{v}_2 = (0,1)^\top \quad \text{and} \quad \mathbf{v}_3 = (1,0)^\top.$$

For this integral operator, there exists a one-point integration formula Q_1 which integrates all constant and linear polynomials in $\mathbb{P}^2$. Let the single abscissa $\mathbf{x}_1$ be chosen as $\mathbf{x}_1 = (1/3, 1/3)^\top$, and let the basis $\{b_1, b_2, b_3\}$ of $\mathbb{P}_1^2$ be chosen as

$$b_1(x_1, x_2) = 1, \quad b_2(x_1, x_2) = x_1, \quad \text{and} \quad b_3(x_1, x_2) = x_2.$$

It can easily be seen that

$$\mathrm{I}b_1 = 1/2, \quad \mathrm{I}b_2 = 1/6, \quad \text{and} \quad \mathrm{I}b_3 = 1/6$$

and that

$$b_1(\mathbf{x}_1) = 1, \quad b_2(\mathbf{x}_1) = 1/3, \quad \text{and} \quad b_3(\mathbf{x}_1) = 1/3.$$

The unique solution of the corresponding moment equations (6.12),

$$\begin{aligned} c_1 &= 1/2, \\ (1/3)c_1 &= 1/6, \\ (1/3)c_1 &= 1/6, \end{aligned}$$

is $c_1 = 1/2$, and so the weight c_1 is uniquely determined by the choice of the abscissa $\mathbf{x}_1 = (1/3, 1/3)^\top$. The resulting formula,

$$Q_1 f = \frac{f(1/3, 1/3)}{2},$$

is interpolatory.

The condition in Definition 6.2.3 that the weights $c_1, \ldots, c_n$ must be the unique solution of the moment equations (6.12) can be explained as follows. Let

$$\mathbf{x}_1^*, \mathbf{x}_2^*, \ldots, \mathbf{x}_n^* \; : \; \mathbb{P}_d^N \to \mathbb{R}$$

be the linear functionals defined in (6.6). Using this notation, (6.12) can be written as

$$\sum_{i=1}^{n} c_i \mathbf{x}_i^*(b_j) = \mathrm{I}_w b_j, \quad j = 1, 2, \ldots, \dim(d, N). \tag{6.13}$$

These equations can be solved if and only if the linear functional I_w over $\mathbb{P}_d^N$ is a linear combination of the linear functionals $\mathbf{x}_1^*, \ldots, \mathbf{x}_n^*$, i.e., if and only if $\mathrm{I}_w \in \operatorname{span}\{\mathbf{x}_1^*, \ldots, \mathbf{x}_n^*\}$. If this is the case, then the weights $c_1, \ldots, c_n$ of Q_n are chosen as the coefficients of a linear combination

$$\mathrm{I}_w = \sum_{i=1}^{n} c_i \mathbf{x}_i^* \quad \text{over} \quad \mathbb{P}_d^N .$$

This shows that the weights $c_1, \ldots, c_n$ are uniquely determined by the abscissas $\mathbf{x}_1, \ldots, \mathbf{x}_n$ if and only if the functionals $\mathbf{x}_1^*, \ldots, \mathbf{x}_n^*$ are linearly independent over $\mathbb{P}_d^N$. If $\mathbf{x}_1^*, \ldots, \mathbf{x}_n^*$ are linearly dependent, then one of the linear functionals, for instance, $\mathbf{x}_n^*$, can be represented as a linear combination of the others,

$$\mathbf{x}_n^* = \sum_{i=1}^{n-1} d_i \mathbf{x}_i^* \quad \text{over} \quad \mathbb{P}_d^N .$$

Thus,

$$\mathrm{I}_w = \sum_{i=1}^{n-1} c_i \mathbf{x}_i^* + c_n \sum_{i=1}^{n-1} d_i \mathbf{x}_i^* = \sum_{i=1}^{n-1} (c_i + c_n d_i) \mathbf{x}_i^* \quad \text{over} \quad \mathbb{P}_d^N .$$

Hence, if the integration formula Q_n is *not* interpolatory, then there is an integration formula Q_{n-1} with $n-1$ abscissas which satisfies the same moment equations; if Q_{n-1} itself is not interpolatory, then there is an integration formula Q_{n-2} with $n-2$ abscissas which again satisfies the same moment equations; etc. Thus, for any noninterpolatory integration formula Q_n which satisfies a set of moment equations (6.11), there is an interpolatory integration formula $\mathrm{Q}_{n'}$ with n' abscissas, $n' < n$, which satisfies the same moment equations.

For reasons of efficiency, the number of function values needed to attain a given accuracy requirement should be kept as low as possible and thus $\mathrm{Q}_{n'}$ is generally preferred to Q_n. This is why, in practice, essentially all cubature formulas derived from moment equations are interpolatory.

Existence of Interpolatory Formulas. This section considers multivariate integration formulas Q_n which satisfy the moment equations (6.11) for $d = D$, where D is the required degree of accuracy. In order to be usable for numerical integration purposes, a solution of (6.11) should have the following properties. (i) The value of n should be as small as possible; in particular, the number of abscissas n should not be larger than the dimension of the polynomial space $\mathbb{P}_d^N$ whose elements are integrated exactly. (ii) The abscissas $\mathbf{x}_1, \ldots, \mathbf{x}_n$ should be located within the integration region B. (iii) The weights $c_1, \ldots, c_n$ should be nonnegative, provided the weight function w is admissible. For dimensions $N \geq 2$, it is not at all evident whether or not there are, in the aforementioned sense, satisfactory solutions of (6.11). Nevertheless, the following theorem by Tchakaloff establishes that appropriate formulas exist for general regions $B \subseteq \mathbb{R}^N$.

THEOREM 6.2.2 (TCHAKALOFF). *Let w be an admissible weight function on a bounded and closed region $B \subseteq \mathbb{R}^N$. Then, for any $D \geq 0$, there exist $n = \dim \mathbb{P}_d^N$ abscissas $\mathbf{x}_1, \ldots, \mathbf{x}_n \in B$ and positive weights $c_1, \ldots, c_n$ such that the corresponding cubature formula Q_n for the integral I_w is of degree D.*

Proof. See Stroud [423].

Note that the proof of this general formulation of Tchakaloff's theorem is existential, i.e., nonconstructive, which makes it useless for the practical construction of cubature formulas. For the special weight function $w(\mathbf{x}) \equiv 1$, however, a constructive proof of Tchakaloff's theorem is given in Davis [91]. Although an algorithm for cubature formula construction based on Davis's proof is derived in Wilson [446], practical methods for constructing cubature formulas with a given degree of accuracy (as described in Sections 6.2.3 and 6.2.4) do *not* rely on Tchakaloff's theorem.

The number of abscissas n in interpolatory cubature formulas with a degree of precision D can generally be bounded as follows.

THEOREM 6.2.3. *Let w be an admissible weight function on a region $B \subseteq \mathbb{R}^N$. Then, the number n of abscissas in any interpolatory integration formula Q_n for the integral I_w with degree of accuracy $D \geq 0$ satisfies*

$$\binom{N + \lfloor D/2 \rfloor}{N} \leq n \leq \binom{N + D}{N}. \tag{6.14}$$

Proof. Since, according to Theorem 6.2.1,

$$\binom{N + D}{N} = \dim \mathbb{P}_D^N,$$

inequality (6.14) can also be written as

$$\dim \mathbb{P}_{\lfloor D/2 \rfloor}^N \leq n \leq \dim \mathbb{P}_D^N.$$

As Q_n is assumed to be interpolatory, the functionals $\mathbf{x}_1^*, \ldots, \mathbf{x}_n^*$ pertaining to the abscissas $\mathbf{x}_1, \ldots, \mathbf{x}_n$ of Q_n are linearly independent. Since the space of all linear functionals over $\mathbb{P}_D^N$ has the same dimension as $\mathbb{P}_D^N$, $\dim \mathbb{P}_D^N$ is the maximum number of linearly independent linear functionals over $\mathbb{P}_D^N$ and hence the upper bound $n \leq \dim \mathbb{P}_D^N$ is established.

Now assume that there is an n-point cubature formula Q_n of degree D for which $n < \dim \mathbb{P}_{\lfloor D/2 \rfloor}^N$. Let

$$\{\, b_j \;:\; j = 1, 2, \ldots, \dim \mathbb{P}_{\lfloor D/2 \rfloor}^N \,\}$$

be a basis for $\mathbb{P}^N_{\lfloor D/2 \rfloor}$. The linear system

$$(6.15) \qquad \sum_{j=1}^{\dim \mathbb{P}^N_{\lfloor D/2 \rfloor}} \gamma_j b_j(\mathbf{x}_i) = 0, \quad i = 1, 2, \ldots, n,$$

has a nontrivial solution since the number $\dim \mathbb{P}^N_{\lfloor D/2 \rfloor}$ of unknowns in this system is larger than the number n of equations. Let the polynomial $g \in \mathbb{P}^N_{\lfloor D/2 \rfloor}$ be defined by

$$g := \sum_{j=1}^{\dim \mathbb{P}^N_{\lfloor D/2 \rfloor}} \gamma_j b_j.$$

Clearly, $g^2 \in \mathbb{P}^N_D$, $g^2 \not\equiv 0$, and, according to (6.15),

$$g^2(\mathbf{x}_1) = \cdots = g^2(\mathbf{x}_n) = 0.$$

Since Q_n is assumed to be exact for all polynomials in $\mathbb{P}^N_D$, the contradiction

$$0 = \sum_{i=1}^{n} c_i g^2(\mathbf{x}_i) = \mathrm{Q}_n g^2 = \int_B w(\mathbf{x}) g^2(\mathbf{x})\, d\mathbf{x} > 0$$

can be derived. Hence, there is no cubature formula of degree D with less than $\dim \mathbb{P}^N_{\lfloor D/2 \rfloor}$ abscissas. □

The lower bound on the number of abscissas in Theorem 6.2.3 can be improved if the integral operator I_w is *centrally symmetric*, i.e., if the integral operator I_w is invariant with respect to the transformation

$$G \;:\; \mathbf{x} \mapsto -\mathbf{x}$$

(see Section 6.2.3). For detailed information on the corresponding lower bounds, see, for instance, Cools [61].

Convergence of Simple Interpolatory Formulas. The significance of the degree of accuracy is due to the following theorem, which relates the error in a cubature formula with degree of accuracy D to the error in the best approximation polynomial $P^*_D \in \mathbb{P}^N_D$ with respect to the L_∞-norm.

THEOREM 6.2.4. *Any interpolatory cubature formula Q_n with degree of accuracy $D \geq 0$ satisfies the inequality*

$$|\mathrm{Q}_n f - \mathrm{I}_w f| \leq \left(\|w\|_1 + \sum_{i=1}^{n} |c_i| \right) e^*_D(f),$$

where

$$e^*_D(f) := \inf\{\|P_D - f\|_\infty : P_D \in \mathbb{P}^N_D\}.$$

If all weights of Q_n are nonnegative, then

$$(6.16) \qquad |\mathrm{Q}_n f - \mathrm{I}_w f| \leq (\|w\|_1 + \mathrm{I}w) e^*_D(f).$$

Proof. The proof of this theorem is completely analogous to the proof of Theorem 5.2.2 for quadrature formulas. □

If the coefficients of a sequence $\{Q_{n_D} : D = 0, 1, 2, \dots\}$ of cubature formulas with degrees of accuracy D are *nonnegative* (for all $D \in \mathbb{N}_0$), then the error bound (6.16) implies the convergence

$$Q_{n_D} f \to I_w f \qquad \text{as} \qquad D \to \infty$$

for all $f \in C(B)$, provided the integration region B is bounded and closed. This convergence result is a consequence of the *Weierstrass approximation theorem* (see Davis [90]), which guarantees

$$e_D^*(f) \to 0 \qquad \text{as} \qquad D \to \infty$$

for all $f \in C(B)$.

The relationship

$$e_D^*(f) \leq e_{D-1}^*(f) \qquad \text{for all} \quad D \in \mathbb{N},$$

along with (6.16), implies that the error in cubature formulas (with nonnegative weights) can be expected to decrease as the degree of accuracy increases.

Known Formulas. The construction of interpolatory cubature formulas is an area of ongoing research. *Minimal point formulas*, i.e., formulas with a number n of abscissas which is equal to the corresponding lower bound on the number of abscissas, have only been found for low dimensions and small degrees of accuracy. The following sections describe two methods for constructing cubature formulas in practice.

A listing of almost all interpolatory cubature formulas for multidimensional integration regions known in 1971 is contained in Stroud [423]. An up-to-date compilation of those cubature formulas developed since that publication can be found in Cools and Rabinowitz [64].

6.2.3. Construction Based on Invariant Theory. For n distinct abscissas $\mathbf{x}_1, \dots, \mathbf{x}_n$, $n \leq \dim \mathbb{P}_d^N$, the weights $c_1, \dots c_n$ of an interpolatory integration formula Q_n are uniquely determined as the solution of the *linear* system of equations (6.12). However, for a given choice of abscissas $\mathbf{x}_1, \dots, \mathbf{x}_n$, the moment equations (6.12) may not be solvable; i.e., a corresponding interpolatory integration formula Q_n may not exist. Thus, when trying to construct n-point interpolatory cubature formulas, the system (6.12) must be solved not only for the weights $c_1, \dots, c_n$, but also for the abscissas $\mathbf{x}_1, \dots, \mathbf{x}_n$. Note that the equations (6.12) are *nonlinear* (polynomial) equations in the unknown abscissas $\mathbf{x}_1, \dots, \mathbf{x}_n$. Each abscissa $\mathbf{x}_i$ introduces, in addition to the weight c_i, the N coordinates of the abscissa $\mathbf{x}_i$ as unknowns. Therefore, the cubature formula Q_n to be constructed must satisfy a system of $\dim(d, N)$ nonlinear equations in $n(N+1)$ unknowns. For nontrivial values of n and N, these nonlinear equations are too complex to be solved directly.

Example (The Complexity of Moment Equations). Consider the construction of cubature formulas for $N = 5$ and $D = 5$. For this moderate dimension and this low degree of accuracy, the number of moment equations is

$$\dim(5,5) = \binom{10}{5} = 252.$$

The number of moment equations for $N = 10$ and $D = 5$ is

$$\dim(5,10) = \binom{15}{5} = 3003.$$

The huge size of these systems of nonlinear equations makes their exact, or even numerical, solution most challenging.

Invariant Regions and Functions. Many integration regions, such as the unit cube C_N, the unit sphere S_N, the unit simplex T_N, etc., exhibit a high degree of symmetry. When constructing interpolatory cubature formulas, it seems reasonable to look for formulas which share the symmetry of the given integration problem.

Example (Symmetric Quadrature Formulas). The unit interval $[-1, 1]$ is centrally symmetric with respect to the origin 0. The abscissas of *all* simple quadrature formulas Q_n (with the unit weight function $w(x) \equiv 1$ on $[-1,1]$) described in Section 5.2 are also symmetric about the origin; i.e., if x_i is an abscissa of Q_n, then $-x_i$ is also an abscissa of Q_n. In addition, the abscissas x_i and $-x_i$ of Q_n are assigned the same weights.

In mathematical terms, a region $B \subseteq \mathbb{R}^N$ exhibits symmetry if it is invariant with respect to bijective *linear transformations* G on $\mathbb{R}^N$, i.e., if $G(B) = B$. If B is invariant with respect to two such transformations G_1 and G_2, then $G_2(G_1(B)) = B$; i.e., B is also invariant with respect to the linear transformation $G_2 \circ G_1$. Similarly, if B is invariant with respect to G, then it is also invariant with respect to G^{-1}. This shows that the set $\mathcal{G}$ of all such *symmetry transformations* of a region B forms a group with respect to the composition $\circ$ of functions.

If a region B is invariant with respect to a group $\mathcal{G}$ of transformations, a function g defined on B is called invariant with respect to $\mathcal{G}$ (or $\mathcal{G}$*-invariant*) if

$$g \circ G^{-1} = g \qquad \text{for all} \quad G \in \mathcal{G}.$$

An integral operator I_w is invariant with respect to a group $\mathcal{G}$ of linear transformations if both the integration region B and the weight function w are invariant with respect to $\mathcal{G}$.

Example (Invariant Integration Problems). The unit interval $[-1, 1]$ is invariant with respect to the group Z_2 consisting of the two transformations

$$G_1 \ : \ x \mapsto x \quad \text{and} \quad G_2 \ : \ x \mapsto -x. \tag{6.17}$$

The Legendre and the Chebyshev weight functions for the interval $[-1, 1]$ are invariant with respect to Z_2, while the Jacobi weight function for $[-1, 1]$ is *not* Z_2-invariant if $\alpha \neq \beta$ (see Section 3.1.2).

The N-dimensional unit cube $C_N = [-1, 1]^N$ is invariant with respect to the group $\mathcal{G}$ of linear transformations

$$G \ : \ (x_1, \ldots, x_N) \mapsto (s_1 x_{\pi(1)}, s_2 x_{\pi(2)}, \ldots, s_N x_{\pi(N)}), \tag{6.18}$$

where π is a permutation of the index set $\{1, \ldots, N\}$ and $(s_1, \ldots, s_N)^\top \in \{-1, 1\}^N$. For $N = 2$, $\mathcal{G}$ is said to be the *dihedral* group, and the symbol $\mathcal{D}_4$ is used to denote $\mathcal{G}$. Less formally, $\mathcal{G}$ can be said to consist of sign changes and coordinate permutations. Since $\{-1, 1\}^N$ contains 2^N elements $(s_1, \ldots, s_N)^\top$, and as there are $N!$ permutations of the index set $\{1, \ldots, N\}$, $\mathcal{G}$ consists of $N!\,2^N$ transformations.

An integral operator

$$\mathrm{I}_w f = \int_{C_N} w(\mathbf{x}) f(\mathbf{x})\, d\mathbf{x}$$

is invariant with respect to the group $\mathcal{G}$ if its weight function w is invariant with respect to $\mathcal{G}$. This is, for instance, the case if

$$w(x_1, \ldots, x_N) = \overline{w}(x_1)\, \overline{w}(x_2) \cdots \overline{w}(x_N),$$

where $\overline{w}$ is a Z_2-invariant, one-dimensional weight function.

Orbits and Generators of Invariant Formulas. If an integral operator I_w is invariant with respect to a group $\mathcal{G}$ of linear transformations, an integration formula Q_n for I_w is said to be invariant with respect to $\mathcal{G}$ if and only if (i) its abscissa set $\{\mathbf{x}_1, \ldots, \mathbf{x}_n\}$ is invariant with respect to $\mathcal{G}$, i.e.,

$$G(\{\mathbf{x}_1, \ldots, \mathbf{x}_n\}) = \{\mathbf{x}_1, \ldots, \mathbf{x}_n\} \qquad \text{for all} \quad G \in \mathcal{G},$$

and (ii) any two integration abscissas $\mathbf{x}_i$ and $\mathbf{x}_j$ which can be mapped onto one another using a transformation $G \in \mathcal{G}$ have equal weights, i.e.,

$$\mathbf{x}_i = G(\mathbf{x}_j) \quad \Rightarrow \quad c_i = c_j.$$

The set

$$\mathcal{G}(\mathbf{x}_i) := \{\, G(\mathbf{x}_i) \ : \ G \in \mathcal{G} \,\}$$

is then called the *$\mathcal{G}$-orbit* of the abscissa $\mathbf{x}_i$. Abscissas of an invariant integration formula pertaining to the same orbit have equal weights. Note that $\mathcal{G}(\mathbf{x}_i) \subseteq \{\mathbf{x}_1, \ldots, \mathbf{x}_n\}$ for all $i = 1, \ldots, n$. A subset $\{\mathbf{x}_{i_1}, \ldots \mathbf{x}_{i_K}\} \subseteq \{\mathbf{x}_1, \ldots, \mathbf{x}_n\}$ is called a *generator set* for the $\mathcal{G}$-invariant integration formula Q_n if and only if

$$\mathcal{G}(\mathbf{x}_{i_j}) \cap \mathcal{G}(\mathbf{x}_{i_k}) = \emptyset \quad \text{for} \quad j \neq k$$

and

$$\mathcal{G}(\mathbf{x}_{i_1}) \cup \cdots \cup \mathcal{G}(\mathbf{x}_{i_K}) = \{\mathbf{x}_1, \ldots, \mathbf{x}_n\}.$$

In other words, for each abscissa $\mathbf{x}_i \in \{\mathbf{x}_1, \ldots, \mathbf{x}_n\}$, there exists a unique *generator* $\mathbf{x}_{i_l} \in \{\mathbf{x}_{i_1}, \ldots \mathbf{x}_{i_K}\}$ and a (not necessarily unique) transformation $G \in \mathcal{G}$ such that $\mathbf{x}_i = G(\mathbf{x}_{i_l})$. If $\{\mathbf{x}_{i_1}, \ldots \mathbf{x}_{i_K}\}$ is a generator set for a $\mathcal{G}$-invariant cubature formula Q_n, then Q_n can be written as

$$(6.19) \qquad Q_n f = \sum_{i=1}^{n} c_i f(\mathbf{x}_i) = \sum_{l=1}^{K} c_{i_l} \sum_{\mathbf{x}_i \in \mathcal{G}(\mathbf{x}_{i_l})} f(\mathbf{x}_i).$$

Example (Invariant Cubature Formulas). The integration subroutine `TOMS/698`, developed by Berntsen, Espelid, and Genz [29], incorporates a $\mathcal{D}_4$-invariant 29-point cubature formula Q_{29} (due to Genz and Malik [172]) for the unit square $C_2 = [-1, 1]^2$. The following table contains the coordinates of the 7 generators of Q_{29} as well as the corresponding weights (rounded to 2 significant digits):

Generator	Weight	x_1-Coordinate	x_2-Coordinate
1	-0.36	0	0
2	0.50	0.41	0
3	0.23	0.90	0
4	-0.10	0.96	0
5	0.25	0.69	0.69
6	0.03	0.96	0.96
7	0.09	0.96	0.41

Figure 6.1 shows the abscissas of this invariant cubature formula. Abscissas pertaining to the same orbit are connected by dotted lines.

Simplification of Moment Equations Using Symmetries. The following theorem by Sobolev shows that nonlinear systems of moment equations can be simplified significantly for invariant cubature formulas.

THEOREM 6.2.5 (SOBOLEV). *If an integral operator I_w is invariant with respect to a group of transformations $\mathcal{G}$ and a cubature formula Q_n for I_w is also invariant with respect to $\mathcal{G}$, then*

$$Q_n P_d = I P_d \qquad \text{for all} \quad P_d \in \mathbb{P}_d^N$$

holds if and only if

$$Q_n P_d = I P_d \qquad \text{for all} \quad P_d \in \mathbb{P}_d^N(\mathcal{G}),$$

where $\mathbb{P}_d^N(\mathcal{G})$ is the space of all $\mathcal{G}$-invariant polynomials of maximum degree d.

Proof. See Sobolev [414].

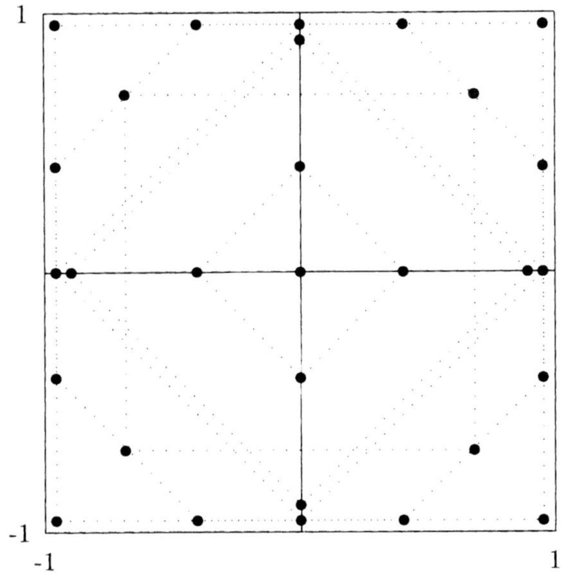

FIG. 6.1. *Abscissas of the $\mathcal{D}_4$-invariant formula* Q_{29} *for* $[-1,1]\times[-1,1]$.

Thus, moment equations for $\mathcal{G}$-invariant cubature formulas Q_n can be written as

$$Q_n b_j = I_w b_j, \quad j = 1, 2, \ldots, \dim \mathbb{P}_d^N(\mathcal{G}), \tag{6.20}$$

where $\{\, b_j \;:\; j = 1, \ldots, \dim \mathbb{P}_d^N(\mathcal{G}) \,\}$ is a basis for $\mathbb{P}_d^N(\mathcal{G})$. A comparison with the moment equations (6.11) for general (nonsymmetric) cubature rules shows that the number of nonlinear equations to be solved is reduced from $\dim \mathbb{P}_d^N$ to $\dim \mathbb{P}_d^N(\mathcal{G})$.

Example (Invariant Polynomials). The group Z_2 of transformations which leave the unit interval $[-1,1]$ invariant is given by (6.17). It can easily be seen that the invariant polynomials with respect to Z_2 are those which can be expressed as polynomials in x^2. Hence, the set of even monomials $\{1, x^2, x^4, \ldots, x^{2\lfloor d/2 \rfloor}\}$ is a basis for $\mathbb{P}_d^1(Z_2)$.

The group of transformations which leave the two-dimensional unit cube C_2 invariant—the dihedral group $\mathcal{D}_4$—is given by (6.18). Obviously, in order for $P(x_1, x_2)$ to be $\mathcal{D}_4$-invariant, all exponents in $P(x_1, x_2) \in \mathbb{P}^2$ have to be even and P must be symmetric in x_1 and x_2. For instance, $x_1^2 + x_2^2$ and $x_1^2 x_2^2$ are $\mathcal{D}_4$-invariant. It can be shown that a polynomial P is $\mathcal{D}_4$-invariant if and only if it can be expressed as a polynomial in $x_1^2 + x_2^2$ and $x_1^2 x_2^2$. Therefore, $\mathbb{P}_d^2(\mathcal{D}_4)$ has the basis $\{1\}$ for $d = 0, 1$, the basis $\{1, x_1^2 + x_2^2\}$ for $d = 2, 3$, the basis $\{1, x_1^2 + x_2^2, x_1^2 x_2^2, (x_1^2 + x_2^2)^2\}$ for $d = 4, 5$, etc.

Structures and Consistency Conditions. Using the orbit representation (6.19) of $\mathcal{G}$-invariant cubature formulas, the moment equations (6.20) can be written as

$$(6.21)\qquad Q_n b_j = \sum_{l=1}^{K} |\mathcal{G}(\mathbf{x}_{i_l})| c_{i_l} b_j(\mathbf{x}_{i_l}) = I b_j, \quad j = 1, 2, \ldots, \dim \mathbb{P}_d^N(\mathcal{G}),$$

where K is the number of generators in Q_n and $|\mathcal{G}(\mathbf{x}_{i_l})|$ denotes the number of points in the orbit $\mathcal{G}(\mathbf{x}_{i_l})$. Each orbit $\mathcal{G}(\mathbf{x}_{i_l})$ of an invariant cubature formula introduces a number of unknowns into the nonlinear equations (6.21). These unknowns are the weight c_{i_l} and a number of parameters determining $\mathbf{x}_{i_l}$. Different types of generators $\mathbf{x}_{i_l}$ differ from each other in the number and the kinds of their parameters.

Example (Types of Generators). $\mathcal{D}_4$-invariant cubature formulas Q_n for the unit square C_2 (for instance, the cubature formula Q_{29} in the example on page 178) have the following types of generators:

Type	Generator $\mathbf{x}_{i_l}$	$\lvert\mathcal{G}(\mathbf{x}_{i_l})\rvert$	Number of unknowns	Unknowns
1	$(0,0)$	1	1	weight
2	$(a,0)$	4	2	weight, a
3	(a,a)	4	2	weight, a
4	$(a,b),\ a>b$	8	3	weight, a, b

For the symmetry group Z_2, there are only two different types of generators:

Type	Generator $\mathbf{x}_{i_l}$	$\lvert\mathcal{G}(\mathbf{x}_{i_l})\rvert$	Number of unknowns	Unknowns
1	(0)	1	1	weight
2	$(a),\ a>0$	2	2	weight, a

The various possible types of generators are usually indexed by $m \in \{1, 2, \ldots, M\}$, where the parameter M depends on the transformation group. Let K_m be the number of generators of type m in an invariant cubature formula Q_n. Then, $\mathbf{K} := (K_1, \ldots, K_M)$ is referred to as the *structure* of Q_n.

Example (Structure of Invariant Cubature Formulas). The first table in the previous example shows that there are $M = 4$ different types of generators of $\mathcal{D}_4$-invariant cubature formulas Q_n for the unit square C_2. The $\mathcal{D}_4$-invariant cubature formula Q_{29} in the example on page 178 has the structure $(1, 3, 2, 1)$.

If there are more equations than unknowns in the moment equations (6.21), the system (6.21) cannot normally be expected to have a solution. This also holds if there is a subsystem of (6.21) which contains more equations than unknowns. Based on this observation, Rabinowitz and Richter [358] introduced the notion of *consistency conditions.* A consistency condition is an inequality for the numbers of generators $K_1, \ldots, K_M$ that is mathematically equivalent to the condition that any subsystem of (6.21) has at least as many unknowns as equations.

Example (Consistency Conditions). Consider the construction of a symmetric, univariate integration rule with degree of accuracy D, i.e., a rule which satisfies the moment equations for $d = D$. Since $\{1, x^2, \ldots, x^{2\lfloor d/2\rfloor}\}$ forms a basis for $\mathbb{P}_d^1(Z_2)$, the system (6.21) contains $\lfloor d/2 \rfloor + 1$ moment equations. The number of unknowns in a symmetric quadrature formula with structure (K_1, K_2) is $K_1 + 2K_2$. Since there can only be one abscissa of type 1, the consistency conditions are

$$\begin{aligned} K_1 &\leq 1, \\ K_1 + 2K_2 &\geq \lfloor d/2 \rfloor + 1. \end{aligned} \tag{6.22}$$

The derivation of the consistency conditions for the dihedral group $\mathcal{D}_4$ is considerably more complicated (see, for instance, Mantel and Rabinowitz [294]).

In order to keep the number of integrand evaluations as low as possible, the invariant cubature formulas Q_n satisfying the moment equations (6.21) should have as few abscissas as possible. Solutions of (6.21) can only be expected to exist if the consistency conditions are satisfied. Hence, as a first step in cubature formula construction, one should try to find a structure $\mathbf{K}$ satisfying the consistency conditions such that the number of abscissas n (which is given by the sum of the components of $\mathbf{K}$ multiplied by the cardinalities of the corresponding $\mathcal{G}$-orbits) is as small as possible. Since $\mathbf{K}$ is an integer vector, this is an *integer programming* problem. Such a choice of $\mathbf{K}$ gives rise to a specific nonlinear system (6.21) of moment equations. If this system cannot be solved, this procedure has to be repeated using other, possibly nonoptimal, structures.

Example (The Construction of Minimum Point Formulas). The number of abscissas in a Z_2-invariant, univariate integration formula of type (K_1, K_2) is $n = K_1 + 2K_2$. Consider the construction of a symmetric quadrature formula of degree $D = 5$. It can easily be seen that the structure $(1, 1)$ satisfies the consistency conditions (6.22) and that the associated number $n = 3$ of abscissas is minimal. The resulting moment equations (6.21) are

$$\begin{aligned} c_1 + 2c_2 &= 2, \\ 2c_2 a^2 &= 2/3, \\ 2c_2 a^4 &= 2/5. \end{aligned}$$

By dividing the third equation by the second one, $a^2 = 3/5$ (i.e., $a = \sqrt{3/5}$) can be obtained. The remaining linear system for c_1 and c_2 yields $c_2 = 5/9$ and $c_1 = 8/9$. This is the Gauss formula G_3 for the unit interval $[-1, 1]$.

The construction of integration formulas for $N \geq 2$ using this approach requires the solution of considerably more complex nonlinear systems. The algebraic manipulations required to solve these nonlinear systems are rather lengthy and do not necessarily lead to a better understanding of this invariant theoretic approach to constructing multivariate integration formulas. This is why no such example is given in this section.

Concluding Remarks. The invariant theoretic approach to constructing cubature formulas has been used by a number of workers in this area of research (see, for instance, Espelid [133], Gatermann [156], Genz [165], Haegemans and Cools [197], Keast and Diaz [228], Maeztu [287], Rasputin [362], Soerevik and

Espelid [416]). However, the invariant theoretical approach is unsatisfactory in a number of ways.

To start with, the invariant theoretic approach does not remove the necessity of solving *nonlinear* systems of equations; it only simplifies these systems. For large values of N and/or D, the resulting nonlinear systems are still too complex to be solved in closed form. This is why some authors (see, for instance, Verlinden et al. [439] and Verlinden and Haegemans [440]) have used numerical methods for solving moment equations.

Then, there exist cubature formulas—so-called *fortuitous* formulas—which do *not* satisfy the consistency conditions, but *do* satisfy the moment equations. For instance, fortuitous rules for the square C_2 can be found in Cools and Schmid [65]. As a result, finding the minimum point rule satisfying the consistency conditions is not necessarily the same as finding the minimum point rule satisfying the moment equations.

Finally, there may be nonsymmetric cubature formulas satisfying the moment equations which have a smaller number of abscissas than the corresponding minimum point, symmetric cubature formula. For instance, invariant cubature formulas over the triangle T_2 of degree $D = 6$ have at least 12 abscissas, whereas nonsymmetric formulas which attain the same degree of precision by using only 10 abscissas are known (Möller [302]).

6.2.4. Construction Based on Ideal Theory. From the representation (6.13) of the moment equations, it can be seen that the moment equations can be solved if and only if the linear functional I_w over $\mathbb{P}_d^N$ is a linear combination of the linear functionals $\mathbf{x}_1^*, \ldots, \mathbf{x}_n^*$, i.e., if and only if $\mathrm{I}_w \in \operatorname{span}\{\mathbf{x}_1^*, \ldots, \mathbf{x}_n^*\}$. This is equivalent (see, for instance, Greub [182]) to

$$\{\mathrm{I}\}_w^\perp \supseteq (\operatorname{span}\{\mathbf{x}_1^*, \mathbf{x}_2^*, \ldots, \mathbf{x}_n^*\})^\perp,$$

where $U^\perp$ denotes the set of polynomials which are annihilated by all linear functionals in U,

$$U^\perp := \{\, P \in \mathbb{P}_d^N \;:\; u(P) = 0 \text{ for all } u \in U \,\}.$$

In other words, the moment equations (6.13) can be solved if and only if

$$\text{(6.23)} \qquad P \in \mathbb{P}_d^N,\ P(\mathbf{x}_i) = 0,\ i = 1, \ldots, n \quad \Rightarrow \quad \mathrm{I}_w(P) = 0.$$

Let $\mathcal{I}^N(\mathbf{x}_1, \ldots, \mathbf{x}_n)$ and $\mathcal{I}_d^N(\mathbf{x}_1, \ldots, \mathbf{x}_n)$ denote the set of all polynomials in $\mathbb{P}^N$ and $\mathbb{P}_d^N$ which vanish at the abscissas $\mathbf{x}_1, \ldots, \mathbf{x}_n$. Using this notation, condition (6.23) is equivalent to

$$\text{(6.24)} \qquad \mathrm{I}_w P = 0 \qquad \text{for all} \quad P \in \mathcal{I}_d^N(\mathbf{x}_1, \mathbf{x}_2, \ldots, \mathbf{x}_n).$$

If P_1 and P_2 vanish at the points $\mathbf{x}_1, \ldots, \mathbf{x}_n$, then so does any linear combination

$$\lambda_1 P_1 + \lambda_2 P_2, \quad \lambda_1, \lambda_2 \in \mathbb{R}.$$

Hence, $\mathcal{I}^N(\mathbf{x}_1, \ldots, \mathbf{x}_n)$ and $\mathcal{I}_d^N(\mathbf{x}_1, \ldots, \mathbf{x}_n)$ form vector spaces over the field of real numbers. They are subspaces of $\mathbb{P}^N$ and $\mathbb{P}_d^N$, respectively.

The vector space structure is not the only algebraic structure with which $\mathbb{P}^N$ can be endowed. $\mathbb{P}^N$ also forms a *ring* with respect to the addition and multiplication of polynomials. If P_1 and P_2 vanish at $\mathbf{x}_1, \ldots, \mathbf{x}_n$, then so does any combination $Q_1 P_1 + Q_2 P_2$, where $Q_1, Q_2 \in \mathbb{P}^N$ are arbitrary polynomials. Thus $\mathcal{I}^N(\mathbf{x}_1, \ldots, \mathbf{x}_n)$ is an *ideal* of the ring $\mathbb{P}^N$.

H-Bases and d-Orthogonality. The concepts of *H-bases* and of *d-orthogonality* form the basis for the application of ideal theoretic methods to cubature formula construction.

A finite subset $\{b_1, \ldots, b_L\}$ of $\mathcal{I}^N(\mathbf{x}_1, \ldots, \mathbf{x}_n)$ is said to be an *H-basis* if, for all polynomials $P \in \mathcal{I}^N(\mathbf{x}_1, \ldots, \mathbf{x}_n)$, there exist polynomials $Q_1, \ldots, Q_L \in \mathbb{P}^N$ such that

$$P = Q_1 b_1 + \cdots + Q_L b_L \quad \text{and} \quad \deg(Q_l b_l) \leq \deg(P), \quad l = 1, 2, \ldots, L.$$

In other words, leading (highest order) terms do not cancel each other when the $Q_1 b_1, \ldots, Q_L b_L$ are added up.

Example (H-Bases). Consider a univariate polynomial $P \in \mathbb{P}_d^1$ which vanishes at n distinct points $x_1, \ldots, x_n$. Then, P is divisible by the polynomial $\omega_n(x) = (x - x_1) \cdots (x - x_n)$:

$$Q := P/\omega_n \in \mathbb{P}_{d-n}.$$

Thus $\{\omega_n\}$ forms an H-basis for $\mathcal{I}^1(x_1, \ldots, x_n)$.

A polynomial P is said to be *d-orthogonal* if $\mathrm{I}_w(QP) = 0$ for all polynomials Q which satisfy $PQ \in \mathbb{P}_d^N$. In other words, P is *orthogonal* to all polynomials $Q \in \mathbb{P}_{d-\deg P}^N$ with respect to the inner product $\langle P, Q \rangle_w := \mathrm{I}_w(PQ)$.[57]

Example (d-Orthogonal Polynomials). Consider the H-basis $\{\omega_n\}$ for $\mathcal{I}^1(x_1, \ldots, x_n)$. Since ω_n is of degree $d = n$, the only possibility of obtaining a polynomial of the form QP of degree $d < n$ is setting $Q = 0$. Thus ω_n is $(n-1)$-orthogonal.

The relevance of d-orthogonal H-bases for the construction of cubature formulas is demonstrated by the following result.

THEOREM 6.2.6. *Let $\mathbf{x}_1, \ldots, \mathbf{x}_n$ denote distinct points in $\mathbb{R}^N$. Then, there exists a cubature formula Q_n with degree of accuracy D with the abscissas $\mathbf{x}_1, \ldots, \mathbf{x}_n$, if and only if, for each H-basis $\{b_1, \ldots, b_L\}$ for the ideal $\mathcal{I}^N(\mathbf{x}_1, \ldots, \mathbf{x}_n)$, the basis polynomials $b_1, \ldots, b_L$ are D-orthogonal.*

Proof. Let $\{b_1, \ldots, b_L\}$ denote an H-basis for $\mathcal{I}^N(\mathbf{x}_1, \ldots, \mathbf{x}_n)$. Assume that there is a cubature formula Q_n of degree D with the abscissas $\mathbf{x}_1, \ldots, \mathbf{x}_n$. For

[57] w is assumed to be an admissible weight function.

$Q \in \mathbb{P}^N_{D-\deg b_l}$, $\mathrm{Q}_n(Qb_l) = \mathrm{I}_w(Qb_l)$ holds because Q_n has the degree of accuracy D. Therefore,

$$\mathrm{I}_w(Qb_l) = \mathrm{Q}_n(Qb_l) = \sum_{i=1}^{n} c_i Q(\mathbf{x}_i) b_l(\mathbf{x}_i) = 0$$

since $b_l \in \mathcal{I}^N(\mathbf{x}_1, \dots, \mathbf{x}_n)$. Consequently, any b_l, $l = 1, \dots, L$, is D-orthogonal.

Now, assume that each b_l, $l = 1, \dots, L$, of an H-basis for $\mathcal{I}^N(\mathbf{x}_1, \dots, \mathbf{x}_n)$ is D-orthogonal. Let P denote an arbitrary polynomial in $\mathcal{I}^N_D(\mathbf{x}_1, \dots, \mathbf{x}_n)$. Since $\{b_1, \dots, b_L\}$ is an H-basis, there exist $Q_1, \dots, Q_L \in \mathbb{P}^N$ such that

$$P = Q_1 b_1 + \cdots + Q_L b_L \quad \text{and} \quad \deg(Q_l b_l) \le \deg(P), \quad l = 1, 2, \dots, L.$$

Hence,

$$\mathrm{I}_w P = \sum_{l=1}^{L} \mathrm{I}_w(Q_l b_l) = 0.$$

According to (6.24), there exists a cubature formula Q_n of degree D with abscissas $\mathbf{x}_1, \dots, \mathbf{x}_n$. □

According to the proof of Theorem 6.2.6, a cubature formula Q_n of degree D with abscissas $\mathbf{x}_1, \dots, \mathbf{x}_n$ exists if there is a *single* D-orthogonal H-basis $\{b_1, \dots, b_L\}$. Therefore, the construction of cubature formulas based on ideal theory consists of finding D-orthogonal H-bases $\{b_1, \dots, b_L\}$ of ideals $\mathcal{I}^N(\mathbf{x}_1, \dots, \mathbf{x}_n)$.

Example (D-Orthogonal H-Bases). Since $\{\omega_n\}$ forms an $(n-1)$-orthogonal H-basis for $\mathcal{I}(x_1, \dots, x_n)$, Theorem 6.2.6 substantiates the existence of a cubature formula of degree $D = n-1$ with abscissas $x_1, \dots, x_n$. If a nonnegative weight function w is included in the integral operator I, then the abscissas $x_1, \dots, x_n$ should be chosen as the n zeros of the orthogonal polynomial P_n with respect to the weight function w. For this choice of $x_1, \dots, x_n$, ω_n is $(2n-1)$-orthogonal (see Section 5.2.7). The associated cubature formula is a Gauss-type formula and has degree $D = 2n - 1$.

Concluding Remarks. The ideal theoretic approach to cubature formula construction was introduced by Möller [300]. Since then, it has proved to be an effective method for performing theoretical analyses of cubature formulas. It has also been useful in the practical construction of bivariate cubature formulas (see, for instance, Möller [301] and Schmid [383], [384]). However, practical results for higher dimensions $N \ge 3$ are yet to be obtained.

Algebraic ideal theory is not the only approach to investigating the relationship between common zeros of sets of polynomials and cubature formulas of polynomial degree of accuracy. An alternative approach, based on studying the properties of multivariate orthogonal polynomials, is, for instance, described in Xu [450].

6.3. Number-Theoretic Formulas

Number-theoretic integration formulas are based on *equidistributed* and *low-discrepancy* sequences of abscissas. These number-theoretic concepts are introduced in Section 6.3.1. Specific one- and multidimensional sequences suitable for numerical integration purposes are dealt with in Sections 6.3.2 and 6.3.3, respectively.

6.3.1. Formula Construction Using Equidistribution.

The number-theoretic notion of *equidistribution* is defined as follows.

DEFINITION 6.3.1 (EQUIDISTRIBUTED INFINITE SEQUENCE). *A sequence of vectors* $\mathbf{x}_1, \mathbf{x}_2, \ldots \in \mathbb{R}^N$ *is said to be equidistributed (or uniformly distributed) in the cube* $C_N^+ = [0,1]^N$ *if*

$$\lim_{n\to\infty} \frac{1}{n} \sum_{i=1}^{n} f(\mathbf{x}_i) = \int_{C_N^+} f(\mathbf{x})\, d\mathbf{x} \tag{6.25}$$

for all properly Riemann integrable functions $f : C_N^+ \to \mathbb{R}$.

Clearly, the characteristic function c_E of a subcube $E \subseteq C_N^+$ is Riemann integrable. Hence, for any equidistributed sequence $\mathbf{x}_1, \mathbf{x}_2, \mathbf{x}_3, \ldots$,

$$\lim_{n\to\infty} \frac{1}{n} \sum_{i=1}^{n} c_E(\mathbf{x}_i) = \lim_{n\to\infty} \frac{A(E;n)}{n} = \int_{C_N^+} c_E(\mathbf{x})\, d\mathbf{x} = \mathrm{vol}(E), \tag{6.26}$$

where

$$A(E;n) := \sum_{i=1}^{n} c_E(\mathbf{x}_i)$$

counts the number of points $\mathbf{x}_i \in E$. This shows that the fraction of points of an equidistributed sequence which lie in any subcube $E \subseteq C_N^+$ is asymptotically equal to the volume of E. It can be shown (see, for instance, Hlawka, Schoißengeier, and Taschner [208]) that (6.26) constitutes a sufficient condition for the sequence $\mathbf{x}_1, \mathbf{x}_2, \mathbf{x}_3, \ldots$ to be equidistributed.

THEOREM 6.3.1. *A sequence* $\mathbf{x}_1, \mathbf{x}_2, \mathbf{x}_3, \ldots \in C_N^+$ *is equidistributed if and only if* (6.26) *is satisfied for all subcubes* $E \subseteq C_N^+$.

Example (Equidistributed Sequences). Let $\{x\}$ denote the fractional part of a real number x, i.e., $\{x\} = x - \lfloor x \rfloor$. If $\theta \in \mathbb{R}$ is *irrational*, then the sequence

$$x_i := \{i\theta\}, \qquad i = 1, 2, 3, \ldots,$$

is equidistributed in $[0,1]$ (Hlawka, Schoißengeier, and Taschner [208]).

In order to obtain similar equidistributed sequences in C_N^+, $N > 1$, it is necessary to choose N irrational numbers $\theta_1, \ldots, \theta_N$ in such a way that $1, \theta_1, \ldots, \theta_N$ are linearly independent over the field of rational numbers. This means that if $\alpha_0, \alpha_1, \ldots, \alpha_N$ are rationals such that

$$\alpha_0 + \alpha_1\theta_1 + \cdots + \alpha_N\theta_N = 0,$$

then

$$\alpha_0 = \alpha_1 = \cdots = \alpha_N = 0.$$

In this case, the sequence $\mathbf{x}_1, \mathbf{x}_2, \ldots$, defined by

$$\mathbf{x}_i := (\{i\theta_1\}, \ldots, \{i\theta_N\}), \quad i = 1, 2, 3, \ldots,$$

is equidistributed in C_N^+ (Hlawka, Schoißengeier, and Taschner [208]).

For any equidistributed sequence $\mathbf{x}_1, \mathbf{x}_2, \ldots$ in C_N^+, a corresponding family of equal-weight integration formulas $\{Q_n\}$ can be defined by

$$Q_n f := \frac{1}{n} \sum_{i=1}^{n} f(\mathbf{x}_i), \quad n = 1, 2, 3, \ldots.$$

According to (6.25), the convergence

$$\lim_{n \to \infty} Q_n f = \int_{C_N^+} f(\mathbf{x})\, d\mathbf{x} \tag{6.27}$$

holds for all Riemann integrable functions f.

The definitions and convergence properties of integration formulas based on equidistributed sequences and integration formulas based on Monte Carlo methods (see Section 6.4) are very similar. This is why computational techniques based on equidistributed sequences are sometimes called *quasi-Monte Carlo methods.* The only difference between the two methods is the origin of the abscissa set $\{\mathbf{x}_i\}$. In Monte Carlo methods the abscissas are supposed to be samples of independent random variables, whereas in equidistributed sequences the abscissas are generated deterministically.

The convergence result (6.27) does not provide any information about the error $Q_n f - If$ for specific values of n. A bound on $Q_n f - If$ in terms of the *discrepancy* of the finite sequence $\mathbf{x}_1, \ldots, \mathbf{x}_n$ and the *variation* of the integrand f is given by the *Koksma–Hlawka inequality* as described in the remainder of this section.

Discrepancy of a Sequence. For an equidistributed sequence $\mathbf{x}_1, \mathbf{x}_2, \ldots$,

$$\frac{A(E; n)}{n} \approx \int_{C_N^+} c_E(\mathbf{x})\, d\mathbf{x} = \text{vol}(E)$$

holds for any measurable subset $E \subseteq C_N^+$, provided n is chosen sufficiently large. If the points of a finite sequence $\mathbf{x}_1, \ldots, \mathbf{x}_n$ are to be "uniformly" distributed in C_N^+, one would expect that $A(E; n)/n$ is a good approximation for vol(E). Hence, the deviation of $A(E; n)/n$ from vol(E) for different subsets $E \subseteq C_N^+$ indicates the deviation of the distribution of $\mathbf{x}_1, \ldots, \mathbf{x}_n$ from the uniform (equal) distribution. The following definition of the *discrepancy* of the points $\mathbf{x}_1, \ldots, \mathbf{x}_n$ gives a precise mathematical meaning to this informal approach to measuring the nonuniformity of the distribution of $\mathbf{x}_1, \ldots, \mathbf{x}_n$.

DEFINITION 6.3.2 (DISCREPANCY OF A FINITE SEQUENCE). *For a nonempty set $\mathcal{M}$ of measurable subsets of C_N^+, the discrepancy $D_n^{\mathcal{M}}$ of the finite sequence $\mathbf{x}_1, \ldots, \mathbf{x}_n \in C_N^+$ with respect to $\mathcal{M}$ is defined by*

$$D_n^{\mathcal{M}}(\mathbf{x}_1, \ldots, \mathbf{x}_n) := \sup_{E \in \mathcal{M}} \left| \frac{A(E;n)}{n} - \int_{C_N^+} c_E(\mathbf{x})\, d\mathbf{x} \right| .$$

Note that $\mathcal{M}_1 \subseteq \mathcal{M}_2$ implies $D_n^{\mathcal{M}_1}(\mathbf{x}_1, \ldots, \mathbf{x}_n) \le D_n^{\mathcal{M}_2}(\mathbf{x}_1, \ldots, \mathbf{x}_n)$.

In conjunction with number-theoretic integration rules, the following two choices of $\mathcal{M}$ are the most important:

$$\mathcal{M} = \mathcal{C} := \left\{ [a_1, b_1) \times [a_2, b_2) \times \cdots \times [a_N, b_N) \subseteq C_N^+ \right\}$$

and

$$\mathcal{M} = \mathcal{C}_0 := \left\{ [0, b_1) \times [0, b_2) \times \cdots \times [0, b_N) \subseteq C_N^+ \right\}.$$

The corresponding discrepancies are denoted $D_n := D_n^{\mathcal{C}}$ and $D_n^* := D_n^{\mathcal{C}_0}$, respectively. D_n^* is referred to as the *star discrepancy*. Since $\mathcal{C}_0 \subset \mathcal{C}$,

$$D_n^*(\mathbf{x}_1, \mathbf{x}_2, \ldots, \mathbf{x}_n) \le D_n(\mathbf{x}_1, \mathbf{x}_2, \ldots, \mathbf{x}_n).$$

Conversely, it can easily be shown that

$$D_n(\mathbf{x}_1, \mathbf{x}_2, \ldots, \mathbf{x}_n) \le 2^N D_n^*(\mathbf{x}_1, \mathbf{x}_2, \ldots, \mathbf{x}_n)$$

(Niederreiter [321]); i.e., D_n and D_n^* differ, independently of the particular finite sequence $\mathbf{x}_1, \ldots, \mathbf{x}_n$, by at most a dimension-dependent factor.

Theorem 6.3.1, along with the definition of D_n, implies that the equidistribution of an infinite sequence is connected with the discrepancies of its finite subsequences as follows.

THEOREM 6.3.2. *An infinite sequence $\mathbf{x}_1, \mathbf{x}_2, \ldots \in C_N^+$ is equidistributed if and only if*

$$\lim_{n \to \infty} D_n(\mathbf{x}_1, \mathbf{x}_2, \ldots, \mathbf{x}_n) = 0.$$

Variation of a Function. The *variation* of a univariate real function $f : [a, b] \to \mathbb{R}$ characterizes the regularity of f on the interval $[a, b]$ (Niederreiter [316]). For a partition $\mathcal{P}$ of the interval $[a, b]$ into n subintervals,

$$\mathcal{P} := \{x_i \; : \; a = x_0 < x_1 < \cdots < x_{n-1} < x_n = b\},$$

the sum

$$V(f; \mathcal{P}) := \sum_{i=1}^{n} |f(x_i) - f(x_{i-1})|$$

measures the *discrete* variation of f with respect to the specific partition $\mathcal{P}$. The *continuous* variation of f can be characterized by the supremum of all such discrete variations $V(f;\mathcal{P})$.

DEFINITION 6.3.3 (VARIATION OF A UNIVARIATE FUNCTION). *The variation of a univariate function* $f : [a,b] \to \mathbb{R}$ *is defined by*

$$V(f) := \sup_{\mathcal{P}} \{V(f;\mathcal{P})\} = \sup_{\mathcal{P}} \left\{ \sum_{i=1}^{n} |f(x_i) - f(x_{i-1})| \right\}.$$

If $V(f)$ *is finite,* f *is said to be of bounded variation on* $[a,b]$.

If f is continuously differentiable then the relationship

$$V(f) = \int_a^b |f'(x)|\, dx$$

follows from the mean value theorem of calculus.

Variation in the Sense of Vitali. The concept of the variation of a univariate function can be generalized to multivariate functions as follows: the term $f(x_i) - f(x_{i-1})$ in the definition of the discrete variation $V(f;\mathcal{P})$ of univariate functions measures the difference between the f-values at the two endpoints (the two vertices) of the subinterval $[x_{i-1}, x_i]$. The absolute value $|f(x_i) - f(x_{i-1})|$ is taken as a discrete measure for the variation of f on $[x_{i-1}, x_i]$. Analogously, the N-dimensional difference operator

$$\Delta(f;C) := \sum_{j_1=0}^{1} \cdots \sum_{j_N=0}^{1} (-1)^{\sum_{k=1}^{N} j_k} f(j_1 a_1 + (1-j_1)b_1, \ldots, j_N a_N + (1-j_N)b_N)$$

measures the 2^{N-1} differences between the f-values at any two vertices of the cube

$$C := [a_1, b_1] \times [a_2, b_2] \times \cdots \times [a_N, b_N]$$

which differ in only one of the N coordinates. The absolute value $|\Delta(f;C)|$ can be taken as a discrete measure for the variation of f on C.

Consequently, for a partition

$$\mathcal{P} := \left\{ [x_{l_1}^{(1)}, x_{l_1+1}^{(1)}] \times \cdots \times [x_{l_N}^{(N)}, x_{l_N+1}^{(N)}] \ : \ l_k = 0, \ldots, m_k - 1, \ k = 1, \ldots, N \right\}$$

of C_N^+, based on the N partitions

$$0 = x_0^{(k)} < x_1^{(k)} < \cdots < x_{m_k}^{(k)} = 1, \qquad k = 1, 2, \ldots, N,$$

of the interval $[0,1]$,

$$V^{(N)}(f;\mathcal{P}) := \sum_{C \in \mathcal{P}} |\Delta(f;C)|$$

measures the *discrete* variation of f with respect to $\mathcal{P}$. The *continuous* variation of f can be characterized by the supremum of all such discrete variations $V(f;\mathcal{P})$.

DEFINITION 6.3.4 (VARIATION IN THE SENSE OF VITALI). *The variation $V^{(N)}(f)$ of a multivariate function $f : C_N^+ \to \mathbb{R}$, in the sense of Vitali, is defined by*

$$V^{(N)}(f) := \sup_{\mathcal{P}} \sum_{C \in \mathcal{P}} |\Delta(f;C)|,$$

where the supremum is extended over all partitions $\mathcal{P}$ of C_N^+. If $V^{(N)}(f)$ is finite, f is said to be of bounded variation on C_N^+ in the sense of Vitali.

If the partial derivative

$$\frac{\partial^N f}{\partial x_1 \partial x_2 \cdots \partial x_N} \tag{6.28}$$

exists and is continuous on C_N^+, then the variation of f in the sense of Vitali can be expressed explicitly as

$$V^{(N)}(f) = \int_{C_N^+} \left| \frac{\partial^N f}{\partial x_1 \partial x_2 \cdots \partial x_N}(x_1, \ldots, x_N) \right| dx_1 \cdots dx_N \tag{6.29}$$

(Niederreiter [316]).

The variation in the sense of Vitali has a severe shortcoming. If f is constant in at least one of its N variables, then $\Delta(f;C) = 0$ for all subcubes $C \subseteq C_N^+$. Hence, $V^{(N)}(f) = 0$, regardless of the variation of f with respect to any other variable.

Variation in the Sense of Hardy and Krause. A more suitable notion of variation than that of Vitali can be obtained by taking into account the behavior of f when its domain is restricted to the *faces* of C_N^+: for $1 \le k \le N$ and $1 \le i_1 < \cdots < i_k \le N$, let $V^{(k)}(f; i_1, \ldots, i_k)$ denote the k-dimensional variation in the sense of Vitali of the restriction of f to the face

$$C_N^{i_1,\ldots,i_k} := \{(x_1, x_2, \ldots, x_N) \in C_N^+ \ : \ x_j = 0 \text{ for all } j \ne i_1, \ldots, i_k\}$$

of C_N^+.

DEFINITION 6.3.5 (VARIATION IN THE SENSE OF HARDY AND KRAUSE). *The variation $V(f)$ of a multivariate function $f : C_N^+ \to \mathbb{R}$ on C_N^+, in the sense of Hardy and Krause, is defined by*

$$V(f) := \sum_{k=1}^{N} \sum_{1 \le i_1 < \cdots < i_k \le N} V^{(k)}(f; i_1, i_2, \ldots, i_k).$$

If $V(f)$ is finite, i.e., if all variations $V^{(k)}(f; i_1, \ldots, i_k)$ in the sense of Vitali are finite, then f is said to be of bounded variation in the sense of Hardy and Krause.

It can easily be seen that a function f is of bounded variation in the sense of Hardy and Krause if and only if *all* its restrictions to arbitrary cross-sections along axes of C_N^+,

$$C_N^{i_1,\dots,i_k}(y_1,\dots,y_k) := \{(x_1,\dots,x_N) \in C \ : \ x_j = y_j \text{ for all } j \neq i_1,\dots,i_k\},$$

are of bounded variation in the sense of Vitali.

If a function $f : C_N^+ \to \mathbb{R}$ is constant in one of its N variables, say x_j, then f is *not* necessarily bounded in the sense of Hardy and Krause. This is due to the fact that the variable x_j is eliminated when f is restricted to a face $C_N^{i_1,\dots,i_k}$ with $j \neq i_1,\dots,i_k$. The corresponding variation $V^{(k)}(f; i_1,\dots,i_k)$ in the sense of Vitali is therefore not necessarily zero.

Koksma–Hlawka Inequality. The relevance of the concepts of discrepancy and variation to numerical integration is established by the Koksma–Hlawka inequality.

THEOREM 6.3.3 (KOKSMA–HLAWKA INEQUALITY). *For every integration formula*

$$\mathrm{Q}_n f := \frac{1}{n}\sum_{i=1}^{n} f(\mathbf{x}_i) \quad \textit{for} \quad \mathrm{I}f = \int_{C_N^+} f(\mathbf{x})\, d\mathbf{x},$$

the error bound

$$|\mathrm{Q}_n f - \mathrm{I}f| \leq V(f)\, D_n^*(\mathbf{x}_1, \mathbf{x}_2, \dots, \mathbf{x}_n) \tag{6.30}$$

holds, where $V(f)$ is the variation of $f : C_N^+ \to \mathbb{R}$ in the sense of Hardy and Krause and $D_n^*(\mathbf{x}_1,\dots \mathbf{x}_n)$ is the star discrepancy of the sequence $\mathbf{x}_1,\dots \mathbf{x}_n \in C_N^+$.

Proof. See Hlawka [207].

In (6.30) $V(f)$ is an indicator of the regularity of the integrand f, whereas the star discrepancy D_n^* characterizes the uniformity of the abscissa distribution.

The error bound (6.30) is sharp in the following sense (see, for instance, Niederreiter [321]). For any finite sequence $\mathbf{x}_1,\dots \mathbf{x}_n$ in C_N^+ and for any $\varepsilon > 0$, there is a function $f \in C^\infty(C_N^+)$ with $V(f) = 1$ such that

$$|\mathrm{Q}_n f - \mathrm{I}f| > D_n^*(\mathbf{x}_1, \mathbf{x}_2, \dots, \mathbf{x}_n) - \varepsilon.$$

The Koksma–Hlawka inequality can be generalized to integrals over arbitrary regions $B \subseteq C_N^+$ (see Niederreiter [321]).

Significance of Low-Discrepancy Sequences. According to the Koksma–Hlawka inequality (6.30), equal-weight integration formulas Q_n with small error bounds for integrands f of bounded variation can be obtained by choosing the abscissas $\mathbf{x}_1,\dots,\mathbf{x}_n$ in such a way that discrepancy of the abscissa set is small. Such *low-discrepancy sequences* are provided and discussed in the following sections.

In practice, it is not usually possible to determine the number of abscissas necessary to meet a given error requirement in advance. Therefore, it is important that the number of abscissas used can be increased in such a way that previously calculated function values are reused. A suitable family $\{Q_n\}$ of *embedded* integration formulas can be obtained by choosing the abscissas of the integration formulas Q_n as the first n elements of an infinite sequence $\mathbf{x}_1, \mathbf{x}_2, \ldots$. This is why, in addition to finite low-discrepancy sequences, *infinite* low-discrepancy sequences are studied in Sections 6.3.2 and 6.3.3.

6.3.2. Univariate Low-Discrepancy Sequences. One-dimensional, low-discrepancy sequences are considered in this section only to provide a basis for the discussion of multidimensional low-discrepancy sequences in Section 6.3.3. Univariate integration formulas based on low-discrepancy sequences are inefficient in comparison to formulas based on polynomial approximation and should therefore not be used in practice.

The discrepancies D_n^* and D_n of a finite sequence $x_1, \ldots, x_n$ in $[0,1]$ can be calculated easily using the following theorem.

THEOREM 6.3.4. *Let* $0 \le x_1 \le x_2 \le \cdots \le x_n \le 1$; *then*

$$D_n^*(x_1, \ldots, x_n) = \frac{1}{2n} + \max\left\{ \left| x_i - \frac{2i-1}{2n} \right| \ : \ i = 1, 2, \ldots, n \right\}$$

and

$$D_n(x_1, \ldots, x_n) = \frac{1}{n} + \max\left\{ \frac{i}{n} - x_i \ : \ i = 1, 2, \ldots, n \right\} - \min\left\{ \frac{i}{n} - x_i \ : \ i = 1, 2, \ldots, n \right\}.$$

Proof. See Niederreiter [321].

Theorem 6.3.4 immediately implies the following lower bounds for the discrepancies $D_n^*(x_1, \ldots, x_n)$ and $D_n(x_1, \ldots, x_n)$:

$$D_n^*(x_1, \ldots, x_n) \ge \frac{1}{2n} \quad \text{and} \quad D_n(x_1, \ldots, x_n) \ge \frac{1}{n}. \tag{6.31}$$

Both lower bounds are attained for the sequence

$$x_i = \frac{2i-1}{2n}, \qquad i = 1, 2, \ldots, n.$$

The corresponding optimal equal-weight formula is the compound one-point Gauss rule (i.e., the compound midpoint rule).

This result shows that, for any $n \in \mathbb{N}$, there exists a sequence $x_1^{(n)}, \ldots, x_n^{(n)}$ such that

$$D_n(x_1^{(n)}, \ldots, x_n^{(n)}) = O(n^{-1}). \tag{6.32}$$

Since the sequences $x_1^{(n)}, \ldots, x_n^{(n)}$ depend on n, this does *not*, however, imply the existence of a *single* infinite sequence $x_1, x_2, \ldots$ such that (6.32) holds. Indeed, it can be shown (Schmidt [385]) that there exists a constant $c > 0$ such that, for any infinite sequence $x_1, x_2, \ldots$,

$$D_n(x_1, \ldots, x_n) \geq c\,\frac{\log n}{n} \qquad \text{for infinitely many} \quad n \in \mathbb{N}. \tag{6.33}$$

Hence, the best possible, i.e., lowest, asymptotic order of magnitude for the discrepancy of one-dimensional infinite sequences is $O(n^{-1}\log n)$. While several kinds of sequences $x_1, x_2, \ldots$ which attain this lower bound are known, only *van der Corput sequences* are considered in this section because they enable the construction of multidimensional low-discrepancy formulas.

Van der Corput Sequences. Any integer $i \in \mathbb{N}_0$ has a unique *digit expansion*

$$i = \sum_{j=0}^{J(i)} d_j(i) b^j, \tag{6.34}$$

where the integer $b \geq 2$ is the *base* and the *digits* $d_j(i)$ satisfy

$$d_j(i) \in \{0, \ldots, b-1\}, \quad j = 0, 1, \ldots, J(i).$$

For a fixed base b, the expansion (6.34) can also be expressed by

$$i = d_{J(i)} \cdots d_2 d_1 d_0.$$

DEFINITION 6.3.6 (RADICAL-INVERSE FUNCTION). *The radical-inverse function* $\varphi_b : \mathbb{N}_0 \to [0, 1)$ *in base* b *is defined by*

$$\varphi_b(i) := \sum_{j=0}^{J(i)} d_j(i) b^{-j-1}, \quad i = 0, 1, 2, \ldots. \tag{6.35}$$

φ_b can be regarded as a reflection of the expansion (6.34) with respect to the decimal point:

$$\varphi_b : d_{J(i)} \cdots d_2 d_1 d_0. \mapsto 0.d_0 d_1 d_2 \cdots d_{J(i)}.$$

Radical-inverse functions can be used to define one-dimensional low-discrepancy sequences as follows.

DEFINITION 6.3.7 (VAN DER CORPUT SEQUENCE). *The van der Corput sequence in base* b *is the infinite sequence* $x_1, x_2, \ldots \in [0, 1]$ *defined by*

$$x_i := \varphi_b(i-1), \quad i = 1, 2, 3, \ldots. \tag{6.36}$$

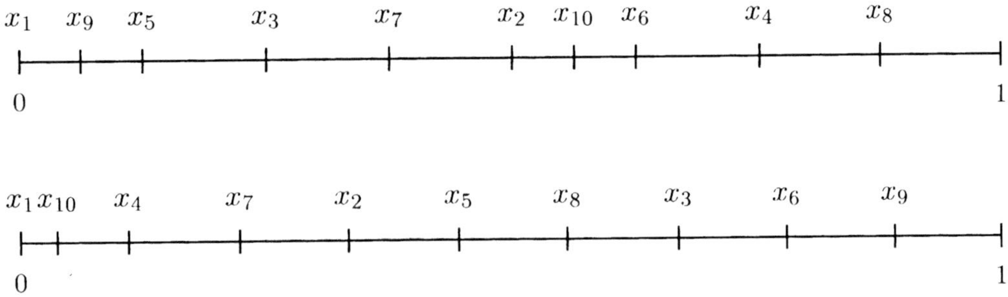

FIG. 6.2. *Van der Corput sequences for $b = 2$ and $b = 3$.*

Figure 6.2 shows the first 10 elements of the van der Corput sequences in bases $b = 2$ and $b = 3$.

It can be proved that the discrepancies of van der Corput sequences in bases b, denoted $D_n^*(b)$ and $D_n(b)$, attain the optimal asymptotic order of magnitude, i.e.,

$$D_n^*(b) = O(n^{-1} \log n) \quad \text{and} \quad D_n(b) = O(n^{-1} \log n).$$

More accurate characterizations of the asymptotic behavior of $D_n^*(b)$ and $D_n(b)$ can be obtained by taking into account the implied constants $c^*(b)$ and $c(b)$ defined by

$$c^*(b) := \limsup_{n\to\infty} \frac{nD_n^*(b)}{\log n} \qquad \text{and} \qquad c(b) := \limsup_{n\to\infty} \frac{nD_n(b)}{\log n}.$$

It can be shown that $c(b)$ equals $c^*(b)$ and that $c(b) = c^*(b)$ attains its minimum at $b = 3$ (see Faure [144]); i.e., the van der Corput sequence in base $b = 3$ has the smallest asymptotic discrepancy.

Slight reductions in the implied constants can be attained by considering *generalized van der Corput sequences*

$$x_i := \sum_{j=0}^{J(i-1)} \sigma(d_j(i-1))b^{-j-1}, \qquad i = 1, 2, 3, \ldots,$$

where σ is a permutation of the numbers $0, \ldots, b-1$.

Descriptions of efficient and numerically stable algorithms for calculating van der Corput sequences can be found in Fox [151] as well as Kolar and O'Shea [239].

Example (Van der Corput Sequences). The integral

$$\text{(6.37)} \qquad \mathrm{I}f = \int_0^1 |x - 1/3|^{1/2}\, dx = 2/3 \left[(2/3)^{3/2} + (1/3)^{3/2}\right] \approx 0.49$$

was approximately calculated using equal-weight quadrature formulas $\mathrm{Q}_n^{(b)}$ based on the first n elements of the van der Corput sequences in bases $b = 2$, $b = 3$, and

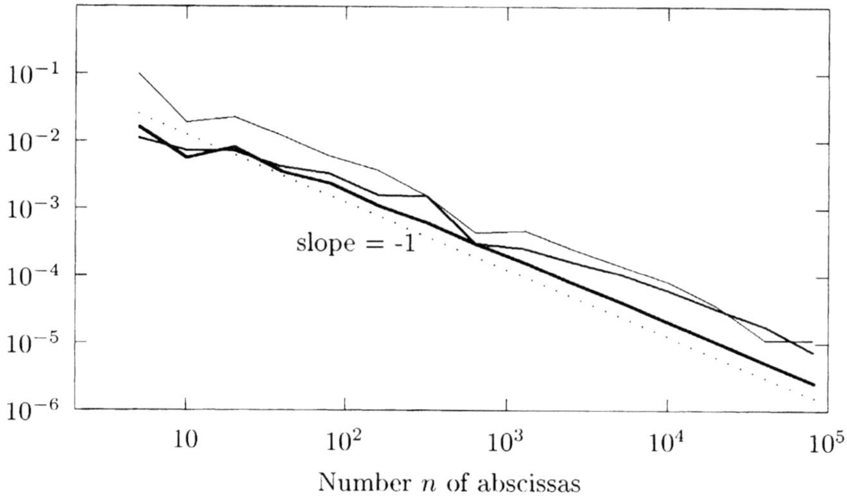

FIG. 6.3. *Absolute errors for quadrature formulas based on van der Corput sequences in bases $b=2$ (—), $b=3$ (—), and $b=5$ (—).*

$b = 5$. Figure 6.3 shows the absolute values of the resulting integration errors using $n \le 100{,}000$ abscissas.

Since Figure 6.3 uses a log-log scale, the order of convergence can be directly seen as the slope of the error curves. The results demonstrate that the rate of convergence of the error $|\mathrm{Q}_n^{(N)} f_N - \mathrm{I}_N f_N|$ for large values of n is almost linear.

The integral (6.37) was also computed using `QUADPACK/dqag`, a globally adaptive integration program based on Gauss–Kronrod formulas (see Section 5.2.9). Using only $n = 225$ function evaluations, `QUADPACK/dqag` approximated $\mathrm{I}f$ with an error $|\mathrm{Q}_n f - \mathrm{I}f| \approx 2.2 \cdot 10^{-6}$. This result is more accurate than all results obtained using the quadrature formulas $\mathrm{Q}_n^{(b)}$ for $n \le 100{,}000$ abscissas. This is why integration formulas based on equidistributed sequences should be used only for $N \ge 2$.

6.3.3. Multivariate Low-Discrepancy Sequences. Theorem 6.3.4 provides a simple method for calculating the discrepancy of any one-dimensional finite sequence $x_1, \dots, x_n$. Moreover, Theorem 6.3.4 implies the sharp lower bound (6.31) for the discrepancy of any one-dimensional finite sequence $x_1, \dots, x_n$; this lower bound is attained for the n-point compound midpoint rule.

By contrast, there is no simple method known for calculating the discrepancy of multidimensional finite sequence $\mathbf{x}_1, \dots, \mathbf{x}_n$. Furthermore, general sharp lower bounds for the discrepancy of multidimensional finite sequences and corresponding optimal finite sequences $\mathbf{x}_1, \dots, \mathbf{x}_n$ have not been established so far. Nevertheless, it is widely assumed that the lower bounds

$$D_n^*(\mathbf{x}_1, \mathbf{x}_2, \dots, \mathbf{x}_n) \ge c(N)\, \frac{(\log n)^{N-1}}{n} \tag{6.38}$$

for any finite N-dimensional sequence $\mathbf{x}_1, \dots, \mathbf{x}_n$ and

$$(6.39) \qquad D_n^*(\mathbf{x}_1, \mathbf{x}_2, \dots, \mathbf{x}_n) \geq c'(N)\, \frac{(\log n)^N}{n} \quad \text{for infinitely many } n,$$

hold for any infinite sequence $\mathbf{x}_1, \mathbf{x}_2, \dots$ (Niederreiter [321]). The constants $c(N)$ and $c'(N)$ depend only on N; i.e., they are independent of the particular sequence under consideration. For $N = 1$, (6.38) coincides with (6.31) and (6.39) with (6.33). For $N = 2$, (6.38) was proved by Schmidt [385], while for $N \geq 3$, (6.38) remains unproven. The best general result thus far established,

$$D_n^*(\mathbf{x}_1, \mathbf{x}_2, \dots, \mathbf{x}_n) \geq c(N)\, \frac{(\log n)^{(N-1)/2}}{n},$$

is due to Roth [378].

Assuming (6.38) and (6.39) hold, the optimal asymptotic order of magnitude for the discrepancy of finite sequences $\mathbf{x}_1, \dots, \mathbf{x}_n$ is

$$D_n^*(\mathbf{x}_1, \mathbf{x}_2, \dots, \mathbf{x}_n) = O\left(n^{-1}(\log n)^{N-1}\right),$$

whereas for infinite sequences $\mathbf{x}_1, \mathbf{x}_2, \dots$,

$$D_n^*(\mathbf{x}_1, \mathbf{x}_2, \dots, \mathbf{x}_n) = O\left(n^{-1}(\log n)^{N}\right)$$

is obtained. There are different ways of constructing sequences for which these presumed lower bounds are attained.

Halton Sequences. *Halton sequences* are direct multidimensional extensions of the univariate van der Corput sequences (6.36).

DEFINITION 6.3.8 (HALTON SEQUENCE). *The infinite Halton sequence $\mathbf{x}_1, \mathbf{x}_2, \dots \in C_N^+$ in the bases $b_1, \dots, b_N$ is defined by*

$$\mathbf{x}_i := (\varphi_{b_1}(i-1), \dots, \varphi_{b_N}(i-1))^\top, \quad i = 1, 2, 3, \dots,$$

where $\varphi_{b_1}, \dots, \varphi_{b_N}$ are the radical-inverse functions (6.35) *in bases $b_1, \dots, b_N$.*

THEOREM 6.3.5. *If the bases $b_1, \dots, b_N$ are pairwise relatively prime, then, for all $n \geq 2$, the discrepancy of the corresponding Halton sequence $\mathbf{x}_1, \mathbf{x}_2, \dots$ satisfies*

$$(6.40) \quad D_n^*(\mathbf{x}_1, \mathbf{x}_2, \dots, \mathbf{x}_n) \leq c(b_1, \dots, b_N)\frac{(\log n)^N}{n} + O(n^{-1}(\log n)^{N-1}),$$

where the coefficient $c(b_1, \dots, b_N)$ of the leading term is given by

$$(6.41) \qquad c(b_1, \dots, b_N) = \prod_{k=1}^{N} \frac{b_k - 1}{2 \log b_k}.$$

Proof. See Niederreiter [321].

Since the function

$$x \mapsto \frac{x-1}{2\log x}$$

increases monotonically for $x \geq 2$, the coefficient $c(b_1, \ldots, b_N)$ in (6.41) attains its minimum when the bases $b_1, \ldots, b_N$ are chosen as the first N prime numbers $p_1, \ldots, p_N$. However, even for this optimal choice of the bases, the leading coefficient increases *superexponentially* as $N \to \infty$ (Niederreiter [321]).

Example (Halton Sequences). The N-dimensional integral

$$\text{(6.42)} \qquad \begin{aligned} \mathrm{I}_N f_N &= c_N \int_{C_N^+} \prod_{k=1}^{N} |x_k - 1/3|^{1/2}\, dx_k \\ &= c_N (2/3)^N ((2/3)^{3/2} + (1/3)^{3/2})^N \\ &\approx c_N 0.49^N \end{aligned}$$

can be regarded as the N-dimensional generalization of the univariate integral (6.37). In the experiments, the parameter c_N was chosen in such a way that $\mathrm{I}_N f_N = 1$.

The integral (6.42) was computed for the dimensions $N = 4$, 8, and 12 using equal-weight cubature formulas $\mathrm{Q}_n^{(N)}$ based on the first n elements of the N-dimensional Halton sequences in the bases $p_1, \ldots, p_N$. Figure 6.4 shows the absolute value of the errors $\mathrm{Q}_n^N f_N - \mathrm{I}_N f_N$ for $n \leq 100{,}000$ abscissas.

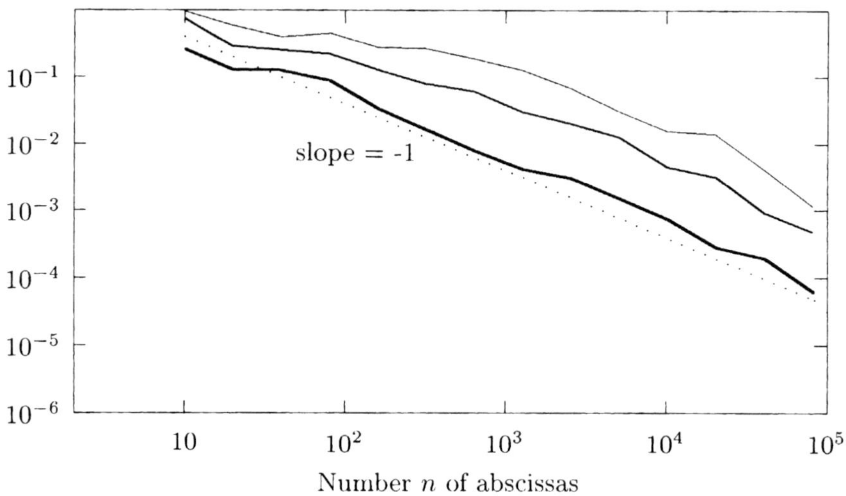

FIG. 6.4. *Absolute values of errors for integration formulas based on Halton sequences for dimensions* $N=4$ (—), $N=8$ (—), *and* $N=12$ (—).

The results demonstrate that the rate of convergence of the error $|\mathrm{Q}_n^{(N)} f_N - \mathrm{I}_N f_N|$ for large values of n is approximately linear.

Software (Halton Sequences). `TOMS/647` contains a subroutine which generates Halton sequences for dimensions $N \leq 40$. In addition, `TOMS/647` includes an implementation of a method by Faure [144].

Hammersley Sequences. In contrast to the infinite Halton sequences, *Hammersley sequences* are finite.

DEFINITION 6.3.9 (HAMMERSLEY SEQUENCE). *The n-point Hammersley sequence $\mathbf{x}_1, \ldots, \mathbf{x}_n$ in the bases $b_1, \ldots, b_{N-1} \in C_N^+$ is defined by*

$$\mathbf{x}_i := \left(\frac{i-1}{n}, \varphi_{b_1}(i-1), \ldots, \varphi_{b_{N-1}}(i-1) \right), \qquad i = 1, 2, \ldots, n.$$

Hammersley and Halton sequences differ only in the definition of the first component of the sequence vectors $\{\mathbf{x}_i\}$. It can be shown that the bound (6.40) for the discrepancy of Halton sequences also holds for Hammersley sequences when N is replaced by $N-1$:

$$D_n^*(\mathbf{x}_1, \mathbf{x}_2, \ldots, \mathbf{x}_n) \leq c(b_1, \ldots, b_{N-1}) \frac{(\log n)^{N-1}}{n} + O\left(n^{-1}(\log n)^{N-2}\right).$$

When the N $(N-1)$ bases in Halton (Hammersley) sequences are optimally chosen as the first N $(N-1)$ prime numbers, the bound on the discrepancy of the n-point Hammersley sequence is smaller than the bound on the discrepancy of the first n elements of the corresponding Halton sequence by a factor $(p_N - 1)/(2 \log p_N) \log n$. This is due to the relationship

$$\frac{c(p_1, \ldots, p_{N-1})}{c(p_1, \ldots, p_N)} = \frac{2 \log p_N}{p_N - 1},$$

which follows from (6.41).

Example (Hammersley Sequences). The integral $\mathrm{I}_N f_N$ in (6.42) was computed for the dimensions $N = 4$, 8, and 12 using equal-weight cubature formulas $\mathrm{Q}_n^{(N)}$ based on n-point Hammersley sequences in the bases $p_1, \ldots, p_{N-1}$. Figure 6.5 shows the absolute values of the errors $\mathrm{Q}_n^{(N)} f_N - \mathrm{I}_N f_N$ for $n \leq 100{,}000$ abscissas.

The errors in formulas based on Hammersley sequences can be seen to be generally smaller than the errors in the corresponding formulas based on Halton sequences (see Figs. 6.4 and 6.5).

Nets and (t, N)-Sequences. Hammersley and Halton sequences attain the presumed optimal orders of magnitude $O(n^{-1}(\log n)^{N-1})$ and $O(n^{-1}(\log n)^N)$ for the discrepancies of finite and infinite N-dimensional sequences, respectively. However, the implied constant $c(p_1, \ldots, p_N)$ in (6.40) increases superexponentially as $N \to \infty$. A method for constructing finite and infinite sequences which attain the presumed optimal orders of magnitude for the discrepancy and whose implied constants *decrease* superexponentially as $N \to \infty$

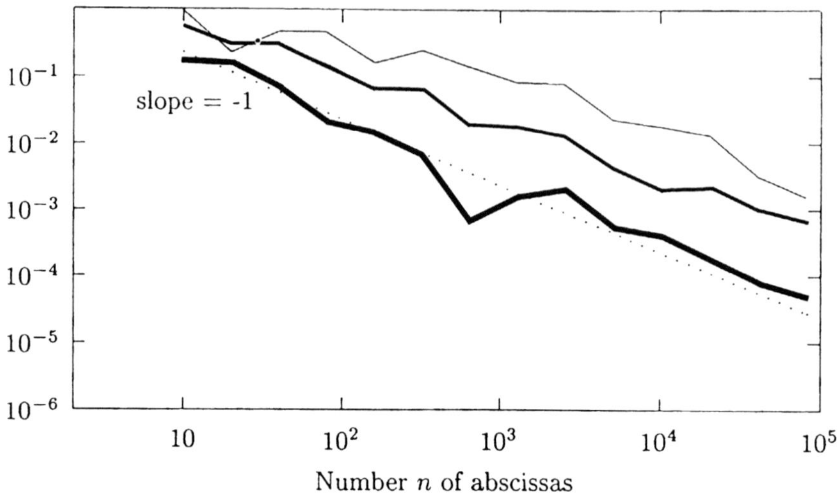

FIG. 6.5. *Absolute errors for integration formulas based on Hammersley sequences for dimensions $N = 4$ (—), $N = 8$ (—), and $N = 12$ (–).*

was developed by Sobol [413]. Niederreiter [317] generalized this approach by introducing the general concepts of *nets* and (t, N)*-sequences*. The best currently available finite and infinite low-discrepancy sequences are based on this method of construction.

The introduction of nets and (t, N)-sequences can be motivated as follows. Let $x_1, x_2, \ldots$ be a van der Corput sequence (6.36) in base $b \geq 2$. For arbitrary integers $k \geq 0$ and $m \geq 1$, consider the b^m integers $n = kb^m, \ldots, (k+1)b^m - 1$. For these n, the m least significant digits $d_{m-1}(n) \cdots d_0(n)$ in the digit expansion of n run through all possible values from $\{0, \ldots, b-1\}^m$, while the remaining digits remain constant. Hence, the reflections of these digit expansions with respect to the decimal point have the following property: the leading m digits assume all possible values from $\{0, \ldots, b-1\}^m$, while the remaining digits stay fixed. This means that each of the b^m intervals

(6.43) $$[jb^{-m}, (j+1)b^{-m}), \qquad j = 0, 1, \ldots, b^m - 1,$$

contains exactly one of the points

(6.44) $$x_n, \quad n = kb^m, kb^m + 1, \ldots, (k+1)b^m - 1,$$

of the van der Corput sequence in base b. Since all *elementary intervals* (6.43) have the same length b^{-m}, the finite sequence (6.44) is perfectly uniformly distributed with respect to these elementary intervals.

The following definitions make it possible to generalize these results to dimensions greater than one.

DEFINITION 6.3.10 (N-DIMENSIONAL ELEMENTARY INTERVAL). *A subcube $E \subseteq C_N^+$ is said to be an N-dimensional elementary interval if*

$$E = \prod_{k=1}^{N} [j_k b^{-e_k}, (j_k + 1)b^{-e_k}), \quad 0 \le j_k < b^{e_k} \quad \text{for} \quad k = 1, 2, \ldots, N,$$

where $e_1, \ldots, e_N$ are nonnegative integers and $b \ge 2$.

DEFINITION 6.3.11 ((t, m, N)-NET). *Let t and m be integers satisfying $0 \le t \le m$. A b^m-point sequence $\mathbf{x}_1, \ldots, \mathbf{x}_{b^m} \in C_N^+$ is said to be a (t, m, N)-net in base b if $A(E; b^m) = b^t$ for all elementary intervals E with* $\mathrm{vol}(E) = b^{t-m}$.

In other words, a (t, m, N)-net $\mathbf{x}_1, \ldots, \mathbf{x}_{b^m}$ is perfectly uniformly distributed with respect to the b^{m-t} elementary intervals E with volume $\mathrm{vol}(E) = b^{t-m}$.

DEFINITION 6.3.12 ((t, N)-SEQUENCE). *An infinite sequence $\mathbf{x}_1, \mathbf{x}_2, \ldots \in C_N^+$ is said to be a (t, N)-sequence in base b if, for all integers $k \ge 0$ and $m \ge t$, the finite sequence $\mathbf{x}_n$, $n = kb^m, \ldots, (k+1)b^{m+1} - 1$, is a (t, m, N)-net in base b.*

If a b^m-point sequence is a (t, m, N)-net, it is also a (u, m, N)-net for all $u = t, t+1, \ldots, m$: each elementary interval E with $\mathrm{vol}(E) = b^{u-m}$ can be represented as a union of b^{u-t} disjoint elementary intervals $E_1, \ldots, E_{b^{u-t}}$ with $\mathrm{vol}(E_j) = b^{t-m}$, $j = 1, \ldots, b^{u-t}$. Since $A(E_j; b^m) = b^t$ for $j = 1, \ldots, b^{u-t}$,

$$A(E; b^m) = \sum_{j=1}^{b^{u-t}} A(E_j; b^m) = b^{u-t} b^t = b^u.$$

Similarly, a (t, N)-sequence is also a (u, N)-sequence for all $t \le u \le m$. Thus smaller values of the parameter t generally result in a more uniform distribution.

Example ((t, m, N)-Nets and (t, N)-Sequences). For $N = 1$, the van der Corput sequence in base b is a (t, N)-sequence for all $t \ge 0$.

The following theorem shows that (t, m, N)-nets and (t, N)-sequences yield low-discrepancy sequences.

THEOREM 6.3.6. *For $m > 0$ the discrepancy D_n^* of a (t, m, N)-net $\mathbf{x}_1, \ldots, \mathbf{x}_n \in C_N^+$, $n = b^m$, in base b satisfies*

$$(6.45) \quad D_n^*(\mathbf{x}_1, \ldots, \mathbf{x}_n) \le c(N, b)\, b^t \,\frac{(\log n)^{N-1}}{n} + O(b^t n^{-1} (\log n)^{N-2}),$$

whenever $n \ge 2$. The constant $c(N, b)$ is given by

$$c(N, b) = \left(\frac{b-1}{2 \log b}\right)^{N-1}$$

if $N = 2$ or if $N = 3, 4$ and $b = 2$; otherwise

$$c(N, b) = \frac{1}{(N-1)!} \left(\frac{\lfloor b/2 \rfloor}{\log b}\right)^{N-1}.$$

The discrepancy D_n^* of the first n elements of a (t, N)-sequence,

$$\mathbf{x}_1, \mathbf{x}_2, \mathbf{x}_3, \ldots \in C_N^+,$$

in base b satisfies

$$(6.46) \quad D_n^*(\mathbf{x}_1, \mathbf{x}_2, \ldots, \mathbf{x}_n) \leq c'(N, b)\, b^t \, \frac{(\log n)^N}{n} + O(b^t n^{-1} (\log n)^{N-1})$$

for all $n \geq 2$. The constant $c'(N, b)$ is given by

$$c'(N, b) = \frac{1}{N} \left(\frac{b-1}{2 \log b} \right)^N$$

if $N = 2$ or if $N = 3, 4$ and $b = 2$; otherwise

$$c'(N, b) = \frac{1}{N!} \frac{b-1}{2\lfloor b/2 \rfloor} \left(\frac{\lfloor b/2 \rfloor}{\log b} \right)^N .$$

Proof. See Niederreiter [321].

Theorem 6.3.6 shows that (t, m, N)-nets and (t, N)-sequences indeed attain the presumed optimal orders of magnitude for the discrepancy of finite and infinite sequences, respectively. Additionally, for given b and N, the best error bounds in (6.45) and (6.46) are attained for the lowest value of t, $t = 0$. This confirms the previously made observation that lower values of t result in a more uniform distribution.

However, Theorem 6.3.6 does *not* imply the *existence* of (t, m, N)-nets or (t, N)-sequences. Niederreiter [321] suggests constructing (t, N)-sequences of the form

$$(6.47) \qquad \mathbf{x}_i := \left(\sum_{j=1}^{\infty} y_{ij}^{(1)} b^{-j}, \ldots, \sum_{j=1}^{\infty} y_{ij}^{(N)} b^{-j} \right), \quad i = 1, 2, 3, \ldots,$$

where the coefficients $y_{ij}^{(k)}$ are defined by

$$(6.48) \qquad \begin{aligned} y_{ij}^{(k)} &:= \sum_{r=0}^{J(i-1)} c_{jr}^{(k)} d_r(i-1) \bmod b, \\ k &= 1, 2, \ldots, N, \quad i, j = 1, 2, 3, \ldots .^{58} \end{aligned}$$

In (6.48) only a finite number of coefficients $c_{jr}^{(k)} \in \{0, \ldots, b-1\}$, $j = 1, 2, \ldots$, are supposed to be nonzero for each $k = 1, \ldots, N$ and $r = 0, 1, \ldots$.

Although this construction scheme looks fairly complicated at first sight, it can be easily explained. Remember that van der Corput sequences in base

[58] The entities $J(i-1)$ and $d_r(i-1)$ are defined in the digit expansion (6.34).

b are $(0, N)$-sequences in base b for $N = 1$. When extending van der Corput sequences to dimensions $N \geq 2$, it is possible to use van der Corput sequences for each coordinate; this is the approach taken in Halton sequences. However, in order to ensure that different coordinates are not *correlated*, relatively prime bases $b_1, \ldots, b_N$ have to be chosen for different coordinates. This eventually leads to the superexponential growth of the constant $c(b_1, \ldots, b_N)$ in (6.40).

In order to avoid this superexponential growth, N different, *uncorrelated* one-dimensional low-discrepancy sequences should be constructed from the digit expansion with respect to a *single* base b. This is achieved in (6.48) by subjecting the digit vectors $(d_0(i-1), \ldots, d_{J(i-1)}(i-1))^{\top}$ to N different linear transformations. The N transformed digit expansions are then used in (6.47) for constructing sequences in C_N^+ in the same way that the original digit expansion is used for constructing the van der Corput sequence.

Niederreiter [321] establishes conditions on the coefficients $c_{jr}^{(k)}$ in order for sequences (6.47) to be (t, N)-sequences. He also describes a method for constructing suitable coefficients $c_{jr}^{(k)}$. Due to space limitations, the construction of (t, m, N)-nets and (t, N)-sequences cannot be discussed in more detail (see Niederreiter [321]).

Example ((t, N)-Sequences). The integral $\mathrm{I}_N f_N$ in (6.42) was computed for $N = 4$, 8, and 12 using equal-weight cubature formulas $\mathrm{Q}_n^{(N)}$ based on the first n elements of the above $(0, N)$-sequences. The prime numbers $b = 5$, $b = 11$, and $b = 17$ were chosen for $N = 4$, $N = 8$, and $N = 12$, respectively. Figure 6.6 shows the absolute values of the errors $\mathrm{Q}_n^{(N)} f_N - \mathrm{I}_N f_N$ for $n \leq 100{,}000$ abscissas.

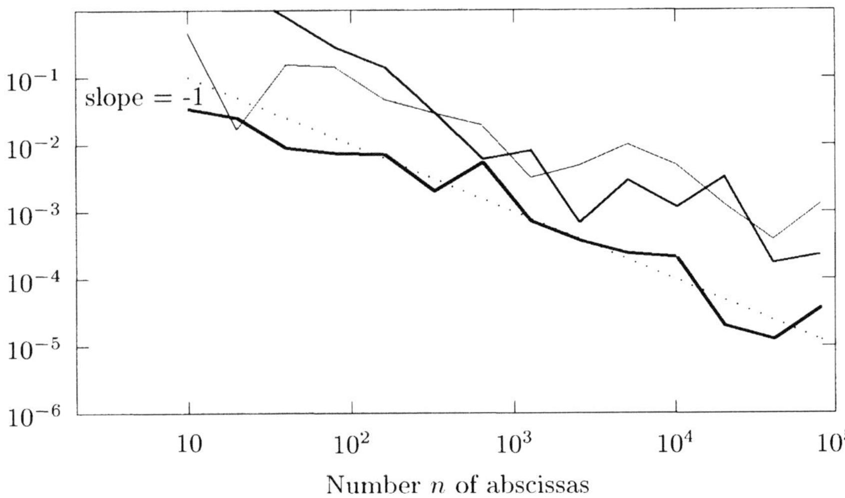

FIG. 6.6. *Absolute errors for integration formulas based on* $(0, N)$*-sequences for dimensions* $N = 4$ (—), $N = 8$ (—), *and* $N = 12$ (—).

The errors in formulas based on $(0, N)$-nets can be seen to be, at least for large values of n, smaller than the errors in the corresponding formulas based on Halton sequences (see Figure 6.4).

Software ((*t*, *N*)-Sequences). `TOMS/659` contains an implementation of Sobol's original method. An implementation of Niederreiter's generalization can be found in `TOMS/738`.

6.3.4. Concluding Remarks.

A major disadvantage of integration formulas based on general low-discrepancy sequences is that the bound on the discretization error (6.30), given in the Koksma–Hlawka inequality, depends only on the variation, but not on the analytical smoothness of the integrand. In particular, the order of convergence of $Q_n f \to If$ as $n \to \infty$ is *not* higher for smooth (for instance, analytic) functions than for discontinuous functions. However, there is a class of integration formulas based on certain low-discrepancy sequences which attain higher rates of convergence for smooth integrands. These formulas, which are known as *lattice rules*, are dealt with in Section 6.5.

6.4. Pseudorandom Formulas

The sampling points $\mathbf{x}_1, \ldots, \mathbf{x}_n$ in integration formulas are usually chosen *deterministically*. In *randomized formulas*, however, these points are derived from random quantities characterized by appropriate probability distributions. In numerical integration, special types of randomized algorithms are used: *Monte Carlo methods*.

6.4.1. Basic Monte Carlo Techniques.

The application of Monte Carlo methods in numerical computation is based on interpreting the deterministic quantity to be computed as a stochastic entity.[59] Random sampling for given probability distributions is used to estimate this unknown stochastic quantity.

In Monte Carlo methods for numerical integration, the integral If to be computed is interpreted as the following *expected (mean) value*:[60]

$$(6.49)\quad If = \int_B f(\mathbf{x})\,d\mathbf{x} = \mathrm{vol}(B)\int_{\mathbb{R}^N} \mathrm{vol}(B)^{-1} c_B(\mathbf{x}) f(\mathbf{x})\,d\mathbf{x} = \mathrm{vol}(B)\,\mu(f).$$

$\mu(f)$ is the mean value of f, interpreted as a random variable, with respect to the uniform distribution on B, i.e., the distribution with probability density $\mathrm{vol}(B)^{-1} c_B$ (where c_B denotes the characteristic function for the region B).

---

[59] In Monte Carlo *optimization* methods, for example, the optimal value to be computed is interpreted as the limit of an appropriate sequence of random variables (see, for instance, Mitchell [299]).

[60] For the sake of simplicity, it is initially assumed that the integral If involves the constant weight function $w \equiv 1$. The generalization of Monte Carlo methods to weighted integrals is discussed in conjunction with *importance sampling* techniques in Section 6.4.2.

A fundamental statistical technique is to estimate an expected value using a *sample mean*. The *Monte Carlo estimate* of the expected value $\mu(f)$ is obtained by taking n independent samples $\mathbf{x}_1, \ldots, \mathbf{x}_n \in \mathbb{R}^N$ with density $\mathrm{vol}(B)^{-1}c_B$ and making the estimate

$$\mu(f) \approx \mathrm{Q}_n f := \frac{1}{n}\sum_{i=1}^{n} f(\mathbf{x}_i). \tag{6.50}$$

Convergence. From a stochastic point of view, this procedure can be justified as follows. The samples $\mathbf{x}_1, \ldots, \mathbf{x}_n$ are regarded as realizations of a sequence $\mathbf{X}_1, \mathbf{X}_2, \ldots$ of independent, identically distributed random variables with the common density $\mathrm{vol}(B)^{-1}c_B$. In this interpretation, the formula $\mathrm{Q}_n f$ in (6.50) is a realization of the random variable

$$M_n f := \frac{1}{n}\sum_{i=1}^{n} f(\mathbf{X}_i).$$

According to the *strong law of large numbers*, the convergence

$$M_n f \quad \to \quad \mu(f) \qquad \text{as} \quad n \to \infty$$

is almost certain, i.e., happens with a probability of 1, which essentially guarantees that any sequence of realizations $\{\mathrm{Q}_n f\}$ of $M_n f$ converges to $\mu(f)$ as $n \to \infty$.

The strong law of large numbers does not provide any information about the error $\mathrm{Q}_n f - \mu(f)$ for a specific sample of size n. A *probabilistic estimate* of the error can be obtained by calculating the expected deviation of $M_n f$ from $\mu(f)$. The expected deviation can be characterized, for instance, by the quadratic mean of the error $M_n f - \mu(f)$, i.e., by the *standard deviation* $\sigma(M_n f)$ of the sample mean $M_n f$,

$$\sigma(M_n f) := \sqrt{\mu\left(M_n f - \mu(f)\right)^2}. \tag{6.51}$$

If f, interpreted as a random variable, has standard deviation $\sigma(f)$, then the standard deviation of the sample mean $M_n f$ is

$$\sigma(M_n f) = \frac{\sigma(f)}{\sqrt{n}}.$$

Therefore, the expected size of the error $|\mathrm{Q}_n f - \mu(f)|$ for a realization $\mathrm{Q}_n f$ of $M_n f$ is given by $\sigma(f)n^{-1/2}$, and for this reason $\mathrm{Q}_n f - \mu(f) = O(n^{-1/2})$ as $n \to \infty$.

This "$1/\sqrt{n}$ law" can also be obtained using the *central limit theorem*:

$$\lim_{n\to\infty} P\left(a \le \frac{M_n f - \mu(f)}{\sigma(f)n^{-1/2}} < b\right) = \frac{1}{\sqrt{2\pi}}\int_a^b \exp(-t^2/2)\, dt. \tag{6.52}$$

The central limit theorem states that the distribution of the *standardized* random variable

$$\frac{M_n f - \mu(f)}{\sigma(f) n^{-1/2}}$$

converges to the normal distribution with mean $\mu = 0$ and variance $\sigma^2 = 1$. From equation (6.52),

$$\lim_{n\to\infty} P\left(|M_n f - \mu(f)| \leq \alpha\sigma(f)n^{-1/2}\right) = \frac{1}{\sqrt{2\pi}} \int_{-\alpha}^{\alpha} \exp(-t^2/2)\, dt \tag{6.53}$$

can be derived for any $\alpha \geq 0$. By setting

$$\varepsilon_n := \alpha\sigma(f)n^{-1/2} \qquad \text{and} \qquad \delta := \frac{1}{\sqrt{2\pi}} \int_{-\alpha}^{\alpha} \exp(-t^2/2)\, dt,$$

(6.53) can be written as

$$\lim_{n\to\infty} P\left(|M_n f - \mu(f)| \leq \varepsilon_n\right) = \delta.$$

Consequently, the rate of convergence $\varepsilon_n = O(n^{-1/2})$ as $n \to \infty$ holds when the likelihood δ, for $|M_n f - \mu(f)| \leq \varepsilon_n$ to be true, is fixed.

The expected convergence rate of the error in integration formulas based on Monte Carlo methods, $\mathrm{Q}_n f - \mu(f) = O(n^{-1/2})$, does *not* depend on the dimension N. Hence, integration formulas based on Monte Carlo methods attain the same order of convergence for any dimension.

If the standard deviation $\sigma(f)$ were available, both (6.51) and (6.53) could be used to provide information about the integration error $\mathrm{Q}_n f - \mu(f)$. Since $\sigma(f)$ is generally not known, practical methods for estimating the integration error are based on calculating approximate values of $\sigma(f)$ (see Section 8.1.4).

Modified Monte Carlo Method for Difficult Regions. The computation of the integral $\mathrm{I}f$ according to (6.49) and (6.50) requires not only the calculation of $\mu(f)$, but also of $\mathrm{vol}(B)$. If $\mathrm{vol}(B)$ cannot be computed easily, the Monte Carlo method can be modified using the following transformation:

$$\begin{aligned} \mathrm{I}f &= \int_B f(\mathbf{x})\, d\mathbf{x} = \int_{B'} c_B(\mathbf{x}) f(\mathbf{x})\, d\mathbf{x} \\ &= \mathrm{vol}(B') \int_{\mathbb{R}^N} \mathrm{vol}(B')^{-1} c_{B'}(\mathbf{x})(c_B(\mathbf{x}) f(\mathbf{x}))\, d\mathbf{x} \\ &= \mathrm{vol}(B')\, \mu(c_B f). \end{aligned} \tag{6.54}$$

The superset B' of B is chosen in such a way that the volume $\mathrm{vol}(B')$ can be computed easily; the characteristic function c_B is interpreted as part of the modified integrand $c_B f$ rather than as part of the density function, as in the basic Monte Carlo method. The expected value of $c_B f$ has to be computed with respect to the density $\mathrm{vol}(B')^{-1} c_{B'}$.

Example (Difficult Integration Regions). A bounded integration region $B \subset \mathbb{R}^N$ is determined by its boundary ∂B. Suppose the $(N-1)$-dimensional manifold ∂B is given by the *implicit* representation

$$\partial B = \{ \mathbf{x} \in \mathbb{R}^N \ : \ G(\mathbf{x}) = 0 \},$$

and assume that $G : \mathbb{R}^N \to \mathbb{R}$ is chosen in such a way that the integration region B is characterized by

$$B = \{ \mathbf{x} \in \mathbb{R}^N \ : \ G(\mathbf{x}) \geq 0 \}. \tag{6.55}$$

The computation of

$$\mathrm{vol}(B) = \int_B 1\, d\mathbf{x}$$

for an integration region B defined by (6.55) is often an intricate task. Analytical or numerical integration methods are usually not applicable if the implicit representation of B cannot be transformed into an explicit representation. Hence, the basic Monte Carlo method (6.50) cannot be applied easily.

By contrast, the evaluation of the characteristic function c_B for integration regions B given by (6.55) is straightforward:

$$c_B(\mathbf{x}) = \begin{cases} 1 & \text{if } G(\mathbf{x}) \geq 0, \\ 0 & \text{else.} \end{cases}$$

Thus, the modified Monte Carlo method (6.54) is applicable, provided a region B', $B \subset B'$, with a simple boundary (for instance, an N-dimensional cube) can be found.

Issues Involved in Monte Carlo Methods. The practical application of Monte Carlo methods gives rise to the following issues.

(i) Estimates of the error based on (6.51) and/or (6.53) are probabilistic.[61] In a specific situation, there is no guarantee that the actual integration error is bounded by such error estimates. Practical techniques for estimating the error in Monte Carlo methods are discussed in Section 8.1.4.

(ii) The order of convergence $O(n^{-1/2})$ does not reflect the smoothness of the integrand. Smooth functions are *not* generally integrated more accurately or efficiently than nonsmooth functions. Modified Monte Carlo methods with an increased rate of convergence for smooth integrands are described in Section 6.4.3.

(iii) The abscissas $\mathbf{x}_1 \ldots, \mathbf{x}_n$ are supposed to be samples of independent random variables. The question arises as to how such "random sequences" can be generated. In practice, so-called *pseudorandom sequences* are used, i.e., deterministic sequences which, in some respects, *simulate* samples of random variables (see Section 6.4.4).

[61] For the sake of simplicity, it is assumed in the remainder of this section that the integration problem is scaled in such a way that $\mathrm{vol}(B) = 1$, which leads to $\mu(f) = If$.

Example (Monte Carlo Method). The integral $\mathrm{I}_N f_N$ in (6.42) was computed using pseudorandom formulas $\mathrm{Q}_n^{(N)}$. The abscissas $\mathbf{x}_1, \ldots \mathbf{x}_n$ of $\mathrm{Q}_n^{(N)}$ were generated using the intrinsic subroutine `RANDOM_NUMBER` in Fortran 90, which generates random numbers uniformly distributed in $[0, 1)$. In order to generate uniform random vectors in C_N^+, any N consecutive sequence elements were assembled in an N-dimensional vector. Figure 6.7 shows the absolute value of the errors $\mathrm{Q}_n^{(N)} f_N - \mathrm{I}_N f_N$ for $N = 12$ using $n \leq 100{,}000$ abscissas.

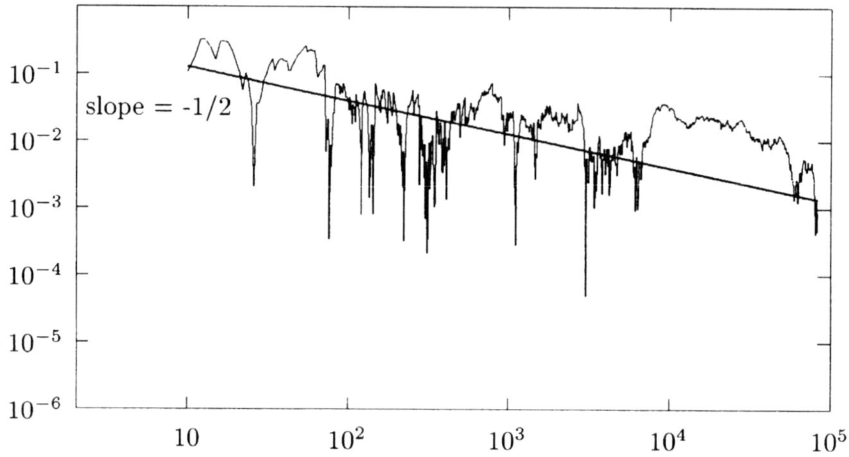

FIG. 6.7. *Absolute errors in 12-dimensional integration formulas based on the Monte Carlo method.*

The observed errors can be seen to follow, to some extent, a straight line with slope $-1/2$, the theoretical order of convergence. However, there are rather large random deviations in the actual errors from this theoretical convergence rate.

A comparison with the corresponding results for number-theoretic integration formulas based on Halton sequences (see Figure 6.4) and $(0, N)$-nets (see Figure 6.6) shows that the number-theoretic formulas are not superior to the pseudorandom formulas for the specific numbers of abscissas used in the experiments. However, since the number-theoretic formulas exhibit a higher rate of error reduction, they can be expected to outperform integration formulas based on the Monte Carlo method for larger values of n.

6.4.2. Variance Reduction Schemes. The expected size $\sigma(f)n^{-1/2}$ of the integration error in Monte Carlo methods suggests two alternative ways to reduce the error: (i) increase the number n of function evaluations and (ii) reduce the variance of the integrand f. The first method is rather inefficient: in order to increase the accuracy of integral approximations by a factor of 10, for instance, the number of function evaluations has to be increased by a factor of 100. This is why substantial efforts have been made to develop methods for reducing the variance. The most important of these techniques are outlined in this section.

Variance Reduction Using Approximation. Integration formulas are often obtained by approximating the integrand f using a suitable model function g and taking the exact value $\mathrm{I}g$ as an approximation of $\mathrm{I}f$ (see Section 6.1.2). In order for the integration error $\mathrm{I}g - \mathrm{I}f$ to be small, g must be a good approximation (with respect to the L_∞ norm) for f.

If the approximation quality of g is insufficient, then a more accurate integral approximation can be obtained by estimating the integration error $\mathrm{I}g - \mathrm{I}f = \mathrm{I}(g-f)$ using an integration formula Q_n: $\mathrm{I}(g-f) \approx \mathrm{Q}_n(f-g)$.[62] Clearly, the error of the modified approximation $\mathrm{I}f \approx \mathrm{I}g + \mathrm{Q}_n(f-g)$ is given by $\mathrm{Q}_n(f-g) - \mathrm{I}(f-g)$.

If the formula Q_n is based on the Monte Carlo method, the expected size of $|\mathrm{Q}_n(f-g) - \mathrm{I}(f-g)|$ becomes, according to (6.51), $\sigma(f-g)n^{-1/2}$. Hence, the error in Q_n is generally smaller for the modified integrand $f-g$ than for the original integrand f if $\sigma(f-g) \le \sigma(f)$. Because $\sigma(f-g) \le \|f-g\|_\infty$, this condition is satisfied whenever g is a sufficiently good approximation for f. In other words, the variance of the integrand subjected to Monte Carlo integration can be reduced by subtracting a suitable approximation function.

In order to compute $\mathrm{Q}_n(f-g)$, it is necessary to evaluate the original integrand f *and* the approximation function g at the abscissas $\mathbf{x}_1, \ldots, \mathbf{x}_n$ of Q_n. Hence, the computational complexity of the modified Monte Carlo method is higher than the complexity of the basic Monte Carlo method. Let t_f and t_{f-g} denote the computing time required for evaluating f and $f-g$ at a single abscissa. Variance reduction using approximation can be expected to be effective, i.e., to lead to a reduction in the overall computing time, only if

$$\frac{t_f}{t_{f-g}} > \frac{\sigma^2(f-g)}{\sigma^2(f)}. \tag{6.56}$$

In other words, variance reduction by a factor c is effective only if the corresponding integrand evaluation time is increased by a factor (significantly) smaller than c.

Importance Sampling. The basic idea behind *importance sampling* is to artificially introduce a weight function w,

$$\mathrm{I}f = \int_B f(\mathbf{x})\, d\mathbf{x} = \int_B w(\mathbf{x}) \frac{f(\mathbf{x})}{w(\mathbf{x})}\, d\mathbf{x}, \tag{6.57}$$

which satisfies $w(\mathbf{x}) > 0$ on B and is normalized according to $\mathrm{I}w = 1$. In other words, the weight function w is a positive probability density on B. Equation

[62] The integration formula Q_n is usually obtained using a construction principle other than approximation.

(6.57) can thence also be written as

$$\mathrm{I}f = \mathrm{I}_w\left(\frac{f}{w}\right) = \mu_w\left(\frac{f}{w}\right),$$

where μ_w denotes the expected value of its argument with respect to the probability density w.

The *Monte Carlo estimate* of the expected value $\mu_w(f/w)$ is obtained by taking n independent samples $\mathbf{x}_1, \ldots, \mathbf{x}_n$ with density w and estimating

$$\text{(6.58)} \qquad \mathrm{I}_w f = \mu_w(f/w) \approx \mathrm{Q}_n(f/w) := \frac{1}{n}\sum_{i=1}^{n}\frac{f(\mathbf{x}_i)}{w(\mathbf{x}_i)}.$$

Each integration formula $\mathrm{Q}_n(f/w)$ can be regarded as a realization of the random variable

$$M_n(f/w) := \frac{1}{n}\sum_{i=1}^{n} f(\mathbf{X}_i)/w(\mathbf{X}_i),$$

where $\mathbf{X}_1, \ldots, \mathbf{X}_n$ is a sequence of independent, identically where $\mathbf{X}_1, \ldots, \mathbf{X}_n$ is a sequence of independent, identically distributed random variables with common density w. As with the basic Monte Carlo method, the standard deviation $\sigma_w(M_n(f/w))$ of $M_n(f/w)$ with respect to the density w is a probabilistic estimate of $|\mathrm{Q}_n(f/w) - \mu_w(f/w)|$. Moreover, if f/w has a standard deviation of $\sigma_w(f/w)$ with respect to the density w, the standard deviation of the sample mean $M_n(f/w)$ is given by $\sigma_w(M_n(f/w)) = \sigma_w(f/w)n^{-1/2}$.

The modified Monte Carlo method outlined so far leads to a variance reduction if $\sigma_w^2(f/w) < \sigma^2(f)$. If f is positive[63] on B, the choice

$$w(\mathbf{x}) := \frac{f(\mathbf{x})}{\mathrm{I}f}, \qquad \mathbf{x} \in B,$$

results in the optimal variance $\sigma_w^2(f/w) = \sigma_w^2(\mathrm{I}f) = 0$. Unfortunately, this choice of w unrealistically presupposes that the integral value $\mathrm{I}f$ to be computed is known in advance. Nevertheless, these considerations suggest that $\sigma_w^2(f/w)$ can be expected to be small if w is chosen in such a way that the function f/w is *approximately constant*, i.e., does not exhibit large variations on B.

The abscissas $\mathbf{x}_1, \ldots, \mathbf{x}_n$ in (6.58) have to be sampled for the probability density w on B. The generation of such *nonuniform* random numbers takes more time than generating samples for the uniform density c_B. Furthermore, calculating $\mathrm{Q}_n(f/w)$ requires the evaluation of the integrand f *and* the density

[63] If f is of mixed sign and has a finite lower or upper bound, $f(\mathbf{x}) > C$ or $f(\mathbf{x}) < C$ for all $\mathbf{x} \in B$, the variance reduction scheme described can be applied to the function $f - C$ or $f + C$.

w at the abscissas $\mathbf{x}_1, \ldots, \mathbf{x}_n$ of $Q_n(f/w)$. Consequently, the computing time required to evaluate n-point integration formulas is larger for importance sampling Monte Carlo methods than for the basic Monte Carlo method; a condition similar to (6.56) must be satisfied in order for importance sampling Monte Carlo methods to be effective.

Constructing suitable weight functions for general integrands is in practice a difficult task. However, if a probability density is already included in the *original* integration problem, i.e., if the integrand f is of the form $\tilde{w}(x)\tilde{f}(x)$, where $\tilde{w}$ is a probability density, then w can easily be chosen as $w := \tilde{w}$. In addition to being straightforward, this choice has the advantage that evaluating the modified integrand $f/w = \tilde{f}$ takes *less* computing time than evaluating the original integrand $\tilde{w}\tilde{f}$. Integration problems containing probability densities as weight functions typically arise in computational statistics (see Section 2.1).

6.4.3. Monte Carlo Methods with Increased Convergence Rate. If an integration formula Q_n can be interpreted as a realization of a random variable $R_n f$ with expectation value If, the expected size of the integration error $|Q_n f - If|$ can be measured by $\sigma(R_n f)$. When $R_n f$ is chosen as the *sample mean* $M_n f$, the relationship $\sigma(M_n f) = \sigma(f) n^{-1/2}$, and therefore the $1/\sqrt{n}$ law, holds. For other possible definitions of $R_n f$, however, this relationship, and therefore the $1/\sqrt{n}$ law, does not generally apply. Indeed, for properly defined $R_n f$, the corresponding sequences of realizations $\{Q_n\}$ can be shown to converge *faster* than $O(n^{-1/2})$. The most important of these sequences are introduced in this section.

Stratified Sampling. In Haber's method of stratified sampling, the integration region $B = C_N^+$ is divided into k^N disjoint subcubes $C_1, \ldots, C_{k^N}$ of side k^{-1} (Haber [188], [189]). The equal-weight integration formula Q_n, $n = k^N$, is defined by selecting the abscissas $\mathbf{x}_1, \ldots, \mathbf{x}_n$ as samples of independent random variables $\mathbf{X}_1, \ldots, \mathbf{X}_n$ uniformly distributed on $C_1, \ldots, C_n$, respectively. If the integrand f is continuously differentiable on C_N^+, then the mean error magnitude is asymptotically given by

$$|Q_n f - If| \approx \frac{c}{n^{1/2+1/N}}, \qquad n \to \infty,$$

where the constant c depends on the first partial derivatives of f. Note that the improvement in the order of convergence by $1/N$ becomes less significant when the dimension is increased. The order of convergence $O(n^{1/2+1/N})$ is also attained when, instead of a single sample, a fixed number L of independent samples is taken in each subcube $C_1, \ldots, C_{k^N}$ for all $k = 1, 2, \ldots$.

Antithetic Variates. The order of convergence attained in stratified sampling can be improved by applying *antithetic variates* as follows. For any abscissa $\mathbf{x}_i \in C_i$, $i = 1, \ldots, n$, the point $\mathbf{x}_i^*$ is chosen such that the midpoint $(\mathbf{x}_i + \mathbf{x}_i^*)/2$ is the center of C_i. The equal-weight rule $\tilde{\mathrm{Q}}_{2n}$ is defined to use both the original abscissas $\mathbf{x}_1, \ldots, \mathbf{x}_n$ and the symmetric points $\mathbf{x}_1^*, \ldots, \mathbf{x}_n^*$. If the integrand f is twice continuously differentiable on C_N^+, then the error magnitude is given by

$$|\tilde{\mathrm{Q}}_{2n} f - \mathrm{I} f| \approx \frac{c^*}{n^{1/2+2/N}}, \qquad n \to \infty,$$

where the constant c^* depends on the second-order partial derivatives of f. The improvement in the order of convergence by $2/N$ is twice as large as in the original stratified sampling method.

Software (Stratified Sampling). The adaptive integration program `NAG/d01gbf` uses integration formulas based on stratified sampling.

6.4.4. Pseudorandom Sequences.

Monte Carlo methods for numerical integration require *pseudorandom sequences* $\mathbf{x}_1, \mathbf{x}_2, \ldots$ which simulate (certain aspects of) the behavior of samples of independent, identically distributed (according to a given distribution) random variables. Integration formulas whose abscissas are based on such pseudorandom sequences can be referred to as *pseudorandom formulas.*

The construction of pseudorandom sequences is commonly, though inaccurately, called *random number generation.* In this section, the fundamentals of random number generation are very briefly outlined. For thorough discussions of this subject, see, for instance, Dagpunar [83], Kalos and Whitlock [225], and Knuth [237]. A survey of random number generators on advanced computer systems can be found in Anderson [5].

A distinction is to be made between *uniform random numbers* and *nonuniform random numbers* (or *random variates*). Uniform random numbers are taken from the continuous uniform distribution on C_N^+, whereas nonuniform random numbers are taken from nonuniform distributions and/or distributions defined on regions B other than C_N^+.

Pseudorandom Sequences and Equidistribution. In order to qualify as a practically useful pseudorandom sequence, a deterministic sequence must pass certain statistical tests, for instance, tests for the uniformity of the samples, for the correlation between the samples, etc. For a description of a number of test methods, see Knuth [237].

Among the various tests for randomness, the test for uniformity of a sequence $x_1, \ldots, x_n$ in $[0, 1]$ based on the *law of the iterated logarithm* (see, for instance, Niederreiter [321]),

$$\limsup_{n\to\infty} \frac{\sqrt{2n}D_n^*(X_1, \ldots, X_n)}{\sqrt{\log\log n}} = 1 \quad \text{almost surely,} \tag{6.59}$$

is especially important. Equation (6.59) implies the asymptotic upper bound

$$O(\sqrt{n^{-1} \log \log n}) \qquad \text{as} \quad n \to \infty$$

on the discrepancy $D_n^*(x_1, \dots x_n)$ of the first n elements of a uniform sample sequence $x_1, x_2, \dots$. As a result, according to Theorem 6.3.2, uniform sample sequences $x_1, x_2, \dots$ can be expected to be equidistributed.

Furthermore, by substituting the bound $O(\sqrt{n^{-1} \log \log n})$, which is essentially equal to $O(n^{-1/2})$, for the discrepancy in the Koksma–Hlawka inequality (6.30), the $1/\sqrt{n}$ law of Monte Carlo methods can be recovered. The discrepancies of the number-theoretic sequences described in Section 6.3, such as the Halton sequences, essentially attain an asymptotic order of $O(n^{-1})$. Hence, (6.59) also shows that the discrepancy of pseudorandom sequences is asymptotically greater than the discrepancies of number-theoretic sequences by a factor of $O(n^{1/2})$. Pseudorandom formulas can therefore be expected to perform significantly worse, at least for large values of the number n of abscissas, than integration formulas based on low-discrepancy sequences. The example on page 206 partially confirms this hypothesis.

Linear Congruential Methods. Most uniform random number generators are based on linear congruential methods. Any such random number generator is characterized by the following three parameters: the *modulus* m is a large positive integer; the *multiplier* $a \in \{0, \dots, m-1\}$ is assumed to be relatively prime to m; whereas the *increment* c is an arbitrary integer, $c \in \{0, \dots, m-1\}$.

For any initial value $y_1 \in \{0, 1, \dots, m-1\}$, the sequence

$$y_1, y_2, y_3, \dots \in \{0, 1, \dots, m-1\}$$

can be defined recursively by

$$\text{(6.60)} \qquad y_{i+1} := a\, y_i + c \mod m, \qquad i = 1, 2, 3, \dots .$$

A sequence of *linear congruential pseudorandom numbers* $x_1, x_2, \dots \in [0, 1)$ is derived from $y_1, y_2, \dots$ by

$$x_i := y_i / m, \qquad i = 1, 2, 3, \dots .$$

Clearly, any sequence $y_1, y_2, \dots$ generated according to (6.60)—and consequently the corresponding sequence $x_1, x_2, \dots$—is periodic and the period T satisfies $T \le m$. This fact raises doubts concerning the suitability of linear congruential methods for random number generation. Indeed, it can be shown that, for the commonly used linear congruential methods,

$$D_T^*(x_1, \dots, x_T) = O(T^{-1})$$

(Niederreiter [321]). This contradicts the law of the iterated logarithm (6.59), which states that $D_T^*(x_1, \ldots, x_T)$ should essentially behave as $O(T^{-1/2})$. In this sense, the distribution of n linear congruential pseudorandom numbers is *too* uniform when n is close to the period T. On the other hand, for the first $n < T$ linear congruential pseudorandom numbers,

$$D_n^*(x_1, \ldots, x_n) = O(n^{-1}T^{1/2}(\log T)^2)$$

holds (Niederreiter [321]). Hence, for $n \ll T$, the discrepancy $D_n^*(x_1, \ldots, x_n)$ is too large when compared with (6.59). For an extensive discussion of the properties of linear congruential methods see, for instance, Knuth [237].

Linear congruential methods are widely used in practice despite their obvious deficiencies. The main advantage of linear congruential methods lies in their simplicity, which makes the generation of random numbers inexpensive.

Other Uniform Random Number Generators. A number of alternative methods exist for the generation of uniform random numbers. *Multiple-recursive methods* (see Niederreiter [315]) are straightforward multistep generalizations of linear congruential methods:

$$y_{i+k} := \sum_{l=0}^{k-1} a_l y_{i+l} \mod m, \qquad i = 1, 2, 3, \ldots,$$

whereas *first-order nonlinear congruential methods* are defined by

$$y_{i+1} := f(y_i) \mod m, \qquad i = 1, 2, 3, \ldots.$$

Properly chosen nonlinear congruential methods generate pseudorandom sequences whose discrepancies are in accordance with the law of the iterated logarithm (Eichenauer-Herrmann and Niederreiter [125]). Other interesting methods are *shift-register* methods (Tausworthe [426]) and *generalized feedback shift-register* methods (Lewis and Payne [262]). A thorough discussion of these alternative methods can be found in Niederreiter [321].

Multivariate Uniform Random Sequences. N-dimensional uniform random numbers can be obtained from one-dimensional uniform random numbers by gathering any N consecutive sequence elements into N-dimensional vectors. More precisely, if $x_1, x_2, \ldots$ is a one-dimensional pseudorandom sequence, then the associated N-dimensional pseudorandom sequence $\mathbf{x}_1, \mathbf{x}_2, \ldots$ is defined by

$$\mathbf{x}_i := (x_{(i-1)N+1}, x_{(i-1)N+2}, \ldots, x_{iN})^\top, \qquad i = 1, 2, 3, \ldots.$$

N-dimensional uniform random numbers can be generated directly (without referring to one-dimensional uniform random sequences) using the *matrix method* (see Grothe [185])

$$\mathbf{y}_{i+1} = A\mathbf{y}_i \mod m, \qquad A \in \mathbb{R}^{N\times N}, \quad \mathbf{y}_i \in \mathbb{R}^N, \quad i = 1, 2, 3, \ldots,$$

which is a straightforward generalization of the linear congruential method for N-dimensional vectors.

Software (Uniform Number Generator). There is a myriad of uniform random number generators available. They are not only contained in all major libraries, but they are also available as standard or intrinsic functions in Fortran 90, ANSI C, and numerous Fortran 77 extensions. Additionally, random number generators are part of the application programming interface of a number of operating systems, such as UNIX. The quality of the various uniform random number generators, however, varies significantly. Therefore, the use of software of proven quality, such as the programs found in the NAG and IMSL libraries, is recommended.

Generation of Nonuniform Random Numbers. Nonuniform random numbers distributed according to a probability measure P on a region $B \subseteq \mathbb{R}^N$ are usually generated by starting from a sequence of uniform random numbers on a cube C_N^+ and transforming this sequence to fit the distribution P. A thorough discussion of this subject can be found in Devroye [108]. For commonly used statistical distributions, special methods for generating random numbers have been developed. Such methods are available, for instance, in the NAG and IMSL libraries.

6.5. Lattice Rules

The error in compound trapezoidal rules T_n can be characterized by the Euler–Maclaurin sum formula (5.31). This error expansion shows that compound trapezoidal rules T_n produce particularly accurate results for $(b-a)$-periodic integrands; additionally, the order of convergence of $T_n f \to \mathrm{I} f$ as $n \to \infty$ crucially depends on the smoothness of the periodic integrand f. These properties also hold for the class of *lattice rules* described in this section. Lattice rules can therefore be regarded as multidimensional generalizations of compound trapezoidal rules.

6.5.1. Error Estimation Using Harmonic Analysis. Multidimensional integration rules which extend compound trapezoidal rules in the aforementioned way can easily be obtained by forming N-fold Cartesian products of trapezoidal rules (see Section 6.1.4). Unfortunately, the resulting product trapezoidal rules are impractical for high dimensions due to the rapid growth in the number of abscissas with increasing dimension. In order to construct efficient integration formulas for smooth, periodic integrands which use smaller numbers of abscissas, it is necessary to obtain a general representation of the error in integration formulas applied to periodic integrands. The assumed periodicity of the function f suggests that such a representation can be obtained using *harmonic analysis*, i.e., by expanding f in terms of its Fourier series.

One-Dimensional Integrands. Consider the expansion of a Riemann integrable, 1-periodic[64] function $f : \mathbb{R} \to \mathbb{R}$ in terms of its (exponential) Fourier series

$$\sum_{m\in\mathbb{Z}} a(m)\, e^{2\pi imx}, \qquad x \in \mathbb{R}. \tag{6.61}$$

The Fourier coefficients $a(m)$ are given by

$$a(m) = \int_0^1 f(x)\, e^{-2\pi imx}\, dx, \qquad m \in \mathbb{Z}.$$

It is assumed that the Fourier series (6.61) is absolutely convergent and converges to $f(x)$ for all $x \in \mathbb{R}$.

Let Q_n denote a quadrature formula for the interval $[0, 1]$. By replacing the integrand in $\mathrm{Q}_n f$ with its Fourier series (6.61) and changing the order of summation (which can be justified by the absolute convergence of the Fourier series), the formula

$$\mathrm{Q}_n f = \sum_{m\in\mathbb{Z}} a(m)\mathrm{Q}_n(e^{2\pi imx})$$

can be derived. Because

$$a(0) = \int_0^1 f(x)\, e^{-2\pi i 0\cdot x}\, dx = \int_0^1 f(x)\, dx = \mathrm{I}f,$$

the general error representation

$$\mathrm{Q}_n f - \mathrm{I}f = \sum_{m\in\mathbb{Z}\setminus\{0\}} a(m)\mathrm{Q}_n(e^{2\pi imx})$$

can be obtained.

Hence, a quadrature formula Q_n can be expected to be efficient for a given periodic integrand f if Q_n annihilates (i.e., integrates to zero) those *trigonometric polynomials* $e^{2\pi imx}$ for which the absolute value of the corresponding Fourier coefficient $a(m)$ of f is large. For smooth functions the most significant Fourier coefficients $a(m)$ are usually those with small absolute index values $|m|$ because the Fourier coefficients of smooth functions decay rapidly as $|m| \to \infty$. For instance, the asymptotic bound $a(m) = o(m^{-k})$ holds for $f \in C^{(k)}[0, 1]$ (Lyness [271]).

In order to be efficient for smooth, periodic integrands, a quadrature formula Q_n should annihilate as many low-frequency trigonometric polynomials $e^{\pm 2\pi ix}, e^{\pm 4\pi ix}, e^{\pm 6\pi ix}, \ldots$ as possible. For a given number of abscissas, the maximum number of consecutive low-frequency terms that can be annihilated by

[64]The following discussion can easily be generalized to general, $(b-a)$-periodic functions.

any n-point quadrature formula Q_n is $2(n-1)$ (Beckers and Cools [18]). The maximum possible number $2(n-1)$ of consecutive annihilated terms is obtained for the (displaced) trapezoidal rules

$$T_n(v)f = \frac{1}{n}\sum_{i=0}^{n-1} f\left(\left\{\frac{i}{n}+v\right\}\right), \qquad v \in \mathbb{R}.$$ [65]

Moreover, it can be shown (see Theorem 6.5.2) that these rules annihilate *all* trigonometric polynomials $e^{2\pi i m x}$ except for $m = \pm kn$, $k = 0, 1, 2, \ldots$. As a result, the error in the trapezoidal rules $T_n(v)$ is given by

$$T_n(v)f - \mathrm{I}f = \sum_{k\in\mathbb{Z}\setminus\{0\}} a(kn)T_n(v)(e^{2\pi i knx}).$$

Multidimensional Integrands. Let the function $f : \mathbb{R}^N \to \mathbb{R}$ be Riemann integrable and 1-periodic in each independent variable, i.e.,

$$f(\mathbf{x}) = f(\mathbf{x}+\mathbf{z}) \qquad \text{for all} \quad \mathbf{x} \in \mathbb{R}^N \text{ and } \mathbf{z} \in \mathbb{Z}^N,$$

and assume that f can be expanded into an absolutely convergent Fourier series,

(6.62) $$f(\mathbf{x}) = \sum_{\mathbf{m}\in\mathbb{Z}^N} a(\mathbf{m})\, e^{2\pi i \mathbf{m}\cdot\mathbf{x}}, \qquad \mathbf{x} \in \mathbb{R}^N,$$

with

(6.63) $$a(\mathbf{m}) = \int_{C_N^+} f(\mathbf{x})\, e^{-2\pi i \mathbf{m}\cdot\mathbf{x}}\, d\mathbf{x}, \quad \mathbf{m} \in \mathbb{Z}^N,$$

where $\mathbf{m}\cdot\mathbf{x}$ denotes the inner product

$$\mathbf{m}\cdot\mathbf{x} := m_1x_1 + m_2x_2 + \cdots + m_Nx_N.$$

Let Q_n denote a cubature formula for the N-dimensional unit cube C_N^+. Replacing the integrand in Q_nf with its Fourier series (6.62) and changing the order of summation (which again can be justified by the absolute convergence of the Fourier series) yields

$$Q_nf = \sum_{\mathbf{m}\in\mathbb{Z}^N} a(\mathbf{m})Q_n(e^{2\pi i \mathbf{m}\cdot\mathbf{x}}),$$

from which the error representation

(6.64) $$Q_nf - \mathrm{I}f = \sum_{\mathbf{m}\in\mathbb{Z}^N\setminus\{\mathbf{0}\}} a(\mathbf{m})Q_n(e^{2\pi i \mathbf{m}\cdot\mathbf{x}})$$

can be obtained.

When trying to construct cubature formulas Q_n for which the right-hand sum in (6.64) is small, one encounters the following two fundamental problems.

[65]The unusual form of the trapezoidal rule $T_n = T_n(0)$ is the result of the relationship $(f(0)+f(1))/(2n) = f(0)/n$, which follows from the assumed periodicity of f.

(i) A cubature formula can be expected to be efficient for a given periodic integrand f if it annihilates those *trigonometric polynomials* $e^{2\pi i\mathbf{m}\cdot\mathbf{x}}$ for which the absolute value of the corresponding Fourier coefficient $a(\mathbf{m})$ of f is large. For univariate smooth integrands, $|a(m_1)| < |a(m_2)|$ can be expected to hold if $|m_1| < |m_2|$ and $|m_1|$ and $|m_2|$ are sufficiently large. The relative size of the Fourier coefficients of multivariate functions, however, cannot be characterized so easily: the precise form of asymptotic bounds on $a(\mathbf{m})$ is determined by the specific set of existing partial derivatives of the function f. Different multivariate functions f give rise to different sets of existing partial derivatives, resulting in different asymptotic bounds on the corresponding Fourier coefficients $a(\mathbf{m})$.

As a result, there is no generally applicable ordering, $<$, of the index set $\mathbb{Z}^N$ such that, for all smooth integrands f, $|a(\mathbf{m}_1)| < |a(\mathbf{m}_2)|$ can be expected to hold if $\mathbf{m}_1 < \mathbf{m}_2$. Nevertheless, several reasonable ordering schemes characterizing the asymptotic behavior of Fourier coefficients for relevant classes of integrands exist.

(ii) The construction of multidimensional rules which annihilate as many important (according to a given ordering) functions $e^{2\pi i\mathbf{m}\cdot\mathbf{x}}$ as possible is significantly more difficult than the corresponding one-dimensional task.

6.5.2. Basic Lattice Rule Concepts. In univariate integration the set of integers $\{m\}$ whose corresponding trigonometric polynomials $e^{2\pi imx}$ are *not* annihilated by the optimal quadrature rule $T_n(v)$ is given by

$$\{m = kn \ : \ k = 0, \pm 1, \pm 2, \ldots\}. \tag{6.65}$$

This set forms a *lattice* in the following sense.

DEFINITION 6.5.1 (LATTICE). *A lattice L in $\mathbb{R}^N$ is a subset of $\mathbb{R}^N$ which satisfies the following conditions.*

1. *If $\mathbf{x}_1$ and $\mathbf{x}_2$ belong to L, then so do $\mathbf{x}_1 + \mathbf{x}_2$ and $\mathbf{x}_1 - \mathbf{x}_2$.*
2. *L contains N linearly independent elements.*
3. *$\mathbf{0}$ is an isolated point of L; i.e., there exists an open sphere whose intersection with L contains only $\mathbf{0}$.*[66]

It follows that every lattice forms a discrete additive subgroup of $\mathbb{R}^N$ and that lattices have no limit points: all lattice points are isolated.

The lattice property of the set (6.65) is induced by the fact that the abscissa set of $T_n(v)$ is obtained by shifting the lattice

$$L = \left\{ \frac{i}{n} \ : \ i \in \mathbb{Z} \right\}$$

[66] From the first two conditions, $\mathbf{0} \in L$ can be immediately derived.

and performing a *modulo* 1 operation on the components of the shifted lattice points. L can be seen to contain the integer lattice $\mathbb{Z}$. Such lattices are generally referred to as *integration lattices*.

DEFINITION 6.5.2 (INTEGRATION LATTICE). *An integration lattice L in $\mathbb{R}^N$ is a lattice that contains the integer lattice $\mathbb{Z}^N$ as a sublattice, i.e., $\mathbb{Z}^N \subseteq L$.*

The task of constructing efficient cubature formulas for periodic integrands is significantly facilitated by considering only formulas which satisfy the following condition: the set of integer vectors $\{\mathbf{m}\}$ whose corresponding trigonometric polynomials $e^{2\pi i \mathbf{m}\cdot\mathbf{x}}$ are *not* annihilated by Q_n is a lattice in $\mathbb{R}^N$. The abscissas of cubature formulas Q_n meeting this requirement form a lattice; see the following definition.

DEFINITION 6.5.3 (LATTICE RULE). *A lattice rule for the numerical integration of functions $f\colon [0,1)^N \to \mathbb{R}$ is an equal-weight integration formula*[67]

$$Q_n f := \frac{1}{n}\sum_{i=0}^{n-1} f(\mathbf{x}_i),$$

whose abscissas $\mathbf{x}_0, \ldots, \mathbf{x}_{n-1}$ are the points of an integration lattice L which are located in the cube $C_N^q := [0,1)^N$.

In other words, the abscissa set of a lattice rule is given by

$$A(L) := \{\mathbf{x}_0, \ldots, \mathbf{x}_{n-1}\} = L \cap C_N^q.$$

$A(L)$ is nonempty, because $\mathbf{0} \in L$, and finite, because L has no limit points.

Example (Product Trapezoidal Rule). The most obvious integration lattice is the cubic lattice

$$L = \left\{ \left(\frac{i_1}{k}, \ldots, \frac{i_N}{k}\right)^{\top} : i_j \in \mathbb{Z},\ 1 \le j \le N \right\} \tag{6.66}$$

with $k \in \mathbb{N}$. The corresponding lattice rule is the product trapezoidal rule

$$T_n f = \frac{1}{n}\sum_{i_1=0}^{k-1} \cdots \sum_{i_N=0}^{k-1} f\left(\frac{i_1}{k}, \ldots, \frac{i_N}{k}\right), \tag{6.67}$$

whose number of abscissas is given by $n := k^N$.

Example (The Method of Good Lattice Points). The *method of good lattice points* is based on integration formulas of the form

$$Q_n f = \frac{1}{n}\sum_{i=0}^{n-1} f\left(\left\{\frac{i}{n}\mathbf{p}\right\}\right), \tag{6.68}$$

---

[67]As the construction of lattice rules is related to the concept of remainder classes, it is convenient to let the summation index run from $i = 0$ to $i = n - 1$.

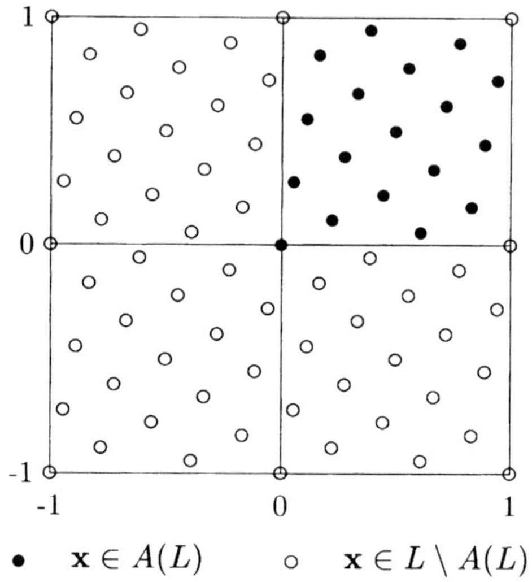

FIG. 6.8. *Integration lattice in* $\mathbb{R}^2$.

where $\mathbf{p} = (p_1, \ldots, p_N)^\top \in \mathbb{Z}^N$ is an integer vector. Braces $\{\cdot\}$ enclosing a vector indicate that each component of the vector is to be replaced by its fractional part (located in the interval $[0, 1)$). The greatest common divisor of $p_1, \ldots, p_N$ and n must be 1 in order to ensure that all n abscissas are distinct. (6.68) can easily be seen to be a lattice rule corresponding to the lattice

$$L = \left\{ \left\{ \frac{i}{n}\mathbf{p} \right\} + \mathbf{z} \;:\; i \in \mathbb{Z},\, \mathbf{z} \in \mathbb{Z}^N \right\}. \tag{6.69}$$

Figure 6.8 shows the integration lattice L in $\mathbb{R}^2$ proposed by Beckers and Cools [18] and the corresponding abscissa set $A(L)$. L is obtained by choosing the parameter vector $\mathbf{p} = (1, 5)^\top$ in (6.69).

The displaced trapezoidal rules $T_n(v)$ are as well suited to the integration of periodic functions as the basic trapezoidal rules T_n. This fact is the main motivation for defining *displaced* lattice rules as follows.

DEFINITION 6.5.4 (DISPLACED LATTICE RULE). *The displaced lattice rule for a given integration lattice L and a vector* $\mathbf{v} \in \mathbb{R}^N$ *is defined by*

$$\mathrm{Q}_n(\mathbf{v})f := \frac{1}{n}\sum_{i=0}^{n-1} f\left(\{\mathbf{x}_i + \mathbf{v}\}\right),$$

where $\{\mathbf{x}_0, \ldots, \mathbf{x}_{n-1}\} = A(L)$.

Clearly, the choice $\mathbf{v} \in L$ leads to a conventional lattice rule, $\mathrm{Q}_n(\mathbf{v}) = \mathrm{Q}_n(\mathbf{0})$. The displacement $\mathbf{v}$ can, for instance, be chosen in such a way that all cubature abscissas are located in the interior of C_N^q.

6.5.3. Practical Representation of Lattice Rules. A useful representation of the abscissas in lattice rules can be obtained by generalizing the representations (6.66) for product trapezoidal rules and (6.69) for the method of good lattice points as follows: for any combination of k integer vectors $\mathbf{z}_1, \ldots, \mathbf{z}_k \in \mathbb{Z}^N$ and k numbers $n_1, \ldots, n_k \in \mathbb{N}$, the set

$$L := \left\{ \frac{i_1}{n_1}\mathbf{z}_1 + \cdots + \frac{i_k}{n_k}\mathbf{z}_k + \mathbf{z} \ : \quad i_j = 0, 1, \ldots, n_j - 1, \quad j = 1, 2, \ldots, k, \ \mathbf{z} \in \mathbb{Z}^N \right\} \tag{6.70}$$

forms an integration lattice. The abscissa set of the corresponding lattice rule is given by

$$A(L) = \left\{ \left\{ \frac{i_1}{n_1}\mathbf{z}_1 + \cdots + \frac{i_k}{n_k}\mathbf{z}_k \right\} \ : \ i_j = 0, \ldots, n_j - 1, \ j = 1, \ldots, k \right\}. \tag{6.71}$$

This representation of the abscissa set $A(L)$ of a lattice rule is called the *k-cycle form* of L (Lyness and Sloan [281]). In numerical integration programs, lattice rules can be compactly represented simply by storing the k integer vectors $\mathbf{z}_1, \ldots, \mathbf{z}_k$ and the k numbers $n_1, \ldots, n_k$. The abscissas can easily be computed from this data by forming all of the linear combinations in (6.71).

However, the following problems prevent the immediate application of the k-cycle representation in numerical integration programs. (i) It is not clear whether *all* integration lattices can be written in the form (6.70), and (ii) the form (6.71) of lattice rules is possibly *repetitive*; i.e., abscissas may occur several times. The following theorem shows that these problems can be overcome.

THEOREM 6.5.1. *Any N-dimensional lattice rule Q_n can be expressed in nonrepetitive form as*

$$\mathrm{Q}_n f = \frac{1}{n_1 \cdots n_m} \sum_{i_1=0}^{n_1-1} \cdots \sum_{i_m=0}^{n_m-1} f\left(\frac{i_1}{n_1}\mathbf{z}_1 + \cdots + \frac{i_m}{n_m}\mathbf{z}_m \right),$$

where m, the rank *of Q_n, and $n_1, \ldots, n_m$, the* invariants *of Q_n, are uniquely determined natural numbers satisfying $1 \le m \le N$ and*

$$n_{j+1} \mid n_j, \quad j = 1, 2, \ldots, m-1; \quad n_m > 1.$$

The vectors $\mathbf{z}_1, \ldots, \mathbf{z}_m$ are linearly independent.

Proof. See Sloan and Joe [402].

The practical relevance of this theorem is that it shows that all lattice rules can be represented in a nonrepetitive k-cycle form. In addition, it provides an upper bound for the amount of storage required: for a rank m lattice rule, $m(N+1)$ integers must be stored (in order to store the m vectors $\mathbf{z}_1, \ldots, \mathbf{z}_m \in \mathbb{Z}^N$ and the m invariants $n_1, \ldots, n_m \in \mathbb{N}$).

Example (Rank and Invariants of Lattice Rules). Product trapezoidal rules (6.67) have the maximum possible rank, $m = N$, and their invariants are $n_1 = n_2 = \cdots = n_N = k = n^{1/N}$. Integration rules originating from the method of good lattice points have the minimum possible rank, $m = 1$, and their single invariant is $n_1 = n$.

6.5.4. Error Analysis of Lattice Rules. The usefulness of a class of integration formulas crucially depends on the behavior of the discretization error $\mathrm{Q}_n f - \mathrm{I}f$ for relevant classes of integrands $\mathcal{F}$. In this section, fundamental theorems characterizing the discretization error in lattice rules for various kinds of integrands are discussed. Based on these error estimates, various indicators of the quality of lattice rules can be derived.

For periodic integrands, error bounds can be established which show that the rate of convergence of $\mathrm{Q}_n f \to \mathrm{I}f$ as $n \to \infty$ depends on the smoothness of f. More precisely, the asymptotic bound

$$\mathrm{Q}_n f - \mathrm{I}f = O(\rho_n^{-\alpha}) \quad \text{as} \quad n \to \infty$$

(see Theorem 6.5.4), where ρ_n is a quality index of the lattice rule and α reflects the smoothness of f, can be derived. For a given sequence $\{\mathrm{Q}_n\}$ of lattice rules, larger values of α, i.e., smoother integrands f, result in faster rates of convergence of $\mathrm{Q}_n f \to \mathrm{I}f$. In contrast, for *nonperiodic* integrands, the corresponding asymptotic error bound $\mathrm{Q}_n f - \mathrm{I}f = O(\rho_n^{-1})$ (see Theorem 6.5.5) is independent of the smoothness of f. In this case, larger values of α, i.e., smoother integrands, do not generally result in faster convergence.

Periodic Integrands. Let the function $f : \mathbb{R}^N \to \mathbb{R}$ be continuous and 1-periodic in each independent variable, and assume that f can be expanded into an absolutely convergent Fourier series. Then, the representation (6.64) of the integration error holds and, for lattice rules Q_n, can be written as

$$\mathrm{Q}_n f - \mathrm{I}f = \sum_{\mathbf{m} \in \mathbb{Z}^N \setminus \{\mathbf{0}\}} a(\mathbf{m}) \frac{1}{n} \sum_{i=0}^{n-1} e^{2\pi \mathrm{i} \mathbf{m} \cdot \mathbf{x}_i}, \tag{6.72}$$

where $a(\mathbf{m})$ are the Fourier coefficients of f.

While the error representation (6.72) is true for any equal-weight rule Q_n, the relationship

$$\frac{1}{n} \sum_{i=0}^{n-1} e^{2\pi \mathrm{i} \mathbf{m} \cdot \mathbf{x}_i} = \begin{cases} 1 & \text{if } \mathbf{m} \cdot \mathbf{x}_i \in \mathbb{Z} \quad \text{for all} \quad i = 0, 1, \ldots, n-1, \\ 0 & \text{otherwise}, \end{cases} \tag{6.73}$$

holds only if the abscissa set of Q_n is derived from an integration lattice, $\{\mathbf{x}_1, \ldots, \mathbf{x}_n\} = A(L)$ (Sloan and Kachoyan [403]). The set

$$L^\perp := \left\{ \mathbf{m} \in \mathbb{R}^N : \mathbf{m} \cdot \mathbf{x} \in \mathbb{Z} \text{ for all } \mathbf{x} \in L \right\}$$

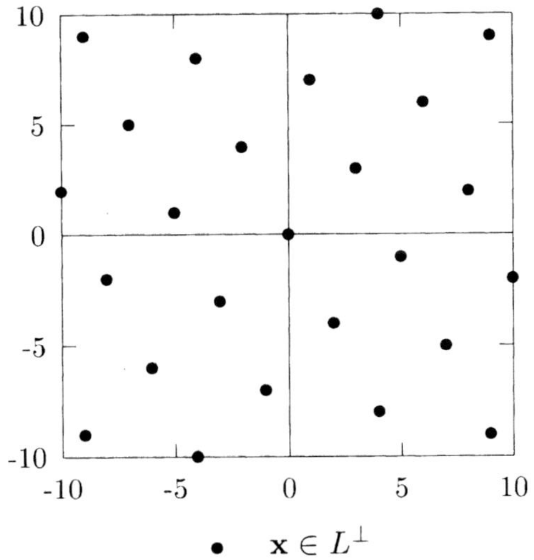

FIG. 6.9. *Dual lattice* $L^{\perp}$.

can easily be seen to be a lattice contained in $\mathbb{Z}^N$.

DEFINITION 6.5.5 (DUAL LATTICE). *$L^{\perp}$ is called the dual lattice of L.*

Example (Dual Lattice). The dual lattice $L^{\perp}$ of the integration lattice L in Figure 6.8 is given by

$$L^{\perp} = \{a_1\mathbf{z}_1 + a_2\mathbf{z}_2 \ : \ a_1, a_2 \in \mathbb{Z}\},$$

where

$$\mathbf{z}_1 = (18, 0)^{\top} \quad \text{and} \quad \mathbf{z}_2 = (-5, 1)^{\top}.$$

$L^{\perp}$ is shown in Figure 6.9.

The results obtained so far lead to the following fundamental theorem given by Sloan and Kachoyan [403].

THEOREM 6.5.2. *Let L be an integration lattice. Then, the error in the corresponding lattice rule Q_n is given by*

$$\mathrm{Q}_n f - \mathrm{I}f = \sum_{\mathbf{m} \in L^{\perp} \setminus \{\mathbf{0}\}} a(\mathbf{m}), \tag{6.74}$$

and the errors in the displaced lattice rules $\mathrm{Q}_n(\mathbf{v})$ are given by

$$\mathrm{Q}_n(\mathbf{v}) f - \mathrm{I}f = \sum_{\mathbf{m} \in L^{\perp} \setminus \{\mathbf{0}\}} e^{2\pi i \mathbf{m}\cdot\mathbf{v}} a(\mathbf{m}), \tag{6.75}$$

where $L^{\perp}$ is the dual lattice of L.

Proof. Equality (6.74) can be immediately obtained by combining (6.72) and (6.73). Equality (6.75) follows from the identities $\mathrm{Q}_n(\mathbf{v})f = \mathrm{Q}_n\tilde{f}$ and $\mathrm{I}f = \mathrm{I}\tilde{f}$, where $\tilde{f}(\mathbf{x}) := f(\mathbf{x}+\mathbf{v})$, because, from (6.63), the Fourier coefficients $\tilde{a}(\mathbf{m})$ of $\tilde{f}$ are given by

$$\tilde{a}(\mathbf{m}) = e^{2\pi i \mathbf{m}\cdot\mathbf{v}} a(\mathbf{m}), \qquad \mathbf{m} \in \mathbb{Z}^N. \qquad \square$$

Periodizing Multivariate Integrands. Integrands occurring in practice can generally not be expected to be periodic.[68] Such nonperiodic integrands must be subjected to periodizing transformations in order to accelerate the convergence of sequences $\{\mathrm{Q}_n\}$ of lattice rules.

This is usually achieved by transforming each independent variable separately using one of the univariate periodizing transformations given in Section 5.3.4; i.e., the independent variables $x_1, \ldots, x_N$ are transformed by

$$x_1 = \psi_1(\overline{x}_1), \; x_2 = \psi_2(\overline{x}_2), \; \ldots \;, \; x_N = \psi_N(\overline{x}_N),$$

where $\psi_1, \ldots, \psi_N$ are univariate periodizing transformations. For the corresponding special N-dimensional transformation $\psi = (\psi_1, \ldots, \psi_N)^\top$, the general multivariate transformation rule (1.7) is simplified:

$$\begin{aligned}
&\int_{C_N^q} f(x_1, \ldots, x_N)\, dx_1 \cdots dx_N \\
&= \int_{C_N^q} f(\psi_1(\overline{x}_1), \ldots, \psi_N(\overline{x}_N))\, \psi_1'(\overline{x}_1) \cdots \psi_N'(\overline{x}_N)\, d\overline{x}_1 \cdots d\overline{x}_N.
\end{aligned}$$

Note that applying periodizing transformations can only be expected to improve the *asymptotic* behavior of the discretization error $\mathrm{Q}_n f - \mathrm{I}f$ as $n \to \infty$. The actual integration error for practical numbers n of abscissas may exhibit a rather different behavior. In particular, periodizing an integrand does *not* necessarily lead to more accurate integral approximations.

Example (The Effect of Periodizing Transformations). Joe and Sloan [221] discuss the construction of sequences of embedded lattice rules $\mathrm{Q}_{n_0}, \mathrm{Q}_{n_1}, \ldots, \mathrm{Q}_{n_N}$. Each rule Q_{n_j}, $j = 0, 1, \ldots, N$, has $n_j = 2^j r$ distinct points, where r is an odd number. The following experiments are based on the particular sequence of lattice rules corresponding to the choice $r = 619$ found in Joe and Sloan [221].

First, the sequence was used for the numerical calculation of the integral

$$\int_{[0,1]^6} \cos(-5.5 + 2\|\mathbf{x}\|_1)\, d\mathbf{x} = \cos(0.5)\sin^6(1) \approx 0.312. \tag{6.76}$$

[68] It is assumed that previous transformations have converted the original integration problem into one with the integration region C_N^q.

The integrand in (6.76) is analytic on $[0,1]^6$, but the 1-periodic extension of f is discontinuous on the boundary of $[0,1]^6$. An integrand with a continuous 1-periodic extension was obtained using the polynomial transformation

$$x_k = 3\overline{x}_k^2 - 2\overline{x}_k^3, \quad \overline{x}_k \in [0,1], \qquad k = 1,2,\ldots,6,$$

and the tanh-transformation

$$(6.77) \quad x_k = \frac{1}{2}\left(1 + \tanh\left(\frac{1}{2}\left(\frac{1}{1-\overline{x}_k} - \frac{1}{\overline{x}_k}\right)\right)\right), \quad \overline{x}_k \in (0,1), \qquad k = 1,2,\ldots,6.$$

The periodic extension of the transformed integrand is analytic when the tanh-transformation is used, whereas the periodic extension is not differentiable at the boundary of $[0,1]^6$ for the polynomial transformation.

Figure 6.10 shows the absolute values of the relative integration errors of the sequence of lattice rules for the original integrand as well as for both transformed integrands. The lattice rules uniformly perform best when applied to the integrand obtained using the polynomial transformation. The largest *error reduction* (for $Q_{619} \to Q_{39616}$) is attained for the integrand obtained using the tanh-transformation.

Figure 6.11 shows the analogous results for the integral

$$(6.78) \qquad \int_{[0,1]^6} \exp(\|\mathbf{x}\|_1/6)\, d\mathbf{x} = (6(\exp(1/6)-1))^6 \approx 1.66.$$

The lattice rules Q_{n_j}, $j = 0,1,2$, perform best for the untransformed integrand, while the rules Q_{n_j}, $j = 3,4,5,6$, attain the highest accuracy for the integrand obtained using the tanh-transformation. The tanh-transformation also achieves the largest error reduction. Note that in both examples there is hardly any error reduction for the original, nonperiodic integrand.

These examples demonstrate that the choice of an appropriate periodizing transformation may have a strong impact on the accuracy of lattice rules.

Korobov Spaces. Based on Theorem 6.5.2, accuracy indices of lattice rules can be defined for classes $\mathcal{F}$ of periodic integrands if the Fourier coefficients $a(\mathbf{m})$ of functions $f \in \mathcal{F}$ exhibit, at least asymptotically, a similar behavior. This section introduces important instances of suitable classes of periodic integrands $\mathcal{F}$: *Korobov spaces*.

Korobov spaces arise naturally from periodizing integrands $f : C_N^+ \to \mathbb{R}$ which are sufficiently smooth[69] but *nonperiodic*; i.e., f cannot be extended periodically into $\mathbb{R}^N$ without loss of continuity at the boundary of C_N^+. Multivariate periodizing transformations generally use the same univariate transformation function $\psi_1 = \cdots = \psi_N =: \psi$ for each independent variable $x_1, \ldots, x_N$. If a polynomial transformation (5.33) or a sine transformation (5.34) with parameter $\alpha \in \mathbb{N}$, $\alpha > 1$, is chosen for ψ, then

$$(6.79) \qquad \psi^{(k)}(0) = \psi^{(k)}(1) = 0 \quad \text{for} \quad k = 1,2,\ldots,\alpha-1.$$

[69]For the sake of simplicity, it is assumed that f is infinitely many times differentiable.

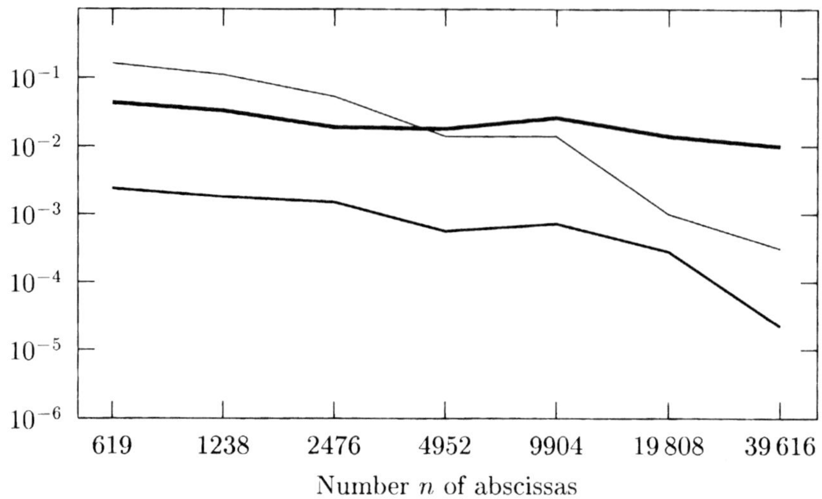

FIG. 6.10. *Absolute errors for lattice rules applied to the original integrand (—) in (6.76), the integrand subjected to the polynomial transformation (—), and the tanh-transformation (—).*

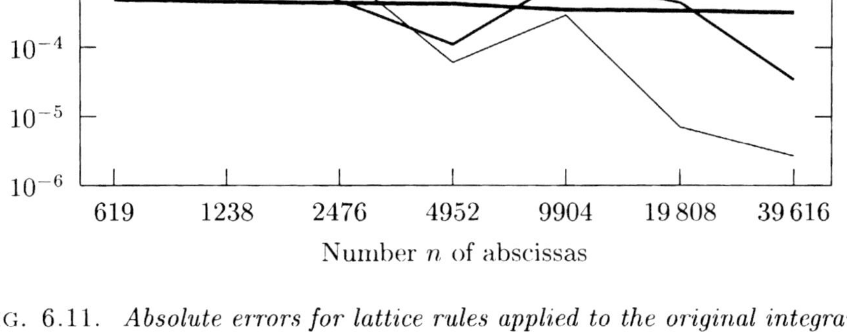

FIG. 6.11. *Absolute errors for lattice rules applied to the original integrand (—) in (6.78), the integrand subjected to the polynomial transformation (—), and the tanh-transformation (—).*

If property (6.79) holds for a transformation function ψ, then the Fourier coefficients $a(\mathbf{m})$ of the transformed integrand

$$f(\psi(\overline{x}_1), \ldots, \psi(\overline{x}_N)) \cdot \psi'(\overline{x}_1) \cdots \psi'(\overline{x}_N) \tag{6.80}$$

can be shown[70] to satisfy the inequality

$$|a(\mathbf{m})| \leq \frac{c}{(\overline{m}_1 \cdots \overline{m}_N)^{\alpha}} \quad \text{for all} \quad \mathbf{m} \in \mathbb{Z}^N, \tag{6.81}$$

where the constant $c \in \mathbb{R}_+$ is independent of $\mathbf{m} \in \mathbb{Z}^N$ and

$$\overline{m} := \begin{cases} |m| & \text{if } |m| \geq 1, \\ 1 & \text{if } |m| < 1. \end{cases}$$

Hence, periodizing a class of nonperiodic but smooth integrands f using the same transformation ψ (for which (6.79) is supposed to hold) gives rise to a set of transformed integrands (6.80) whose Fourier coefficients all satisfy the same kind of asymptotic error bound $a(\mathbf{m}) = O((\overline{m}_1 \cdots \overline{m}_N)^{-\alpha})$. This fact motivates the following definition.

DEFINITION 6.5.6 (KOROBOV SPACES). *Let $\alpha > 1$ and $c > 0$ be real numbers. Then, $E_N^{\alpha}(c)$ denotes the set of all functions f whose Fourier coefficients $a(\mathbf{m})$ satisfy the inequality* (6.81). *E_N^{α} denotes the union of all sets $\{E_N^{\alpha}(c) \; : \; c \in \mathbb{R}_+\}$,*

$$E_N^{\alpha} := \bigcup_{c \in \mathbb{R}_+} E_N^{\alpha}(c).$$

$E_N^{\alpha}(c)$ and E_N^{α} are said to be Korobov spaces.

When IMT transformations (5.36) or IMT-type transformations (5.37) or (5.38) are used for periodization purposes, the transformed integrand (6.80) is contained in all Korobov spaces E_N^{α}, $\alpha > 1$.

Irrespective of the use of transformation functions, periodic functions can be shown to belong to certain Korobov spaces $E_N^{\alpha}(c)$, provided they satisfy appropriate regularity conditions.

THEOREM 6.5.3. *Let the function $f : \mathbb{R}^N \to \mathbb{R}$ be 1-periodic and assume that for an $\alpha \in \mathbb{N}$, $\alpha > 1$, all partial derivatives*

$$\frac{\partial^{\alpha_1 + \cdots + \alpha_N} f}{\partial x_1^{\alpha_1} \cdots \partial x_N^{\alpha_N}}, \qquad \alpha_j = 0, 1, \ldots, \alpha - 1, \quad j = 1, 2, \ldots, N, \tag{6.82}$$

exist and are of bounded variation on C_N^+ in the sense of Hardy and Krause. Then, there exists a constant $c \in \mathbb{R}_+$ such that the Fourier coefficients of f satisfy the inequality (6.81); *i.e., $f \in E_N^{\alpha}(c)$.*

Proof. See Niederreiter [316].

[70] Inequality (6.81) can easily be proved by iterating the right-hand side integral in (6.63) and applying integration by parts to each resulting one-dimensional integral.

Figure of Merit (Zaremba Index). A cubature formula Q_n can be expected to be efficient for integrands f contained in a Korobov space $E_N^\alpha(c)$ if Q_n annihilates those trigonometric polynomials $e^{2\pi i \mathbf{m}\cdot\mathbf{x}}$ for which $\overline{m}_1 \cdots \overline{m}_N$ is small. The size of the smallest value of $\overline{m}_1 \cdots \overline{m}_N$ for which the associated trigonometric polynomial $e^{2\pi i \mathbf{m}\cdot\mathbf{x}}$ is *not* annihilated by Q_n can therefore be regarded as an indicator of the quality of Q_n.

DEFINITION 6.5.7 (FIGURE OF MERIT, ZAREMBA INDEX). *The figure of merit or Zaremba index $\rho(L)$ of an integration lattice L is defined by*

$$\rho(L) = \min_{\mathbf{m}\in L^\perp \setminus \{\mathbf{0}\}} \overline{m}_1 \overline{m}_2 \cdots \overline{m}_N .$$

The figure of merit indicates how many of the most significant Fourier coefficients are annihilated by a given lattice rule. The figure of merit can be interpreted graphically as follows: *Zaremba crosses* (Beckers and Cools [19]) Z_d, $d \in \mathbb{N}$, are defined by

$$Z_d := \left\{ \mathbf{m} \in \mathbb{R}^N \ : \ \overline{m}_1 \overline{m}_2 \cdots \overline{m}_N = d \right\} .$$

Using the Zaremba crosses, the Zaremba index of an integration lattice L can expressed as

$$\rho(L) = \min \left\{ d \ : \ L^\perp \cap Z_d \neq \emptyset \right\} ;$$

i.e., the Zaremba index is given by the smallest Zaremba cross containing a point of the dual lattice.

Example (Zaremba Index and Zaremba Crosses). Figure 6.12 shows the dual lattice $L^\perp$ (depicted in Figure 6.9) of the integration lattice L (depicted in Figure 6.8) together with the Zaremba crosses Z_2, Z_5, and Z_8. It can be seen that Z_5 is the smallest Zaremba cross containing a point of the dual lattice, i.e., $\rho(L) = 5$.

The intuitive use of the figure of merit as an indicator of the accuracy of lattice rules for functions $f \in E_N^\alpha(c)$ can be justified by the following theorem.

THEOREM 6.5.4. *Let L be an integration lattice. Then, the error in the corresponding lattice rules $Q_n(\mathbf{v})$, $\mathbf{v} \in \mathbb{R}^N$, for integrands $f \in E_N^\alpha(c)$ can be bounded by*

$$|Q_n(\mathbf{v})f - If| \leq c\, d(N, \alpha)\, \frac{(1 + \log \rho)^{N-1}}{\rho^\alpha}, \tag{6.83}$$

where $d(N,\alpha)$ depends only on N and α (but not on L).

Proof. See Sloan and Kachoyan [403].

This theorem shows that the asymptotic behavior of the integration error as $\rho \to \infty$ is essentially described by $O(\rho^{-\alpha})$. The improvement in the error bound (6.83) due to the use of better integration lattices, with larger values of ρ, is most pronounced for large values of α, i.e., for smooth integrands.

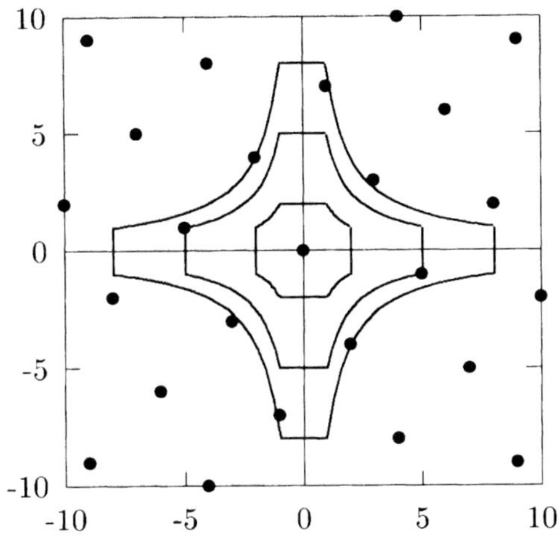

FIG. 6.12. *Zaremba crosses Z_2, Z_5, and Z_8, and dual lattice $L^\perp$.*

Worst-Case Index. For $f \in E_N^\alpha(c)$, the error estimate

$$|\mathrm{Q}_n(\mathbf{v})f - \mathrm{I}f| \leq \sum_{\mathbf{m}\in L^\perp\setminus\{\mathbf{0}\}} \frac{c}{(\overline{m}_1 \cdots \overline{m}_N)^\alpha} \tag{6.84}$$

can be derived from Theorem 6.5.2 and (6.81). This error bound can easily be seen to be sharp; i.e., there is an $f \in E_N^\alpha(c)$ for which equality is attained in (6.84). Let $f_\alpha \in E_N^\alpha(1)$ be defined by

$$f_\alpha(\mathbf{x}) := \sum_{\mathbf{m}\in\mathbb{Z}^N} \frac{1}{(\overline{m}_1 \cdots \overline{m}_N)^\alpha} e^{2\pi i \mathbf{m}\cdot\mathbf{x}}, \qquad \mathbf{x} \in \mathbb{R}^N, \text{[71]}$$

and let $P_\alpha(\mathrm{Q}_n)$ denote the error in Q_n for f_α,

$$P_\alpha(\mathrm{Q}_n) := \mathrm{Q}_n f_\alpha - \mathrm{I}f_\alpha = \mathrm{Q}_n f_\alpha - 1.$$

From Theorem 6.5.2, the equality

$$P_\alpha(\mathrm{Q}_n) = \mathrm{Q}_n f_\alpha - \mathrm{I}f_\alpha = \sum_{\mathbf{m}\in L^\perp\setminus\{\mathbf{0}\}} \frac{1}{(\overline{m}_1 \cdots \overline{m}_N)^\alpha}$$

follows. Hence, the error bound (6.84) can be written as

$$|\mathrm{Q}_n f - \mathrm{I}f| \leq c P_\alpha(\mathrm{Q}_n) \quad \text{for all} \quad f \in E_N^\alpha(c). \tag{6.85}$$

[71]Note that the right-hand side Fourier series is absolutely convergent because $\alpha > 1$.

Equality is attained in (6.85) for $f = cf_\alpha$, which shows that $cP_\alpha(Q_n)$ is the best possible error bound for integrands $f \in E_N^\alpha(c)$.

Therefore, $P_\alpha(Q_n)$ characterizes the worst-case integration error of an n-point lattice rule Q_n for integrands $f \in E_N^\alpha$; i.e., $P_\alpha(Q_n)$ can be regarded as a worst-case index of the accuracy of lattice rules for integrands $f \in E_N^\alpha(c)$. For even values $\alpha = 2, 4, \ldots$, f_α can be expressed explicitly in terms of Bernoulli polynomials (Sloan [401]), and therefore $P_\alpha(Q_n) = Q_n f_\alpha - 1$ can be evaluated efficiently for a given Q_n.

Theorem 6.5.4 implies that the worst-case index $P_\alpha(Q_n)$ of a lattice rule Q_n can be bounded in terms of the Zaremba index $\rho(L)$ of the corresponding integration lattice L:

$$P_\alpha(Q_n) = O((1 + \log \rho(L))^{N-1} \rho(L)^{-\alpha}) \quad \text{as} \quad n \to \infty.$$

Conversely, the bound $\rho(L)^{-\alpha} = O(P_\alpha(Q_n))$ can be established (Niederreiter [321]). Hence, the worst-case index $P_\alpha(Q_n)$ can be expected to decrease in direct proportion to $\rho(L)^{-\alpha}$, at least for large values of n.

Nonperiodic Integrands. Bounds on the error in lattice rules Q_n for nonperiodic integrands f can be obtained from the Koksma–Hlawka inequality (6.30) provided that f is of bounded variation in the sense of Hardy and Krause (see Section 6.3.1):

$$(6.86) \qquad |Q_n f - If| \le V(f)\, D_n^*(\mathbf{x}_0, \mathbf{x}_1, \ldots, \mathbf{x}_{n-1}).$$

In (6.86) $V(f)$ denotes the variation of f in the sense of Hardy and Krause, and $D_n^*(\mathbf{x}_0, \ldots \mathbf{x}_{n-1})$ is the star discrepancy of the abscissas $\mathbf{x}_0, \ldots \mathbf{x}_{n-1}$ of Q_n. Note that (6.86) also holds when the star discrepancy D_n^* is replaced by the discrepancy D_n (see Section 6.3.1).

The discrepancy of the abscissas $\mathbf{x}_0, \ldots, \mathbf{x}_{n-1}$ in a lattice rule Q_n can be bounded in terms of the corresponding figure of merit.

THEOREM 6.5.5. *Let L be an integration lattice in $\mathbb{R}^N$. Then, the discrepancy D_n of the n abscissas $\mathbf{x}_0, \ldots, \mathbf{x}_{n-1}$ in the corresponding lattice rule Q_n satisfies*

$$D_n(\mathbf{x}_0, \ldots, \mathbf{x}_{n-1}) < \frac{N}{n} + \frac{1}{\rho(L)} \left(\frac{2}{\log 2}\right)^{N-1} \left((\log n)^N + \frac{3}{2}(\log n)^{N-1}\right),$$

where $\rho(L)$ is the figure of merit of L.

Proof. See Niederreiter and Sloan [322].

Now, clearly, $\rho(L) \le n$, and it thus follows that

$$D_n(\mathbf{x}_0, \ldots, \mathbf{x}_{n-1}) \le c(N) \frac{(\log n)^N}{\rho(L)} \quad \text{for all} \quad n \ge 3.$$

A lower bound for the discrepancy of a lattice rule is provided in the following theorem.

THEOREM 6.5.6. *Let L be an integration lattice in $\mathbb{R}^N$. Then, the discrepancy D_n of the n abscissas $\mathbf{x}_0, \ldots, \mathbf{x}_{n-1}$ in the corresponding lattice rule Q_n satisfies*

$$D_n(\mathbf{x}_0, \ldots, \mathbf{x}_{n-1}) \geq \frac{1}{d(N)\rho(L)},$$

where $d(2) = 4$, $d(3) = 27$, and $d(N) = \frac{2}{\pi}((\pi + 1)^N - 1)$ as $N \geq 4$.

Proof. See Niederreiter and Sloan [322].

The asymptotic magnitudes of the upper and lower bounds in Theorems 6.5.5 and 6.5.6 as $n \to \infty$ differ only by a factor $O(\log^N n)$. As explained in Section 6.5.5, there are sequences of lattice rules Q_n for which the figures of merit ρ_n grow much faster than $O(\log^N n)$. For such sequences of lattice rules, the discrepancy D_n is asymptotically given by $O(\rho_n^{-1})$. Low-discrepancy formulas are obtained when the figure of merit is large. As a result, the figure of merit is an indicator of the accuracy of lattice rules not only for smooth, periodic integrands, but also for nonperiodic integrands of bounded variation. This fact supports the choice of the figure of merit as a measure of the quality of lattice rules.

6.5.5. Efficiency of Lattice Rules. In Section 6.5.4 fundamental theorems characterizing the discretization error in lattice rules are discussed. For a given integrand f, the discretization error $\mathrm{Q}_n f - \mathrm{I}f$ can be bounded in terms of different indicators of the accuracy of the lattice rule Q_n. For $f \in E_N^\alpha(c)$, for instance, Theorem 6.5.4 establishes a bound on the discretization error in terms of the figure of merit ρ, while the bound (6.85) is given in terms of the worst-case index $P_\alpha(\mathrm{Q}_n)$.

In order to assess the *efficiency* of an integration rule Q_n, the accuracy of Q_n has to be related to its computational cost, i.e., to the number of f-evaluations required. In this section, general theoretical results concerning the existence of efficient lattice rules are provided. The theorems presented prove the existence of sequences of lattice rules $\{\mathrm{Q}_n\}$ such that the order of convergence of $\mathrm{Q}_n f \to \mathrm{I}f$ as $n \to \infty$ depends on the smoothness of f, but is essentially independent of the dimension N. Formulas with sufficiently large numbers of abscissas in such sequences can be expected to be efficient. This does not, however, necessarily imply the existence of efficient formulas for practical numbers of abscissas.

Historically, the theoretical foundations of the method of good lattice points (i.e., rank-1 lattice rules; see Section 6.5.3) were developed much earlier than the general theory of lattice rules. Therefore, the existence of efficient integration formulas was first established for the method of good lattice points and only later extended to general lattice rules.

Method of Good Lattice Points. For the method of good lattice points, a number of theorems of the following general form have been proved.

THEOREM 6.5.7. *Let $\alpha > 1$ be a given real number. For any $n \in \mathbb{N}$ there exists an n-point, rank-1 rule Q_n such that*

$$P_\alpha(\mathrm{Q}_n) \le d(N,\alpha)\frac{(\log n)^{c(N,\alpha)}}{n^\alpha}, \tag{6.87}$$

where c and d depend only on the dimension N and α, but not on the number of abscissas n.

For a survey of the best results concerning the constants c and d, see Niederreiter [320].

Theorem 6.5.7, along with (6.85), shows that there is a sequence of rank-1 lattice rules $\{\mathrm{Q}_n\}$ such that, for all $f \in E_N^\alpha$,

$$\mathrm{Q}_n f - \mathrm{I}f = O\left(\frac{(\log n)^{c(N,\alpha)}}{n^\alpha}\right) \quad \text{as} \quad n \to \infty.$$

Because $(\log n)^{c(N,\alpha)} = o(n^\alpha)$ as $n \to \infty$, the order of convergence of $\mathrm{Q}_n f \to \mathrm{I}f$ does *not* deteriorate significantly when the dimension N is increased. The essential order of convergence $O(n^{-\alpha})$ depends only on the smoothness of the integrand.

General Lattice Rules. Niederreiter [320] shows that inequality (6.87) holds for lattice rules of arbitrary rank $m = 2, \ldots, N$. The method used by Niederreiter to prove this result is based on extending appropriate rank-1 rules which satisfy (6.87) so that they become rank-m lattice rules.

THEOREM 6.5.8. *Let $\alpha > 1$. Then, for any rank $m = 1, \ldots, N$ and any set of invariants $n_1, \ldots, n_m$, there exists a lattice rule Q_n, $n = n_1 \cdots n_m$, with the prescribed rank and invariants such that*

$$P_\alpha(\mathrm{Q}_n) \le d(N,\alpha)\frac{(\log n_1)^{c(N,\alpha)}}{n_1^\alpha}, \tag{6.88}$$

where c and d depend only on the dimension N and α, but not on the rank or the invariants.

Note that the error bound (6.88) for general lattice rules is obtained from the error bound (6.87) for the method of good lattice points by replacing the number of abscissas n by the first, largest invariant n_1.[72]

Theorem 6.5.8 along with (6.85) shows that, for any rank m and for any invariants $n_2, \ldots, n_m$, there is a sequence of lattice rules $\{\mathrm{Q}_n\}$ with the

[72]For rank-1 lattice rules, $n_1 = n$ holds and the two error bounds coincide.

prescribed rank and invariants[73] such that, for all $f \in E_N^\alpha$,

$$Q_n f - If = O\left(\frac{(\log n)^{c(N,\alpha)}}{n^\alpha}\right) \quad \text{as} \quad n \to \infty. \tag{6.89}$$

6.5.6. Practical Lattice Rules.

While Theorems 6.5.7 and 6.5.8 establish the existence of efficient lattice rules (at least for large numbers of abscissas), they do not provide methods to actually generate efficient lattice rules because they are based on *nonconstructive* proof techniques.

Search for Lattice Rules. Since there is currently no general method for constructing efficient lattice rules, such lattice rules have been found mostly by computer search. The large number of lattice rules for most practically relevant values of n and N prohibits a straightforward *exhaustive* computer search.[74] Therefore, computer searches have to be restricted to certain classes of lattice rules, such as *projection regular rules* (see Niederreiter [319], Sloan and Lyness [404], [405], and Sloan and Walsh [407]), *copy rules* (see Disney and Sloan [110] and Sloan and Lyness [404]), *component scaled rules* (see Lyness and Soerevik [282]), or *embedded sequences of lattice rules* (see Joe and Sloan [221], [222]). Searches for good rank-1 rules (6.68) are generally restricted to lattice points $\mathbf{p}$ of the *Korobov form*

$$\mathbf{p} = (1, q, q^2, \ldots, q^{N-1})^\top \mod p, \tag{6.90}$$

where p is a prime number and q can be chosen from the set $\{0, \ldots, p-1\}$. The main problem with restricting searches is finding appropriate classes of lattice rules, i.e., *small* classes which are likely to contain efficient lattice rules.

Known Formulas. Tables of efficient lattice rules have been published in a number of papers. In Maisonneuve [291], rank-1 lattice rules can be found for dimensions N ranging from $N = 3$ to $N = 10$. Extensions of these results to larger numbers of abscissas for three-dimensional integrals are contained in Kedem and Zaremba [229]. Extensions for dimension $N = 4$ are given in Bourdeau and Pitre [38]. Extensive tables of rank-1 lattice rules for dimensions $N = 2$ to $N = 15$ are provided in the appendix of the book by Hua and Wang [212].

Based on an exhaustive search, a table containing the most efficient three-dimensional general lattice rules was published in Lyness and Soerevik [284]. Optimal lattice rules of nonprime order n for dimension $N = 4$ are described

[73] Note that the first invariant n_1 in Q_n is not prescribed; n_1 increases proportionally to the number of abscissas n.

[74] Characterizations of the number of lattice rules can be found in Joe and Hunt [220], Lyness [272], and Lyness and Soerevik [283].

in Lyness and Soerevik [285]. Rank-2 lattice rules for dimensions $N = 3$ to $N = 8$ using approximately $n = 1000$, 10,000, or 100,000 abscissas are given in Sloan and Walsh [407]. Rank-N lattice rules and embedded sequences of lattice rules are provided in Joe and Sloan [221] (for dimensions $N = 6$, 14, and 20) and in Sloan and Joe [402] (for dimensions $N = 2, 3, \dots, 12$). High-order three- and four-dimensional lattice rules based on component scaling can be found in Lyness and Soerevik [282].

Software (Searches for Good Lattice Rules). For a given dimension N and a number $n = p$ of abscissas, the subroutine `NAG/d01gyf` calculates a vector $\mathbf{p}$ of the Korobov form (6.90) such that the corresponding rank-1 lattice rule (6.68) is optimal (with respect to the P_2 index) among all n-point, rank-1 lattice rules whose generating vectors are of Korobov form.

Similarly, `NAG/d01gzf` computes optimal lattice points $\mathbf{p}$ of Korobov form for numbers of abscissas which are products of two distinct primes $n = p_1 p_2$. The P_2 values of the resulting formulas are generally larger than the P_2 values of the formulas obtained from `NAG/d01gyf` using a comparable (prime) number of abscissas. The main advantage of `NAG/d01gzf` is the smaller computational complexity as compared to `NAG/d01gyf`.

Software (Integration Algorithms Based on Lattice Rules). The routine `NAG/d01gcf` uses rank-1 lattice rules to calculate iterated integrals of any dimension $N \leq 20$. The generating vector of the lattice rule to be applied can be chosen from a set of six precalculated vectors of Korobov form or may be supplied by the user. The polynomial transformation (5.33) with $\alpha = 2$ can be used optionally to periodize the given integrand.

6.5.7. Concluding Remarks.

The foregoing discussion is limited to the fundamentals of the theory of lattice rules. An up-to-date, comprehensive account of the subject can be found in Sloan and Joe [402]. An interesting recent development is the introduction of lattice rules with *modified vertex weights* (see Niederreiter and Sloan [323], [324] and Price and Sloan [344]). These vertex-modified rules can be shown to be more accurate for nonperiodic integrands than equal-weight lattice rules. Lattice rules have also been studied recently from the viewpoint of integration rules with *trigonometric degrees of accuracy*. Work in this area can be found in Beckers and Cools [19], Cools and Sloan [66], and in the (mostly Russian) literature quoted therein.

6.6. Miscellaneous Formulas

Several special multidimensional integration methods, for instance, the *method of averaging*, are related to, but cannot strictly be subsumed under, any of the integration techniques discussed in this chapter (see, for instance, Davis and Rabinowitz [92]). The main reason for omitting them, apart from space limitations, is that no corresponding high-quality software products have been developed. A theoretical discussion of these methods would therefore be of little relevance to most readers.

A notable exception is the method of Sag and Szekeres [380]. In this method, the integration region B is transformed to the unit sphere S_N in such a way that the transformed integrand and all its derivatives vanish at the surface of S_N. The resulting integral over S_N is computed using a product trapezoidal rule.

Software (Method of Sag and Szekeres). The subroutine `NAG/d01fdf` uses the method of Sag and Szekeres to compute definite integrals over *arbitrary* product regions for dimensions $N \leq 30$. The user can set certain parameters of the transformation function and thus tailor the transformation to the particular behavior of the given integrand. While `NAG/d01fdf` computes an approximate integral value, it does *not* provide any information regarding the integration error.

The subroutine `NAG/d01jaf` computes definite integrals for the dimensions $N = 2$, 3, and 4 over N-dimensional spheres using the method of Sag and Szekeres. The user can choose between different transformations. `NAG/d01jaf` implements an integration *algorithm*; i.e., it attempts to approximate the integral value with a user-specified accuracy and returns, in addition to the approximate integral value, an error estimate.

Chapter 7

Methods for Special Integration Problems

In practical computations, one often has to deal with particular kinds of integration problems, involving, for instance, oscillatory or singular integrands. Substantial efforts have been made towards the development of tailor-made computational techniques which take advantage of the structure and properties of specific integrands and integration regions, thus enabling the efficient and reliable solution of special integration problems. A number of such methods are described in this chapter: Section 7.1 discusses techniques tailored to highly oscillatory integration problems on bounded regions, whereas Section 7.2 is dedicated to oscillatory and nonoscillatory integrals on unbounded regions. Sections 7.3, 7.4, and 7.5 outline methods for weakly singular, Cauchy singular, and hypersingular integration problems, respectively.

7.1. Oscillatory Integrals on Bounded Regions

A function $f : B \subseteq \mathbb{R}^N \to \mathbb{R}$ is said to be *oscillatory* if it possesses several local extrema (minima and maxima) within B. The term *highly* or *rapidly* oscillatory function is used when the number of extrema is large (i.e., it significantly exceeds, say, 10). This section focuses on *univariate* oscillatory integration problems. Multivariate (usually, bi- or trivariate) oscillatory integrals are usually iterated (possibly after appropriate transformations of the independent variables), and the resulting univariate oscillatory integration problems are dealt with successively.

7.1.1. Practical Problem Settings

Multiple Oscillatory Integrals. Oscillatory integrands occur, for instance, in integral transforms (and their inverse transforms) such as the Fourier transform or the Hankel transform (see Section 2.2). In integral transforms, as well as in other applications, a given function f must be integrated with respect to a *family* of oscillatory weight functions $\{k(x, y_m) : m = 1, 2, \ldots, M\}$:

$$(7.1) \qquad \mathrm{I}_m f := \int_a^b k(x, y_m) f(x)\, dx, \quad m = 1, 2, \ldots, M.$$

In (7.1) the factor $k(x, y_m)$ is oscillatory in the variable x, and the number of oscillations grows as the value of y_m increases. The function f, by contrast, is not generally assumed to oscillate and is *independent* of the parameter y_m.

When dealing with the multiple integration problems in (7.1), it is crucial that the same (or, at least, highly overlapping) sets of abscissas be used to approximately calculate the different integrals because this allows the reuse of previously computed f-values.

Problem Conditioning and Error Requirements. Practical kernel functions $k(., y_m)$ usually assume positive and negative values at consecutive extrema alternately. Therefore, $|\mathrm{I}_m f|$ can be expected to be small in comparison to $\|f\|_\infty$, at least for large values of y_m. According to Section 3.4.1, such integration problems have a bad *relative* conditioning, and so calculating $\mathrm{I}_m f$ to a prescribed relative accuracy can become increasingly difficult (due to the effects of data and roundoff errors) for large values of m.

However, if the values of the functions $k(., y_m)$ are of comparable size (as is the case, for instance, for the Fourier kernel), it is more sensible to require the values $\mathrm{I}_m f, m = 1, 2, \ldots, M$, to have the same *absolute* accuracy. Since, in this case, the absolute condition number $K_{f,k(.,y_m)} = \|k(., y_m)\|_\infty$ does not usually increase as y_m grows, the effects of data and roundoff errors are generally not critical even for large values of m.

Problem Classes. Special techniques for numerically integrating oscillatory functions over finite intervals $[a, b]$ are usually targeted at one of the three following classes of kernel functions:

(i) the Fourier kernel functions

$$k(x, y) = \exp(\mathrm{i}\omega x) = \cos(\omega x) + \mathrm{i} \sin(\omega x),$$

where the circular frequency $\omega = \omega(y)$ depends on the parameter y, but not on the integration variable x;

(ii) the *nonlinear* Fourier kernel functions

$$k(x, y) = \exp(\mathrm{i}\omega(x) x) = \cos(\omega(x) x) + \mathrm{i} \sin(\omega(x) x),$$

where $\omega(x) = \omega(x; y)$ depends on both the parameter y and the integration variable x;

(iii) other specific types of oscillatory kernel functions k.

Calculating integrals with (linear) Fourier kernel functions—a task mostly associated with the computation of Fourier coefficients—has received by far the greatest attention amongst these three problem classes. Recently, there has also been increasing interest in calculating integrals involving nonlinear Fourier kernel functions (see, for instance, Ehrenmark [123]). Other kernel functions have rarely been deemed sufficiently important to warrant the effort needed to develop tailor-made integration methods for them. A notable exception is the kernel function

$$k(x,y) := J_n(xy), \qquad x \in [0,1], \tag{7.2}$$

where J_n is the Bessel function of the first kind of order n (see, for instance, Lewanowicz [261]). This kernel function appears, for instance, in the calculation of the *Fourier–Bessel transform*,

$$\mathrm{I}_m f := \int_0^1 x J_n(y_m x) f(x), \qquad m = 1,2,3,\ldots, \tag{7.3}$$

where $0 < y_1 < y_2 < \cdots$ are the zeros of J_n.

7.1.2. Integration between Zeros. Suppose the exact locations of the $k-1$ interior zeros $\bar{x}_1 < \bar{x}_2 < \cdots < \bar{x}_{k-1}$ of an oscillatory integrand f are known. For the sake of simplicity, assume that the interval endpoints a and b are also zeros of f. Then, a straightforward way of exploiting this knowledge is to use these zeros as abscissas of properly chosen integration formulas Q_n: since f is known to be zero at these abscissas, the corresponding integrand evaluations do not actually have to be performed, which reduces the computational effort required to compute $\mathrm{Q}_n f$.

Suitable *compound* integration formulas $\mathrm{Q}_n = \mathrm{Q}^1_{n_1} + \cdots + \mathrm{Q}^k_{n_k}$ can easily be obtained as follows. The zeros of f are used to define a subdivision

$$a = \bar{x}_0 < \bar{x}_1 < \cdots < \bar{x}_{k-1} < \bar{x}_k = b$$

of $[a,b]$, and, on each subinterval $[\bar{x}_{j-1}, \bar{x}_j]$, a *closed* basic rule $\mathrm{Q}^j_{n_j}$ (i.e., a rule which contains $\bar{x}_{j-1}$ and $\bar{x}_j$ as abscissas) is applied. Clearly, the overall number of function evaluations required to evaluate $\mathrm{Q}_n f$ is given by $n_1 + \cdots + n_k - 2k$ rather than the usual $n_1 + \cdots + n_k - k + 1$, which amounts to a saving of $k+1$ function evaluations.

On each subinterval $[\bar{x}_{j-1}, \bar{x}_j]$, the corresponding basic rule $\mathrm{Q}^j_{n_j}$ should be chosen in such a way that the highest possible degree of accuracy is attained for the given number n_j of abscissas to be used. According to Section 5.2.8, this means that $\mathrm{Q}^j_{n_j}$ should be the n_j-point *Lobatto formula*, which attains the degree of accuracy $D = 2n_j - 2$. For reasons of computational convenience, it is customary to use properly transformed copies of a *single* Lobatto formula on the different subintervals, unless the distances between consecutive zeros (i.e., the lengths of subintervals) vary significantly, in which case the number of abscissas in each subinterval should be adjusted correspondingly.

Example (Integration between Zeros). The integral

$$\mathrm{I}_m f := \int_0^1 \sin(m\pi x)e^{-x}\,dx = \frac{m\pi(1+(-1)^{m+1}/e)}{1+m^2\pi} \tag{7.4}$$

was computed for $m = 10$, 30, and 90 using compound Gauss and Lobatto formulas, $k \times G_{l-2}$ and $k \times L_l$, $k = m$, respectively. For any given value of l, both compound formulas require the same number $k(l-2)$ of integrand evaluations. The following table shows the errors $E_l^G := (k \times G_{l-2})f - \mathrm{I}_m f$ and $E_l^L := (k \times L_l)f - \mathrm{I}_m f$ for $l = 3$, 6, and 9.

m	E_3^G	E_3^L	E_6^G	E_6^L	E_9^G	E_9^L
10	$1.1 \cdot 10^{-2}$	$9.4 \cdot 10^{-4}$	$-1.5 \cdot 10^{-7}$	$-1.3 \cdot 10^{-9}$	$1.6 \cdot 10^{-14}$	$4.6 \cdot 10^{-17}$
30	$3.8 \cdot 10^{-3}$	$3.2 \cdot 10^{-4}$	$-5.3 \cdot 10^{-8}$	$-4.4 \cdot 10^{-10}$	$5.9 \cdot 10^{-15}$	$1.7 \cdot 10^{-17}$
90	$1.3 \cdot 10^{-3}$	$1.1 \cdot 10^{-4}$	$-1.8 \cdot 10^{-8}$	$-1.5 \cdot 10^{-10}$	$2.0 \cdot 10^{-15}$	$5.8 \cdot 10^{-18}$

It can clearly be seen that the compound Lobatto formulas yield significantly more accurate results—by up to a factor of around 10^3—than the corresponding compound Gauss formulas. This can be attributed to the fact that the basic Lobatto rule L_l has the degree of accuracy $D = 2l - 3$, while the basic Gauss rule G_{l-2} attains only $D = 2l - 5$.

The method of integrating between the zeros of oscillatory integrands has one severe drawback: when a family of integrals (7.1), instead of a single integral, must be calculated, the zeros of different kernels $k(x, y_m)$ do *not* usually coincide entirely. Hence, the integration formulas applied for different values of m use different sets of abscissas, which prevents the reuse of previously calculated f-values.

7.1.3. Filon-Type Integration.

Filon-type formulas are based on the general method of *product integration* (see Section 5.1.2): the integrand f is approximated using a piecewise polynomial model function g, and the integral $\mathrm{I}_m g$ of g with respect to the oscillatory kernel function $k(., y_m)$ is taken as an approximation of $\mathrm{I}_m f$. Such methods presuppose that the integrals $\mathrm{I}_m g$ can be calculated analytically. This presupposition holds, for instance, for the (linear) Fourier kernel functions.

Filon's original formulas $F_{2k+1}^{C;\omega}$ and $F_{2k+1}^{S;\omega}$ (Filon [149]) are targeted at integrals with cosine and sine Fourier kernels, respectively, and are based on Simpson's approach to constructing piecewise parabolic approximation functions: the integration interval $[a, b]$ is subjected to an equidistant partitioning

$$\bar{x}_j = a + (j-1)h, \qquad j = 0, 1 \ldots, k, \quad h = (b-a)/k,$$

and on each subinterval $[\bar{x}_{j-1}, \bar{x}_j]$, f is approximated using the parabolic polynomial which interpolates it at the points $\bar{x}_{j-1}$, $(\bar{x}_{j-1} + \bar{x}_j)/2$, and $\bar{x}_j$.

The resulting integration formulas can be found, for instance, in Davis and Rabinowitz [92].

Thus, Simpson's and Filon's methods differ only in the weight functions associated with the respective integration problems, but not in the approximating model functions or the approximation technique used. The analogy between Simpson's and Filon's rules also extends to their rates of convergence; i.e.,

$$F_{2k+1}^{C;\omega} f - \int_a^b \cos(\omega x) f(x)\, dx = O(k^{-3}) \quad \text{and}$$

$$F_{2k+1}^{S;\omega} f - \int_a^b \sin(\omega x) f(x)\, dx = O(k^{-3})$$

as $k \to \infty$, provided that $f \in C^4[a, b]$. However, in contrast to Simpson's rules, Filon's rules are *not* generally exact for cubic polynomials.

Example (Filon's Rules). The integral

$$\mathrm{I}_m f := \int_0^1 \sin(\omega_m x) e^{-x}\, dx, \quad \omega_m = (m - 1/4)\pi, \tag{7.5}$$

which is comparable in difficulty to (7.4), was computed for $m = 10$, 30, and 90 using Filon's rules $F_{m(l-2)+1}^{S;\omega_m}$, $l = 3, 6, 9$. Note that the number of f-evaluations required to evaluate $F_{m(l-2)+1}^{S;\omega_m} f$ is the same as the number required to evaluate $m \times L_l$ for the integrand in (7.4).

m	$F_{m+1}^{S;\omega_m} f - \mathrm{I}_m f$	$F_{4m+1}^{S;\omega_m} f - \mathrm{I}_m f$	$F_{7m+1}^{S;\omega_m} f - \mathrm{I}_m f$
10	$1.9 \cdot 10^{-5}$	$2.3 \cdot 10^{-9}$	$2.3 \cdot 10^{-10}$
30	$7.0 \cdot 10^{-7}$	$2.9 \cdot 10^{-11}$	$3.0 \cdot 10^{-12}$
90	$2.6 \cdot 10^{-8}$	$3.6 \cdot 10^{-13}$	$3.7 \cdot 10^{-14}$

Filon's rules prove to be superior to the corresponding compound Lobatto formulas for $l = 3$ (where the precision achieved is relatively low), while the reverse is true for $l = 9$ (where the precision attained is high).

Numerous generalizations of Filon's rule can be and have been obtained by using different sets of piecewise polynomial model functions g and/or by imposing different interpolation/approximation conditions on g. By using piecewise linear model functions, for instance, *Filon-trapezoidal* rules are obtained, by using cubic splines, *Filon-spline* rules can be derived, etc. The main advantage of *all* Filon-type formulas, and of all other product integration methods, is that the approximating model function g is independent of the kernel functions $k(x, y_m)$. This enables the same set of f-values needed for determining g to be used for calculating $\mathrm{I}_m g \approx \mathrm{I}_m f$ for all $m = 1, \ldots, M$.

Example (Filon's Rules). The integral (7.5) was computed for $m = 10^j$, $j = 0, 2, 4, 6, 8$, using Filon's rules $F_{2k+1}^{S;\omega_m}$, $k = 3$, 6, and 12. The following table shows the resulting relative errors $E_{2k+1}^{\mathrm{rel}} f := (F_{2k+1}^{S;\omega_m} f - \mathrm{I}_m f)/\mathrm{I}_m f$.

j	$E_7^{\text{rel}} f$	$E_{13}^{\text{rel}} f$	$E_{15}^{\text{rel}} f$
0	$-2.3 \cdot 10^{-5}$	$-1.5 \cdot 10^{-6}$	$-9.2 \cdot 10^{-8}$
2	$-1.1 \cdot 10^{-5}$	$-2.4 \cdot 10^{-6}$	$-7.3 \cdot 10^{-7}$
4	$-1.2 \cdot 10^{-7}$	$-2.3 \cdot 10^{-8}$	$-6.8 \cdot 10^{-9}$
6	$-1.2 \cdot 10^{-9}$	$-2.3 \cdot 10^{-10}$	$-6.8 \cdot 10^{-11}$
8	$-1.2 \cdot 10^{-11}$	$-2.3 \cdot 10^{-12}$	$-6.8 \cdot 10^{-13}$

The results indeed demonstrate that a fixed set of f-values can be used to approximately calculate the integral (7.5) for strongly varying values of m.

7.1.4. Modified Clenshaw–Curtis Integration. Similar to Filon-type integration, modified Clenshaw–Curtis integration adheres to the general principle of product integration (see Section 5.1.2). In contrast to Filon-type integration, however, Clenshaw–Curtis integration is based on approximating f on the whole of $[a, b]$ using a *single* polynomial P_d which interpolates f at the zeros or extrema of an appropriate Chebyshev polynomial (see Section 5.2.6). According to (5.21), evaluating the resulting integration formula with respect to the kernel function $k(., y_m)$ requires (in addition to computing the coefficients $a_0, a_1, \ldots, a_d$ of the Chebyshev expansion of P_d) calculating the *modified* moments $\mathrm{I}_{k(.,y_m)}T_0, \mathrm{I}_{k(.,y_m)}T_1, \ldots, \mathrm{I}_{k(.,y_m)}T_d$.

For a number of oscillatory kernel functions k, these moments can be calculated efficiently using appropriate recurrence relations. Most importantly, a suitable recurrence relation for Fourier kernel functions is given in Piessens and Branders [335]. For mixed kernels which contain both Fourier and algebraic/logarithmic endpoint singularity components, similar recurrence relations have been derived recently in Piessens and Branders [338], whereas a recurrence relation for Bessel function kernels is provided in Piessens and Branders [336].

Example (Modified Clenshaw–Curtis Integration). The integral (7.5) was computed for $m = 10^j$, $j = 0, 2, 4, 6, 8$, using n-point modified Clenshaw–Curtis formulas for $n = 2$, 4, and 6. The following table shows the resulting relative errors E_n^{rel}.

j	$E_2^{\text{rel}} f$	$E_4^{\text{rel}} f$	$E_6^{\text{rel}} f$
0	$-1.3 \cdot 10^{-3}$	$1.4 \cdot 10^{-6}$	$2.4 \cdot 10^{-10}$
2	$-1.3 \cdot 10^{-4}$	$-7.3 \cdot 10^{-7}$	$-1.2 \cdot 10^{-9}$
4	$-1.4 \cdot 10^{-6}$	$-8.9 \cdot 10^{-9}$	$-2.0 \cdot 10^{-11}$
6	$-1.4 \cdot 10^{-8}$	$-8.9 \cdot 10^{-11}$	$-2.0 \cdot 10^{-13}$
8	$-1.4 \cdot 10^{-10}$	$-8.9 \cdot 10^{-13}$	$-2.0 \cdot 10^{-15}$

The same integrals were previously calculated with comparable accuracy by applying Filon's rules using a significantly larger number of abscissas. For smooth integrands, modified Clenshaw–Curtis formulas are usually superior to Filon's formulas.

Software (Modified Clenshaw–Curtis Integration). The subroutines `CMLIB/dqc25f` $\approx$ `QUADPACK/dqc25f` $\approx$ `SLATEC/dqc25f` evaluate the modified 25-point Clenshaw–Curtis formula with respect to the Fourier kernel functions for a given integrand f. Additionally, an embedded 13-point Clenshaw–Curtis formula is used to calculate an error estimate.

7.1.5. Euler Expansion Methods for Calculating Fourier Coefficients. An efficient method (proposed by Lyness [268]) for calculating the Fourier coefficients of smooth functions $f \in C^p[a,b]$, $p > 1$, is based on the *Euler expansion*,

$$(7.6) \qquad f(x) = h_{p-1}(x) + g_p(x), \quad h_{p-1}(x) := \sum_{q=1}^{p-1} \frac{\lambda_q}{q!} B_q\left(\frac{x-a}{b-a}\right),$$

where B_q denotes the qth Bernoulli polynomial (see Abramowitz and Stegun [1]) and

$$\lambda_q := (b-a)^{q-1}(f^{(q-1)}(b) - f^{(q-1)}(a)), \quad q = 1, 2, \ldots, p-1.$$

From the elementary properties of Bernoulli polynomials, the relationships

$$f^{(q-1)}(b) - f^{(q-1)}(a) = h_{p-1}^{(q-1)}(b) - h_{p-1}^{(q-1)}(a), \quad q = 1, 2, \ldots, p-1,$$

follow, and consequently

$$g_p^{(q-1)}(a) = g_p^{(q-1)}(b), \quad q = 1, 2, \ldots, p-1.$$

This shows that the remainder term g_p in (7.6) has a smooth $(b-a)$-periodic extension.

Clearly, the Fourier coefficients of f can be obtained by superimposing the corresponding Fourier coefficients of h_{p-1} and g_p, respectively. While analytical formulas for the Fourier coefficients of h_{p-1} in terms of the quantities $\lambda_1, \ldots, \lambda_{p-1}$ can be computed easily (see, for instance, Giunta and Murli [175]), the Fourier coefficients of g_p can be calculated numerically with high precision using compound trapezoidal rules T_l (see Section 5.3.2). More precisely, it can be shown that the error in T_l when used to calculate specific Fourier coefficients of g_p converges like $O(l^{-p})$ as $l \to \infty$ and that the error in each specific trapezoidal rule T_l can be bounded uniformly for *all* Fourier coefficients. Furthermore, significant computational savings can be made by taking advantage of fast Fourier transform methods in the evaluation of T_l.

The derivatives of f at the interval endpoints a and b required for calculating $\lambda_1, \ldots, \lambda_{p-1}$ do not usually have to be computed with high precision and are generally approximated using finite difference methods.

Software (Calculation of Fourier Coefficients Using Euler Expansions). `TOMS/649` supplies a package of subroutines for evaluating Fourier coefficients based on Euler expansions. Two top-level routines deal with integrands given as function subprograms and integrands provided at equally spaced abscissas, respectively. The user can determine the number of Fourier coefficients to be computed, prescribe a uniform absolute accuracy requirement, and choose the maximum value of l to be used in the trapezoidal formulas. Other routine arguments enable the user, for instance, to select the differentiation method to be applied.

7.1.6. Levin's Method. The methods discussed in this section so far do not readily apply to integrals involving nonlinear Fourier weight functions, for example,

$$\mathrm{I}f = \int_a^b \exp(\mathrm{i}\omega(x))f(x)\,dx.$$

Suppose f satisfies the identity

$$f(x) = \mathrm{i}\omega'(x)g(x) + g'(x). \tag{7.7}$$

Then, calculating $\mathrm{I}f$ is straightforward:

$$\mathrm{I}f = g(b)\exp(\mathrm{i}\omega(b)) - g(a)\exp(\mathrm{i}\omega(a)). \tag{7.8}$$

Now, (7.7) can be interpreted as a differential equation for the unknown function g. The representation of the general solution of (7.7),

$$g(x) = \exp(-\mathrm{i}\omega(x))\left(\int_a^x \exp(\mathrm{i}\omega(t))f(t)\,dt + C\right),$$

shows that g can generally be expected to be as oscillatory as the original integrand.[75] However, Levin [259] proves that, even if the weight $\exp(\mathrm{i}\omega(x))$ oscillates rapidly, there exists a slowly oscillatory *particular* solution of (7.7), provided f and ω' are only slowly oscillatory.

Since a slowly oscillatory particular solution is known to exist, it is sensible to expect that a solution of (7.7) can be approximated with reasonable accuracy using an n-dimensional linear space $\mathcal{M}_n$ of slowly oscillating model functions, for instance, a suitable space of (piecewise) polynomials. Given an appropriate space $\mathcal{M}_n$, an approximate solution $g_n \in \mathcal{M}_n$ can most easily be constructed using a *collocation* approach, i.e., by requiring g_n to satisfy the differential equation (7.7) at n properly chosen points $a = x_1 < x_2 < \cdots < x_n = b$:

$$f(x_i) = \mathrm{i}\omega'(x_i)g_n(x_i) + g_n'(x_i), \qquad i = 1, 2, \ldots, n. \tag{7.9}$$

Expanding g_n in terms of the basis $\{b_1, b_2, \ldots b_n\}$ for $\mathcal{M}_n$,

$$g_n(x) = a_1 b_1(x) + a_2 b_2(x) + \cdots + a_n b_n(x),$$

[75] This general representation is not suitable for calculating g because it involves solving an integration problem which is even more intricate than the original one.

and substituting this representation for g_n in the collocation equations (7.9) yields a system of linear equations for $a_1, a_2, \ldots, a_n$. Once these coefficients have been computed, an approximate integral value can be obtained from (7.8):

$$\text{(7.10)} \qquad \mathrm{I}f \approx \sum_{j=1}^{n} a_j (b_j(b) \exp(\mathrm{i}\omega(b)) - b_j(a) \exp(\mathrm{i}\omega(a))).$$

Levin's method often proves highly effective (i.e., it requires a small number of f-evaluations) when integrals involving nonlinear Fourier kernels are to be calculated with relatively low precision. Note, however, that Levin's method requires calculating the derivatives $\omega'(x_i)$, which have to be approximated using finite differences if ω' is not available in closed form.

Example (Levin's Method). The integral

$$\mathrm{I}_m f = \int_0^1 e^{-x+imx^2}\, dx$$

was computed for $m = 20$, 30, and 40 using Levin's method. $\mathbb{P}_{n-1}$ was chosen to be the n-dimensional space of model functions $\mathcal{M}_n$, and the collocation points $x_1, \ldots, x_n$ were chosen as the properly transformed Chebyshev extrema. The following table shows the resulting relative errors $E_n^{\text{rel}} f$ in both the real and imaginary parts of $\mathrm{I}_m f$ for $n = 10$, 15, and 20.

m	$E_{10}^{\text{rel}} f$	$E_{15}^{\text{rel}} f$	$E_{20}^{\text{rel}} f$
20	$-4.7 \cdot 10^{-4}$ $3.2 \cdot 10^{-3}$	$-4.2 \cdot 10^{-6}$ $5.1 \cdot 10^{-6}$	$-3.2 \cdot 10^{-10}$ $1.1 \cdot 10^{-9}$
30	$-5.2 \cdot 10^{-3}$ $-1.8 \cdot 10^{-3}$	$3.0 \cdot 10^{-5}$ $3.0 \cdot 10^{-5}$	$-2.0 \cdot 10^{-8}$ $2.2 \cdot 10^{-7}$
40	$-2.9 \cdot 10^{-3}$ $-1.0 \cdot 10^{-2}$	$-8.3 \cdot 10^{-5}$ $5.5 \cdot 10^{-5}$	$4.9 \cdot 10^{-7}$ $-3.3 \cdot 10^{-7}$

7.1.7. Approximation of the Compound Integrand. Product integration methods (such as Filon-type or modified Clenshaw–Curtis methods) presuppose that the integral $\mathrm{I}(kg)$, involving the oscillatory kernel k, can be calculated analytically for suitable model functions g approximating the integrand f. If this presupposition does not hold, the product integration approach fails. In this case, a possible alternative is to approximate the compound integrand kf using special model functions g which are tailored to the specific structure of kf, in particular, to the oscillatory nature of k. This approach has been applied recently to integrals involving nonlinear Fourier kernels.

The basic idea in Ehrenmark [124] is that the compound integrands $\cos(\omega(x)x)f(x)$ and $\sin(\omega(x)x)f(x)$ can be approximated using the simple functions 1, $\cos(\overline{\omega}x)$, and $\sin(\overline{\omega}x)$ if $\omega'(x)$ varies only moderately on $[a, b]$. In this case, $\overline{\omega}$ is chosen as an estimate of the average value of $\omega'(x)$ on $[a, b]$. The resulting model function, composed of the functions 1, $\cos(\overline{\omega}x)$, and $\sin(\overline{\omega}x)$, can then be integrated accurately using a closed three-point formula Q_3 with the abscissas $a, (a+b)/2$, and b. If the above assumption concerning the variation of $\omega'(x)$ on $[a, b]$ does not hold, then the integration interval can be divided into sufficiently short subintervals. The compound integrands are then approximated, and the resulting model functions are integrated on each subinterval separately.

Evans [142], [143] discusses extensions of Ehrenmark's approach and proposes a different method, which is found to be superior. In Evans's approach both f *and* ω are approximated using piecewise linear (or quadratic) model functions which interpolate f and ω, respectively, at equidistant interpolation points. The integral of the resulting compound approximation function is taken as the approximate integral value. For quadratic approximation functions, the integral of the resulting compound approximation function is *not* elementary and requires the evaluation of *Fresnel integrals*.

7.2. Integrals on Unbounded Regions

This section describes computational methods which are tailored to the solution of integration problems on unbounded regions. The discussion is limited to *univariate* integrals on infinite and semi-infinite intervals. Multivariate integrals on unbounded regions are usually iterated (possibly after appropriate transformations of the independent variables) in such a way that unbounded variables are integrated separately in univariate inner integrals. These univariate (semi-)infinite integration problems are then dealt with successively.

7.2.1. Conversion to Integrals on Bounded Regions. A large number of numerical integration methods exist for integrals on bounded regions. It follows that a reasonable strategy for dealing with integrals on unbounded regions is to convert them into (almost) equivalent integrals on bounded regions. This can be achieved by either *transforming* or *truncating* the independent variable appropriately.

Transformations to Bounded Regions. The transformation rule (1.6) describes the changes in the integration interval and the integrand resulting from replacing the independent variable x with another independent variable $\overline{x}$ based on the relationship $x = \psi(\overline{x})$. Using appropriate transformation functions ψ, the resulting integration interval can, for instance, be made to be bounded. The following list contains a number of transformations frequently used for converting the semi-infinite interval $[0, \infty)$ to the standard, finite integration interval $[0, 1)$.

$\psi(\overline{x})$	$\psi'(\overline{x})$
$-\alpha \log(1-\overline{x}),\ \alpha > 0$	$\alpha/(1-\overline{x})$
$\overline{x}/(1-\overline{x})$	$1/(1-\overline{x})^2$
$(\overline{x}/(1-\overline{x}))^2$	$2\overline{x}/(1-\overline{x})^3$

Similarly, the infinite integration interval $(-\infty, \infty)$ can be transformed into the standard integration interval $(-1, 1)$ using one of the following transformations:

$\psi(\overline{x})$	$\psi'(\overline{x})$
$\overline{x}/(1-\lvert\overline{x}\rvert)$	$1/(1-\lvert\overline{x}\rvert)^2$
$\tan(\pi\overline{x}/2)$	$(\pi/2)(1+\tan^2(\pi\overline{x}/2))$

More useful transformations can be found, for instance, in Genz [169]. Additionally, a thorough knowledge and understanding of the integration problem at hand may enable the development of tailor-made transformations (see Section 3.6.1).

Note that ψ' is generally singular at the points $\overline{x}$ for which $\psi(\overline{x}) = \pm\infty$, and this may entail corresponding singularities in the transformed integrand $(f \circ \psi)\psi'$. In this case, one of the methods described in Section 7.3 can be applied.

Example (Singularities due to Transformation). Transforming the convergent integral

$$\int_0^\infty \frac{1}{1+x^\alpha}\, dx, \qquad \alpha > 1,$$

using the substitution

$$x = \psi(\overline{x}) = \overline{x}/(1-\overline{x}), \qquad 0 \le \overline{x} < 1,$$

yields

$$\int_0^1 \frac{1}{(1-\overline{x})^2 + (1-\overline{x})^{2-\alpha}\overline{x}^\alpha}\, d\overline{x}.$$

For $\alpha < 2$, the transformed integrand is singular at $\overline{x} = 1$.

Truncation to Bounded Region. Rather than integrating a function f over an unbounded integration interval, say $[a, \infty)$, one can simply truncate $[a, \infty)$ to a bounded subinterval $[a, \beta] \subset [a, \infty)$, and integrate f over $[a, \beta]$. Truncating $[a, \infty)$ is appropriate when $f(x)$ decreases rapidly as $x \to \infty$ and/or an a priori bound on the integral of the truncated tail, $\mathrm{I}(f; [\beta, \infty))$, can be determined. While, for reasons of mathematical rigor, transforming unbounded integration intervals is usually preferable to simply truncating them, truncation can be used to avoid numerical difficulties (typically, due to overflows) with evaluating f (or suitable transformation functions ψ) for large abscissa values.

Proceeding to the Limit. If there is no a priori bound on $\mathrm{I}(f;[\beta,\infty))$ available, truncating $[a,\infty)$ can be made more reliable by approximately calculating a sequence of truncated integrals:

$$\begin{aligned} \mathrm{Q}(f;[a,\beta_1]) &\approx \mathrm{I}(f;[a,\beta_1]), \\ \mathrm{Q}(f;[a,\beta_1]) + \mathrm{Q}(f;[\beta_1,\beta_2]) &\approx \mathrm{I}(f;[a,\beta_2]), \\ \mathrm{Q}(f;[a,\beta_1]) + \mathrm{Q}(f;[\beta_1,\beta_2]) + \mathrm{Q}(f;[\beta_2,\beta_3]) &\approx \mathrm{I}(f;[a,\beta_3]), \\ &\vdots \end{aligned}$$

where $\beta_1 < \beta_2 < \beta_3 < \cdots$. The size of the contributions $\mathrm{Q}(f;[\beta_{j-1},\beta_j])$ can then be used to determine an appropriate cut-off point β. Additionally, it may be possible to accelerate the convergence of the sequence of truncated integrals using extrapolation methods (see Section 7.2.4).

7.2.2. Trapezoidal Rules for Infinite Intervals. Compound trapezoidal rules for (semi-)infinite intervals are obtained in essentially the same way as compound trapezoidal rules for finite intervals: given a step size $h > 0$, the semi-infinite interval $[a,\infty)$, for instance, is divided into the subintervals

$$[a + (i-1)h,\ a + ih], \quad i = 1, 2, 3, \ldots,$$

and a properly transformed (simple) trapezoidal rule is applied to each subinterval. However, the number of subintervals $[a + (i-1)h, a + ih]$ in the partitioning of $[a,\infty)$ is (countably) infinite, and thus the resulting compound trapezoidal rule for $[a,\infty)$ becomes an infinite series:

$$T_h f := h\left(\frac{1}{2}f(a) + \sum_{i=1}^{\infty} f(a+ih)\right).$$

A similar compound trapezoidal rule can be obtained for the infinite interval $(-\infty,\infty)$:

$$T_h f := h \sum_{i=-\infty}^{\infty} f(ih).$$ [76]

Note that the infinite sum defining $T_h f$ may *diverge* even when the integral of f over $[a,\infty)$ or $(-\infty,\infty)$ converges. Nevertheless, the Euler–Maclaurin sum formula (see Theorem 5.3.3) can easily be seen to remain valid for infinite compound trapezoidal rules.

---

[76] The choice of $x = 0$ as an integration abscissa in $T_h f$ is somewhat arbitrary, and more general compound trapezoidal rules for $(-\infty,\infty)$, with essentially the same analytical properties as $T_h f$, can be defined by $T_h(v)f := h\sum_{i=-\infty}^{\infty} f(ih+v)$, $v \in [0,h)$. However, the symmetry of $T_h f$ makes it possible to take computational advantage of symmetries in f.

THEOREM 7.2.1 (EULER–MACLAURIN SUM FORMULA FOR INFINITE INTERVALS). *Let $a \in \mathbb{R}$ or $a = -\infty$, and $b = +\infty$. If f is (improperly) Riemann integrable over $[a, b]$ and $f \in C^{2k+1}[a, b]$, then*

$$\begin{aligned} T_h f - \mathrm{I}f &= \frac{B_2}{2!} h^2 [f'(b) - f'(a)] + \frac{B_4}{4!} h^4 [f^{(3)}(b) - f^{(3)}(a)] \\ &\quad + \cdots + \frac{B_{2k}}{(2k)!} h^{2k} [f^{(2k-1)}(b) - f^{(2k-1)}(a)] \\ &\quad + h^{2k+1} \int_a^b \overline{P}_{2k+1}\left(\frac{x-a}{h}\right) f^{(2k+1)}(x)\, dx, \end{aligned} \tag{7.11}$$

where the entities $B_2, B_4, \ldots, B_{2k}$ and $\overline{P}_{2k+1}$ are defined in Theorem 5.3.3.

Proof. If $a \in \mathbb{R}$ and $b = +\infty$, set $b = a + lh$ in (5.31) and proceed to the limit $l \to \infty$. Proceed analogously if $a = -\infty$ and $b = +\infty$. □

The error expansion (7.11) shows, among other things, that infinite compound trapezoidal rules T_h give particularly accurate results for smooth integrands whose odd derivatives (up to the order $2k-1$) vanish at the (finite or infinite) endpoints of the integration interval:

$$f^{(j)}(a) = f^{(j)}(b) = 0, \qquad j = 1, 2, \ldots, 2k-1.$$

In particular, if

$$\int_a^b |f^{(2k+1)}(x)|\, dx < \infty,$$

then it can be obtained from (7.11) that $|T_h f - \mathrm{I}f| = O(h^{2k+1})$ as $h \to 0+$. For functions which are analytic in a strip

$$D_d := \{\, z \in \mathbb{C} \;:\; \mathrm{Im}(z) < d \,\}, \quad d > 0, \tag{7.12}$$

of the complex plane, even the exponential convergence

$$|T_h f - \mathrm{I}f| = O(e^{-cd/h}), \quad c > 0,$$

can be established (see Davis and Rabinowitz [92]). *Periodizing* transformations similar to the ones described in Section 5.3.4 can be used to accelerate the convergence of infinite compound trapezoidal rules and/or to make these rules applicable to (nonperiodic) integrands defined on *finite* intervals.

Truncation of Infinite Compound Trapezoidal Rules. For practical computations, the series defining infinite compound trapezoidal rules must be truncated. For the semi-infinite interval $[a, \infty)$, for instance,

$$T'_{h;n} f := h\left(\frac{1}{2} f(a) + \sum_{i=1}^{n} f(a + ih)\right)$$

or

$$T''_{h;n}f := h\left(\frac{1}{2}f(a) + \sum_{i=1}^{n-1} f(a+ih) + \frac{1}{2}f(a+nh)\right)$$

is calculated instead of $T_h f$.[77] A reasonable strategy is to choose the cut-off index n in such a way that the truncation error $T_{h;n}f - T_h f$ is comparable in size to the integration error $T_h f - \mathrm{I}f$:

$$|T_{h;n}f - T_h f| \approx |T_h f - \mathrm{I}f|.$$

The truncation error can be estimated using a "proceeding to the limit" strategy, as described in Section 7.2.1. In order to estimate the integration error, several values $T_{h_1}f, T_{h_2}f, T_{h_3}f, \ldots$ must be calculated for $h_1 > h_2 > h_3 > \cdots$ (see Section 8.1.3). The cut-off indices n_k, $k = 2, 3, \ldots$, used in the approximations $T_{h_1;n_1}f, T_{h_2;n_2}f, T_{h_3;n_3}f, \ldots$ are then determined according to estimated discretization errors.

Clearly, $T_{h;n}f$ cannot be expected to converge to $\mathrm{I}f$ as $h \to 0+$ unless the relationship $nh \to \infty$ as $h \to 0+$ holds between the step size h and the cut-off index n. However, the precise nature of the functional relationship $n = n(h)$ or $h = h(n)$ required to ensure that $T_{h;n}f$ converges (rapidly) to $\mathrm{I}f$ is not obvious. This question is, among other things, dealt with in the framework of *sinc methods*.

Sinc Methods. This section provides a coarse outline of sinc methods and their application to numerical integration. Extensive and thorough discussions—in particular, technical details and proofs—can be found in Lund and Bowers [267] and Stenger [417].

The *sinc function* (see Fig. 7.1) is defined by

$$\text{(7.13)} \qquad \operatorname{sinc}(x) = \begin{cases} \dfrac{\sin(\pi x)}{\pi x} & \text{for } x \neq 0, \\ 1 & \text{for } x = 0. \end{cases}$$

Note that the sinc function is improperly Riemann integrable over $(-\infty, +\infty)$ and that

$$\int_{-\infty}^{+\infty} \operatorname{sinc}(x)\,dx = 1.$$

Furthermore, for any $h > 0$, the set of shifted and scaled sinc functions

$$\text{(7.14)} \qquad \left\{ \frac{1}{\sqrt{h}} \operatorname{sinc}\left(\frac{x - ih}{h}\right) \; : \; i \in \mathbb{Z} \right\}$$

[77] The symbol $T_{h;n}f$ is used to denote either of the truncated series.

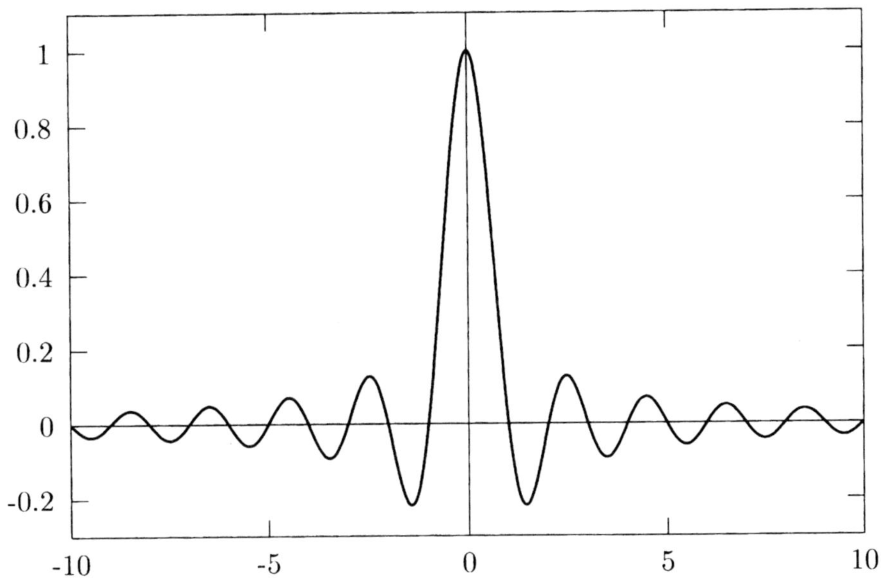

FIG. 7.1. *Sinc function* $\sin(\pi x)/(\pi x)$.

is an *orthonormal* system with respect to the inner product

$$\langle f, g\rangle := \int_{-\infty}^{\infty} f(x)g(x)\,dx.$$

Now, for any function $f : \mathbb{R} \mapsto \mathbb{R}$, the series

$$C_h f(x) := \sum_{i=-\infty}^{\infty} f(ih)\,\mathrm{sinc}\left(\frac{x-ih}{h}\right) \tag{7.15}$$

is said to be the *Whittaker cardinal function* of f. The sinc expansion $C_h f$ interpolates f at the points $\{ih : i \in \mathbb{Z}\}$ and can be regarded as an approximation of f on $\mathbb{R}$. The connection between $C_h f$ and the trapezoidal rule $T_h f$ is established by the fundamental relationship

$$\int_{-\infty}^{\infty} C_h f(x)\,dx = T_h f.$$

In other words, $T_h f$ can be interpreted not only as the integral of the piecewise linear function which interpolates f at the points $\{ih : i \in \mathbb{Z}\}$, but also as the integral of the smooth function $C_h f$ which satisfies the same interpolation conditions. This suggests that the discretization errors $T_h f - \mathrm{I}f$ and $T_{h;n} f - \mathrm{I}f$ depend on the quality of the approximations $C_h f \approx f$ and

$$C_{h;n} f(x) := \sum_{i=-n}^{n} f(ih)\,\mathrm{sinc}\left(\frac{x-ih}{h}\right) \approx f,$$

respectively. Based on this kind of analysis, it can be shown that $T_h f$ converges exponentially to $\mathrm{I}f$ if f is (among other things) analytic in a strip D_d, $d > 0$, defined in (7.12). Moreover, if f decays exponentially as $|x| \to \infty$, i.e., if

$$|f(x)| = O\left(e^{-\alpha|x|}\right), \quad \alpha > 0,$$

then the step size h and the truncation index n in $T_{h;n}$ should be related by

$$h = \sqrt{\frac{\pi d}{\alpha n}}.$$

For this choice of interrelationship between h and n, it can be shown that

$$|T_{h;n}f - \mathrm{I}f| \leq C(f, d, \alpha)e^{-\sqrt{\pi d \alpha n}},$$

i.e., that $T_{h;n}f - \mathrm{I}f$ converges as $O(e^{-c\sqrt{n}})$, $c > 0$.

Sinc integration methods are obtained by transforming given integration problems $\mathrm{I}(f; [a, b])$, using conformal transformation functions ψ, in such a way that the transformed integration region $\overline{I} = \psi^{-1}([a, b])$ becomes $\overline{I} = (-\infty, +\infty)$ and the corresponding integrand $\overline{f} = (f \circ \psi)\psi'$ satisfies the aforementioned analyticity and growth conditions. As explained in Section 3.6.1, applying a truncated compound trapezoidal rule $T_{n;h}$ to $\overline{f}$ can be regarded as applying a corresponding transformed formula $\overline{T}_{n;h}$ to the original integration problem. Formulas $\overline{T}_{n;h}$ obtained in this way are referred to as *sinc integration formulas*. Numerous examples can be found in Lund and Bowers [267] and Stenger [417].

Software (Sinc Integration). The quadrature algorithm implemented in TOMS/614 is based on sinc formulas which can be shown to be optimal for certain function spaces. Mathematical and algorithmic details can be found in Sikorski and Stenger [397], [398].

7.2.3. Simple Interpolatory (Product) Integration Formulas. Simple integration formulas for finite intervals are based on approximating f in the integration interval using a single polynomial $p \in \mathbb{P}$ (see Section 5.1.2). When it comes to extending this approach to (semi-)infinite intervals, the following problem arises. Polynomials p (except for the trivial polynomial $p \equiv 0$) are *not* improperly Riemann integrable over (semi-)infinite intervals. Therefore, polynomials cannot be used as model functions for approximating improperly Riemann integrable functions on (semi-)infinite intervals.

One possible way of overcoming this problem is to use more suitable classes of model functions g instead of polynomials. In this context, "suitable" means (i) f can be approximated with sufficient accuracy on the given unbounded integration interval using easily computable model functions, and (ii) the model functions can be efficiently integrated in closed form. In general, it is not possible to construct suitable classes of model functions unless information about the asymptotic behavior of $f(x)$ as $x \to +\infty$ and/or $x \to -\infty$ (in particular, about the rate of decay) is available. Asymptotic information about the integrand usually states that the integrand can be expressed in the form wf, where (i) w is nonnegative and (ii) $f(x)$ does *not* converge to 0 as $x \to \infty$ and/or $x \to -\infty$. Note that property (ii) in addition to the improper integrability of wf implies that $w(x) \to 0$ as $x \to \infty$ and/or $x \to -\infty$.

Example (Product Form of Integrands). The integrand in

$$\int_0^\infty \frac{\sin(x)}{x}\, dx$$

can be split into $w(x) = 1/x$ and $f(x) = \sin(x)$, whereas the integrand in

$$\int_0^\infty p(x)e^{-x}\, dx, \quad p \in \mathbb{P},$$

is the product of $w(x) = e^{-x}$ and $f(x) = p(x)$.

Once a product form of an integrand has been established, it is natural to use the product integration approach (see Section 5.1.2) to construct corresponding integration formulas. Note, however, that it is not usually possible to choose practical finite-dimensional spaces of model functions $\mathcal{M}$ in such a way that f can be approximated *uniformly* on the entire, unbounded integration interval by a function $g \in \mathcal{M}$. Therefore, the error bound (5.4), which usually serves as a simple justification for the validity of product integration formulas, does *not* apply, and more refined techniques are necessary to establish approximation properties of product formulas for unbounded intervals.

Informally speaking, in order for a product formula on an unbounded interval to be accurate, it is important that (i) $w(x)$ decrease rapidly as $x \to \infty$ and/or $x \to -\infty$ and that (ii) f can be approximated sufficiently accurately, in relation to the rate of decrease in w, by model functions $g \in \mathcal{M}$ on any bounded subinterval $[\alpha, \beta]$ of the integration region. Properties (i) and (ii) are meant to ensure that the contributions $\mathrm{I}_w(g - f; [\beta, +\infty])$ (and/or $\mathrm{I}_w(g - f; [-\infty, \alpha])$) and $\mathrm{I}_w(g - f; [\alpha, \beta])$, respectively, to the overall integration error are small.

Generally, product integration is of less importance for integrals on unbounded regions than it is for other special kinds of integration problems (see Sections 7.1.3 and 7.1.4, for example). The formulas of primary practical relevance are the Gauss formulas for nonnegative weight functions w, in particular, for Hermite and (generalized) Laguerre weight functions as well as for rational weight functions (see, for instance, Gautschi [157]). These are discussed in detail in Section 5.2.7.

7.2.4. Extrapolation Methods.

Section 7.2.1 describes truncating an interval of integration and proceeding to the limit as one particular way of dealing with unbounded integration regions. The convergence of this approximation process can often be accelerated using appropriate extrapolation methods, the most important of which are discussed in this section.

Convergence Acceleration of Alternating Series. Consider the integral

$$\mathrm{I}(gh) = \int_a^\infty g(x)h(x)\,dx,$$

and assume that g is periodic and h is nonnegative on $[b, \infty)$, $b \geq a$. Furthermore, if $2q$ denotes the smallest period of g, then shifting g by the half-period q is supposed to leave the function values invariant except for a sign change:

$$g(x+q) = -g(x) \quad \text{for all } x \geq b. \tag{7.16}$$

Now, partitioning the semi-infinite interval $[a, \infty)$ at the points

$$\bar{x}_0 = a, \quad \bar{x}_j = b + (j-1)q, \quad j = 1, 2, 3, \ldots,$$

and defining

$$\mathrm{I}_j(gh) = \int_{\bar{x}_j}^{\bar{x}_{j+1}} g(x)h(x)\,dx, \; j = 0, 1, 2, \ldots,$$

yields the series representation

$$\mathrm{I}(gh) = \sum_{j=0}^{\infty} \mathrm{I}_j(gh). \tag{7.17}$$

By using the transformation $x = \overline{x} + b + (j-1)q$ and taking into account (7.16), the terms in this series can be expressed as

$$\mathrm{I}_j(gh) = (-1)^{j-1} \int_0^q g(\overline{x} + b)h(\overline{x} + b + (j-1)q)\,d\overline{x}.$$

If the sign of the remaining integral is independent of j, then the series (7.17) is *alternating*. This presupposition is satisfied, for instance, when g is of one sign in $[b, b+q]$. Alternatively, if h has a *Poincaré-type* asymptotic expansion,

$$h(x) \sim x^\gamma \sum_{k=1}^{\infty} \frac{d_k}{x^{k-1}}, \tag{7.18}$$

with $\gamma < 0$,[78] then the representation

$$\mathrm{I}_j f = (-1)^{j-1} \sum_{k=1}^{\infty} \frac{D_k}{(c+j)^{-\gamma+k-1}}, \tag{7.19}$$

[78] It is assumed that this series is absolutely convergent to $h(x)$ for any $x \in [b, \infty)$.

where $c = b/q$, can easily be derived. The set of functions in $C^\infty[a, \infty)$ which have Poincaré-type asymptotic expansions (7.18) with $d_1 \neq 1$ is denoted $A^{(\gamma)}$, $\gamma \in \mathbb{R}$. The representation (7.19) shows that the series (7.17) becomes alternating for sufficiently large values of j irrespective of whether g is of one sign in $[b, b+q]$ or not, provided (7.18) holds for h.

The representation (7.19) also shows that the sequence of partial sums $\{s_m\}$ of the series (7.17),

$$s_m := \sum_{j=0}^{m} \mathrm{I}_j(gh), \quad m = 1, 2, 3, \ldots,$$

is *logarithmically* convergent (see Section 3.7.7). Therefore, the Euler transform (3.33) can be used to accelerate the convergence of $\{s_m\}$. Since the transformed sequences can be seen to again converge logarithmically, the Euler transform can be applied iteratively to accelerate the convergence of $\{s_m\}$. Modifications of the Euler transform which are particularly efficient for series (7.17) with (7.19) are described in Espelid and Overholt [140].

If h does not have a Poincaré-type asymptotic expansion, (modified) Euler transforms may not be suitable for accelerating the convergence of (7.17). Note that there is no *universal* acceleration algorithm for the partial sums of alternating series (Brezinski and Redivo Zaglia [41]). Nevertheless, some of the extrapolation methods described in Section 3.7 (in particular, Levin's u-transform and the ε-algorithm) have been shown, both on theoretical and experimental grounds, to be effective for broad classes of alternating series.

Software (Numerical Integration Based on Acceleration of Alternating Series). The subroutines `TOMS/639`, `QUADPACK/dqawf`, and `NUMERALGO/na6` are based on methods for accelerating the convergence of alternating series and use the Euler transform, the ε-algorithm, and modified Euler transforms, respectively. Both `QUADPACK/dqawf` and `NUMERALGO/na6` use *adaptive* integration algorithms to compute approximations to the integrals I_j, $j = 0, 1, \ldots$, and are therefore dealt with in Section 8.3.3. By contrast, `TOMS/639` uses a fixed quadrature formula Q_n (either a compound trapezoidal rule or a Gauss formula) with a user-specified number of abscissas n for approximating the integrals I_j, $j = 0, 1, \ldots$. In addition to the problem parameters a, g, and h, the user must provide the algorithmic parameters b and q, and, if a Gauss formula is to be used, a routine which generates the abscissas and weights of the Gauss formula.

***D*- and *W*-Transformation Methods.** Levin and Sidi [260] demonstrated that for a large class of improperly Riemann integrable functions $f : [a, \infty) \to \mathbb{R}$ the "tail" integral $G(\beta) := \mathrm{I}(f; [a, \infty)) - \mathrm{I}(f; [a, \beta)) = \mathrm{I}(f; [\beta, \infty))$ satisfies a differential equation of the form

$$(7.20) \quad b_J(\beta) G^{(J)}(\beta) + b_{J-1}(\beta) G^{(J-1)}(\beta) + \cdots + b_1(\beta) G'(\beta) + G(\beta) = 0,$$

where

(7.21) $\qquad b_j \in A^{(\sigma_j)}, \quad \sigma_j \in \mathbb{N}, \ \sigma_j \le j \quad \text{for } j = 1, 2, \ldots, J.$

This result can be proven assuming that f itself satisfies (7.20) for a suitable set of coefficient functions $\tilde{b}_1, \ldots, \tilde{b}_J$ and that

$$\lim_{x\to\infty} \tilde{b}_j^{(i-1)}(x) f^{(j-i)}(x) = 0$$

for $j = i, i+1, \ldots, J$ and $i = 1, \ldots, J$. Using the relationship $G^{(j)} = -f^{(j-1)}$, $j = 1, 2, \ldots$, and the representation (7.18) of the coefficient functions $b_1, \ldots, b_J$,

$$b_j(x) = x^{\sigma_j} \sum_{k=1}^{\infty} \frac{B_{j,k}}{x^{k-1}},$$

equation (7.20) can be written as

$$G(\beta) = \sum_{j=1}^{J} \beta^{\sigma_j} f^{(j-1)}(\beta) \sum_{k=1}^{\infty} \frac{B_{j,k}}{\beta^{k-1}}. \tag{7.22}$$

From the representation (7.22), the asymptotic expansion

$$G(\beta) \sim \sum_{j=1}^{J} \beta^{\sigma_j} f^{(j-1)}(\beta) \sum_{k=1}^{K_j} \frac{B_{j,k}}{\beta^{k-1}} \quad \text{as} \quad \beta \to \infty \tag{7.23}$$

follows for any choice of $\mathbf{K} = (K_1, \ldots, K_J)^\top \in \mathbb{N}^J$. This asymptotic expansion can be used to construct corresponding sequence acceleration methods as follows. Let $\beta_1 < \beta_2 < \beta_3 < \cdots$ be a monotonic sequence in (a, ∞) with $\beta_m \to \infty$ as $m \to \infty$, and let the sequence $\{s_m\}$ be defined by

$$s_m := \mathrm{I}(f; [a, \beta_m]), \quad m = 1, 2, 3, \ldots.$$

Then, the elements of the transformed sequence $\{s^{(m;\mathbf{K})}\}$ are obtained by solving the linear systems of equations

$$\begin{aligned} s^{(m;\mathbf{K})} - s_l &= \sum_{j=1}^{J} \beta_l^{\sigma_j} f^{(j-1)}(\beta_l) \sum_{k=1}^{K_j} \frac{B_{j,k}}{\beta_l^{k-1}}, \\ l &= m, m+1, \ldots, m + \|\mathbf{K}\|_1 \end{aligned} \tag{7.24}$$

for $m = 1, 2, \ldots$. This sequence transformation is known as the *D-transform*. Note that, for $J = 1$, setting

$$t_m := 1/\beta_m, \ m = 1, 2, 3, \ldots, \quad \text{and} \quad g(t) := (1/t)^{\sigma_j} f(1/t)$$

converts (7.23) into the form (3.29); i.e., the D-transform becomes a particular instance of the W-transform described in Section 3.7.6. Hence, for $J = 1$, the D-transform can be efficiently computed using the recursive W-algorithm (3.30), whereas for $J > 1$ a generalization of the W-algorithm—the $W^{(J)}$-*algorithm* described in Ford and Sidi [150]—can be used to calculate $\{s^{(m;\mathbf{K})}\}$.

Example (D-Transform). The integrals

$$\mathrm{I}f_q = \int_0^\infty e^{-x^{q+1}}\, dx, \quad q = 1, 2, 3,$$

were approximately calculated using the D-transform based on the sequence of equidistant points

$$\beta_m = m/10, \quad m = 1, 2, 3, \ldots.$$

The integrands f_q and the corresponding tail integrals G_q can be seen to satisfy the differential equation (7.20) for $J = 1$ and $\sigma_1 = 0$. The sequence elements $s_1, \ldots, s_{10}$ were approximately computed with a relative accuracy $\varepsilon_{\text{rel}} = 10^{-14}$ using the subroutine NAG/d01auf. The following table shows the relative errors of $s_{10} = s^{(10;0)}$ and of the corresponding extrapolated values $s^{(10-k;k)}$, $k = 1, 3, 5, 7, 9$, for f_1, f_2, and f_3.

k	$E^{\text{rel}} f_1$	$E^{\text{rel}} f_2$	$E^{\text{rel}} f_3$
0	$1.6 \cdot 10^{-1}$	$9.6 \cdot 10^{-2}$	$6.8 \cdot 10^{-2}$
1	$6.2 \cdot 10^{-2}$	$5.7 \cdot 10^{-2}$	$5.1 \cdot 10^{-2}$
3	$4.3 \cdot 10^{-3}$	$8.6 \cdot 10^{-4}$	$3.3 \cdot 10^{-3}$
5	$3.5 \cdot 10^{-4}$	$1.3 \cdot 10^{-3}$	$1.6 \cdot 10^{-3}$
7	$6.2 \cdot 10^{-5}$	$6.2 \cdot 10^{-4}$	$4.9 \cdot 10^{-4}$
9	$1.1 \cdot 10^{-4}$	$4.2 \cdot 10^{-4}$	$1.8 \cdot 10^{-4}$

The basic D-transform can be modified in different ways in order to decrease the computational effort required for the calculation of $\{s^{(m;\mathbf{K})}\}$ and/or to make the D-transform easier to use. (i) The effort required for determining the exponents $\sigma_1, \ldots, \sigma_J$ in (7.21) can be avoided by setting $\sigma_1 = 1, \ldots, \sigma_J = J$, as suggested by Sidi [393], without significantly decreasing the accuracy of the extrapolated values. (ii) The functions $\beta \mapsto \beta^{\sigma_j} f^{(j-1)}(\beta)$ in (7.23) can be replaced with easily computable functions. (iii) If f is oscillatory, then choosing the elements of the sequence $\{\beta_m\}$ as the consecutive zeros of one or more functions $f, f', \ldots, f^{(J-1)}$ reduces the computational effort required for the calculation of $\{s^{(m;\mathbf{K})}\}$. Detailed analyses of the D-transform and its modifications can be found in Ford and Sidi [150], Levin and Sidi [260], and Sidi [389], [390], [393], [395].

Sidi [392] proposed a convergence acceleration technique, closely related to the D-transform, which is applicable to a class of oscillatory integrands. These integrands are improperly Riemann integrable over $[a, \infty)$ and can be written in the form

$$f(x) = u(\theta(x))e^{\varphi(x)}h(x),$$

where

(i) u is any linear combination of sin and cos;

(ii) $\theta \in A^{(l_\theta)}$, $l_\theta \in \mathbb{N}$;

(iii) $\varphi \in A^{(l_\varphi)}$, $l_\varphi \in \mathbb{N}_0$, and, if $l_\varphi \geq 1$, $\varphi(x) \to -\infty$ as $x \to \infty$; and

(iv) $h \in A^{(\gamma)}$, $\gamma \in \mathbb{R}$.

If $\bar{\theta}$ and $\bar{\varphi}$ denote the polynomial parts of θ and φ, respectively,[79] the representation

$$(7.25)\qquad \int_{\beta}^{\infty} f(x)\,dx = \beta^{\sigma+\gamma} e^{\bar{\varphi}(x)} \left[\cos(\bar{\theta}(x)) b_{\mathrm{c}}(x) + \sin(\bar{\theta}(x)) b_{\mathrm{s}}(x)\right]$$

can be derived, where b_{c}, $b_{\mathrm{s}} \in A^{(0)}$, and

$$\sigma := 1 - \max\{l_{\theta}, l_{\varphi}\}.$$

This asymptotic expansion can be used to construct corresponding sequence acceleration methods as follows. Let $\beta_1 < \beta_2 < \beta_3 < \cdots$ be a monotonic sequence of zeros of $\sin(\bar{\theta}(x))$ in (a, ∞) such that consecutive elements of the sequence $\{\cos(\bar{\theta}(\beta_m)) \in \{-1, 1\}\}$ are alternating in sign.[80] For the sake of simplicity, assume $\cos(\bar{\theta}(\beta_1)) = -1$. Then,

$$(7.26)\qquad \int_{a}^{\infty} f(x)\,dx = \int_{a}^{\beta_m} f(x)\,dx + g(\beta_m) b_{\mathrm{c}}(\beta_m), \qquad m = 1, 2, 3, \ldots,$$

is obtained from (7.25), where

$$(7.27)\qquad g(\beta_m) = (-1)^m \beta_m^{\sigma+\gamma} e^{\bar{\varphi}(\beta_m)}.$$

Setting $s := \mathrm{I}(f; [a, \infty))$, $s_m := \mathrm{I}(f; [a, \beta_m))$,

$$t_m := 1/\beta_m,\ m = 1, 2, 3, \ldots, \qquad \text{and} \quad b_{\mathrm{c}}(\beta_m) = \sum_{k=1}^{\infty} \frac{B_k}{\beta_m^{k-1}}$$

converts the asymptotic expansion (7.26) into the form (3.29). Hence, the W-algorithm (3.30) can be used to accelerate the convergence of $\{s_m\}$. Specifically, the results

$$s^{(m;k)} - s = O(\beta_m^{\sigma+\gamma-k} e^{\bar{\varphi}(\beta_m)}) \quad \text{as} \quad m \to \infty,$$

and

$$s^{(m;k)} - s = o(k^{-\mu}) \quad \text{as} \quad k \to \infty \quad \text{for any} \quad \mu > 0$$

are established in Sidi [392].

Note that evaluating the function g in (7.27) requires knowledge of the constants σ and γ and of the polynomial $\bar{\varphi}$, which may be difficult or impossible to obtain. This problem can be remedied by replacing (7.27) with

$$g(\beta_m) := \int_{\beta_m}^{\beta_{m+1}} f(x)\,dx.$$

---

[79]The polynomial parts are obtained by omitting the terms with negative exponents in the respective Poincaré-type expansions (7.18).

[80]Clearly, the roles of sin and cos in this method are interchangeable.

Sidi [394] showed that the corresponding *modified* W-transform essentially retains the previously stated rates of convergence. The modified W-transform is also useful in that it often results in accelerated convergence when applied to more general integrands of the form

$$f(x) = u(\theta(x))h(x),$$

where u and θ have the aforementioned properties and h is a general nonoscillatory function.

Example (Modified W-Transform). The integrals

$$\mathrm{I}f_q = \int_0^\infty \sin(x)e^{-x}x^{2^{q-1}}\,dx, \quad q = 1, 2, 3, 4,$$

were approximately calculated using the modified W-algorithm based on the zeros $\beta_m = m\pi$, $m = 1, 2, \ldots$, of $\sin(\bar{\theta}(x)) = \sin(x)$. The sequence elements $s_1, \ldots, s_6$ were approximately computed with a relative accuracy $\varepsilon_{\text{rel}} = 10^{-14}$ using the subroutine `NAG/d01auf`. The following table shows the relative errors in $s_6 = s^{(6;0)}$ and of the corresponding extrapolated values $s^{(6-k;k)}$, $k = 1, \ldots, 5$, for $\mathrm{I}f_q$, $q = 1, \ldots, 4$.

k	$E^{\text{rel}} f_1$	$E^{\text{rel}} f_2$	$E^{\text{rel}} f_3$	$E^{\text{rel}} f_4$
0	$1.3 \cdot 10^{-7}$	$2.6 \cdot 10^{-6}$	$5.1 \cdot 10^{-4}$	$6.2 \cdot 10^{-2}$
1	$1.5 \cdot 10^{-10}$	$6.9 \cdot 10^{-9}$	$1.2 \cdot 10^{-6}$	$1.5 \cdot 10^{-3}$
2	$4.0 \cdot 10^{-12}$	$8.2 \cdot 10^{-12}$	$5.7 \cdot 10^{-8}$	$1.8 \cdot 10^{-4}$
3	$2.1 \cdot 10^{-13}$	$1.1 \cdot 10^{-11}$	$4.6 \cdot 10^{-9}$	$2.2 \cdot 10^{-6}$
4	$2.1 \cdot 10^{-14}$	$2.3 \cdot 10^{-12}$	$1.0 \cdot 10^{-10}$	$7.8 \cdot 10^{-6}$
5	$4.1 \cdot 10^{-15}$	$7.1 \cdot 10^{-13}$	$3.0 \cdot 10^{-10}$	$4.2 \cdot 10^{-6}$

The (modified) W-transform requires the generation of suitable sequences $\beta_1 < \beta_2 < \beta_3 < \cdots$ of zeros of $\sin(\bar{\theta}(x))$. Such sequences can generally be determined as follows. Let β_1 be the smallest zero of $\sin(\bar{\theta}(x))$ greater than a. This means that β_1 is a root of the polynomial equation $\bar{\theta}(x) = q\pi$ for a particular $q \in \mathbb{N}$. Then, depending on whether $\theta(x) \to \infty$ or $\theta(x) \to -\infty$ as $x \to \infty$, determine β_m, $m = 2, 3, \ldots$, as the *greatest* root of the polynomial equation $\bar{\theta}(x) = (q+m-1)\pi$ or $\bar{\theta}(x) = (q-m+1)\pi$, respectively. The elements of the resulting sequence ultimately become consecutive zeros of $\sin(\bar{\theta}(x))$ and can easily be seen to have the necessary properties.

Confluent Extrapolation Methods. D-transform sequences $\{s^{(m;\mathbf{K})}\}$, as defined by (7.24), are obtained from the asymptotic expansion (7.23) by replacing the asymptotic equality relationship with exact equality and requiring the resulting equation to hold for as many distinct points β_l as there are unknown quantities in (7.23), giving rise to a system of $\|\mathbf{K}\|_1 + 1$ linear equations. An alternative way to obtain systems of linear equations from which the unknown quanities in (7.23) can be approximately determined is to differentiate the asymptotic expansion (7.23) with respect to β:

$$(7.28) \quad G^{(l)}(\beta) = -f^{(l-1)}(\beta) \sim \sum_{j=1}^{J} \sum_{k=1}^{K_j} B_{j,k}((\cdot)^{\sigma_j - k+1} f^{(j-1)})^{(l)}(\beta) \quad \text{as} \quad \beta \to \infty$$

for $l = 1, 2, \ldots$. Now, the $\|\mathbf{K}\|_1 + 1$ unknowns in (7.23) can be approximately determined by requiring that (7.23) and the expansions (7.28), for $l = 1, \ldots, \|\mathbf{K}\|_1$, hold strictly as equations:

$$(7.29) \quad \begin{aligned} s^{(\mathbf{K})}(\beta) - F(\beta) &= \sum_{j=1}^{J} \beta^{\sigma_j} f^{(j-1)}(\beta) \sum_{k=1}^{K_j} \frac{B_{j,k}}{\beta^{k-1}}, \\ -f^{(l-1)}(\beta) &= \sum_{j=1}^{J} \sum_{k=1}^{K_j} B_{j,k}((\cdot)^{\sigma_j - k+1} f^{(j-1)})^{(l)}(\beta), \\ & \quad l = 1, 2, \ldots, \|\mathbf{K}\|_1, \end{aligned}$$

where $F(\beta) := \mathrm{I}(f; [a, \beta])$ and $s^{(\mathbf{K})}(\beta)$ is an extrapolated approximation of $\mathrm{I}(f; [a, \infty))$.

The extrapolation procedure described above is fundamentally different from the extrapolation procedures described previously in this section and from those in Section 3.7 in two ways.

(i) The argument β in the extrapolated value $s^{(\mathbf{K})}(\beta)$ can vary continuously in the interval $[a, \infty)$, and thus the above extrapolation procedure is said to be a *continuous prediction* algorithm. Continuous prediction algorithms are *function transformations* rather than sequence transformations in that they transform functions into other functions (for example, F into $s^{(\mathbf{K})}$) such that the transformed functions can be expected to converge faster than the original ones.

(ii) The computation of the approximation $s^{(\mathbf{K})}(\beta)$ requires only a *single* value, $F(\beta)$, of the function to be transformed, but also requires the values of a number of derivatives of F, i.e., f, f', $\ldots$, at β. By contrast, computing each element of a transformed sequence involves two or more (consecutive) elements of the original sequence and usually requires (weighted) differences between elements of the original and intermediate sequences.

The analogy between the differences of consecutive sequence elements used in sequence transformations and the functional derivatives used in function transformations mentioned in (ii) points to the following alternative way of obtaining $s^{(\mathbf{K})}(\beta)$. Let the sequence $\{\beta_m\}$ be defined by $\beta_m = \beta + mh$, $m =$

$1, 2, \ldots$. Then, the elements of the D-transform sequence $\{s^{(m;\mathbf{K})}\}$ defined in (7.24) can be seen to converge to $s^{(\mathbf{K})}(\beta)$ as $h \to 0+$,

$$\lim_{h\to 0+} s^{(m;\mathbf{K})} = s^{(\mathbf{K})}(\beta) \quad \text{for any} \quad m \in \mathbb{N}.$$

Continuous prediction methods which are obtained from sequence transformations by such a limiting process are said to be *confluent extrapolation methods*.

The above confluent version of the D-transform is introduced and analyzed in Gray and Wang [181]. In a number of numerical examples presented in this paper, the confluent D-transform compares quite favorably to the original D-transform.

7.2.5. Calculation of Fourier Transforms. The Fourier transform and its inverse (see Section 2.2) require integrals of the form

$$\mathrm{I}_y f = \int_{-\infty}^{+\infty} e^{\mathrm{i}xy} f(x)\, dx. \tag{7.30}$$

Often, $\mathrm{I}_y f$ is represented as $\mathrm{I}_y f = \mathrm{I}_y^{\mathrm{C}} f_{\mathrm{e}} + \mathrm{i}\mathrm{I}_y^{\mathrm{S}} f_{\mathrm{o}}$, where

$$f_{\mathrm{e}}(x) := f(x) + f(-x), \qquad f_{\mathrm{o}}(x) := f(x) - f(-x),$$

and

$$\mathrm{I}_y^{\mathrm{C}} f_{\mathrm{e}} = \int_0^{\infty} \cos(xy) f_{\mathrm{e}}(x)\, dx \tag{7.31}$$

and

$$\mathrm{I}_y^{\mathrm{S}} f_{\mathrm{o}} = \int_0^{\infty} \sin(xy) f_{\mathrm{o}}(x)\, dx \tag{7.32}$$

are the *Fourier cosine* and *sine transforms* of f_{e} and f_{o}, respectively.

Miscellaneous Integration Methods. The integrals (7.30), (7.31), or (7.32) usually have to be evaluated for a number of values $y = y_1, \ldots, y = y_M$. In this case, it is important to use the same (or, at least, highly overlapping) sets of abscissas in calculating the integrals in order to be able to reuse f-values. If the same set of n values $S(f) = \{f(x_1), \ldots, f(x_n)\}$ is used in all integrals, then the number of arithmetic operations and exponential function evaluations required to approximately calculate $\mathrm{I}_{y_1} f, \ldots, \mathrm{I}_{y_M} f$ for a given sample $S(f)$ is generally of the order $O(nM)$. This computational effort is not significant if M is small compared to n or if the computational cost of an f-evaluation is large in comparison to that of an arithmetic operation and an exponential function evaluation.

In applications which can be subsumed under this general problem scenario, f can usually be evaluated at any point $x \in \mathbb{R}$. Hence, any method for dealing with integrals on unbounded regions described in this section—in particular, any method tailored to oscillatory integrands—can, in principle, be used to calculate Fourier transforms. For instance, Ooura and Mori [326] compute Fourier transforms using infinite compound trapezoidal rules in conjunction with certain periodizing transformations (see Section 7.2.2); Piessens and Branders [337] employ a product integration approach based on Chebyshev series expansions (see Section 7.2.3); and Lund [266] proposes a sinc method (see Section 7.2.2). An extrapolation method based on accelerating the convergence of alternating sequences is used in the subroutine `QUADPACK/dqawf` (see Section 7.2.4).

Note, however, that in most of these methods the set of abscissas used is dependent upon the parameter y, which makes it difficult (or even impossible) to reuse previously computed f-values if multiple Fourier transforms have to be calculated. A notable exception are product integration methods (see, for instance, Piessens and Branders [337]).

Discrete Fourier Transform. The problem scenario described previously does *not* apply to a large number of applications of the Fourier transform, for instance, applications concerned with signal representation and manipulation. In these applications, obtaining the sample $S(f)$ requires little computational effort, or no computational effort at all, but f is often available only at a potentially large number of equidistant sampling points. For instance, $S(f)$ may be obtained from the output of a Digital/Analog converter.

Functions given at equidistant points are most easily integrated using compound trapezoidal rules (see Section 7.2.2). Suppose the samples available correspond to the values of f at the points $x_j = jh$, $j = 0, \dots, n-1$, where $h > 0$ is the sampling rate. Then,

$$T_y f := h \sum_{j=0}^{n-1} e^{\mathrm{i}jhy} f(jh)$$

can be used as an approximation for $\mathrm{I}_y f$. Furthermore, for reasons whose thorough discussion is beyond the scope of this book (see, for instance, Ueberhuber [435]), the Fourier transform is usually to be computed for the values

$$y_k := (2\pi k)/(nh), \quad k = 0, 1, \dots, n-1.$$

Setting

$$f_j := f(jh), \quad j = 0, 1, \dots, n-1,$$

one obtains

$$(7.33) \qquad T_{y_k} f = hT_k f, \quad T_k f := \sum_{j=0}^{n-1} f_j e^{2\pi \mathrm{i}jk/n}, \quad k = 0, 1, \dots, n-1.$$

The linear transformation $(f_0, \ldots, f_{n-1})^\top \to (T_0 f, \ldots, T_{n-1} f)^\top$ defined in (7.33) is known as the *discrete Fourier transform* (DFT). The discrete Fourier transform can be analyzed, and a number of interesting properties can be established, completely independently of its connection to the Fourier integral transform (see, for instance, Van Loan [437]).

Clearly, a naive implementation of the discrete Fourier transform requires $O(n^2)$ floating-point multiplications and additions for n data values. However, unless n is a prime number, the particular structure of the discrete Fourier transform makes it possible to significantly reduce the number of required floating-point operations by eliminating redundant operations (see Van Loan [437] for details). The resulting algorithms are called *fast Fourier transforms* (FFTs). Specifically, if n can be factored as

$$n = n_1 n_2 \cdots n_K, \qquad n_1, n_2, \ldots, n_K \in \mathbb{N},$$

the number of floating-point operations can be reduced to $O(n(n_1 + \cdots + n_K))$. In particular, if $n = 2^K$, FFT algorithms require $O(n \log n)$ operations.

Software (FFT). There exists a large number of software products for calculating FFTs. (For a more comprehensive survey, see Ueberhuber [435].) Highly recommendable is the FFTPACK package developed by P. N. Swarztrauber of the National Center of Atmospheric Research. FFTPACK can be obtained from NETLIB but has also been incorporated into the CMLIB, IMSL, and SLATEC libraries. Alternative high-quality FFT implementations are provided by the NAG Fortran 77 library in Chapter C06.

7.2.6. Calculation of the Inverse Laplace Transform. The Laplace transform (2.8) and the inverse Laplace transform (2.9) are powerful mathematical tools used primarily to solve ordinary or partial differential equations (see Section 2.2). The principal difficulty in using them is that it is often not possible to calculate Bromwich's integral,

$$f(x) = \frac{1}{2\pi \mathrm{i}} \int_{\gamma - \mathrm{i}\infty}^{\gamma + \mathrm{i}\infty} e^{xs} F(s)\, ds, \tag{7.34}$$

which is generally used to invert the Laplace transform analytically. Choosing the contour in (7.34) as $s = \gamma + \mathrm{i}t$, $t \in (-\infty, +\infty)$, yields the Fourier integral

$$f(x) = \frac{e^{\gamma x}}{2\pi} \int_{-\infty}^{\infty} e^{\mathrm{i}xt} F(\gamma + \mathrm{i}t)\, dt. \tag{7.35}$$

This shows that Bromwich's integral is ill conditioned with respect to changes in F, and therefore its numerical calculation is likely to be beset with considerable difficulties (see Section 3.4.1).

Indeed, the numerical inversion of the Laplace transform has been found so difficult that the topic has been covered extensively in a large number of publications. Surveys of and comparisons between different methods can be found in Davies and Martin [89] and, more recently, in Duffy [119]. Clearly, it is not possible to give a comprehensive survey of numerical methods for inverting the Laplace transform in this book. Instead, we focus on methods for which high-quality implementations are readily available. These methods generally assume that F is provided as a function subprogram and that this function subprogram can be used to evaluate F at any complex value z for which $F(z)$ is defined. Methods which are applicable when F is available only for real abscissas or only for an equidistant set of real abscissas can, for instance, be found in Ang, Lund, and Stenger [7] and Rodriguez and Seatzu [374], respectively.

Accelerated Trapezoidal Rules for Bromwich's Integral. A simple method (due to Crump [73]) for dealing with Bromwich's integral is applying a truncated trapezoidal rule $T_{h;n}$ to (7.35), which yields

$$f_{\gamma;h;n}(x) = \frac{he^{\gamma x}}{\pi}\left[\frac{F(\gamma)}{2} + \sum_{j=1}^{n} \quad (\cos(jhx)\,\mathrm{Re}F(\gamma + \mathrm{i}jh) \right.$$
$$\left. - \sin(jhx)\,\mathrm{Im}F(\gamma + \mathrm{i}jh))\right],$$

and accelerating the convergence of $f_{\gamma;h;n}(x)$ as $n \to \infty$ using extrapolation methods (see Section 3.7). In particular, the ε-algorithm (see Section 3.7.4) and Levin's u-transform (see Section 3.7.5) have been found useful for this purpose.

Software (Accelerated Trapezoidal Rules for Bromwich's Integral). `NAG/c06laf` implements Crump's method using the ε-algorithm for convergence acceleration purposes. The routine simultaneously inverts the Laplace transform to a user-requested relative accuracy for m user-specified points $x_1, \ldots, x_m$. The user can also specify the step size h and an upper bound on the cut-off index n to be used. Additionally, in order for the routine to function reliably, it is crucial that the user provide a correct upper bound on the abscissa of convergence γ_0. The routine calculates the approximations $f_{\gamma;h;n}(x_j)$, $j = 1, \ldots, m$, for two or more values of $\gamma > \gamma_0$ in order to obtain corresponding error estimates.

`TOMS/619` provides a simpler, but significantly less flexible and reliable alternative implementation of Crump's method based on the ε-algorithm.

Talbot's Method. The basic idea in Talbot's method (Talbot [425]) is changing the contour in Bromwich's integral (7.34) in such a way that trapezoidal rules are significantly more accurate for the transformed integral than for the original integral (7.35). Specifically, if C denotes the new contour, then it is desirable that as $|s| \to \infty$, $s \in C$, $\mathrm{Im}\, s$ remains bounded, which limits the oscillations in the factor e^{xs}, and that $\mathrm{Re}\, s \to -\infty$, which leads to a rapid decay in e^{xs}.

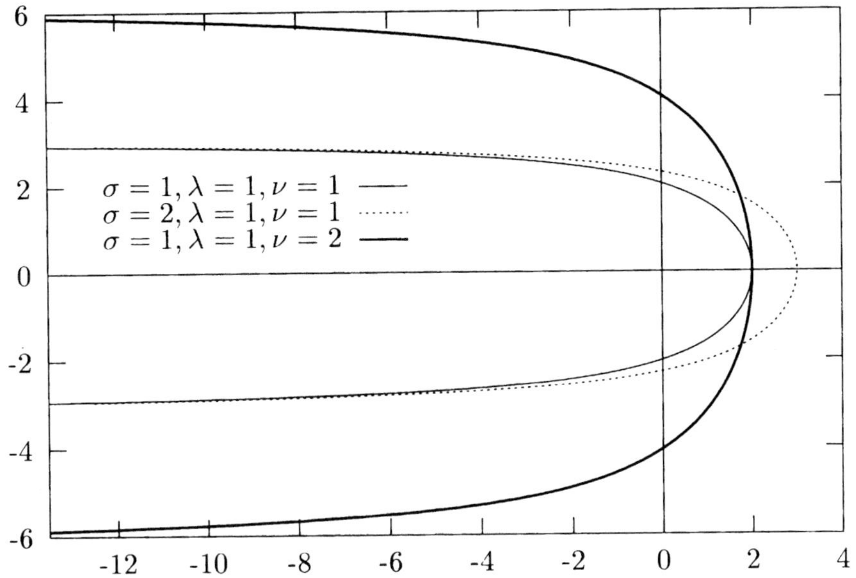

FIG. 7.2. *Talbot's contours.*

The particular contour proposed by Talbot (see Fig. 7.2) can be parameterized using

$$(7.36)\quad s(t) = \sigma + \lambda s_\nu(t) \quad \text{with} \quad s_\nu(t) := t\cot(t) + \mathrm{i}\nu t, \quad t \in (-\pi, \pi).$$

This transforms (7.34) into

$$(7.37)\qquad f(x) = \frac{\lambda e^{\sigma x}}{2\pi i}\int_{-\pi}^{\pi} e^{\lambda x s_\nu(t)} F(\sigma + \lambda s_\nu(t)) s_\nu'(t)\, dt.$$

(7.37) yields the same value as (7.34) if (i) $|F(s)| \to 0$ as $|s| \to \infty$ uniformly for $\{s : \mathrm{Re}(s) < \gamma_0\}$, and (ii) the region bounded by the original and by Talbot's contour does not contain any singularities in F. Note that $e^{\lambda x s_\nu(t)} \to 0$ as $t \to \pm\pi$, and thus applying (finite) compound trapezoidal rules T_l to (7.37) can be expected to yield highly accurate results.

The geometric parameters σ, λ, and ν must be chosen with particular care since they can have a significant impact on the accuracy of the method (see Talbot [425]).

Software (Talbot's Method). Talbot's method is implemented in `TOMS/682` (Murli and Rizzardi [309]). Two top-level routines are available. The first routine determines the geometric parameters σ, λ, and ν as well as an appropriate value of l (in T_l) given

(i) the abscissa x for which the Laplace transform is to be inverted,

(ii) the locations and kinds of the singularities of F, and

(iii) the accuracy required.

The second routine applies the trapezoidal rule T_l to (7.37) for a user-supplied Laplace transform F, given specific values of x, σ, λ, ν, and l.

Weeks's Method. Weeks's method is based on the following series representation of the inverse Laplace transform f:

$$f(x) = e^{\gamma x} \sum_{k=0}^{\infty} a_k e^{-bx/2} L_k(bx), \quad \gamma > \gamma_0, \ b > 0, \tag{7.38}$$

where L_k is the kth Laguerre polynomial. The coefficients $a_k = a_k(\gamma, b)$, $k = 0, 1, \dots$, in this expansion can be shown to be the Taylor coefficients of

$$G(z) = G(z; \gamma, b) = \frac{b}{1-z} F\left(\frac{b}{1-z} + \gamma - \frac{b}{2}\right), \tag{7.39}$$

where G is analytic in a circle $\{z : |z| < R_{\gamma,b}\}$, $R_{\gamma,b} \geq 1$ (Weeks [445]). Now, an approximation to f can be obtained by truncating the series in (7.38), say, after the index K, and calculating approximate values $\tilde{a}_k$ for the remaining coefficients a_k, $k = 0, 1, \dots, K$:

$$f(x) \approx \tilde{f}_{\gamma,b,K} = e^{\gamma x} \sum_{k=0}^{K} \tilde{a}_k e^{-bx/2} L_k(bx).$$

Clearly, the accuracy of $\tilde{f}_{\gamma,b,K}$ depends on rate of convergence of (7.38) and the accuracy of the approximate Taylor coefficients.

Suitable approximations $\tilde{a}_k$ can be obtained from the Cauchy integral representation of the Taylor coefficients of G:

$$a_k = \frac{1}{2\pi \mathrm{i}} \int_C \frac{G(z)}{z^{k+1}}\, dz, \tag{7.40}$$

where C is a contour which contains the origin, but does not contain any singularity of G. Choosing C as the circle with radius r yields

$$a_k = \frac{1}{2\pi r^k} \int_0^{2\pi} G(re^{\mathrm{i}t}) e^{-\mathrm{i}kt}\, dt. \tag{7.41}$$

Since the integrand in (7.41) is periodic, compound trapezoidal rules T_l can be used to calculate the Taylor coefficients a_k with high accuracy (Lyness and Giunta [276]). Applying T_l (l even) to (7.41) gives

$$\tilde{a}_k = \frac{2}{r^k l} \sum_{j=1}^{l/2} \left(\cos(jkh) \mathrm{Re}\, G(re^{\mathrm{i}jh}) + \sin(jkh) \mathrm{Im}\, G(re^{\mathrm{i}jh})\right), \quad h = 2\pi/l.$$

The algorithmic parameters γ, b, and r must be chosen carefully since they can have a significant impact on the accuracy of the method. See Garbov et al. [155] and Giunta, Laccetti, and Rizzardi [174] for detailed discussions.

A particular advantage of Weeks's method is that, once the coefficients $\tilde{a}_0, \ldots, \tilde{a}_K$ have been obtained, the approximate inverse Laplace transform $\tilde{f}_{\gamma,b,K}(x)$ can be evaluated quite cheaply for any number of x values.

Software (Weeks's Method). An implementation of Weeks's method, as described above, is provided in `TOMS/662` (Garbov et al. [155], [154]) $\approx$ `NAG/c06lbf` and `NAG/c06lcf`. For a given Laplace transform F, `NAG/c06lbf` calculates the approximate Taylor coefficients $\tilde{a}_0, \ldots, \tilde{a}_K$ to a user-specified accuracy. The user must provide the abscissa of convergence γ_0, the algorithmic parameters γ and b, and an upper bound for the truncation index K. The information produced by `NAG/c06lbf` can then be used by `NAG/c06lbc` to evaluate $\tilde{f}_{\gamma,b,K}$ at m user-specified points $x_1, \ldots, x_m$.

7.3. Weakly Singular Integrals

A function $f\colon B \subseteq \mathbb{R}^N \to \mathbb{R}$ is said to have a *singularity* or *singular point* at a limit point $\mathbf{x}_0$ of B if f is unbounded in any neighborhood of $\mathbf{x}_0$. An integrand f is called *singular* if f has one or more singularities within or at the boundary of the integration region B. Most common integration formulas Q_n are based on approximating f using model functions g—(piecewise) polynomials, for instance—which are *bounded* on any bounded domain. Since the maximum norm $\|g - f\|_\infty$ of the resulting approximation error is infinite, the integral $\mathrm{I}g$ cannot generally be expected to yield a reasonably accurate approximation to $\mathrm{I}f$. This is why special computational techniques are required for the accurate calculation of singular integrals.

This section deals with *weakly* singular integrands which are improperly Riemann integrable over B. *Strongly* singular integrals, i.e., Cauchy singular and hypersingular integrals, are treated in Sections 7.4 and 7.5, respectively.

7.3.1. Practical Problem Settings.

The choice of an appropriate computational method for solving a given integration problem generally depends heavily on the particular setting of the problem at hand. This section highlights problem settings which are particularly relevant to singular integrals.[81]

Unknown Singular Integrands. In *quadrature methods* for solving integral equations, numerical integration methods are used for discretizing integral operators whose arguments depend on the unknown solutions f of the integral equations to be solved.

Example (Fredholm Linear Integral Equation of the Second Kind). *Nyström's* method for solving the *Fredholm linear integral equation of the second kind*,

$$f(y) - \lambda \int_a^b k(x,y) f(x)\, dx \;=\; g(y), \qquad y \in [a,b], \tag{7.42}$$

[81] Some of the problem settings may arise in connection with nonsingular integrals as well.

is based on discretizing the integral operator in (7.42) using a properly chosen linear combination of function values,

$$f(y) - \lambda \sum_{i=1}^{n} c_i k(x_i, y) f(x_i) = g(y), \qquad y \in [a, b], \tag{7.43}$$

and requiring equation (7.43) to hold for the points $y = x_1, y = x_2, \ldots, y = x_n$. In other words, equation (7.43) is approximately solved by applying a collocation method which uses the quadrature abscissas as collocation points.

Since the integration abscissas and weights must be chosen before f can be (approximately) determined, *adaptive* integration methods (see Section 7.3.3), which are otherwise highly effective in dealing with singular integration problems, cannot readily be applied.

Known Analytical Properties of Singular Integrands. In many cases, tailor-made, highly efficient integration methods can be developed for singular integrands by exploiting known auxiliary information about the singular points. This information concerns the exact locations of the singularities $\mathbf{x}_0$ and/or the asymptotic behavior of $f(\mathbf{x})$ as $\mathbf{x} \to \mathbf{x}_0$.

Example (Fredholm Linear Integral Equation of the Second Kind). A typical kernel occurring in the integral equation (7.42) is

$$k(x, y) = |x - y|^{\alpha}, \qquad \alpha > -1.$$

For $\alpha < 0$, the resulting integral operator in (7.42) is singular at $x_0 = y$. Nevertheless, the solution f can be shown to be nonsingular on $[a, b]$ and smooth on (a, b), provided the right-hand side function g is sufficiently smooth (see, for instance, Hackbusch [194]). As a result, the natural representation $k(x, y)f(x)$ splits the integrand into the factor k, which is singular at $x_0 = y$, and the nonsingular, mostly smooth factor f. The factor k, which is given analytically, therefore completely determines the asymptotic behavior of $k(x, y)f(x)$ for x close to a singularity $x_0 = y \in (a, b)$.

Multiple Singular Integrals. The numerical treatment of singular integrals is made more complicated by the fact that in many applications a family of singular integrals of the form

$$\mathrm{I}_m f := \int_B k(\mathbf{x}, \mathbf{y}_m) f(\mathbf{x})\, d\mathbf{x}, \qquad m = 1, 2, \ldots, M, \tag{7.44}$$

must be discretized. The kernel in (7.44) is usually singular at $\mathbf{x} = \mathbf{y}$, but otherwise smooth. For this kind of problem, it is advantageous to use the same set of integration abscissas (but different integration weights) for discretizing $\mathrm{I}_m f$, $m = 1, \ldots, M$. (i) If f is fixed, then all f-values required for calculating $\mathrm{I}_1 f$ can be reused for calculating $\mathrm{I}_m f$, $m = 2, \ldots, M$. (ii) If f is the unknown solution of an integral equation involving the operator $f \mapsto \mathrm{I}(kf)$, then using different sets of abscissas for different indices $m = 1, \ldots, M$ makes it more difficult to construct systems of equations from which approximations of f can be obtained.

7.3.2. Conversion to Nonsingular Integrals.

Singular integration problems can be converted into (almost) equivalent, nonsingular problems by applying an appropriate *transformation technique*, *truncating* the independent variable, or *subtracting the singularity*.

Transformation to Nonsingular Integrals. Often, one of the transformation techniques described in Section 1.1.5 can be used to remove singularities in integration problems. If f is given in closed form, symbolic computation systems can greatly facilitate this task (see the example on page 110). Of particular importance in this context are the periodizing transformations discussed in Sections 5.3.4 and 6.5.4. These transformations convert integrands (defined on N-dimensional cubes) with boundary singularities into integrands which have smooth, periodic extensions. As a result, lattice rules, in particular, compound trapezoidal rules, usually prove to be particularly efficient for the transformed integrands (see Sections 5.3.2 and 6.5).

Singularities of N-dimensional integrands ($N \geq 2$) can often be removed by using N-dimensional spherical coordinates (1.8). In particular, the transformation

$$\int_B \frac{f(\mathbf{x})}{\|\mathbf{x}\|_2^\alpha}\, d\mathbf{x}$$
$$= \int_{\overline{B}} f(\psi(r, \varphi_1, \ldots, \varphi_{N-1}))\, r^{N-\alpha-1} q(\varphi_1, \ldots, \varphi_{N-1})\, d(r, \varphi_1, \ldots, \varphi_{N-1}),$$

where the trigonometric polynomial q is given by (1.9), shows that using spherical coordinates removes singularities of maximum order $\alpha \leq N - 1$.

Similarly, *Duffy's coordinates* can be used to remove singularities located at the vertices of simplices or cubes (Duffy [120]). In particular, let T denote the triangle with vertices $\mathbf{v}_1 = (0, 0)^\top$, $\mathbf{v}_2 = (1, 0)^\top$, and $\mathbf{v}_3 = (0, 1)^\top$. Then, the transformation

$$x_1 = \overline{x}_1 \quad \text{and} \quad x_2 = \overline{x}_1 \overline{x}_2 \tag{7.45}$$

yields

$$\int_T \frac{f(x_1, x_2)}{\sqrt{x_1^2 + x_2^2}^{\,\alpha}}\, d(x_1, x_2) = \int_{C_2} \frac{f(\overline{x}_1, \overline{x}_1\overline{x}_2)}{\overline{x}_1^{\alpha-1} \sqrt{1 + \overline{x}_2^2}^{\,\alpha}}\, d(\overline{x}_1, \overline{x}_2).$$

This shows that using Duffy's *triangular* coordinates removes singularities of maximum order $\alpha \leq 1$.

Truncation to Nonsingular Integrals. Singular points $\mathbf{x}_0$ of integrands f can be avoided by truncating the corresponding integration regions B

to $B \setminus B(\mathbf{x}_0)$ (i.e., by integrating f over $B \setminus B(\mathbf{x}_0)$ rather than over B), where $B(\mathbf{x}_0)$ is the intersection of a suitable neighborhood of $\mathbf{x}_0$ with B. Truncating integration regions about singular points is very similar to truncating unbounded integration regions (as described in Section 7.2.1), and consequently similar considerations apply. In particular, the reliability of this technique can be increased by "proceeding to the limit," i.e., by calculating sequences of truncated integrals:

$$\begin{aligned} \mathrm{Q}(f; B \setminus B_1(\mathbf{x}_0)) &\approx \mathrm{I}(f; B \setminus B_1(\mathbf{x}_0)), \\ \mathrm{Q}(f; B \setminus B_1(\mathbf{x}_0)) + \mathrm{Q}(f; B_1(\mathbf{x}_0) \setminus B_2(\mathbf{x}_0)) &\approx \mathrm{I}(f; B \setminus B_2(\mathbf{x}_0)), \\ &\vdots \end{aligned}$$

for $B \supset B_1(\mathbf{x}_0) \supset B_2(\mathbf{x}_0) \supset \cdots$.

Subtracting the Singularity. Suppose a singular integrand is given in the product form wf, where w is singular at $\mathbf{x}_0 \in B$, f is continuous at $\mathbf{x}_0$, and $\mathrm{I}\,w$ can be computed analytically. Then, it is often advantageous to split the integral $\mathrm{I}(wf) = \mathrm{I}_w f$ by writing

$$\int_B w(\mathbf{x}) f(\mathbf{x})\, d\mathbf{x} = \int_B w(\mathbf{x}) \left(f(\mathbf{x}) - f(\mathbf{x}_0)\right) d\mathbf{x} + f(\mathbf{x}_0) \int_B w(\mathbf{x})\, d\mathbf{x}, \tag{7.46}$$

or more concisely,

$$\mathrm{I}_w f = \mathrm{I}_w(f - f(\mathbf{x}_0)) + f(\mathbf{x}_0)\mathrm{I}\,w.$$

The remaining integral $\mathrm{I}_w(f - f(\mathbf{x}_0))$ is usually easier to deal with than the original integral $\mathrm{I}_w f$: Since $f(\mathbf{x}) \to f(\mathbf{x}_0)$ as $\mathbf{x} \to \mathbf{x}_0$, the function $w(\mathbf{x})(f(\mathbf{x}) - f(\mathbf{x}_0))$ is not singular, or at least less singular than $w(\mathbf{x})f(\mathbf{x})$, at $\mathbf{x}_0$. For instance, if f is continuously differentiable at $\mathbf{x}_0$ and

$$w(\mathbf{x}) = O(\|\mathbf{x} - \mathbf{x}_0\|^\alpha), \quad -1 < \alpha < 0, \qquad \text{as} \quad \mathbf{x} \to \mathbf{x}_0,$$

then $w(\mathbf{x})(f(\mathbf{x}) - f(\mathbf{x}_0))$ is Hölder continuous at $\mathbf{x}_0$.

While, consequently, standard formulas Q_n can be used for approximately integrating $w(f - f(\mathbf{x}_0))$, the convergence of $\mathrm{Q}_n(w(f - f(\mathbf{x}_0)))$ as $n \to \infty$ is likely to be slow because $w(f - f(\mathbf{x}_0))$ is not generally differentiable at $\mathbf{x}_0$. This problem can be remedied by subtracting from f the Taylor polynomial $T_d(f; \mathbf{x}_0)$ (1.28) of f of degree d, $d \geq 1$, at $\mathbf{x}_0$, rather than just $f(\mathbf{x}_0)$ (which can be regarded as the Taylor polynomial of degree $d = 0$ at $\mathbf{x}_0$). In particular, for univariate integrals,

$$\int_a^b w(x) f(x)\, dx = \int_a^b w(x)(f(x) - T_d(f; x_0)(x))\, dx + \sum_{l=0}^{d} \frac{f^{(l)}(x_0)}{l!} \int_a^b w(x)(x - x_0)^l\, dx \tag{7.47}$$

is obtained. Generally, this method presupposes that f is at least d-times differentiable at $\mathbf{x}_0$ and that the moments $\mathrm{I}(w(\mathbf{x})(\mathbf{x}-\mathbf{x}_0)^{\mathbf{l}})$ can be computed analytically for all monomials $(\mathbf{x}-\mathbf{x}_0)^{\mathbf{l}} \in \mathbb{P}_d^N$. If f is $(d+1)$-times continuously differentiable at $\mathbf{x}_0$, then

$$|f(\mathbf{x}) - T_d(f;\mathbf{x}_0)(\mathbf{x})| = O(\|\mathbf{x}-\mathbf{x}_0\|^{d+1}) \qquad \text{as} \quad \mathbf{x} \to \mathbf{x}_0.$$

Hence, if

$$w(\mathbf{x}) = O(\|\mathbf{x}-\mathbf{x}_0\|^{\alpha}),\ -1 < \alpha < 0 \qquad \text{as} \quad \mathbf{x} \to \mathbf{x}_0,$$

then $w(\mathbf{x})(f(\mathbf{x}) - T_d(f;\mathbf{x}_0)(\mathbf{x}))$ is d-times differentiable at $\mathbf{x}_0$ and the dth derivative is Hölder continuous at $\mathbf{x}_0$.

Unfortunately, in order for this approach to be practical, the required derivatives must be computable accurately and inexpensively. This assumption is not usually true in practice, particularly in the case of high-order derivatives.

Example (Subtracting the Singularity). The integral

$$\mathrm{I}(wf) = \int_0^1 \frac{e^x}{(x-1/3)^{2/3}}\,dx \approx 7.0863 \tag{7.48}$$

was approximately calculated using the Gauss rules G_1, G_3, G_5, G_7, and G_9. These formulas were applied to the original integrand wf and to the regularized integrands $w(f - T_d(e^x;1/3))$ for $d = 0, 1, 2$, and 3. The following table shows the resulting errors $E_n := G_n(wf) - \mathrm{I}(wf)$ and $E_n^d := G_n(w(f - T_d(e^x;1/3))) + \mathrm{I}(wT_d(e^x;1/3)) - \mathrm{I}(wf)$.

n	E_n	E_n^0	E_n^1	E_n^2	E_n^3
1	-1.6	$3.1 \cdot 10^{-1}$	$-9.0 \cdot 10^{-2}$	$-1.5 \cdot 10^{-2}$	$-2.5 \cdot 10^{-3}$
3	-2.8	$6.1 \cdot 10^{-2}$	$3.0 \cdot 10^{-3}$	$-3.1 \cdot 10^{-4}$	$-2.4 \cdot 10^{-6}$
5	-2.6	$-1.9 \cdot 10^{-2}$	$1.6 \cdot 10^{-3}$	$1.8 \cdot 10^{-5}$	$-2.6 \cdot 10^{-6}$
7	-1.5	$-4.4 \cdot 10^{-2}$	$-7.8 \cdot 10^{-5}$	$2.4 \cdot 10^{-5}$	$2.9 \cdot 10^{-8}$
9	4.3	$2.9 \cdot 10^{-2}$	$-6.7 \cdot 10^{-4}$	$-8.3 \cdot 10^{-7}$	$2.3 \cdot 10^{-7}$

While there is apparently no convergence of the Gauss rules for the initial, singular integrand, the Gauss rules clearly converge for the regularized integrands. It is also evident that the resulting integration errors strongly depend on the smoothness of the regularized integrands.

7.3.3. Adaptive Integration. Adaptive integration algorithms are designed to minimize the computational effort (the number of f-evaluations) required to approximately calculate a given integral by placing abscissas predominantly in those parts of B in which the numerical integration of f is most difficult. When applied to singular functions, adaptive integration algorithms usually concentrate the sampling points in regions close to the singular point(s). Using adaptive integration algorithms is often the only effective way of dealing with singular integration problems when the locations and/or the kinds of singularities are *not* known. A general discussion of adaptive integration algorithms and descriptions of relevant software products can be found in Section 8.2.2.

Adaptive integration methods prove particularly efficient for singular integration problems when they are combined with convergence acceleration techniques (see Section 7.3.4). See Section 8.3.2 for a thorough discussion of adaptive extrapolation methods.

7.3.4. Extrapolation Methods. The Euler–Maclaurin summation formula (5.31) provides an asymptotic expansion for the error in compound integration formulas $k^N \times \mathrm{Q}_n$, $N \geq 1$, for smooth integrands. It can be shown that similar asymptotic error expansions are valid for certain classes of singular functions. These *generalized* Euler–Maclaurin sum formulas form the basis for the application of extrapolation methods to singular integration problems.

Univariate Singular Integrands. The following theorem provides a typical example of an asymptotic error expansion for a specific kind of univariate singular integrand.

THEOREM 7.3.1. *Let* Q_n *be an n-point quadrature formula with degree of accuracy* $D \geq 0$ *for the interval* $[a, b]$ *which assigns the same quadrature weight* c *(possibly* $c = 0$*) to the endpoint abscissas* a *and* b*. If the function* $f : [a, b] \to \mathbb{R}$ *has an algebraic singularity at* $x = a$*, i.e., if* f *can be represented as*

$$f(x) = (x - a)^\alpha h(x), \quad -1 < \alpha < 0, \; h \in C^{p+1}[a, b], \; p \in \mathbb{N},$$

then

$$(k \times \mathrm{Q}_n)f - \mathrm{I}f = \sum_{q=1}^{p} \frac{a_q^{(1)}}{k^{\alpha+q}} + \sum_{q=1}^{p} \frac{a_q^{(2)}}{k^q} + O\left(\frac{1}{k^{p+1}}\right). \tag{7.49}$$

Proof. See Lyness and Ninham [279].

Similar, though more complicated, expansions can be obtained for integrands with algebraic-logarithmic endpoint singularities and for integrands with interior algebraic singularities (see, for instance, Lyness and Ninham [279]).

Usually, the choice of an appropriate extrapolation method for accelerating the convergence of a sequence $\{s_m := (k_m \times \mathrm{Q}_n)f\}$ of compound quadrature formulas depends on the specific kind of error expansion valid for f. For instance, if the exponent α in (7.49) is known, then a modified version of Richardson's extrapolation (see Section 3.7.3) can be used to accelerate the convergence of $\{s_m\}$. However, it may be that the exact locations of the singularities x_0 of f and/or the asymptotic behavior of $f(x)$ as $x \to x_0$ are not precisely known. The remainder of this section therefore focuses on extrapolation methods which are applicable to various kinds of singular functions.

Such *flexible* extrapolation methods depend on the chosen sequence $k_1 < k_2 < \cdots$ of subdivision factors. When $\{k_m\}$ is chosen as a *geometric* sequence, $k_m = k_1^m$, $m = 1, 2, \ldots$, the following crucial observation can be made for *all* error expansions applicable to the aforementioned kinds of singular integrands (see Piessens et al. [339]): the sequence of integration errors $\{s_m - s\}$, $s = \mathrm{I}f$, satisfies—asymptotically—*homogeneous linear difference equations*:

$$(7.50) \qquad b_0(s_m - s) + b_1(s_{m+1} - s) + \cdots + b_K(s_{m+K} - s) = 0$$

for $m = 1, 2, \ldots$ with $K \in \mathbb{N}$. The significance of this result is that, whenever the relationship (7.50) holds for a sequence $\{s_m\}$, the ε-algorithm (see Section 3.7.4) can be used to calculate its limit s: $s = \varepsilon_{2K}^{(m)}$, $m = 1, 2, \ldots$ (Brezinski and Redivo Zaglia [41]). When (7.50) holds *asymptotically*, the transformed sequence $\{\varepsilon_{2K}^{(m)}\}$ usually converges faster than the original sequence $\{s_m\}$.[82]

The ε-algorithm is used, for instance, in a number of programs in the subroutine package QUADPACK [339]. Here and in other programs, the ε-algorithm is combined with adaptive subdivision strategies. A thorough discussion of adaptive extrapolation methods and descriptions of relevant software products can be found in Section 8.3.2.

On the other hand, when $\{k_m\}$ is chosen as an *harmonic* sequence, $k_m = m$, $m = 1, 2, \ldots$, then Levin's u-transform (see Section 3.7.5) can be used to accelerate the convergence of the resulting sequences $\{s_m\}$. See Carino, Robinson, and de Doncker [52] for details.

Multivariate Singular Integrands. The following theorem provides a typical example of an asymptotic error expansion for a specific kind of multivariate singular integrand f.

THEOREM 7.3.2. *Let Q_n be an n-point integration formula with degree of accuracy $D \geq 0$ for the N-dimensional cube C_N. If the function $f : C_N \to \mathbb{R}$ has a homogeneous vertex singularity at $\mathbf{x} = 0$, i.e., if f can be represented as*

$$f(\mathbf{x}) = r^\alpha h(\mathbf{x}), \quad r = \|\mathbf{x}\|_2, \quad -N < \alpha < 0, \ h \in C^\infty(C_N),$$

then

$$(7.51) \quad (k^N \times \mathrm{Q}_n)f - \mathrm{I}f = \sum_{q=1}^{p} \frac{a_q^{(1)} + a_q^{(2)} \log k}{k^{\alpha+q+N-1}} + \sum_{q=1}^{p} \frac{a_q^{(3)}}{k^q} + O\left(\frac{1}{k^{p+1}}\right)$$

for any $p \in \mathbb{N}$.

Proof. See Lyness [269].

[82] The application of the ε-algorithm in this context was first described by Kahaner [223].

Similar error expansions for compound formulas $k^N \times Q_n$ for N-dimensional cubes C_N can be obtained for other kinds of singular functions. In particular, error expansions for functions with *face singularities*,

$$f(\mathbf{x}) = x_{j_1}^{\alpha_1} \cdots x_{j_L}^{\alpha_L} h(\mathbf{x}), \quad \text{with}$$

$$j_1 < j_2 < \cdots < j_L, \; L \leq N, \quad -1 < \alpha_j < 0, \; j = 1, 2, \ldots, L, \quad h \in C^\infty(C_N),$$

can be found in de Doncker [101]. For squares C_2, additional error expansions have been established for other kinds of singularities; see Lyness and de Doncker [274] and [275] for details. Error expansions valid for N-dimensional cubes C_N are usually matched by analogous expansions valid for N-dimensional simplices T_N (see, for instance, Lyness and Monegato [278]). A review of asymptotic error expansions for multivariate singular functions can be found de Doncker [101].

Generally speaking, the asymptotic error expansions obtained for the aforementioned classes of multivariate singular integration problems are structurally very similar to asymptotic error expansions obtained for univariate singular integration problems. As a result, choosing geometric sequences of subdivision factors, $k_m = k_1^m$, $m = 1, 2, \ldots$, gives rise to sequences $\{s_m := (k_m^N \times Q_n)f\}$ of compound quadrature formulas which asymptotically satisfy homogeneous linear difference equations (7.50). The convergence of these sequences can be accelerated by using the ε-algorithm (see Section 3.7.4).

Note that the calculation of an approximate result s_{m+1} requires k_1^N as many f-evaluations as calculating the preceding approximation s_m. This rapid growth in the number of required f-evaluations for increasing values of N can be reduced by combining the ε-algorithm with adaptive subdivision strategies (see Section 8.3.2). Alternatively, the number of required f-evaluations can be reduced to a minimum by choosing $\{k_m\}$ as an harmonic sequence, $k_m = m$, $m = 1, 2, \ldots$. In this case, Levin's u-transform (see Section 3.7.5) can be used to accelerate the convergence of the resulting sequences $\{s_m\}$. Details of this method can be found in Carino [51] and Carino, Robinson, and de Doncker [52] and [53].

7.3.5. Ignoring the Singularity. The asymptotic error expansion (7.49) for integrands with algebraic endpoint singularities and similar expansions for other kinds of singular integrands show that compound quadrature formulas $k \times Q_n$ converge (as $k \to \infty$) not only for Riemann integrable functions, but even for certain classes of singular (improperly Riemann integrable) functions. The convergence in the presence of endpoint singularities can be established more generally for monotonic functions f and for functions f whose absolute value is bounded by a monotonic, improperly Riemann integrable function. For these classes of singular functions, the convergence $Q_n f \to If$, $n \to \infty$ can also be established for certain sequences of simple quadrature formulas $\{Q_n\}$, for instance, for sequences of Gauss, Radau, and Lobatto formulas. A general criterion for suitable sequences $\{Q_n\}$ can be found in Myerson [310].

Similar results can be obtained for functions with interior singularities x_0, provided the abscissas of Q_n closest to x_0 are omitted. See Lubinsky and Rabinowitz [264] and Rabinowitz [348] for details. See also the following section for similar results obtained for product integration formulas.

It is important to emphasize that the convergence achieved by ignoring the singularity is slow in comparison to the convergence achieved by other methods described in this section. Therefore, singularities should *not* be ignored unless it is, for application-specific reasons, impossible to take them into account.

7.3.6. Product Integration. Following the general idea of product integration (see Section 5.1.2), singular integration problems can be dealt with by splitting integrands into the form wf, where w incorporates the singularities and f the smooth part of the integrand.[83] The smooth part f is then approximated using suitable model functions $g \in \mathcal{M}$ (usually, (piecewise) polynomials), and the exact integral $I(wg)$ is used as an approximation to $I(wf)$. The accuracy of $I(wg)$ essentially depends on the accuracy of the approximation of f by g (which is usually high if f is sufficiently smooth), whereas the fact that the singular part w cannot be approximated accurately does *not* affect the accuracy of $I(wg)$. In order for product integration to be practical, w must be available in closed form, and it must be possible to calculate $I(wg)$, $g \in \mathcal{M}$, analytically.

Product Gauss Formulas. If the singular factor w is nonnegative, then the results in Chapters 5 and 6 concerning integration formulas for admissible weight functions apply. In particular, the *Gauss formulas* G_n^w with respect to w attain the maximum degree of accuracy, $D = 2n - 1$, possible for a given number of abscissas n and are therefore highly efficient if f is smooth (see Section 5.2.7).

Example (Product Gauss Formulas). The integral (7.48) was approximately calculated using product Gauss formulas G_n^w, $n = 1, 2, 3, 4, 5, 6$, for the singular weight function $w(x) = (x - 1/3)^{-2/3}$. The following table shows the resulting integration errors $E_n := G_n^w f - I_w f$.

n	1	2	3	4	5	6
E_n	$-1.5 \cdot 10^{-1}$	$-1.0 \cdot 10^{-3}$	$-2.2 \cdot 10^{-6}$	$-2.2 \cdot 10^{-9}$	$-1.8 \cdot 10^{-12}$	$-7.7 \cdot 10^{-16}$

Due to the analyticity of $f(x) = e^x$, the product Gauss formulas converge extremely rapidly—much faster than the formulas obtained by subtracting the singularity used in the example on page 269.

[83]Often, singular integration problems arise naturally in this product form.

There are two major practical problems with using product Gauss formulas: to begin with, if the weight function w is nonclassical, the routines described in Section 5.2.7 for calculating the abscissas and weights of Gauss formulas require users to provide information about the orthogonal polynomials associated with w, for instance, the recurrence coefficients in (5.24). Users may not find it straightforward to generate this information. And then, the abscissas of product Gauss formulas G_n^w depend on the corresponding weight function w. This feature usually makes them unsuitable for families of singular integrals (7.44) in which the weight function w depends on an additional parameter, $w = k(., y_m)$, $m = 1, 2, \dots, M$.

Simple Interpolatory Product Formulas. The aforementioned problems with product Gauss formulas can be overcome by choosing the quadrature abscissas $x_1, \dots, x_n$ *independently* of the weight function w. Clearly, $x_1, \dots, x_n$ should be chosen in such a way that the corresponding interpolation polynomials $p \in \mathbb{P}_{n-1}$ of smooth functions f, $f(x_i) = p(x_i)$, $i = 1, \dots, n$, approximate f accurately on the entire integration interval $[a, b]$. A good choice in this respect are the extrema or zeros of Chebyshev polynomials (see Section 5.2.6), or, more generally, the zeros of Jacobi polynomials. For application-specific reasons, Radau- and Lobatto-type sets of abscissas, which include the interval endpoints a and/or b, are frequently used as well.

An interesting problem in this context is that of *optimal* selection, given $n \in \mathbb{N}$, of the quadrature abscissas $x_1, \dots, x_n$ for specific classes of admissible singular weight functions w and smooth functions f. This problem is studied in a worst-case setting in Smith and Paget [410].

Example (Interpolatory Product Formulas). The integral (7.48) was approximated using interpolatory product formulas (i) $C_n^{z,1}$, based on the zeros of the Chebyshev polynomials of the first kind; (ii) $C_n^{z,2}$, based on the zeros of the Chebyshev polynomials of the second kind; and (iii) C_n^{e}, based on Chebyshev extrema. The numbers n = 2, 4, 6, 8, 10, and 12 of abscissas were chosen in such a way that the resulting formulas attained the same degrees of accuracy as the Gauss formulas $G_{n/2}^w$ used in the previous example. The following table shows the integration errors in the various formulas.

n	$C_n^{z,1} - If$	$C_n^{z,2} - If$	$C_n^{e} - If$
2	$2.8 \cdot 10^{-1}$	$4.2 \cdot 10^{-2}$	$7.6 \cdot 10^{-1}$
4	$2.3 \cdot 10^{-4}$	$-1.2 \cdot 10^{-5}$	$-1.7 \cdot 10^{-3}$
6	$-9.1 \cdot 10^{-7}$	$-9.3 \cdot 10^{-7}$	$-4.3 \cdot 10^{-7}$
8	$2.1 \cdot 10^{-9}$	$1.0 \cdot 10^{-9}$	$3.1 \cdot 10^{-9}$
10	$-1.4 \cdot 10^{-12}$	$6.8 \cdot 10^{-13}$	$-2.8 \cdot 10^{-12}$
12	$3.4 \cdot 10^{-16}$	$6.9 \cdot 10^{-17}$	$1.0 \cdot 10^{-15}$

On average, the accuracy of the interpolatory product formulas is comparable to the accuracy of the corresponding product Gauss formulas (which use only half the number of abscissas). Note also that the formulas based on Chebyshev zeros are generally more accurate than the formulas based on Chebyshev extrema.

Convergence of Interpolatory Product Formulas. For sufficiently smooth functions f, the convergence (and specific rates of convergence of) $\mathrm{Q}_n f \to \mathrm{I}f$, $n \to \infty$, can easily be established for suitable sequences of interpolatory product formulas $\{\mathrm{Q}_n\}$ using bounds on the approximation errors $\|p_n - f\|_\infty$ in the corresponding interpolation polynomials $p_n \in \mathbb{P}_{n-1}$ (see, for instance, Section 5.2.2). However, subject to certain conditions on the singular weight function w, suitable sequences of interpolatory product formulas $\{\mathrm{Q}_n\}$ can be shown to converge for far more comprehensive classes of functions. In particular, if $\{\mathrm{Q}_n\}$ is based on using the zeros of Jacobi polynomials as quadrature abscissas, then the convergence $\mathrm{Q}_n f \to \mathrm{I}f$, as $n \to \infty$, holds for all Riemann integrable functions $f \in \mathcal{R}[a,b]$, provided

$$w(x)(x-a)^{-\max\{(2\alpha+1)/4,0\}}(b-x)^{-\max\{(2\beta+1)/4,0\}} \in L_p[a,b], \quad p > 1,$$

where α and β are the exponents in the Jacobi weight function (Sloan and Smith [406]). The above condition on w becomes least stringent when $\alpha, \beta \leq -1/2$, in which case $w \in L_p[a,b]$, $p > 1$, must be satisfied. This condition is also sufficient to establish the convergence of sequences $\{\mathrm{Q}_n\}$ based on the Chebyshev extrema for all $f \in \mathcal{R}[a,b]$. Rabinowitz [347] obtains similar conditions on w which ensure the convergence of $\{\mathrm{Q}_n\}$ based on the zeros of *generalized* Jacobi weight functions for continuous functions.

Interpolatory product formulas may even converge for singular functions, making it possible to "ignore" the singularities in f. This is important, for instance, when some of the singularities of a given integrand cannot be located accurately and thus cannot be incorporated into w. This topic is dealt with in a number of recent papers for various sets of quadrature abscissas and classes of singular functions. Detailed information can be found in Criscuolo, Mastroianni, and Monegato [72], Mastroianni and Monegato [295], Lubinsky and Sidi [265], Rabinowitz [355], Rabinowitz and Smith [360], [361], and the references cited therein.

While the above convergence results serve as a theoretical justification for using simple interpolatory product integration formulas, the practical usefulness of these formulas may depend on whether reliable error estimates can be computed efficiently. Suitable error estimation procedures are usually based on *extending* the set of abscissas used in a given product formula. Kronrod-type extensions (see Section 5.2.9) for interpolatory product formulas are discussed in Dagnino [74], and practical algorithms for computing these Kronrod-type extension formulas are described in Calio and Marchetti [48], Dagnino and Fiorentino [79], and Palamara Orsi [328].

Compound Product Formulas. Product formulas can also be obtained by using piecewise polynomials, in particular, polynomial splines, to interpolate f at an equidistant set of interpolation points. The resulting compound product formulas can be shown to converge for smooth functions and for certain singular functions. For details, see Dagnino [76], Dagnino and Palamara Orsi [80], Rabinowitz [354], and the references cited therein.

Noninterpolatory Product Formulas. Results for noninterpolatory product formulas, based, for instance, on truncated series expansions, can be found in Dagnino [75], Dagnino, Demichelis, and Santi [78], and Rabinowitz [349].

Product Integration on (Semi-)Infinite Intervals. Most literature on product integration for singular integrands only deals with finite integration intervals. As to product integration of singular integrands over (semi-)infinite integrals, the interested reader is referred to Lubinsky and Sidi [265], Mastroianni and Monegato [296], and Smith, Sloan, and Opie [411].

7.4. Cauchy Singular Integrals

This section outlines computational techniques for approximately integrating Cauchy singular functions, i.e., strongly singular functions whose Cauchy principal value integrals are well defined (see Section 1.3). First, Section 7.4.1 reviews some of the particular problem settings affecting the calculation of Cauchy principal value integrals. Next, Sections 7.4.2 and 7.4.3 describe two particular ways of converting univariate Cauchy singular integration problems into equivalent weakly singular or even nonsingular problems. While other conversion methods exist (for instance, truncating the integration region), they have been discussed in sufficient detail in Section 7.3.2 and are therefore not considered in this section. Then, an alternative to conversion-based integration methods—product integration formulas for Cauchy singular integrands—is described in Section 7.4.4, and references to miscellaneous methods are provided in Section 7.4.5. Finally, the calculation of multivariate Cauchy singular integrals is dealt with in Section 7.4.6.

7.4.1. Practical Problem Settings. Most of the observations made in Section 7.3.1 concerning the particular setting(s) in which weakly singular integration problems are posed also apply to Cauchy singular integration problems. In particular, methods for integrating Cauchy singular functions are frequently used to solve *Cauchy singular integral equations*. In these cases, it often happens that families of Cauchy singular integrals of the form

$$\mathrm{I}_{y_m} f := ⨍_B w(\mathbf{x}) k(\mathbf{x}, \mathbf{y}_m) f(\mathbf{x})\, d\mathbf{x}, \quad m = 1, 2, \ldots, M, \tag{7.52}$$

have to be discretized. The kernel $k(\mathbf{x}, \mathbf{y})$ in (7.52) possesses a singularity of order N at $\mathbf{x} = \mathbf{y}$, but is otherwise smooth. The additional weight function w is (improperly) integrable, but usually not smooth; for instance, it might be weakly singular. Both the kernel function k and the weight function w are known analytically.

In the following sections, computational methods for univariate Cauchy principal value integrals are described using the Hilbert kernel $k(x, y) = 1/(x - y)$, which is (by far) the most frequently studied univariate kernel function. The techniques described can also be applied to any other kernel, for instance, the cotangent kernel (Gautschi, Kovacevic, and Milovanovic [159])

$$k(x, y) = \cot\left(\frac{x - y}{2}\right),$$

which behaves like the Hilbert kernel as $x \to y$ (i.e., for which $k(x, y) \sim c/(x - y)$ as $x \to y$).

Quadrature Methods for Cauchy Singular Integral Equations. The need to calculate univariate Cauchy principal value integrals with Hilbert kernels typically arises in Nyström's (quadrature) method for solving Fredholm-type integral equations of the first kind,

$$(7.53) \qquad ⨍_a^b \frac{f(x)}{x - y}\, dx + \int_a^b \tilde{k}(x, y) f(x)\, dx = g(y), \qquad y \in (a, b),$$

and of the second kind,

$$(7.54) \quad h(y) f(y) + \tilde{h}(y) ⨍_a^b \frac{f(x)}{x - y}\, dx + \int_a^b \tilde{k}(x, y) f(x)\, dx = g(y), \quad y \in (a, b),$$

where $\tilde{k}$, g, h, and $\tilde{h}$ are given functions. Often, information about the asymptotic behavior of the unknown solution f at the interval endpoints a and b, for instance, $f(x) \sim \sqrt{x - a}\sqrt{b - x}$, can be derived analytically. This information is incorporated into suitable weight functions w.

Similar to (7.43), the Cauchy singular integral operators in (7.53) and (7.54) are discretized using quadrature formulas $\mathrm{Q}_n^{(y)}$, and the resulting discretized equation is required to hold for a set of collocation points in (a, b). This method gives rise to particular requirements: to begin with, the collocation points and quadrature abscissas must be chosen carefully in order for the resulting systems of linear equations not to be underdetermined. Also, *uniform* convergence,

$$\sup\{|\mathrm{Q}_n^{(y)} f - \mathrm{I}_y f| \; : \; y \in (a, b)\} \to 0 \quad \text{as} \quad n \to \infty,$$

is usually required to establish the convergence of the resulting approximate solutions.

7.4.2. Symmetrizing the Integrand. Suppose the Cauchy principal value integral

$$\mathrm{I}f = ⨍_{-b}^{b} \frac{f(x)}{x}\,dx$$

exists. Let φ be defined by

$$\varphi(x) := \frac{f(x) - f(-x)}{2x},$$

and note that φ is an even (symmetric) function. Then, according to definition (1.11),

$$\begin{aligned}
\mathrm{I}f &= \lim_{\delta\to 0+}\left(\int_{-b}^{-\delta} \frac{f(x)}{x}\,dx + \int_{\delta}^{b} \frac{f(x)}{x}\,dx\right) \\
&= \lim_{\delta\to 0+} \int_{\delta}^{b} \frac{f(x) - f(-x)}{x}\,dx \qquad (7.55)\\
&= 2\int_0^b \varphi(x) = \int_{-b}^{b} \varphi(x)\,dx.
\end{aligned}$$

This shows that φ is sufficiently regular at $x = 0$ in order for the (possibly improper) Riemann integral of φ over $[-b, b]$ to exist. More precisely, if f is differentiable at 0, then

$$\lim_{x\to 0} \varphi(x) = \lim_{x\to 0} \frac{f(x) - f(-x)}{2x} = f'(0)$$

shows that φ is continuous at $x = 0$; and, more generally, if f is k times differentiable at $x = 0$, then φ is $(k-1)$ times continuously differentiable at $x = 0$. Note that φ is defined in the form of a difference quotient. Hence, the numerical evaluation of $\varphi(x)$ for arguments $x \approx 0$ is numerically unstable and should therefore be avoided.

Integrating the Symmetrized Integrand. Since φ is even, the integral of φ over $[-b, b]$ should be approximated using symmetric quadrature formulas because in this case only $\lfloor (n+1)/2 \rfloor$ evaluations of φ are needed to compute $\mathrm{Q}_n\varphi$. Furthermore, $x = 0$ should not be used as a quadrature abscissa because φ may not be continuous at $x = 0$ (if f is not differentiable at $x = 0$), or $\varphi(0)$ may have to be evaluated numerically (if f' is not known analytically). This implies that an even number $n = 2m$ of quadrature abscissas should be used.

Applying a symmetric formula Q_{2m} to φ yields

$$\mathrm{Q}_{2m}\varphi = \sum_{i=1}^{m} 2c_i\varphi(x_i) = \sum_{i=1}^{m} c_i \frac{f(x_i) - f(-x_i)}{x_i}$$

$$= \sum_{i=1}^{m} \frac{c_i}{x_i} f(x_i) + \sum_{i=1}^{m} \frac{c_i}{-x_i} f(-x_i)$$

$$= \sum_{i=1}^{2m} \frac{c_i}{x_i} f(x_i) =: \mathrm{Q}_{2m}^{(1/x)} f.$$

This shows that $\mathrm{Q}_{2m}\varphi$ can also be interpreted as an odd quadrature formula for f, $\mathrm{Q}_{2m}^{(1/x)} f$, with the abscissas x_i and coefficients c_i/x_i, $i = 1, \ldots, 2m$. The condition number of $\mathrm{Q}_{2m}^{(1/x)}$ is given by

$$\|\mathrm{Q}_{2m}^{(1/x)}\|_1 = 2 \sum_{i=1}^{m} \left| \frac{c_i}{x_i} \right|$$

and can be significantly larger than the condition number of Q_{2m} if some of the x_i are chosen close to 0.

Clearly, if Q_{2m} has the degree of accuracy D, then the resulting formula $\mathrm{Q}_{2m}^{(1/x)}$ is exact for all polynomials of degree $d \le D + 1$. Moreover, since $\mathrm{Q}_{2m}^{(1/x)}$ is odd, it is also accurate for all even polynomials, which increases its degree of accuracy to $D + 2$, provided D is even. For a given value of m, $\mathrm{Q}_{2m}^{(1/x)}$ attains the maximum possible accuracy when Q_{2m} is chosen as the Gauss formula G_{2m}, and the resulting rule $G_{2m}^{(1/x)}$ has the degree of accuracy $D = 4m$. Additionally, the rules $G_{2m}^{(1/x)}$ are well conditioned for modest values of m because the abscissas of G_{2m} cluster towards the interval endpoints. For $m = 100$, for example, $\|G_{100}^{(1/x)}\|_1 \approx 10.8$.

Example (Symmetrizing the Integrand). The Cauchy principal value integral

$$\mathrm{I}f := ⨍_{-1}^{1} \frac{e^{5x}}{x}\, dx \tag{7.56}$$

was approximately calculated using the Gauss formulas $G_{2m}^{(1/x)}$, $m = 1, \ldots, 5$. The following table shows the resulting relative integration errors $E_{2m}^{\mathrm{rel}} := |(G_{2m}^{(1/x)} - \mathrm{I}f)/\mathrm{I}f|$.

m	E_{2m}^{rel}
1	$2.3 \cdot 10^{-1}$
2	$2.6 \cdot 10^{-3}$
3	$5.3 \cdot 10^{-6}$
4	$3.4 \cdot 10^{-9}$
5	$8.8 \cdot 10^{-13}$

Centering the Singularity. The method of symmetrizing the integrand in a Cauchy principal value integral described in this section is based on the assumption that the singularity y is located at the center of the integration interval $[a, b]$. If y is not centered, for instance, $y > (a + b)/2$, the interval

$[a,b]$ is subdivided into the intervals $[a, 2y-b]$ and $[2y-b, b]$. The second interval contains the singularity y at its center and thus can be dealt with using the symmetrization method. The integral over the first interval, which is not singular, can be dealt with using a variety of methods. Monegato [303], for instance, proposed an approach based on subtracting the singularity (see Section 7.4.3), the convergence properties of which were subsequently studied in Chiodo and Criscuolo [58].

The set of abscissas obtained by subdividing the interval and symmetrizing the integrand in the subinterval containing the singularity depends on the value of $y \in (a,b)$. The symmetrization method is therefore *not* suitable for applications in which families of Cauchy principal value integrals (7.52) must be computed.

7.4.3. Subtraction of the Singularity. Subtraction of the singularity is described in Section 7.3.2 as a method for removing singular points in weakly singular integrals. It can also be used to regularize Cauchy principal value integrals as follows:

$$\text{(7.57)} \qquad \mathrm{I}_y(wf) = ⨍_a^b w(x)\frac{f(x)}{x-y}\,dx$$

$$\text{(7.58)} \qquad = ⨍_a^b w(x)\frac{f(x)-f(y)}{x-y}\,dx + f(y)⨍_a^b \frac{w(x)}{x-y}\,dx,$$

i.e., $\mathrm{I}_y(wf) = \mathrm{I}_y(w(f-f(y))) + f(y)\mathrm{I}_y w$. Since w is assumed to be known analytically, $\mathrm{I}_y w$ can be computed with high precision, either analytically or numerically, independently of f. Gautschi and Wimp [160], for instance, discuss the calculation of $\mathrm{I}_y w$ for Jacobi weight functions w.

The remaining function $\varphi_y(x) := (f(x)-f(y))/(x-y)$ usually proves to be sufficiently regular at $x = y$ in order for the (possibly improper) Riemann integral of $w\varphi_y$ over $[a,b]$ to exist. More precisely, if f is differentiable at $x = y$, then

$$\lim_{x\to y}\varphi_y(x) = \lim_{x\to 0}\frac{f(x)-f(y)}{x-y} = f'(y)$$

shows that φ_y is continuous at $x = y$; and, more generally, if f is k-times differentiable at $x = y \in (a,b)$, then φ_y is $(k-1)$-times continuously differentiable at $x = y$. Note that φ_y is defined in terms of a difference quotient. Hence, the numerical evaluation of $\varphi_y(x)$ for arguments $x \approx y$ is numerically unstable and should therefore be avoided.[84]

[84]The functions φ_y obtained by subtracting the singularity are in many ways very similar to the functions φ obtained by symmetrizing the integrand.

Integrating the Regularized Integrand. The equality $\mathrm{I}_y(w(f - f(y))) = \mathrm{I}(w\varphi_y) = \mathrm{I}_w\varphi_y$ shows that an approximate value for $\mathrm{I}_y(w(f - f(y)))$ can be obtained by applying a formula Q_n for the integral operator I_w to φ_y. The sum of $\mathrm{Q}_n\varphi_y$ and $f(y)\mathrm{I}_y w$ then yields an approximation for the original integral $\mathrm{I}_y(wf)$. If y is *not* used as an abscissa in Q_n, $y \notin \{x_1, \ldots, x_n\}$, then this sum evaluates to

$$(7.59) \qquad \begin{aligned} \mathrm{Q}_{n+1}^{(y)} f &:= f(y)\mathrm{I}_y w + \mathrm{Q}_n\varphi_y \\ &= f(y)\mathrm{I}_y w + \sum_{i=1}^{n} c_i \frac{f(x_i) - f(y)}{x_i - y} \\ &= \left(\mathrm{I}_y w - \sum_{i=1}^{n} \frac{c_i}{x_i - y}\right) f(y) + \sum_{i=1}^{n} \frac{c_i}{x_i - y} f(x_i). \end{aligned}$$

Otherwise, if $y = x_j$, then

$$\begin{aligned} \mathrm{Q}_{n+1}^{(y)} f &:= f(y)\mathrm{I}_y w + \mathrm{Q}_n\varphi_y \\ &= f(y)\mathrm{I}_y w + c_j f'(y) + \sum_{i=1, i\neq j}^{n} c_i \frac{f(x_i) - f(y)}{x_i - y} \\ &= \left(\mathrm{I}_y w - \sum_{i=1, i\neq j}^{n} \frac{c_i}{x_i - y}\right) f(y) + c_j f'(y) + \sum_{i=1, i\neq j}^{n} \frac{c_i}{x_i - y} f(x_i). \end{aligned}$$

This shows that the coefficients of $\mathrm{Q}_{n+1}^{(y)}$ can easily be obtained once the coefficients of Q_n have been calculated. Note that the evaluation of $\mathrm{Q}_{n+1}^{(y)}$ becomes ill conditioned if (i) an abscissa $x_i \neq y$ is chosen close to y and/or (ii) $x_j = y$ and $f'(y)$ is approximately calculated based on a finite difference scheme which uses abscissas close to y.

Clearly, if Q_n has the degree of accuracy D, then the derived formula $\mathrm{Q}_{n+1}^{(y)}$ is exact for all polynomials of degree $d \leq D + 1$. Hence, if Q_n is an *interpolatory* formula for I_w, then $\mathrm{Q}_{n+1}^{(y)}$ is also interpolatory: $\mathrm{Q}_{n+1}^{(y)}$ is exact for all polynomials of degree $d \leq n$, and, equivalently, $\mathrm{Q}_{n+1}^{(y)} f$ is the *exact* Cauchy principal value integral of the unique polynomial of degree $d \leq n$ which interpolates f at the abscissas $x_1, \ldots, x_n$ and, additionally, interpolates f or f' at y. As a result, these formulas $\mathrm{Q}_{n+1}^{(y)}$ are *interpolatory product formulas* as defined in Section 7.4.4. What distinguishes the $\mathrm{Q}_{n+1}^{(y)}$ from other interpolatory product formulas is the fact that, in addition to a fixed set of points $\{x_1, \ldots, x_n\}$, the singularity y is chosen as a quadrature abscissa in $\mathrm{Q}_{n+1}^{(y)}$.

Example (Subtracting the Singularity). The Cauchy principal value integrals

$$(7.60) \qquad \mathrm{I}_y f := ⨍_{-1}^{1} \frac{e^{5x}}{x - y}\, dx, \qquad y = -3/4,\ -1/4,\ 1/4,\ 3/4,$$

were approximately calculated by applying the Gauss formulas G_{2m}, $m = 1, \ldots, 5$, to the Riemann integral obtained by subtracting the singularity at y. The resulting formulas $G_{2m+1}^{(y)}$ are numerically reasonably stable because the numbers of abscissas used are low and the abscissas of G_{2m} and the singular points y are well separated. The following table shows the resulting relative integration errors $E_{y;2m+1}^{\text{rel}} := |(G_{2m+1}^{(y)} - \mathrm{I}_y f)/\mathrm{I}_y f|$.

m	$E_{3/4;2m+1}^{\text{rel}}$	$E_{1/4;2m+1}^{\text{rel}}$	$E_{-1/4;2m+1}^{\text{rel}}$	$E_{-3/4;2m+1}^{\text{rel}}$
1	$1.6 \cdot 10^{-1}$	$1.9 \cdot 10^{-1}$	$2.6 \cdot 10^{-1}$	$3.0 \cdot 10^{-1}$
2	$1.4 \cdot 10^{-3}$	$2.0 \cdot 10^{-3}$	$3.1 \cdot 10^{-3}$	$3.9 \cdot 10^{-3}$
3	$2.6 \cdot 10^{-6}$	$4.0 \cdot 10^{-6}$	$6.5 \cdot 10^{-6}$	$8.6 \cdot 10^{-6}$
4	$1.5 \cdot 10^{-9}$	$2.5 \cdot 10^{-9}$	$4.2 \cdot 10^{-9}$	$5.7 \cdot 10^{-9}$
5	$3.9 \cdot 10^{-13}$	$6.5 \cdot 10^{-13}$	$1.1 \cdot 10^{-12}$	$1.5 \cdot 10^{-12}$

The accuracy of these results is comparable to those obtained by symmetrization. In particular, there is no apparent loss of accuracy due to numerical instability.

Suitability for Integral Families. Calculating $\mathrm{Q}_{n+1}^{(y)} f$ requires f to be evaluated at the abscissas $x_1, \ldots, x_n$ of Q_n and, additionally, f or f' to be evaluated at the singular point y. Therefore, the set of function (including derivative) values required to calculate $\mathrm{Q}_{n+1}^{(y)}$ is largely independent of the location of y. In particular, calculating a family of Cauchy principal value integrals (7.52) requires only $n + M$ evaluations of the integrand and/or its derivative, instead of the $O(nM)$ evaluations required if the symmetrization method is used.

If the $\mathrm{Q}_{n+1}^{(y)}$ are to be used in quadrature methods for Fredholm-type integral equations (see Section 7.4.1), it is crucial that the collocation points $y_1, \ldots, y_n$ be chosen in such a way that the coefficients of $f(y_1), \ldots, f(y_n)$ in the resulting equations vanish. For the Fredholm integral equation of the first kind (7.53), for instance, this means that the coefficient

$$c(y) := \mathrm{I}_y w - \sum_{i=1}^{n} \frac{c_i}{x_i - y}$$

of $f(y)$ in $\mathrm{Q}_{n+1}^{(y)}$ must vanish for $y = y_1, \ldots, y = y_n$. Interestingly, it can be shown that $c : y \mapsto \mathbb{R}$ has at least $n - 1$ zeros in (a, b), provided that the quadrature coefficients of Q_n are positive. See Monegato [303] for a more extensive discussion of this issue.

Accuracy and Convergence. Statements concerning the accuracy and convergence of quadrature formulas derived by subtracting the singularity can easily be obtained using the relationship

$$\mathrm{Q}_{n+1}^{(y)} f - \mathrm{I}_y(wf) = \mathrm{Q}_n f - \mathrm{I}_w(\varphi_y).$$

For instance, if f is differentiable on (a, b) and continuous on $[a, b]$, then the corresponding function φ_y is continuous on $[a, b]$ for any $y \in (a, b)$. Consequently, $\{\mathrm{Q}_{n+1}^{(y)} f\}$ converges whenever the associated sequence of quadrature formulas $\{\mathrm{Q}_n\}$ converges for continuous integrands. Furthermore, since $f \in C^k[a, b]$, $k \geq 1$, implies $\varphi_y \in C^{k-1}[a, b]$ for any $y \in (a, b)$, the above relationship can also be used to establish specific rates of convergence of $\{\mathrm{Q}_{n+1}^{(y)} f\}$ for smooth functions f. For instance, Theorem 5.2.2 along with Jackson's theorems (see Section 5.2.2) shows that

$$\mathrm{Q}_{n+1}^{(y)} f = O(n^{-k+1}) \qquad \text{as} \quad n \to \infty$$

for any $f \in C^k[a, b]$, $k \geq 2$, provided that all quadrature weights in the sequence of interpolatory formulas $\{\mathrm{Q}_n\}$ are nonnegative.

These results can be used, for instance, to establish the convergence, and specific rates of convergence, of sequences $\{\mathrm{Q}_{n+1}^{(y)}\}$ based on choosing Q_n as the Gauss formula G_n^w with respect to the admissible weight function w. Clearly, the resulting quadrature formulas $\{\mathrm{Q}_{n+1}^{(y)}\}$, sometimes called *Hunter's rules* (Hunter [213]), can be expected to be particularly accurate for smooth functions. Detailed convergence analyses of Hunter's formulas for various classes of integrands f can be found in a number of articles: Tsamasphyros and Theocaris [434] establish convergence rates of Hunter's formulas in the case of Jacobi weight functions w for functions $f \in C^k[a, b]$, $k \geq 1$, for which $f^{(k)}$ is Hölder continuous. Ioakimidis [215] proves the *uniform* convergence (for $y \in (a, b)$) of Hunter's formulas for the same class of integrands. This result applies to admissible weight functions w which are Hölder continuous on any closed subinterval of (a, b) and may have algebraic or algebraic-logarithmic endpoint singularities. Criscuolo and Mastroianni [71] study the convergence of formulas obtained from Hunter's rules by omitting the term corresponding to the abscissa x_j closest to the singularity y. The aim of this omission is to avoid the numerical instability caused by this term.

Note that G_n^w can be regarded as a special product rule for the weight function w. Depending on the particular requirements of a given application, other classes of product formulas, in particular, product Radau, Lobatto, and Clenshaw–Curtis formulas, may provide more suitable choices for Q_n. Kronrod extensions of product Gauss, Radau, and Lobatto formulas may be used to obtain practical error estimates (see Rabinowitz [346]).

Alternatively, product formulas based on piecewise polynomial interpolation may be chosen for Q_n. Convergence results of the formulas $\mathrm{Q}_{n+1}^{(y)}$ are obtained in Rabinowitz [350] (for product compound trapezoidal rules), in Dagnino and Santi [81] (for product rules based on cubic spline model functions), and in Potra and Venturino [341] (for product compound trapezoidal, midpoint, and Simpson's rules).

7.4.4. Product Formulas. Following the general idea of product integration (see Section 5.1.2), Cauchy singular integration problems can be dealt with by replacing f with suitable model functions $g \in \mathcal{M}$ and using the exact integral $\mathrm{I}_y(wg)$ as an approximation to $\mathrm{I}_y(wf)$. Note, however, that the evaluation of Cauchy principal value integrals is ill conditioned with respect to changes in f (see Section 3.4.1). Consequently, the error bound (5.4), which usually serves as a simple justification for the validity of product integration formulas, does *not* apply, and more thorough analyses are necessary to establish approximation properties of product formulas for Cauchy principal value integrals.

Interpolatory Product Formulas. Interpolatory product formulas for Cauchy singular integrals are mostly based on using (i) the extrema of Chebyshev polynomials or (ii) the zeros of the orthogonal polynomials $\{P_n^v\}$ associated with (generalized) Jacobi weight functions v, as quadrature abscissas. The weight function v may, or may not, be identical to the weight function w in the integral operator. Convergence analyses have been performed for different problem settings, i.e., for different weight functions and different sets of abscissas, in a number of publications.[85]

In Elliott and Paget [128], [129], convergence analyses are carried out for Jacobi weight functions w and the zeros of $\{P_n^w\}$, whereas Chawla and Kumar [55] use the extrema of Chebyshev polynomials as abscissas. An analysis of using the zeros of $\{P_n^w\}$ for more general weight functions w is given by Elliott and Paget [130]. Monegato [303] investigates, among other things, the convergence of interpolatory product rules based on the zeros of Jacobi polynomials for more general admissible weight functions w. Criscuolo and Mastroianni [69], [70] use various weight functions w in connection with the zeros of $\{P_n^v\}$, where v is a generalized Jacobi weight function. Rabinowitz [351] studies Gauss, Radau, and Lobatto points of generalized Jacobi weight functions v in connection with the weight function $w = v$ as well as weight functions w of mixed sign. Convergence analyses for weakly singular integrands f are given in Rabinowitz [353].

Example (Interpolatory Product Formulas). The Cauchy principal value integrals (7.60) were approximately calculated using interpolatory product formulas $\mathrm{Q}_{2m+1}^{(y)}$, $m = 1, \ldots, 5$, based on the zeros of the Legendre polynomials of degree $2m+1$. The following table shows the resulting relative integration errors $E_{y;2m+1}^{\mathrm{rel}} := |(\mathrm{Q}_{2m+1}^{(y)} - \mathrm{I}_y f)/\mathrm{I}_y f|$.

[85] The following list of references is not meant to be complete, but to provide the reader with starting points for further investigations.

m	$E^{\text{rel}}_{3/4;2m+1}$	$E^{\text{rel}}_{1/4;2m+1}$	$E^{\text{rel}}_{-1/4;2m+1}$	$E^{\text{rel}}_{-3/4;2m+1}$
1	$8.4 \cdot 10^{-1}$	$3.6 \cdot 10^{-1}$	$3.1 \cdot 10^{-1}$	$7.3 \cdot 10^{-1}$
2	$6.0 \cdot 10^{-3}$	$3.4 \cdot 10^{-2}$	$2.5 \cdot 10^{-2}$	$3.4 \cdot 10^{-3}$
3	$4.0 \cdot 10^{-2}$	$8.8 \cdot 10^{-3}$	$6.1 \cdot 10^{-3}$	$1.7 \cdot 10^{-2}$
4	$7.4 \cdot 10^{-4}$	$1.7 \cdot 10^{-3}$	$1.1 \cdot 10^{-3}$	$2.7 \cdot 10^{-4}$
5	$1.8 \cdot 10^{-4}$	$1.2 \cdot 10^{-4}$	$7.6 \cdot 10^{-5}$	$5.9 \cdot 10^{-5}$

The accuracy of these results is low in comparison with the accuracy of the results for the formulas $G^{(y)}_{2m+1}$ obtained by subtracting the singularity in the example in Section 7.4.3. This is not particularly surprising, as $G^{(y)}_{2m+1}$ is accurate for all polynomials of degree $d \leq 4m$, whereas $Q^{(y)}_{2m+1}$ only integrates the polynomials of degree $d \leq 2m$ correctly.

Compound Product Formulas. Recently, product formulas based on piecewise polynomial approximation have received considerable interest. Dagnino and Santi [82], for instance, employ cubic spline interpolation based on equally distributed interpolation points, whereas Dagnino, Demichelis, and Santi [77] use quasi-interpolating splines. The uniform convergence (for $y \in (a, b)$) of product formulas based on certain piecewise polynomial approximation schemes is, among other things, established in Rabinowitz [356]. These publications also provide further references to relevant literature.

Practical Considerations. The main advantage of product formulas for Cauchy principal value integrals is that the set of abscissas used can be chosen independently of the singularity y. Product integration formulas are therefore particularly suitable for dealing with families of Cauchy principal value integrals (7.52), in particular, when these occur in quadrature methods for Fredholm-type integral equations. A drawback of product formulas is that they cannot be generated as easily as formulas based on subtracting the singularity from existing formulas for Riemann integrals. See Monegato [303] for a thorough discussion of computational aspects of interpolatory product formulas.

7.4.5. Miscellaneous Methods. There are a number of methods for dealing with Cauchy principal value integrals which cannot be subsumed under the techniques described in the foregoing sections. While a detailed discussion of these methods is beyond the scope of this book, the reader may find the following references useful.

Sinc integration methods (see Section 7.2.2) for Cauchy principal value integrals are described in Venturino [438] and Bialecki [33], [35], whereas extrapolation-based methods are discussed in Lyness [270]. Gerasoulis [173], Hasegawa and Torii [201], and Rabinowitz and Santi [359] use the following hybrid form of subtracting the singularity and approximating f using model functions g:

$$\text{(7.61)} \qquad \begin{aligned} \mathrm{I}_y(wf) &= \fint_a^b w(x)\frac{f(x)}{x-y}\,dx \\ &\approx \fint_a^b w(x)\frac{g(x)-g(y)}{x-y}\,dx + f(y)\fint_a^b \frac{w(x)}{x-y}\,dx. \end{aligned}$$

Rabinowitz and Lubinsky [357] and Rabinowitz [352] investigate methods based on expanding $\mathrm{I}_y(wv)$ in terms of the *functions of the second kind*,

$$Q_n^v(y) := \fint_a^b w(x)\frac{P_n^w(x)}{x-y}\,dx,$$

where $\{P_n^v\}$ are the orthogonal polynomials associated with the admissible weight function v.

7.4.6. Calculating Multivariate Cauchy Principal Value Integrals. Multivariate (usually bivariate) Cauchy principal value integrals occur mostly in boundary integral methods (see Section 2.4). It is beyond the scope of this book to describe the large variety of methods (some of which have a rather limited scope) that have been proposed to deal with (certain kinds of) Cauchy singular integrals in this particular area of application. Information concerning such methods can be found, for instance, in Guiggiani [186], Hackbusch and Sauter [195], Rosen and Cormack [376], and the references cited therein. Instead, this section describes two very general methods for converting multivariate Cauchy principal value integrals into easier-to-handle integration problems.

Subtracting the Singularity. As in Section 7.4.3, the method of subtracting the singularity can be used to regularize multivariate Cauchy principal value integrals as follows:

$$\text{(7.62)} \qquad \begin{aligned} \mathrm{I}_{\mathbf{y}} f &= \fint_B k(\mathbf{x},\mathbf{y})f(\mathbf{x})\,d\mathbf{x} \\ &= \fint_B k(\mathbf{x},\mathbf{y})(f(\mathbf{x})-f(\mathbf{y}))\,d\mathbf{x} + f(\mathbf{y})\fint_B k(\mathbf{x},\mathbf{y})\,d\mathbf{x}. \end{aligned}$$

It is assumed that the second integral in (7.62) can be calculated either analytically or numerically with high precision. The remaining function

$$\varphi_{\mathbf{y}}(\mathbf{x}) := k(\mathbf{x},\mathbf{y})(f(\mathbf{x})-f(\mathbf{y}))$$

usually proves to be sufficiently regular at $\mathbf{x} = \mathbf{y}$ in order for the improper Riemann integral of $\varphi_{\mathbf{y}}$ to exist over B. More precisely, if f is differentiable at $\mathbf{x} = \mathbf{y}$, then the singularity in $\varphi_{\mathbf{y}}(\mathbf{x})$ at $\mathbf{x} = \mathbf{y}$ is of maximum order $N-1$, and thus $\varphi_{\mathbf{y}}$ is only *weakly singular*. Note, however, that $\|\mathbf{x}-\mathbf{y}\|_2^{N-1}\varphi_{\mathbf{y}}$ is only bounded, but not generally continuous at $\mathbf{x}=\mathbf{y}$. In other words, if $\varphi_{\mathbf{y}}$ is represented in the form

$$\varphi_{\mathbf{y}}(\mathbf{x}) = \frac{g(\mathbf{x})}{\|\mathbf{x}-\mathbf{y}\|_2^{N-1}},$$

then g often proves to be discontinuous at $\mathbf{x}=\mathbf{y}$ for practical kernels k. Typically, the *directional limits*

$$g_{\mathbf{x}} = \lim_{t\to 0+} g(t\mathbf{x})$$

exist, but yield different values for different directions $\mathbf{x}/\|\mathbf{x}\|_2$. This particular behavior must be taken into account when using one of the methods described in Section 7.3 to evaluate $\mathrm{I}\varphi_{\mathbf{y}}$.

Theoretical and practical aspects of subtracting the singularity in boundary element methods are thoroughly discussed, for instance, in Kieser, Schwab, and Wendland [233] and Schwab and Wendland [387].

Cartesian Product Formulas in Spherical Coordinates. Kernel functions $k(\mathbf{x},\mathbf{y})$ can usually be written in the form

$$k(\mathbf{x},\mathbf{y}) = \frac{g_{\mathbf{y}}((\mathbf{x}-\mathbf{y})/\|\mathbf{x}-\mathbf{y}\|_2)}{\|\mathbf{x}-\mathbf{y}\|_2^N} + h_{\mathbf{y}}(\mathbf{x}), \tag{7.63}$$

where $g_{\mathbf{y}}$ is smooth and $h_{\mathbf{y}}$ is (improperly) Riemann integrable over B, typically with a singularity of order $N-1$ at $\mathbf{y}$. As a result, the equality

$$⨍_B k(\mathbf{x},\mathbf{y})f(\mathbf{x})\,d\mathbf{x} = \int_B h_{\mathbf{y}}(\mathbf{x})f(\mathbf{x})\,d\mathbf{x} + ⨍_B \frac{g_{\mathbf{y}}((\mathbf{x}-\mathbf{y})/\|\mathbf{x}-\mathbf{y}\|_2)}{\|\mathbf{x}-\mathbf{y}\|_2^N} f(\mathbf{x})\,d\mathbf{x}$$

holds. The remaining Cauchy principal value integral can be transformed using spherical coordinates (1.8) to parameterize the difference $\mathbf{x}-\mathbf{y}$. For notational simplicity, the following discussion is limited to bivariate, Cauchy singular integrals ($N=2$). However, all results can easily be generalized to higher dimension.

Let $\tilde{f}(r,\varphi)$ and $\tilde{g}_{\mathbf{y}}(\varphi)$ be the representations of $f(\mathbf{x})$ and $g_{\mathbf{y}}((\mathbf{x}-\mathbf{y})/\|\mathbf{x}-\mathbf{y}\|_2)$ obtained by parameterizing the difference $\mathbf{x}-\mathbf{y}$ in terms of the spherical coordinates (r,φ). Note that $\tilde{g}_{\mathbf{y}}$ is independent of the radial coordinate r. Furthermore, let the boundary of B have an explicit representation $R:\Phi=[0,2\pi]\to\mathbb{R}_+$ in the angular coordinate φ:

$$\partial B = \{(r,\varphi)\,:\,\varphi\in\Phi,\, r=R(\varphi)\}.$$

Then, by applying the multivariate transformation rule (1.7) with (1.9) and (1.10), one obtains

$$\lim_{\delta\to 0+}\int_{B\setminus S_N^{\|\cdot\|_2}(\mathbf{y};\delta)} \frac{g_{\mathbf{y}}((\mathbf{x}-\mathbf{y})/\|\mathbf{x}-\mathbf{y}\|_2)}{\|\mathbf{x}-\mathbf{y}\|_2^N} f(\mathbf{x})\,d\mathbf{x}$$
$$= \lim_{\delta\to 0+}\int_{\Phi} \tilde{g}_{\mathbf{y}}(\varphi)\sin(\varphi) q(\varphi,\delta)\,d\varphi,$$

where

$$q(\varphi,\delta) := \int_{\delta}^{R(\varphi)} \frac{\tilde{f}(r,\varphi)}{r}\,dr. \tag{7.64}$$

Since the limits on the outer integral with respect to φ are independent of δ, one is tempted to interchange the outer integration and the limit $\delta \to 0+$. Unfortunately, since the integrand in (7.64) is generally *strongly singular* at $r = 0$, $q(\varphi, \delta)$ does *not* usually converge as $\delta \to 0+$. However, if f is *Hölder continuous* at $\mathbf{y}$, i.e., if $\tilde{f}$ is Hölder continuous in r at $r = 0$, then

$$\int_{\delta}^{R(\varphi)} \frac{\tilde{f}(r,\varphi)}{r}\,dr = \int_{\delta}^{R(\varphi)} \frac{\tilde{f}(r,\varphi)-f(\mathbf{y})}{r}\,dr + f(\mathbf{y})\log R(\varphi) - f(\mathbf{y})\log\delta$$

establishes the following splitting of q into a convergent and a divergent part:

$$q(\varphi,\delta) = a_0(\varphi,\delta) + a_1\log\delta,$$

with

$$a_0(\varphi,\delta) := \int_{\delta}^{R(\varphi)} \frac{\tilde{f}(r,\varphi)-f(\mathbf{y})}{r}\,dr + f(\mathbf{y})\log R(\varphi) \quad \text{and} \quad a_1 := -f(\mathbf{y}).$$

It is important to note that

$$\hat{a}_0(\varphi) := \lim_{\delta\to 0+} a_0(\varphi,\delta) = \int_{0}^{R(\varphi)} \frac{\tilde{f}(r,\varphi)-f(\mathbf{y})}{r}\,dr + f(\mathbf{y})\log R(\varphi)$$

is the *finite part integral* of $\tilde{f}(\cdot,\varphi)$, i.e.,

$$\hat{a}_0(\varphi) = \int\!\!\!\!\!\!=_{0}^{R(\varphi)} \frac{\tilde{f}(r,\varphi)}{r}\,dr$$

(see Section 1.4.1). Now, the assumption that the sought-after Cauchy principal value integral exists at least for constant functions, i.e., that

$$\fint_{B} k(\mathbf{x},\mathbf{y})\,d\mathbf{x}$$

exists, implies

$$\lim_{\delta\to 0+}\log\delta\int_{\Phi}\tilde{g}_{\mathbf{y}}(\varphi)\sin(\varphi)\,d\varphi = 0,$$

i.e.,

$$\int_\Phi \tilde{g}_{\mathbf{y}}(\varphi)\sin(\varphi)\,d\varphi = 0.$$

This, along with the fact that $\hat{a}_0(\varphi)$ can be shown to be integrable provided R and f are sufficiently regular, proves that

$$\begin{aligned}\lim_{\delta\to 0+}\int_\Phi \tilde{g}_{\mathbf{y}}(\varphi)\sin(\varphi)q(\varphi,\delta)\,d\varphi &= \lim_{\delta\to 0+}\int_\Phi \tilde{g}_{\mathbf{y}}(\varphi)\sin(\varphi)a_0(\varphi,\delta)\,d\varphi \\ &= \int_\Phi \tilde{g}_{\mathbf{y}}(\varphi)\sin(\varphi)\lim_{\delta\to 0+} a_0(\varphi,\delta)\,d\varphi \\ &= \int_\Phi \tilde{g}_{\mathbf{y}}(\varphi)\sin(\varphi)\hat{a}_0(\varphi)\,d\varphi. \qquad (7.65)\end{aligned}$$

As a result, the sought-after Cauchy principal value integral can be approximated by applying a Cartesian product rule to the right-most iterated integral in (7.65). Since $\hat{a}_0$ is smooth whenever R and f are smooth, standard quadrature formulas (Gauss formulas, for instance) can be applied to the outer integral. The inner integral can be approximated using quadrature formulas appropriate for finite part integrals (see Section 7.5).

The main purpose of splitting the kernel $k(\cdot,\mathbf{y})$ according to (7.63) into a strongly singular, homogeneous part and a weakly singular part is to simplify the subsequent mathematical analysis of the method proposed in this section. However, the analysis given above can easily be extended to kernel representations of the form

$$k(\mathbf{x},\mathbf{y}) = \frac{g_{\mathbf{y}}(\mathbf{x}-\mathbf{y})}{\|\mathbf{x}-\mathbf{y}\|_2^N}.$$

In this case, parameterizing the difference $\mathbf{x}-\mathbf{y}$ in terms of the spherical coordinates $(r,\varphi)^\top$ gives rise to a representation $\tilde{g}_{\mathbf{y}}(r,\varphi)$ of $g_{\mathbf{y}}(\mathbf{x}-\mathbf{y})$ which depends on both the radial and the angular coordinates. If the functions $r\mapsto \tilde{g}_{\mathbf{y}}(r,\varphi)$, $\varphi\in\Phi$, are Hölder continuous (with exponent $\lambda>0$) at $r=0$ and the function

$$\varphi \mapsto \lim_{r\to 0+}\tilde{g}_{\mathbf{y}}(r,\varphi)$$

is smooth, the formula

$$\oint_B k(\mathbf{x},\mathbf{y})f(\mathbf{x})\,d\mathbf{x} = \int_\Phi \sin(\varphi)\hat{a}_0(\varphi)\,d\varphi \qquad (7.66)$$

with

$$\hat{a}_0(\varphi) = \oint_0^{R(\varphi)} \frac{\tilde{g}_{\mathbf{y}}(r,\varphi)\tilde{f}(r,\varphi)}{r}\,dr \qquad (7.67)$$

can be derived. Hence, the sought-after Cauchy principal integral can be approximated by discretizing the right-hand side integrals in (7.66) and (7.67) using appropriate quadrature formulas.

Thorough analyses of convergence properties and discussions of computational aspects of particular Cartesian product rules obtained by the above method can be found in Kieser, Schwab, and Wendland [233], Schwab and Wendland [387], and Monegato [306].

7.5. Finite Part Integrals

This section outlines computational techniques for the numerical integration of hypersingular functions, i.e., strongly singular functions whose integrals are defined in the finite part sense (see Section 1.4). Under certain conditions (see Section 1.4.3), the Cauchy principal value integral of a strongly singular function is well defined, and in this case finite part and Cauchy principal value integrals coincide. Most of the fundamental issues in computational methods for Cauchy principal value integrals (see Section 7.4) are also relevant, more generally, for finite part integrals. Areas of commonality include, in particular, the practical setting of integration problems, the ill conditioning of the discretized integral operators, and the instability of numerical integration methods. In order to avoid excessive duplication of presented material, this section emphasizes differences between numerical methods for finite part integrals and Cauchy principal value integrals; when common areas arise, the reader is referred to the appropriate parts in Section 7.4. Also, readers who wish to further deepen their understanding of the methods presented in this section should consult the excellent survey given in Monegato [307].

Throughout this section, it is tacitly assumed that the integration region B, the integrand f, and the kernel function k meet the assumptions made about them in the various definitions in Section 1.4. In particular, it is assumed that $k(\mathbf{x}, \mathbf{y})$ has a singularity of order $\alpha + N$, $\alpha \geq 0$, at $\mathbf{x} = \mathbf{y}$, that f is $k := \lceil \alpha \rceil$-times continuously differentiable in a neighborhood of $\mathbf{y}$, and that, for $\alpha \in \mathbb{N}_0$, $f^{(k)}$ is Hölder continuous at $\mathbf{y}$. Additional weight functions w are assumed to be (improperly) integrable on B, and asymptotic expansions of type (1.32) must be valid for the corresponding weighted integrals.

7.5.1. Differentiating Formulas for Cauchy Singular Integrals. Let $\{ \mathrm{Q}_n^{(y)} : y \in (a, b) \}$,

$$\mathrm{Q}_n^{(y)} f = \sum_{i=1}^{n} c_i(y) f(x_i(y)),$$

be a family of n-point quadrature formulas for the Hilbert transform integral

$$\mathrm{I}_y f = ⨍_a^b \frac{f(x)}{x - y}\, dx,$$

and assume that the $c_1, \dots, c_n$ and the $x_1, \dots, x_n$ are smooth functions of y. Then, the relationship

$$\text{(7.68)} \qquad \mathrm{I}_{y;\alpha} f := \int_a^b \frac{f(x)}{(x-y)^{\alpha+1}}\,dx = \frac{1}{\alpha!}\frac{d^\alpha}{dy^\alpha} ⨍_a^b \frac{f(x)}{x-y}\,dx,$$

derived from Theorem 1.4.1, suggests that $\mathrm{I}_{y;\alpha} f$ can be approximated by differentiating $\mathrm{Q}_n^{(y)} f$ with respect to the parameter y:

$$\text{(7.69)} \quad \mathrm{I}_{y;\alpha} f \approx \mathrm{Q}_{n_\alpha}^{(y;\alpha)} f := \frac{1}{\alpha!}\frac{d^\alpha\left(\mathrm{Q}_n^{(y)} f\right)}{dy^\alpha} = \frac{1}{\alpha!}\sum_{i=1}^{n} \frac{d^\alpha(c_i(y) f(x_i(y)))}{dy^\alpha}.$$

The number of function and derivative values required to evaluate $\mathrm{Q}_{n_\alpha}^{(y;\alpha)}$ can be bounded by $n \le n_\alpha \le \alpha n$. For $\alpha = 1$, for instance, the formula

$$\text{(7.70)} \qquad \mathrm{Q}_{n_1}^{(y;1)} f := \sum_{i=1}^{n} \frac{dc_i}{dy}(y) f(x_i(y)) + c_i(y)\frac{dx_i}{dy}(y) f'(x_i(y))$$

can be derived. The formulas $\{\mathrm{Q}_{n_\alpha}^{(y;\alpha)}\}$ become particularly simple when the abscissas $x_1, \dots, x_n$ in $\{\mathrm{Q}_n^{(y)}\}$ do *not* depend on y, because in this case $n_\alpha = n$ and

$$\text{(7.71)} \qquad \mathrm{Q}_n^{(y;\alpha)} f = \frac{1}{\alpha!}\sum_{i=1}^{n} \frac{d^\alpha c_i}{dy^\alpha}(y) f(x_i).$$

Formulas with this property usually arise from product integration methods (see Section 7.4.4). In formulas obtained by subtracting the singularity (see Section 7.4.3), on the other hand, one abscissa, say x_j, is identical to y, while the others are independent of y. In this case, $n_\alpha = n + \alpha$ and

$$\text{(7.72)} \quad \mathrm{Q}_{n+\alpha}^{(y;\alpha)} f = \frac{1}{\alpha!}\left(\sum_{i=1, i\ne j}^{n} \frac{d^\alpha c_i}{dy^\alpha}(y) f(x_i) + \sum_{l=0}^{\alpha}\binom{\alpha}{l}\frac{d^l c_i}{dy^l}(y) f^{(\alpha-l)}(y)\right).$$

Clearly, if the formulas $\{\mathrm{Q}_n^{(y)}\}$ are exact for a particular class of functions $\mathcal{F}$, then, by virtue of (7.68) and (7.69), the formulas $\{\mathrm{Q}_{n_\alpha}^{(y;\alpha)}\}$, $\alpha \in \mathbb{N}$, are also exact for functions in $\mathcal{F}$. In particular, differentiating with respect to the parameter y preserves the accuracy of $\{\mathrm{Q}_n^{(y)}\}$. For instance, if $\{\mathrm{Q}_n^{(y)}\}$ is a family of interpolatory product formulas and the abscissas $x_1, \dots, x_n$ in $\{\mathrm{Q}_n^{(y)}\}$ are independent of y, then $\{\mathrm{Q}_n^{(y;\alpha)}\}$ in (7.71) is a family of interpolatory product formulas for any $\alpha \in \mathbb{N}$. This shows that the interpolatory product formulas described in Section 7.5.3 are (for $\alpha \in \mathbb{N}$ and $y \in (a, b)$) a particular instance of formulas obtained by differentiating formulas for Cauchy principal value integrals. Similarly, formulas (7.72) can be seen to be identical to formulas generated by subtracting the singularity in finite part integrals (see Section 7.5.2) if the underlying formulas $\{Q_n^{(y)}\}$ are defined by (7.59), i.e., if the $\{Q_n^{(y)}\}$ are obtained by subtracting the singularity in Cauchy principal value integrals.

Differentiation with respect to y can also be used in connection with classes of formulas $\{Q_n^{(y)}\}$ other than the ones mentioned above. For instance, Bialecki [34] applies this method to sinc quadrature formulas, whereas Hasegawa and Torii [202] use it for formulas obtained from (7.61) by choosing g as the polynomial which interpolates f at the Chebyshev extrema.

7.5.2. Subtracting the Singularity. The method of subtracting the singularity is used in Sections 7.3.2 and 7.4.3 to obtain numerical approximations of weakly and Cauchy singular integrals, respectively. The definition of univariate finite part integrals given in (1.21) shows that this method can also be applied to finite part integrals. Specifically, by approximating the Riemann integral of the regularized integrand

$$\varphi_{y;\alpha}(x) := \frac{f(x) - T_k(f;y)(x)}{(x-y)^{\alpha+1}}, \qquad k := \lceil \alpha \rceil,$$

using a quadrature formula Q_{n-1},[86] the formula

$$(7.73)\quad \int_a^b \frac{f(x)}{(x-y)^{\alpha+1}}\,dx \;\approx\; Q_{n+k}^{(y;\alpha)} f$$
$$:= Q_{n-1}\psi_{y;\alpha} + \sum_{l=0}^{k} \frac{f^{(l)}(y)}{l!} \int_a^b \frac{1}{(x-y)^{\alpha+1-l}}\,dx$$

can be derived. Analytical expressions for the remaining finite part integrals in (7.73) are given in (1.19) and (1.20). An equivalent way of deriving the formulas $Q_{n+k}^{(y;\alpha)}$ for $\alpha \in \mathbb{N}$ is differentiating $Q_n^{(y;0)}$ with respect to y (see Section 7.5.1).

If $\alpha \in \mathbb{N}_0$, then, as with Cauchy principal value integrals, the smoothness of the regularized function $\varphi_{y;\alpha}$ depends only on the smoothness of f. On the other hand, if $\alpha \notin \mathbb{N}_0$ (in which case $y = a$), $\varphi_{y;\alpha}$ is generally at most Hölder continuous (but not differentiable) at y, even if f is analytic. This particular fact must be taken into account when choosing suitable formulas Q_{n-1}.

For $\alpha \in \mathbb{N}_0$ (in which case $k = \alpha$), the formulas $Q_{n+k}^{(y;\alpha)}$ can easily be seen to have the degree of accuracy $D + \alpha + 1$ whenever the degree of accuracy of Q_{n-1} is D. In particular, if Q_{n-1} is an interpolatory formula, then $Q_{n+k}^{(y;\alpha)}$ is also interpolatory in the following sense: $Q_{n+k}^{(y;\alpha)} f$ is the *exact* finite part integral of the unique polynomial of degree $d \leq n + \alpha - 1$ which interpolates f at

[86] It is assumed that the abscissas in Q_{n-1} do not coincide with y.

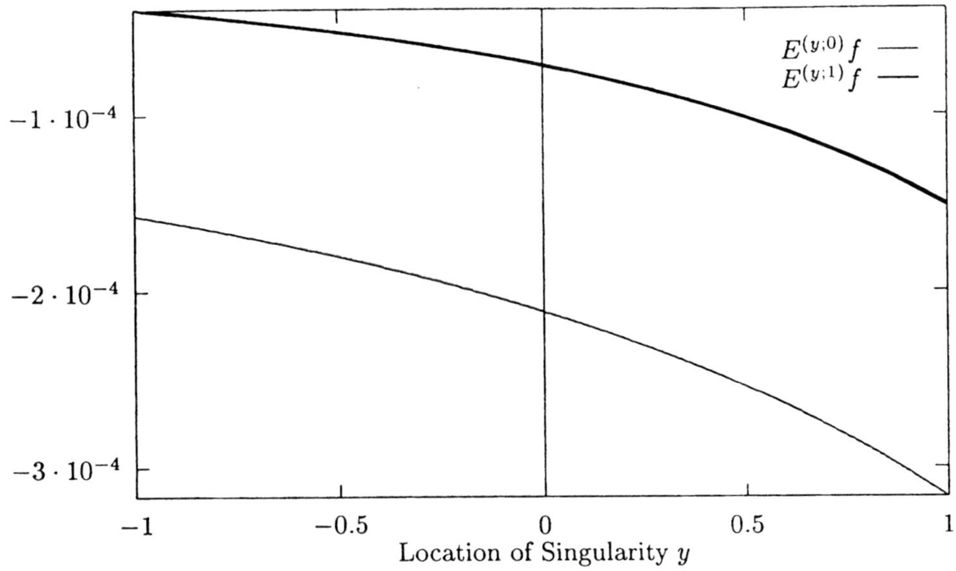

FIG. 7.3. *Errors of formulas* $G_{7+\alpha}^{(y;\alpha)}$.

the abscissas $x_1, \ldots, x_{n-1}$ of Q_{n-1} and, additionally, interpolates $f, f', \ldots, f^{(k)}$ at y.

Example (Subtracting the Singularity). The finite part integrals

$$\text{(7.74)} \qquad I_{y;\alpha}f := \int_{-1}^{1} \frac{e^{5x}}{(x-y)^{\alpha+1}}\, dx, \quad \alpha = 0, 1,$$

were approximately calculated by applying the Gauss formula G_6 to the Riemann integrals obtained by subtracting the singularity at y. Figure 7.3 shows the errors of the resulting formulas $G_{7+\alpha}^{(y;\alpha)}$,

$$E^{(y;\alpha)}f := G_{7+\alpha}^{(y;\alpha)}f - I_{y;\alpha}f,$$

as a function of y. It can be seen that the integration errors $\{E^{(y;\alpha)}f \,:\, y \in (-1,1)\}$ differ by at most a factor of about four, i.e., that the approximation quality of the formulas $\{\, G_{7+\alpha}^{(y;\alpha)} \,:\, y \in (-1,1)\,\}$ is highly uniform.

It is important to note that (i) the evaluation of $Q_{n+k}^{(y;\alpha)}$ becomes ill conditioned when one or more abscissas x_i of Q_{n-1} are chosen close to y and (ii) the conditioning generally becomes worse as α increases. Moreover, if $\alpha > 0$, then the formulas $Q_{n+k}^{(y;\alpha)}$ have the disadvantage of requiring the values of the derivatives $f'(y), \ldots, f^{(k)}(y)$.

For $\alpha \in \mathbb{N}_0$, choosing Q_{n-1} as the Gauss formula G_{n-1} (or as the Gauss formula G_{n-1}^{w} if an additional admissible weight function w is present) yields quadrature formulas $Q_{n+k}^{(y;\alpha)}$ which are highly efficient for smooth integrands (see the previous example). The approximation properties of such formulas for $a < y < b$ are analyzed, for instance, in Paget [327] and Ioakimidis [215], whereas Monegato [306] deals with formulas for $y = a$ resulting from choosing Q_{n-1} as the Gauss or the Gauss–Radau formula with respect to the Jacobi weight function w. Alternatively, Q_{n-1} can be chosen as a compound quadrature formula. The approximation properties of the resulting formulas $\mathrm{Q}_{n+k}^{(y;\alpha)}$ are studied, for instance, in Stolle and Strauss [419] and Diethelm [109].

7.5.3. Product Formulas. Product formulas for finite part integrals are obtained by replacing f with suitable model functions $g \in \mathcal{M}$ and using the exact integrals $\mathrm{I}_{y;\alpha}g =: \mathrm{Q}_n^{(y;\alpha)}f$ as approximations to $\mathrm{I}_{y;\alpha}f$. If the g are independent of the singularity y, then, for $y \in (a,b)$, the $\mathrm{Q}_n^{(y;\alpha)}f$, $\alpha \in \mathbb{N}$, can also be obtained by differentiating $\mathrm{Q}_n^{(y;0)}f$ with respect to y (see Section 7.5.1). As with product formulas for Cauchy principal value integrals (see Section 7.4.4), thorough analyses are necessary in order to establish approximation properties of the resulting formulas because the evaluation of finite part integrals is ill conditioned with respect to changes in f (see Section 3.4.1).

Interpolatory product formulas for finite part integrals are mostly based on using the zeros of the orthogonal polynomials $\{P_n^v\}$ associated with (generalized) Jacobi weight functions v as quadrature abscissas. The weight function v may, or may not, be identical to the weight function w in the integral operator. Monegato [305], for instance, analyzes convergence properties of interpolatory product formulas based on the zeros of $\{P_n^w\}$, where w is a Jacobi weight function, while Criscuolo and Mastroianni [70] study interpolatory product formulas for different classes of weight functions w based on the zeros of $\{P_n^v\}$, where v is a generalized Jacobi weight function. It is, of course, also possible to construct product formulas using piecewise polynomial approximation functions (see, for example, Khan [231]).

Example (Subtracting the Singularity). The finite part integrals (7.74) were approximately calculated using interpolatory product formulas $\mathrm{Q}_8^{(y;\alpha)}$ based on the zeros of the Legendre polynomial of degree 8. Figure 7.4 shows the resulting errors $E^{(y;\alpha)}f := \mathrm{Q}_8^{(y;\alpha)}f - \mathrm{I}_{y;\alpha}f$ as a function of y. It can be seen that the integration errors $\{E^{(y;\alpha)}f \,:\, y \in (-1,1)\}$ vary significantly and clearly tend to infinity as $|y| \to 1$. Furthermore, the integration errors are, in particular for $\alpha = 1$, significantly larger than the integration errors in the formulas $G_{7+\alpha}^{(y;\alpha)}$ shown in Figure 7.3.

7.5.4. Miscellaneous Methods. Elliott [127] and Delbourgo and Elliott [107] study quadrature formulas for finite part integrals based on approximating f using Bernstein polynomials $B_n(f)$. The rate of convergence of the resulting formulas for smooth integrands is $O(n^{-1})$. Modified formulas attain a rate of convergence of $O(n^{-2})$, but require f to be evaluated outside

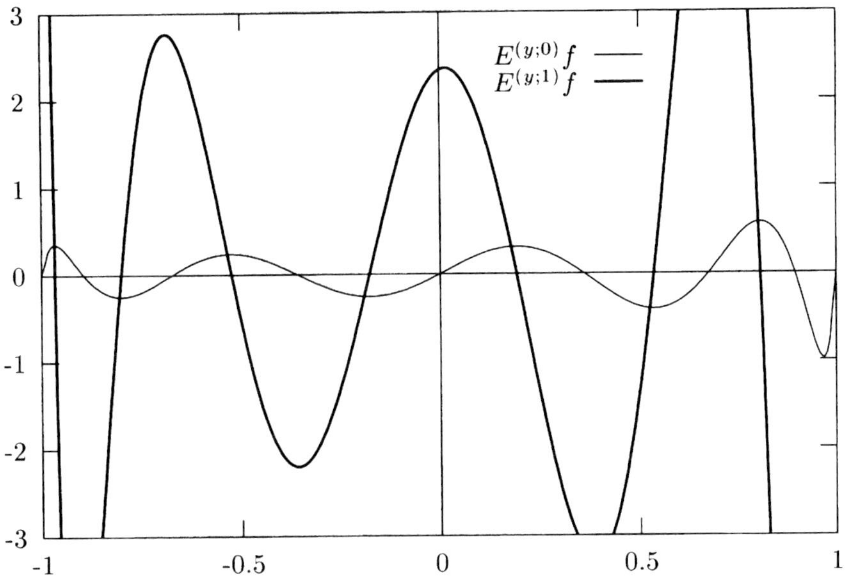

FIG. 7.4. *Errors in product formulas* $Q_8^{(y;\alpha)}$.

the integration region and, for $\alpha \in \mathbb{N}$, certain derivatives of f to be evaluated at the singularity y.

The method proposed in Zhang [451] is based on (i) regularizing the given integration problem using an appropriate central finite difference scheme (instead of a Taylor polynomial) and (ii) applying a sinc method to the regularized integral. Extrapolation-based methods for finite part integrals are discussed in Lyness [273].

7.5.5. Calculating Multivariate Finite Part Integrals. Section 7.4.6 describes two general methods for converting multivariate Cauchy principal value integrals into easier-to-handle integration problems. This section shows how these methods can be adapted to multivariate (usually bivariate) finite part integrals, which occur mostly in boundary integral methods (see Section 2.4). Information regarding special methods for dealing with (certain kinds of) finite part integrals in this particular area of application can be found in Guiggiani et al. [187], Hackbusch and Sauter [195], Rosen and Cormack [376], and the references cited therein.

Subtracting the Singularity. As with Section 7.5.2, the method of subtracting the singularity can be used to regularize multivariate finite part integrals as follows:

$$(7.75) \quad \mathrm{I}_{\mathbf{y}} f \;=\; \fint_B k(\mathbf{x},\mathbf{y}) f(\mathbf{x})\, d\mathbf{x}$$

$$= \int_B k(\mathbf{x},\mathbf{y})(f(\mathbf{x}) - T_k(f;\mathbf{y})(\mathbf{x}))\,d\mathbf{x} + \int_B k(\mathbf{x},\mathbf{y})T_k(f;\mathbf{y})(\mathbf{x})\,d\mathbf{x},$$

where $k := \lceil\alpha\rceil$. It is assumed that the second integral on the right-hand side in (7.75) can be calculated either analytically or numerically with high precision. The remaining function $\varphi_{\mathbf{y}}(\mathbf{x}) := k(\mathbf{x},\mathbf{y})(f(\mathbf{x}) - T_k(f;\mathbf{y})(\mathbf{x}))$ is sufficiently regular at $\mathbf{x} = \mathbf{y}$ for the improper Riemann integral of $\varphi_{\mathbf{y}}$ over B to exist.

Theoretical and practical aspects of subtracting the singularity in boundary element methods are thoroughly discussed, for instance, in Kieser, Schwab, and Wendland [233] and Schwab and Wendland [387].

Cartesian Product Formulas in Spherical Coordinates. In Section 7.4.6, numerical approximations of multivariate Cauchy principal value integrals are obtained by transforming these integrals using spherical coordinates and by applying Cartesian product formulas to the resulting iterated integrals. This method can easily be generalized to finite part integrals by representing the kernel in the form

$$k(\mathbf{x},\mathbf{y}) = \frac{g_{\mathbf{y}}(\mathbf{x}-\mathbf{y})}{\|\mathbf{x}-\mathbf{y}\|_2^{\alpha+N}}.$$

For notational simplicity, assume $N = 2$, and let $\tilde{f}(r,\varphi)$ and $\tilde{g}_{\mathbf{y}}(r,\varphi)$ be the representations of $f(\mathbf{x})$ and $g_{\mathbf{y}}(\mathbf{x}-\mathbf{y})$ obtained by parameterizing the difference $\mathbf{x}-\mathbf{y}$ in terms of the spherical coordinates $(r,\varphi)^\top$. Then, if $\tilde{f}(r,\varphi)$ and $\tilde{g}_{\mathbf{y}}(r,\varphi)$ satisfy appropriate regularity conditions, the formula

$$\text{(7.76)} \qquad \int_B k(\mathbf{x},\mathbf{y})f(\mathbf{x})\,d\mathbf{x} = \int_\Phi \sin(\varphi)\hat{a}_0(\varphi)\,d\varphi$$

with

$$\text{(7.77)} \qquad \hat{a}_0(\varphi) = \int_0^{R(\varphi)} \frac{\tilde{g}_{\mathbf{y}}(r,\varphi)\tilde{f}(r,\varphi)}{r^{\alpha+1}}\,dr$$

can be derived. Hence, the sought-after finite part integral can be approximated by discretizing the right-hand side integrals in (7.76) and (7.77) using appropriate quadrature formulas. In particular, different kinds of formulas for the univariate finite part integral (7.77) are described in the foregoing sections.

Thorough analyses of convergence properties and discussions of computational aspects of particular Cartesian product rules obtained by the above method can be found in Kieser, Schwab, and Wendland [233], Schwab and Wendland [387], and Monegato [307].

Chapter 8

Integration Algorithms

In Chapters 5, 6, and 7, numerous methods of computing approximate values, $\mathrm{Q}f$, of definite integrals, $\mathrm{I}f$, are discussed. In certain situations, theoretical error analyses show that the numerical integration method satisfies the error requirements specific to the given problem. This is, for instance, often true when a particular application requires that a certain class $\mathcal{F}$ of functions be integrated exactly, $\mathrm{Q}f - \mathrm{I}f = 0$ for $f \in \mathcal{F}$, or that a specific order of convergence be attained,

$$\mathrm{Q}(f;B) - \mathrm{I}(f;B) = O(h^{\alpha}), \quad \alpha > 0, \qquad \text{as} \quad h = \mathrm{diam}(B) \to 0.$$

If, however, a solution to the numerical problem (3.1), i.e., an approximate integral value $\mathrm{Q}f$ satisfying

$$(8.1) \qquad |\,\mathrm{Q}f - \mathrm{I}f\,| \le \varepsilon,$$

for a given error tolerance ε, is sought, it is not usually possible to theoretically prove the suitability of an integration method. The *discretization error* in numerical integration methods depends on certain analytic properties of the integrand which are generally not known *a priori* and are usually difficult to estimate or bound computationally. For instance, the remainder terms in Newton–Cotes formulas (see Sections 5.2.4 and 5.2.5) can be accurately expressed in terms of values of certain derivatives $f^{(k)}$ at points $\xi \in [a,b]$. Unfortunately, since the location of ξ is unknown, calculating bounds on the discretization error $E(f)$ requires calculating bounds M_k on $|f^{(k)}|[a,b]$. This necessitates the application of nonstandard computational methods, such as the use of *interval arithmetic* calculations (see Section 8.1.5). Furthermore, when integration algorithms are performed using finite precision arithmetic, *rounding errors* occur which can induce significant perturbations in numerical results. This is especially true when the integration problem is badly conditioned (see Section 3.4).

What is needed for the solution of numerical problems of the form (8.1) are *integration algorithms*. These incorporate (in addition to a basic method for calculating approximate integral values) *error estimation* procedures for estimating or bounding the error in numerical integration results. Moreover, in order to cope with situations in which the accuracy of an initial integral approximation turns out to be insufficient, most integration algorithms also possess a means of generating sequences $\{Q_{n_j} f\}$ of approximate integral values of (presumably) increasing accuracy. Since such sequences are generally obtained by increasing the number of samples taken of f, this feature is also referred to as *discretization refinement*.

Sections 8.1 and 8.2 discuss various error estimation and discretization refinement techniques, respectively. The various parts of Section 8.3 deal with a number of miscellaneous features of integration algorithms. Specifically, Section 8.3.1 describes measures for increasing the reliability of integration algorithms, Section 8.3.2 discusses the incorporation of acceleration methods into integration algorithms, and Section 8.3.4 deals with the simultaneous integration of multiple integrands.

8.1. Error Estimation

An approximation of the error in a numerically calculated quantity is said to be an *error estimate*. In this section, error estimates for integration formulas—the fundamental building blocks of all integration algorithms—are discussed. Unless explicitly stated otherwise, it is assumed that the level of error induced by rounding is significantly lower than the discretization error. Methods for detecting and/or avoiding a loss of accuracy due to rounding errors are discussed in Section 8.3.1.

Error Estimation Using Samples. Error estimates are usually obtained by comparing two or more approximations, calculated in different ways, for the quantity which is to be computed. The difference(s) between the various approximate results are used to estimate the actual approximation error.

As with numerical integration, the integration formulas to be compared are chosen in such a way that they can be expected to attain significantly differing levels of accuracy. This can be achieved by choosing simple integration formulas with strongly varying quality indices, for instance, polynomial formulas of different degrees of accuracy (see Sections 8.1.1 and 8.1.2). Integration formulas with different accuracy levels can also be obtained by choosing compound integration formulas corresponding to subdivisions of the integration region with varying fineness (see Section 8.1.3). For certain classes of integration formulas, the discretization error can also be estimated by comparing *randomized* formulas of *similar* accuracy (see Section 8.1.4).

When comparing the results of different integration formulas, the set of abscissas used to calculate the integration formula with the highest expected accuracy should, for reasons of efficiency, contain all abscissas previously used when the results of lower accuracy were calculated. If this is the case, the sampled function values required to compute the most accurate integral approximation are reused for error estimation purposes; no additional integrand evaluations are required for calculating results of lower accuracy. However, different linear combinations of the integrand values must be computed to obtain results with differing degrees of accuracy. These additional linear combinations constitute the main effort necessary to calculate error estimates in numerical integration programs. This is usually negligible compared with the computational cost of integrand evaluation.

Inherent Unreliability of Sampling Methods. Numerical integration methods which use only the *discrete* sample information

$$S(f) = (f(\mathbf{x}_1), f(\mathbf{x}_2), \ldots, f(\mathbf{x}_n))$$

are inherently unreliable. This is easily understood by considering the set

$$V(f; \mathcal{R}, S) := \{\tilde{f} \in \mathcal{R} \; : \; S(\tilde{f}) = S(f)\}$$

of those Riemann integrable functions which cannot be distinguished, based on the sample information $S(f)$, from f. For instance, let the functions $\tilde{f}_c$, $c \in \mathbb{R}$, be defined by

$$\tilde{f}_c(\mathbf{x}) := f(\mathbf{x}) + c \prod_{i=1}^{n} \|\mathbf{x} - \mathbf{x}_i\|_2^2, \quad c \in \mathbb{R}.$$

These functions have the property that they cannot be distinguished from f at the sample points $\mathbf{x}_1, \ldots, \mathbf{x}_n$,

$$\tilde{f}_c(\mathbf{x}_i) = f(\mathbf{x}_i), \quad i = 1, 2, \ldots, n.$$

The corresponding integrals

$$\mathrm{I}\tilde{f}_c = \mathrm{I}f + c \int_B \prod_{i=1}^{n} \|\mathbf{x} - \mathbf{x}_i\|_2^2 \, d\mathbf{x},$$

however, can assume *any* real value, and consequently $V(f; \mathcal{R}, S) = \mathbb{R}$.

Since the set $\{\mathrm{I}\tilde{f} \; : \; \tilde{f} \in V(f; \mathcal{R}, S)\}$ is unbounded, error estimation procedures which only use discrete sampling information are bound to fail for infinitely many functions $\tilde{f} \in V(f; \mathcal{R}, S)$. Only by taking into account additional, global information about f, such as bounds on the absolute value of its derivatives, can reliable error estimates be obtained (see Section 8.1.5). Unfortunately, such information is not available in many practical integration problems.

8.1.1. Using Pairs of Formulas with Different Levels of Accuracy. A straightforward way to obtain error estimates is to compare the results of pairs of integration formulas Q_{n_1} and Q_{n_2} in which Q_{n_2} can be *expected* to give more accurate results than Q_{n_1} for most admissible integrands $f \in \mathcal{F}$.

Example (Polynomial Rules). If the formula Q_{n_1} has the degree of accuracy D_1 (see Sections 5.2.1 and 6.2.2), then Q_{n_1} integrates all polynomials $P \in \mathbb{P}^N_{d_1}$, $d_1 = D_1$, accurately. A formula Q_{n_2} with degree D_2, $D_2 > D_1$, is exact for a *larger* set of functions $\mathbb{P}^N_{d_2}$, $d_2 = D_2$, $\mathbb{P}^N_{d_2} \supset \mathbb{P}^N_{d_1}$; i.e., Q_{n_2} gives better results for polynomials $P \in \mathbb{P}^N_{d_2} \setminus \mathbb{P}^N_{d_1}$ than Q_{n_1}.

The assumption that Q_{n_2} usually produces significantly better results than Q_{n_1},

$$|Q_{n_2}f - If| \ll |Q_{n_1}f - If|, \tag{8.2}$$

implies that the estimate

$$Q_{n_1}f - If \quad \approx \quad Q_{n_1}f - Q_{n_2}f \tag{8.3}$$

should be true for "most" $f \in \mathcal{F}$. Thus, $Q_{n_2}f$ can be regarded as the *exact value* when compared with the less accurate value $Q_{n_1}f$.

The estimate (8.3) is unsatisfactory in that it provides information about the error in the *less* accurate value $Q_{n_1}f$ only. The more accurate value $Q_{n_2}f$, which generally requires more computational effort, is used only to obtain an error estimate. For this reason, (8.2) is often used to conclude

$$|Q_{n_2}f - If| \leq |Q_{n_1}f - Q_{n_2}f|, \tag{8.4}$$

which can be guaranteed to hold if

$$|Q_{n_2}f - If| \leq \frac{1}{2}|Q_{n_1}f - If| \tag{8.5}$$

is satisfied. Based on (8.4), $Q_{n_2}f$ can be taken as the approximate integral value, and $|Q_{n_1}f - Q_{n_2}f|$ as the corresponding error estimate.

While the resulting error estimates are generally overly pessimistic, there is in practice no guarantee that, for a particular integrand, $Q_{n_2}f$ is indeed more accurate than $Q_{n_1}f$. The error estimate (8.4) may fail when (8.5) does *not* hold. In such cases an integration algorithm may fail, i.e., produce a total error which is considerably larger than the maximum admissible error.

Example (Polynomial Rules). A higher degree of accuracy D_2 of Q_{n_2} can only be expected to result in a more precise estimate of If if a polynomial of degree $d_2 = D_2$ approximates f more accurately on B than a polynomial of the lower degree $d_1 = D_1$. This is often the case if f is sufficiently smooth. However, this is usually *not* true, for instance, if f, or a low-order derivative of f, has a discontinuity in the integration region.

Tests for Asymptotic Behavior. If Q_{n_1} and Q_{n_2} are chosen as polynomial integration formulas of degrees of accuracy D_1 and D_2, respectively, then tests for *asymptotic error behavior* are often used to increase the reliability and accuracy of error estimation procedures based on comparing $Q_{n_1}f$ and $Q_{n_2}f$. The purpose of such tests is to determine whether or not the formula Q_{n_2} can really be expected to yield a more precise integral estimate than Q_{n_1} for given f and B, i.e., whether or not f is sufficiently smooth and $\operatorname{diam} B$ is sufficiently small for this expectation to be met. In this context the term *asymptotic error behavior* refers to the fact that, under the aforementioned conditions, the magnitude of the integration error can be expected to be determined by the asymptotic order of convergence of $Q_{n_{1,2}}(f;B) \to I(f;B)$ as $\operatorname{diam} B \to 0$. Specific knowledge concerning the order of convergence, such as

$$|Q_{n_1}(f;B) - I(f;B)| \le C_1\,(\operatorname{diam} B)^{D_1+2},$$
$$|Q_{n_2}(f;B) - I(f;B)| \le C_2\,(\operatorname{diam} B)^{D_2+2}$$

(see Theorem 5.2.6), can then be utilized to obtain more precise and/or more reliable error estimates.

Tests for asymptotic error behavior are highly heuristic and usually depend on the underlying integration formulas. Thus, crucial algorithmic parameters are often based on experimental data rather than on mathematical reasoning.

Example (Gauss–Kronrod Formula Pairs). The subroutines `QUADPACK/dqkn` implement the Gauss–Legendre formulas G_n along with the corresponding Kronrod formulas K_{2n+1} for $n = 7, 21, 31, 41, 51, 61$. The approximate value of the respective Kronrod formula is used as the numerical integration result, and the difference between the Gauss and the Kronrod formulas serves as an error estimate.

The error estimation process for Gauss–Kronrod formula pairs in these routines includes the following test for asymptotic error behavior. The *mean value* of f on $[a,b]$ is computed numerically:

$$M(f;[a,b]) := \frac{1}{b-a} K_{2n+1}(f;[a,b]).$$

The absolute deviation $K_{2n+1}(|f-M|;[a,b])$ of f from $M(f;[a,b])$ on $[a,b]$ provides a measure for the smoothness of f on $[a,b]$. The test for asymptotic behavior compares $|G_n f - K_{2n+1} f|$ to $K_{2n+1}(|f-M|;[a,b])$. If this ratio is small, the difference between the two quadrature formulas is small compared to the variation of f on $[a,b]$; i.e., the discretization of f in the quadrature formulas G_n and K_{2n+1} is fine with respect to its variation. In this case, K_{2n+1} can indeed be expected to yield a better approximation for If than G_n.

Phase Factor Effects. Even when f belongs to the class of functions for which the integration formula Q_{n_2} can be expected to be more accurate than Q_{n_1}, the reliability of error estimation procedures may suffer from *phase factor* effects (Lyness and Kaganove [277]). While for such integrands *bounds* on the error $Q_{n_2}f - If$ are generally smaller than bounds on $Q_{n_1}f - If$, the actual

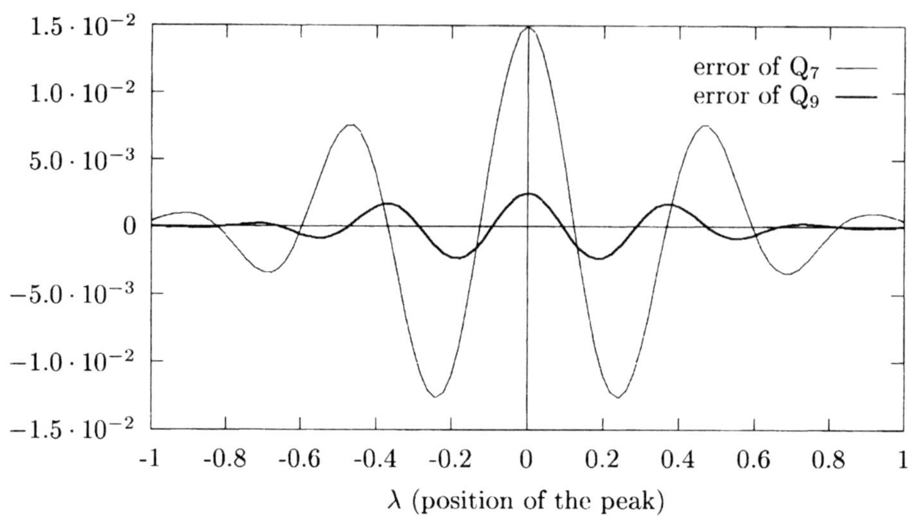

FIG. 8.1. *Error curves for practical Clenshaw–Curtis formulas.*

error does not necessarily behave in this way. This can accidentally lead to values of $|Q_{n_1}f - Q_{n_2}f|$ that are *too small* compared with $|Q_{n_2}f - If|$, which results in a significant *underestimation* of the true error.

Example (Phase Factor Effects). For any choice of the parameters λ, $\mu \in \mathbb{R}$, the function

$$f_{\lambda,\mu}(x) = \frac{10^{-\mu}}{(x-\lambda)^2 + 10^{-2\mu}}$$

has a peak of height 10^{μ} at the point $x = \lambda$. Integration of $f_{\lambda,\mu}$ over the interval $[-1, 1]$ gives

$$\int_{-1}^{1} f_{\lambda,\mu}(x)\,dx = \arctan(10^{\mu}(1-\lambda)) - \arctan(10^{\mu}(-1-\lambda)).$$

This integral was approximated numerically using the two (practical) Clenshaw–Curtis formulas Q_7 and Q_9. Figure 8.1 shows the errors $Q_7 f_{\lambda,\mu} - If_{\lambda,\mu}$ and $Q_9 f_{\lambda,\mu} - If_{\lambda,\mu}$ for the parameter $\mu = 0.35$ as a function of the location of the peak $\lambda \in [-1, 1]$. It can be seen that, on average, $Q_9 f_{\lambda,\mu}$ approximates $If_{\lambda,\mu}$ significantly better than $Q_7 f_{\lambda,\mu}$ does. However, since the two error curves oscillate about zero with different frequencies and phases, the two error curves generally intersect at points other than the zeros of $Q_9 f_{\lambda,\mu} - If_{\lambda,\mu}$. As a result, if the parameter λ is chosen randomly (with uniform distribution) in $[-1, 1]$, $|Q_7 f_{\lambda,\mu} - Q_9 f_{\lambda,\mu}|$ is smaller than $|Q_9 f_{\lambda,\mu} - If_{\lambda,\mu}|$ for about 6% of all samples.

8.1.2. Using Null Rules. The construction of error estimation procedures based on pairs of integration formulas with different degrees of accuracy is difficult because such procedures depend to a large degree on the underlying integration formulas. Extensive experiments are needed for *each* pair of integration formulas to ensure that the error estimates produced are both reliable and accurate. Using sequences of *null rules* instead of pairs of integration formulas is one way of overcoming this difficulty.

DEFINITION 8.1.1 (NULL RULE). *A sum*

$$\mathrm{N}_n f := \sum_{i=1}^{n} u_i f(\mathbf{x}_i) \tag{8.6}$$

is said to be a null rule if at least one of the weights $u_1, \ldots, u_n$ *is nonzero and*

$$\sum_{i=1}^{n} u_i = 0. \tag{8.7}$$

According to (8.7), any null rule N_n annihilates all constant polynomials. Generally, null rules are expected to annihilate polynomials of higher degrees.

DEFINITION 8.1.2 (DEGREE OF NULL RULES). *The degree of an n-point null rule* N_n *is D if*

$$\begin{aligned} \mathrm{N}_n \mathbf{x}^{\mathbf{d}} &= 0 \quad \textit{for all } \mathbf{d} \in \mathbb{N}_0^N \textit{ with } \|\mathbf{d}\|_1 \le D, \\ \mathrm{N}_n \mathbf{x}^{\mathbf{d}} &\ne 0 \quad \textit{for at least one } \mathbf{d} \in \mathbb{N}_0^N \textit{ with } \|\mathbf{d}\|_1 = D+1, \end{aligned}$$

i.e., if N_n *annihilates all polynomials of degree* $d \le D$ *and if there exists at least one polynomial of degree* $d = D+1$ *which is not annihilated by* N_n.

Example (Null Rules). If the integration formulas Q_{n_1} and Q_{n_2} are of degrees of accuracy D_1 and D_2, respectively with $D_1 < D_2$, then $\mathrm{N}_n f := \mathrm{Q}_{n_1} f - \mathrm{Q}_{n_2} f$ defines a null rule of degree D_1. The number n of abscissas in N_n depends on the number of abscissas Q_{n_1} and Q_{n_2} have in common.

In conjunction with null rules, the following definitions are important.[87]

DEFINITION 8.1.3 (INNER PRODUCT, EUCLIDEAN NORM, AND ORTHOGONALITY OF NULL RULES). *The inner product of two n-point null rules*

$$\mathrm{N}_n^{\mathbf{u}} f = \sum_{i=1}^{n} u_i f(\mathbf{x}_i) \quad \textit{and} \quad \mathrm{N}_n^{\mathbf{v}} f = \sum_{i=1}^{n} v_i f(\mathbf{x}_i),$$

using the same set of abscissas, is defined by

$$\langle \mathrm{N}_n^{\mathbf{u}}, \mathrm{N}_n^{\mathbf{v}} \rangle := \mathbf{u}^\top \cdot \mathbf{v} = u_1 v_1 + u_2 v_2 + \cdots + u_n v_n.$$

Correspondingly, the Euclidean norm of a null rule is given by

$$\|\mathrm{N}_n^{\mathbf{u}}\|_2 := \sqrt{\langle \mathrm{N}_n^{\mathbf{u}}, \mathrm{N}_n^{\mathbf{u}} \rangle} = \|\mathbf{u}\|_2,$$

[87]Since the following definitions are given in terms of the weights of (null) rules only, they are also applicable to integration formulas.

and the orthogonality of two null rules $\mathrm{N}_n^{\mathbf{u}}$ *and* $\mathrm{N}_n^{\mathbf{v}}$ *is defined by*

$$\mathrm{N}_{\mathbf{u}} \perp \mathrm{N}_{\mathbf{v}} \quad :\Leftrightarrow \quad \langle \mathrm{N}_n^{\mathbf{u}}, \mathrm{N}_n^{\mathbf{v}} \rangle = 0 \quad \Leftrightarrow \quad \mathbf{u}^\top \cdot \mathbf{v} = 0 \quad \Leftrightarrow \quad \mathbf{u} \perp \mathbf{v}.$$

For the sake of simplicity, the remainder of this section deals only with univariate null rules. However, the basic ideas and concepts introduced can easily be generalized to higher dimensions (see Espelid [134]).

Spaces of Null Rules. In order for a univariate n-point null rule N_n, using abscissas $x_1, \ldots, x_n$, to be of degree D, the weights $u_1, \ldots, u_n$ must be chosen such that N_n annihilates the elements of a basis $\{b_0, b_1, \ldots, b_D\}$ for $\mathbb{P}_D$, for example, the monomials $\{1, x, x^2, \ldots, x^D\}$. Thus, the weights can be obtained by solving the following homogeneous system of linear equations:

$$(8.8) \qquad \begin{array}{rcl} u_1 \quad + u_2 \quad + \cdots + u_n & = & 0, \\ u_1 x_1 + u_2 x_2 + \cdots + u_n x_n & = & 0, \\ & \vdots & \\ u_1 x_1^D + u_1 x_1^D + \cdots + u_n x_n^D & = & 0. \end{array}$$

For any $D \in \{0, \ldots, n-1\}$, the coefficient matrix of (8.8) has rank $D+1$, and the solutions of (8.8) form an $(n-D-1)$-dimensional vector space, denoted S_D. Consequently, the relationship

$$\{\mathbf{0}\} = S_{n-1} \subset S_{n-2} \subset \cdots \subset S_1 \subset S_0$$

holds between the spaces $S_0, \ldots, S_{n-1}$. Since S_{n-1} contains only the trivial solution $u_1 = u_2 = \cdots = u_n = 0$, which does *not* define a null rule (see Definition 8.1.1), there is no null rule which annihilates *all* polynomials of $\mathbb{P}_{n-1}$. The maximum degree of any n-point null rule N_n is therefore $D = n-2$.

Asymptotic Behavior of Null Rules. Theorem 5.2.6 establishes a relationship between the degree of accuracy of a quadrature rule and the asymptotic rate of convergence

$$\mathrm{Q}_n(f; [a,b]) \to \mathrm{I}(f; [a,b]) \qquad \text{as} \quad h := b - a \to 0+.$$

The following theorem shows that an analogous relationship also holds between the degree of a null rule and the asymptotic rate of convergence

$$\mathrm{N}_n(f; [a,b]) \to 0 \qquad \text{as} \quad h \to 0+.$$

THEOREM 8.1.1. *Let* N_n *be a null rule of degree* $D \geq 0$, *and assume the integrand* f *is* $(D+1)$*-times continuously differentiable on* $[a,b]$, $f \in C^{(D+1)}[a,b]$. *Then, there is a constant* C *such that the error bound*

$$(8.9) \qquad |\mathrm{N}_n(f; [\alpha, \beta])| \leq C\,(\beta - \alpha)^{D+2}$$

holds for any subinterval $[\alpha, \beta] \subseteq [a,b]$.

Hence, provided that f is sufficiently smooth and $h = b - a$ is sufficiently small, the magnitude of an n-point null rule $\mathrm{N}_n^{(1)}$ can be expected to be significantly smaller than the magnitude of an n-point null rule $\mathrm{N}_n^{(2)}$ if the degree D_1 of $\mathrm{N}_n^{(1)}$ is larger than the degree D_2 of $\mathrm{N}_n^{(2)}$. Furthermore, it follows from Theorem 5.2.6 that, for any interpolatory quadrature formula Q_n,

$$\mathrm{Q}_n(f;[a,b]) - \mathrm{I}(f;[a,b]) = O(h^{n+1}).$$

As a result, it can be expected that the magnitude of any n-point null rule is asymptotically greater than the magnitude of the error in any n-point interpolatory quadrature formula. This is why null rules can be used to estimate the error in interpolatory quadrature formulas if "asymptotic behavior" is observed.

Test for Asymptotic Behavior. For any set of n distinct points $x_1, \ldots, x_n$, there exists a unique corresponding interpolatory formula Q_n. Tests for asymptotic behavior of Q_n can be implemented using sequences

$$\mathrm{N}_n^{(1)}, \mathrm{N}_n^{(2)}, \ldots, \mathrm{N}_n^{(n-1)}$$

of null rules (with the same abscissas as Q_n) in which each rule

$$\mathrm{N}_n^{(j)}, \quad j = 1, 2, \ldots, n-1,$$

is of degree $D = j - 1$. According to Theorem 8.1.1,

$$(8.10) \quad E_j := |\mathrm{N}_n^{(j)} f| = O(h^{j+1}) \quad \text{as} \quad h := b - a \to 0, \quad j = 1, 2, \ldots, n-1,$$

if the integrand f is sufficiently smooth. This suggests

$$r_j := E_{j+1}/E_j \to 0 \quad \text{as} \quad h \to 0, \quad j = 1, 2, \ldots, n-2.$$

As a result, one way to test for asymptotic error behavior is to check whether all quantities $r_1, r_2, \ldots, r_{n-2}$ are simultaneously small, i.e., whether

$$r := \max\{r_1, r_2, \ldots, r_{n-2}\}$$

is small.

Nonasymptotic Error Estimates. Any null rule N_n with the same abscissas $x_1, \ldots, x_n$ as Q_n is said to be *equally strong* (as Q_n) if

$$\|\mathrm{Q}_n\|_2 = \|\mathrm{N}_n\|_2.$$

Using equally strong null rules for estimating the error in quadrature formulas is crucial to obtaining reliable error bounds in the case of *nonasymptotic* behavior. In this case the deviation of the samples $f(x_1), \ldots, f(x_n)$ from the mean value $\bar{f}$ of f on $[a, b]$,

$$\bar{f} := \mathrm{I}f/(b-a),$$

is often erratic and apparently random. The deviations $f(x_i) - \bar{f}$, $i = 1, \ldots, n$, can therefore be modeled by n independent, identically distributed random variables with mean value 0 and standard deviation σ. Because

$$\mathrm{I}\bar{f} = (b-a)\bar{f} = \mathrm{Q}_n\bar{f},$$

the equality

$$\mathrm{Q}_n f - \mathrm{I}f = \mathrm{Q}_n(f - \bar{f}) = \sum_{i=1}^{n} c_i(f(x_i) - \bar{f})$$

holds. Consequently, the error $\mathrm{Q}_n f - \mathrm{I}f$ can be regarded as a realization of a random variable with mean value 0 and standard deviation $\sigma\,\|\mathrm{Q}_n\|_2$. Similarly,

$$\mathrm{N}_n f = \mathrm{N}_n f - \mathrm{N}_n \bar{f} = \mathrm{N}_n(f - \bar{f}) = \sum_{i=1}^{n} u_i(f(x_i) - \bar{f})$$

shows that $\mathrm{N}_n f$ can be regarded as a realization of a random variable with mean value 0 and standard deviation $\sigma\,\|\mathrm{N}_n\|_2$.

If a null rule N_n is equally strong, $\mathrm{N}_n f$ and the error $\mathrm{Q}_n f - \mathrm{I}f$ are realizations of random variables with the same mean and standard deviation. If several values $\mathrm{N}_n^{(1)} f, \ldots, \mathrm{N}_n^{(m)} f$ of equally strong null rules are available, there is a high probability that the absolute value of at least one null rule is larger than the absolute value of the error $\mathrm{Q}_n f - \mathrm{I}f$. Therefore, a reliable error bound can be obtained, even when asymptotic behavior has not been achieved, by taking the maximum of the values $|\mathrm{N}_n^{(1)} f|, \ldots, |\mathrm{N}_n^{(m)} f|$.

Phase Factor Effects. Even when asymptotic behavior is observed, the occurrence of phase factor effects in null rules may accidentally result in small values of the quantities E_j in (8.10), which can lead to unreliable error estimates. In order to decrease the probability of phase factor effects, Berntsen [24] proposes the construction of the new sequence of error estimates

$$\hat{E}_j := \max\{|\mathrm{N}_n f| \;:\; \mathrm{N}_n \in S_{n-2j-1} \cap S_{n-2j+1}^{\perp},\; \|\mathrm{N}_n\|_2 = \|\mathrm{Q}_n\|_2\},$$

$$j = 1, 2, \ldots, \lfloor (n-1)/2 \rfloor.$$

Each $\hat{E}_j$ yields the *maximum* absolute value of any equally strong null rule in the subspace $S_{n-2j-1} \cap S_{n-2j+1}^{\perp}$. This, in addition to the fact that the subspaces $S_{n-2j-1} \cap S_{n-2j+1}^{\perp}$, $j = 1, \ldots, \lfloor (n-1)/2 \rfloor$, are mutually *orthogonal*, reduces the probability of accidentally small error estimates $\hat{E}_j$ (Berntsen [24]).

The error estimates $\hat{E}_j$ can easily be computed as follows. Let

$$\mathrm{N}_n^{(1)}, \mathrm{N}_n^{(2)}, \ldots, \mathrm{N}_n^{(n-1)}$$

be a sequence of equally strong null rules such that the null rules

$$\mathrm{N}_n^{(D)}, \mathrm{N}_n^{(D+1)}, \ldots, \mathrm{N}_n^{(n-1)}$$

form an orthogonal basis for S_{D-1} for any $D \in \{1, 2, \ldots, n-1\}$. Then, it can be proved (Espelid [134]) that

$$\hat{E}_j = \sqrt{E_{n-2j+1}^2 + E_{n-2j}^2}, \quad j = 1, 2, \ldots, \lfloor (n-1)/2 \rfloor,$$

where the quantities E_j are defined by (8.10). Suitable sequences of equally strong null rules can be obtained by applying the Gram–Schmidt orthogonalization procedure to sequences of equally strong null rules $\mathrm{N}_n^{(1)}, \mathrm{N}_n^{(2)}, \ldots, \mathrm{N}_n^{(n-1)}$ which possess the following property. For any $D = 1, 2, \ldots, n-1$,

$$\mathrm{N}_n^{(D)}, \mathrm{N}_n^{(D+1)}, \ldots, \mathrm{N}_n^{(n-1)}$$

form a (nonorthogonal) basis for S_{D-1}.

Because

$$\hat{E}_j = O(h^{n-2j+1}) \qquad \text{as} \quad h \to 0, \quad j = 1, 2, \ldots, \lfloor (n-1)/2 \rfloor,$$

$$R_j := \hat{E}_j / \hat{E}_{j+1} \to 0 \qquad \text{as} \quad h \to 0, \quad j = 1, 2, \ldots, \lfloor (n-1)/2 \rfloor - 1,$$

holds. Hence, a robust test for asymptotic behavior can be performed by checking whether all quotients $R_1, R_2, \ldots, R_{\lfloor (n-1)/2 \rfloor - 1}$ are simultaneously small.

A somewhat different approach to constructing appropriate sequences of null rules is described in Laurie [254].

8.1.3. Using Compound Formulas. Section 8.1.1 discusses error estimation procedures based on comparing the results of two or more, usually *simple*, integration formulas with *different* degrees of accuracy. This section deals with error estimation techniques based on comparing the results of *compound* formulas with the *same* degree of accuracy which correspond to *different* subdivisions of the integration region B.

Let Q_{n_1} and Q_{n_2} denote compound integration formulas, $\mathrm{Q}_{n_1} = k_1^N \times \mathrm{Q}_n$ and $\mathrm{Q}_{n_2} = k_2^N \times \mathrm{Q}_n$, derived from a simple polynomial integration formula Q_n with degree of accuracy D. If Q_{n_2} corresponds to a finer subdivision of the integration region B than Q_{n_1}, i.e., if $k_2 > k_1$, then $\mathrm{Q}_{n_2} f$ can be expected to be a more accurate integral approximation than $\mathrm{Q}_{n_1} f$, provided

f is sufficiently smooth. More specifically, asymptotic error expansions for compound formulas, as given in Theorem 5.3.4, suggest

$$(8.11)\quad \mathrm{Q}_{n_2}(f;B) - \mathrm{I}(f;B) \sim (k_1/k_2)^{-p}(\mathrm{Q}_{n_1}(f;B) - \mathrm{I}(f;B)) \text{ as } \operatorname{diam} B \to 0$$

for $p = D + 1$, provided f is sufficiently smooth. If asymptotic error behavior is observed, then (8.11) can be used to conclude that

$$\mathrm{Q}_{n_2}f - \mathrm{Q}_{n_1}f \approx (1 - (k_2/k_1)^p)(\mathrm{Q}_{n_2}f - \mathrm{I}f),$$

from which the highly accurate error estimate

$$\mathrm{Q}_{n_2}f - \mathrm{I}f \approx \frac{\mathrm{Q}_{n_2}f - \mathrm{Q}_{n_1}f}{1 - (k_2/k_1)^p}$$

follows. However, this error estimate can be completely incorrect if asymptotic error behavior has not been or cannot be achieved, for instance, when the integrand f has discontinuities in the integration region B.

In order to test for asymptotic error behavior, a third compound formula $\mathrm{Q}_{n_3} = k_3^N \times \mathrm{Q}_n$, $k_3 > k_2$, is required. From (8.11)

$$\begin{aligned} \mathrm{Q}_{n_1}f - \mathrm{I}f &\approx (k_2/k_1)^p(\mathrm{Q}_{n_2}f - \mathrm{I}f), \\ \mathrm{Q}_{n_2}f - \mathrm{I}f &\approx (k_3/k_2)^p(\mathrm{Q}_{n_3}f - \mathrm{I}f) \end{aligned}$$

can be derived. These relationships can be used to conclude that

$$(8.12)\qquad \frac{\mathrm{Q}_{n_2}f - \mathrm{Q}_{n_1}f}{\mathrm{Q}_{n_3}f - \mathrm{Q}_{n_2}f} \approx -\frac{(k_2/k_1)^p - 1}{(k_2/k_3)^p - 1}.$$

The right-hand side expression in (8.12) is simplified for $k_2/k_1 = k_3/k_2 =: k$:

$$(8.13)\qquad \frac{\mathrm{Q}_{n_2}f - \mathrm{Q}_{n_1}f}{\mathrm{Q}_{n_3}f - \mathrm{Q}_{n_2}f} \approx k^p.$$

Integration algorithms can easily examine the validity of (8.12) and, in particular, (8.13).

Example (Cautious Romberg Integration). The Romberg formulas T_m^j have the degrees of accuracy $D = 2j + 1$ (see Section 5.3.6). Hence, when Q_{n_1}, Q_{n_2}, and Q_{n_3} are chosen as consecutive elements of the jth column in Romberg's T-table,

$$\mathrm{Q}_{n_1} = T_m^j, \mathrm{Q}_{n_2} = T_{m+1}^j \quad \text{and} \quad \mathrm{Q}_{n_3} = T_{m+2}^j,$$

the expected order of convergence p and the factor k are given by $p = 2j+2$ and $k = 2$. The reliability of any Romberg integration algorithm can be distinctly improved by continuing calculations only if the left-hand side quotients in (8.13) do not significantly deviate from the expected value $k^p = 2^{2j+2}$. This scheme is referred to as *cautious* Romberg integration.

Software (Cautious Romberg Integration). Cautious Romberg integration is performed by the subroutine `cadre`, a quadrature program developed by de Boor [93]. Cautious Romberg integration is also implemented in `HARWELL/qa05`.

Integrands of Reduced Smoothness. If f (or one of its derivatives) is singular, asymptotic error expansions of the kind (8.11) may still hold (see Section 7.3.4). However, the order of convergence p depends on (usually unknown) analytical properties of f and is generally lower than $D + 1$. If p is unknown, the aforementioned methods of testing for asymptotic error behavior and estimating the discretization error have to be modified. For instance, rather than to test for asymptotic error behavior, the relationships (8.12) or (8.13) can be used to estimate the unknown convergence order based on a sequence of three compound formulas Q_{n_1}, Q_{n_2}, and Q_{n_3}. An additional fourth formula Q_{n_4} is then required to test for asymptotic error behavior.

An alternative way of coping with an unknown order of convergence is to use, in addition to Q_{n_1} and Q_{n_2}, a pair of compound formulas $\tilde{\mathrm{Q}}_{m_1} = k_1 \times \tilde{\mathrm{Q}}_m$ and $\tilde{\mathrm{Q}}_{m_2} = k_2 \times \tilde{\mathrm{Q}}_m$ based on a basic rule $\tilde{\mathrm{Q}}_m$ with the same degree of accuracy D as Q_n. Since Q_n and $\tilde{\mathrm{Q}}_m$ are supposed to have the same degree of accuracy, compound formulas based on Q_n and $\tilde{\mathrm{Q}}_m$, respectively, generally exhibit the same asymptotic order of convergence. Combining the asymptotic relationships between Q_{n_1} and Q_{n_2} as well as $\tilde{\mathrm{Q}}_{m_1}$ and $\tilde{\mathrm{Q}}_{m_2}$ yields the estimate

$$\frac{\mathrm{Q}_{n_2} f - \tilde{\mathrm{Q}}_{m_2} f}{\mathrm{Q}_{n_1} f - \tilde{\mathrm{Q}}_{m_1} f} \approx (k_1/k_2)^p,$$

from which the error estimate

$$\mathrm{Q}_{n_2} f - \mathrm{I} f \approx \frac{(\mathrm{Q}_{n_2} f - \tilde{\mathrm{Q}}_{m_2} f)(\mathrm{Q}_{n_2} f - \mathrm{Q}_{n_1} f)}{\mathrm{Q}_{n_2} f - \mathrm{Q}_{n_1} f - \tilde{\mathrm{Q}}_{m_2} f + \tilde{\mathrm{Q}}_{m_1} f} \tag{8.14}$$

can be derived. The main disadvantage of this error estimation procedure is that the formula $\tilde{\mathrm{Q}}_m$ generally uses one or more abscissas not contained in the abscissa set of Q_n. Computing $\tilde{\mathrm{Q}}_{m_1}$ and $\tilde{\mathrm{Q}}_{m_2}$ therefore requires additional evaluations of f.

However, in Laurie [253] it is shown that the right-hand side expression in (8.14) can be expected to be a *bound* on (rather than an asymptotic estimate of) the discretization error $\mathrm{Q}_{n_2} f - \mathrm{I} f$ if the degree of accuracy of $\tilde{\mathrm{Q}}_m$ is *lower* than the degree of accuracy of Q_n. In *adaptive* integration algorithms (see Section 8.2.2), two or more integration formulas of different degrees of accuracy are usually applied to an integration (sub)region in order to obtain a primary error estimate, and compound formulas are generated automatically as a result of the adaptive subdivision process. Therefore, evaluating the right-hand side of (8.14) does not incur any additional computational cost, which is why (8.14) can be used to improve the reliability and accuracy of error estimates in adaptive integration algorithms.

For a thorough discussion and evaluation of Laurie's method, see Espelid and Soerevik [141]. An extension of Laurie's method to compound formulas derived from *four* different basic rules can be found in Favati, Lotti, and Romani [146].

8.1.4. Using Randomized Formulas. The use of *randomized* integration formulas for error estimation purposes is based on the Monte Carlo principle (see Section 6.4.1): the deterministic result of an integration formula is interpreted as a stochastic entity whose expected value is the desired integral, and a sequence of stochastically independent integral approximations is generated. According to the *strong law of large numbers*, the mean value of this sequence converges to the exact integral value with probability 1, and probabilistic error estimates for the approximate integral value can be obtained using statistical methods. The following discussion is limited to the use of randomized rules in conjunction with *lattice rules* (see Section 6.5). However, the ideas and methods used apply to *pseudorandom* formulas as well (see Section 6.4).

Let the displacement $\mathbf{v}$ of a lattice rule $\mathrm{Q}_n(\mathbf{v})$ be chosen randomly in C_N; i.e., suppose $\mathbf{v}$ can be regarded as a realization of a random variable $\mathbf{V}$ with uniform, multivariate distribution on C_N. In this interpretation, the lattice rule $\mathrm{Q}_n(\mathbf{v})$ and its discretization error $\mathrm{Q}_n(\mathbf{v}) - \mathrm{I}f$ become realizations of the random variables $\mathrm{Q}_n(\mathbf{V})$ and $\mathrm{Q}_n(\mathbf{V}) - \mathrm{I}f$, respectively; and, according to (6.75),

$$\mathrm{Q}_n(\mathbf{V})f - \mathrm{I}f = \sum_{\mathbf{m}\in L^\perp\setminus\{\mathbf{0}\}} a(\mathbf{m})e^{2\pi i\mathbf{m}\cdot\mathbf{V}}$$

holds. The equality

$$\mu(e^{2\pi i\mathbf{m}\cdot\mathbf{V}}) = \int_{C_N} e^{2\pi i\mathbf{m}\cdot\mathbf{v}}d\mathbf{v} = \prod_{j=1}^{N}\int_0^1 e^{2\pi i m_j v_j}dv_j = 0 \qquad \text{for} \quad \mathbf{m} \neq \mathbf{0},$$

along with the linearity of the expectancy operator μ, shows that

$$\mu(\mathrm{Q}_n(\mathbf{V})f - \mathrm{I}f) = 0.$$

Now suppose $\mathbf{V}_1, \ldots, \mathbf{V}_q$ is a sequence of independent random variables uniformly distributed on C_N. The mean

$$M_q(\mathrm{Q}_n f) = \frac{1}{q}\sum_{j=1}^{q} \mathrm{Q}_n(\mathbf{V}_j)f$$

of the sequence $\mathrm{Q}_n(\mathbf{V}_1)f, \ldots, \mathrm{Q}_n(\mathbf{V}_q)f$ is an *unbiased* estimate of $\mathrm{I}f$, i.e.,

$$\mu(M_q(\mathrm{Q}_n f)) = \mathrm{I}f,$$

and the standard deviation $\sigma(M_q(\mathrm{Q}_n f))$ is a measure for the expected size of the error $|M_q(\mathrm{Q}_n f) - \mathrm{I}f|$. An unbiased estimate of the variance $\sigma^2(M_q(\mathrm{Q}_n f))$, i.e., a random variable with expected value $\sigma^2(M_q(\mathrm{Q}_n f))$, is given by

$$V_q(\mathrm{Q}_n f) := \frac{1}{q(q-1)}\sum_{j=1}^{q} (\mathrm{Q}_n(\mathbf{V}_j)f - M_q(\mathrm{Q}_n f))^2$$

(see, for instance, Papoulis [329]). Using the *Chebyshev inequality* (Papoulis [329]), the likelihood that the random variable $M_q(Q_n f)$ differs from its mean value If by more than a given bound ε can be estimated as follows:

$$P\left(|M_q(Q_n f) - If| < \varepsilon\right) \geq 1 - \frac{\sigma^2(M_q(Q_n f))}{\varepsilon^2}. \tag{8.15}$$

Therefore, if the displacements $\mathbf{v}_1, \ldots, \mathbf{v}_q$ used to shift a lattice rule Q_n are independent random samples, then

$$Q_{nq} f := Q_n(\mathbf{v}_1) f + \cdots + Q_n(\mathbf{v}_q) f$$

defines an integration formula for If, and

$$e_q := \sqrt{\frac{1}{q(q-1)} \sum_{j=1}^{q} \left(Q_n(\mathbf{v}_j) f - Q_{nq} f\right)^2}$$

can be used as a probabilistic error estimate for $Q_{nq} f$. Estimates of the likelihood that the actual integration error is *larger* than a given bound ε can be obtained by replacing the exact variance $\sigma^2(M_q(Q_n f))$ in (8.15) with the estimate e_q^2:

$$P\left(|Q_{nq} f - If| < \varepsilon\right) \geq 1 - \frac{e_q^2}{\varepsilon^2} =: s.$$

The quantity s is referred to as the *confidence level* of the error bound ε.

The algorithmic parameter s in error estimation procedures based on randomized formulas is not usually made available to the user; it is automatically chosen by the integration program. There is a trade-off in choosing an appropriate value of s: for small values of s, the actual integration error can easily be underestimated, in which case the reliability of the integration program is decreased. For large values of s, the integration error is most often overestimated, and the efficiency of the integration program is decreased. Extensive numerical experiments are required to obtain satisfactory values of s.

Example (Randomization of Lattice Rules). The method of randomization was used to obtain error estimates for the lattice rules described in the example on page 222. Prior to numerical integration, all integrands were transformed using the *tanh*-transformation (6.77). The following table shows the confidence levels required to prevent an underestimation of the integration error for the integration problem (6.78):

j	n	$q=2$	$q=3$	$q=4$	$q=5$	$q=6$
0	619	0.89	0.96	0.00	0.00	0.00
1	1238	0.00	0.00	0.00	0.00	0.00
2	2476	0.53	0.00	0.00	0.00	0.00
3	4952	0.94	0.00	0.00	0.00	0.00
4	9904	0.99	0.96	0.70	0.16	0.00
5	19 808	0.92	0.00	0.15	0.66	0.18
6	39 616	0.86	0.00	0.00	0.00	0.07

A confidence level $s = 0.00$ indicates that the estimate e_q is larger than the actual integration error. In this case, any confidence level $s \geq 0$ is sufficient to prevent an underestimation of the integration error. In most cases, the values of s decrease as q is increased. The accuracy of the lattice rule used has no apparent influence on the size of s. The maximum value of s in all test cases is $s = 0.99$. This maximum value occurs when the actual integration error is 10 times as large as the probabilistic error estimate e_q.

The following table shows the corresponding results when the integral (6.76) was computed:

j	n	$q=2$	$q=3$	$q=4$	$q=5$	$q=6$
0	619	0.97	0.92	0.60	0.81	0.88
1	1238	0.74	0.00	0.00	0.59	0.44
2	2476	0.00	0.00	0.00	0.00	0.00
3	4952	0.94	0.00	0.00	0.00	0.00
4	9904	0.93	0.97	0.87	0.85	0.26
5	19 808	0.67	0.61	0.29	0.48	0.71
6	39 616	0.46	0.79	0.00	0.00	0.07

The values of s are generally larger than in the first table. Nevertheless, a confidence level of 0.99 is still sufficient to prevent an underestimation of the integration error. The higher values of s might be due to the fact that the integral approximations are less accurate than in the first problem.

Software (Randomized Error Estimation). Randomized formulas are used for estimation purposes in the subroutines `NAG/d01gbf`, `NAG/d01gcf`, and `NAG/d01gdf`.

8.1.5. Error Estimation Using Interval Arithmetic. Numerical computations on digital computers have traditionally been based on representing any real number occurring in a mathematical algorithm (input parameters, intermediate results, and output arguments alike) using a *single* floating-point number. This approach gives rise to the following problems.

(i) Since the set of floating-point numbers $\mathbb{F}$ is a proper, finite subset of $\mathbb{R}$, real numbers cannot generally be represented accurately using floating-point numbers. Instead, a *rounding function* $\square : \mathbb{R} \mapsto \mathbb{F}$ is used to map real numbers x onto their machine representations $\square x$.[88] The difference $\square x - x$ between the rounded value $\square x$ and the exact value x is known as the *rounding error*.

Rounding errors usually affect the accuracy of input data in integration algorithms. The abscissas and weights of Gauss formulas, for instance, are generally *irrational* and cannot be represented exactly in any floating-point number system. Furthermore, rounding errors may also

[88] Here and in the remainder of this section, it is tacitly assumed that any real value x in an algorithm satisfies $|x| \leq x_{\max}$, $x_{\max} := \max \mathbb{F}$, i.e., that x is *not* in the overflow range of $\mathbb{F}$. However, the following discussion applies even to values x with $|x| > x_{\max}$ if $\mathbb{F}$ conforms to the IEC/IEEE floating-point standard. Such x-values can be represented by the signed infinite values $\pm\infty$ defined by the IEC/IEEE floating-point standard.

occur whenever arithmetic operations are performed. Even when the operands of an arithmetic operation are floating-point numbers, the mathematically correct result may not be: the result of an arithmetic operation in $\mathbb{F}$ is obtained by *rounding* the exact result.

(ii) If the numerical value of a real quantity x is not accurately known, the idea of representing x by a single number is altogether questionable because a single number cannot convey any information concerning the level of uncertainty and/or the possible variation of x that might be available.

This kind of uncertainty can, for instance, affect the integrand values used in an integration algorithm: if the integrand values result from measurements, then measurement errors usually give rise to a substantial amount of stochastic noise in the data.

Uncertainty also arises in integration algorithms in a totally different context: whenever variants of the *mean value theorem* are used to derive formulas characterizing the discretization error in integration formulas, the resulting remainder terms are expressed in terms of *unknown* points $\xi \in B$ (see, for instance, Section 5.2.4). Using traditional arithmetic, which operates on single *points*, the fact that an unknown point ξ is located in the integration region B cannot be exploited in any way.

Interval Functions. Interval analysis has been developed as an attempt to overcome the aforementioned (and other) problems. Interval analysis is based on extending the range of applicability of arithmetic operations and, more generally, N-variate functions to (N-dimensional vectors of) real intervals $I \in \mathbb{IR} := \{[\underline{x}, \overline{x}] : \underline{x}, \overline{x} \in \mathbb{R}, \underline{x} \leq \overline{x}\}$. Let op denote any arithmetic operation; i.e., op $\in \{+, -, *, /\}$. Then, the corresponding arithmetic operation op_{I} on intervals is defined by

$$I \,\mathsf{op}_{\mathrm{I}}\, J := \{x \,\mathrm{op}\, y : x \in I, y \in J\}, \qquad I, J \in \mathbb{IR},$$

provided that $x \,\mathrm{op}\, y$ is well defined for all $x \in I$ and $y \in J$.[89] Note that op_{I} is *interval-valued*, i.e., $I \,\mathsf{op}_{\mathrm{I}}\, J \in \mathbb{IR}$, because the mapping $(x, y) \mapsto x \,\mathrm{op}\, y$ is continuous in both variables, and thus maps any compact subset of $\mathbb{R}^2$ (in particular, any Cartesian product $I \times J$ of compact intervals) onto a compact interval. Similarly, for any continuous function $f : B \subseteq \mathbb{R}^N \to \mathbb{R}$, the corresponding function f_{I} on intervals,

$$\mathsf{f}_{\mathrm{I}}(I_1, \ldots, I_N) := \{f(x_1, \ldots, x_N) : x_1 \in I_1, \ldots, x_N \in I_N\}, \tag{8.16}$$

[89]The division operation is therefore restricted to intervals J with $0 \notin J$.

is interval-valued. Interval-valued functions of one or more interval arguments, such as op_{I} or f_{I}, are generally referred to as *interval functions*.

The interval functions $\mathsf{f} = \mathsf{f}_{\mathrm{I}}$ (including the interval arithmetic operations op_{I}) defined above have the following two fundamental properties: they are *inclusion monotonic*,

$$(8.17) \quad I_j \subseteq J_j \text{ for } j = 1, 2, \ldots, N \quad \Rightarrow \quad \mathsf{f}(I_1, \ldots, I_N) \subseteq \mathsf{f}(J_1, \ldots, J_N),$$

and they satisfy

$$(8.18) \qquad \mathsf{f}([x, x]) = f(x).$$

Any interval function f satisfying both (8.17) and (8.18) is said to be an *interval extension* of the function $f : B \to \mathbb{R}$. Any interval extension f of f has the following important *inclusion property*:

$$(8.19) \qquad \begin{aligned} &f(x_1, \ldots, x_N) \in \mathsf{f}(I_1, \ldots, I_N) \\ &\text{for all} \quad (x_1, \ldots, x_N)^\top \in I_1 \times \cdots \times I_N \subseteq B. \end{aligned}$$

The basic implication of (8.19) is that whenever the input data $x_1, \ldots, x_N$ of the algorithm represented by f are known to be contained in the intervals $I_1, \ldots, I_N$, the exact result $f(x_1, \ldots, x_N)$ can be *guaranteed* to be enclosed by the interval $\mathsf{f}(I_1, \ldots, I_N)$.

Operational Definitions of Interval Functions. Definition (8.16) of interval extensions f_{I} is of little practical value since computing $\mathsf{f}_{\mathrm{I}}(I_1, \ldots, I_N)$ according to (8.16) would generally require evaluating $f(x_1, \ldots, x_N)$ at an infinite number of arguments

$$(x_1, \ldots, x_N)^\top \in I_1 \times \cdots \times I_N.$$

Operational definitions of f_{I}, which require the evaluation of $f(x_1, \ldots, x_N)$ only at a finite set of arguments, can easily be obtained whenever f is monotonic in each independent variable. In this case it is sufficient to evaluate f at the 2^N corner points of the N-dimensional cube

$$I_1 \times \cdots \times I_N = [\underline{x}_1, \overline{x}_1] \times \cdots \times [\underline{x}_N, \overline{x}_N];$$

more precisely,

$$f\left([\underline{x}_1, \overline{x}_1], \ldots, [\underline{x}_N, \overline{x}_N]\right) = [\min S, \max S]$$

with

$$S := f\left(\{\underline{x}_1, \overline{x}_1\} \times \cdots \times \{\underline{x}_N, \overline{x}_N\}\right).$$

Similar operational definitions can be given for most commonly used, univariate functions, such as $|\cdot|$, $\sqrt{\cdot}$, or $\sin(\cdot)$, even when they are not monotonic.

For more complex functions, operational definitions of f_{I} are not generally available. However, even in these cases it is often possible to obtain operational definitions of other interval extensions. In particular, if f can be expressed in terms of a finite number of simpler operations $g, h, \ldots$, such that operational definitions of $g_{\mathrm{I}}, h_{\mathrm{I}}, \ldots$ are available, then replacing the operations $g, h, \ldots$ with $g_{\mathrm{I}}, h_{\mathrm{I}}, \ldots$, respectively, results in an operational definition of an interval extension of f. Any such interval extension is said to be an *interval evaluation* of f. Note that different, mathematically equivalent ways of expressing f generally give rise to different interval extensions.

Example (Interval Evaluation). The polynomial $f(x) = x(x-1)$ can be expressed in the equivalent form $f(x) = x^2 - x$. The interval evaluation of the first form for $I = [0,1]$ yields

$$[0,1]([0,1]-[1,1]) = [0,1][-1,0] = [-1,0],$$

whereas for the second form,

$$[0,1]^2 - [0,1] = [0,1] - [0,1] = [-1,1]$$

is obtained.

Machine Interval Functions. When implementing interval extensions of real functions on digital computers using appropriate operational definitions (i.e., definitions which reduce the problem of evaluating $\mathsf{f}(I_1, \ldots, I_N)$ to the problem of evaluating $f(S)$, where S is a finite set of abscissas), one is faced with the same kind of problems as when implementing f itself. To begin with, $f(S)$ is not generally contained in $\mathbb{F}$: even when f is a simple arithmetic operation and $S \subseteq \mathbb{F}$, it need not be the case that $f(S) \subseteq \mathbb{F}$. And then, when f is composed of transcendental functions, it is not usually possible to evaluate the elements of $f(S)$ to machine precision using a reasonable computational effort.

These considerations show why interval extensions cannot be implemented accurately on digital computers. However, f can often be implemented in such a way that the implemented function $\Box\mathsf{f}$ retains the fundamental inclusion property (8.19). Most importantly, such implementations of arithmetic operations can be obtained using the *directed rounding* operations[90] $\triangle, \nabla : \mathbb{R} \to \mathbb{F}$, which map real numbers x onto the closest floating-point numbers $\triangle x$ and ∇x satisfying

$$\nabla x \le x \quad \text{and} \quad \triangle x \ge x.$$

For instance, the inclusion

$$[\underline{x}, \overline{x}] + [\underline{y}, \overline{y}] = [\underline{x} + \underline{y}, \overline{x} + \overline{y}] \subseteq [\nabla(\underline{x} + \underline{y}), \triangle(\overline{x} + \overline{y})]$$

[90]These directed rounding operations are defined in the IEC/IEEE floating-point standard, but they are currently not accessible from the standardized part of any portable programming language.

shows that property (8.19) is satisfied for machine interval addition

$$\square([\underline{x},\overline{x}]+[\underline{y},\overline{y}]) := [\nabla(\underline{x}+\underline{y}),\Delta(\overline{x}+\overline{y})].$$

All interval arithmetic operations can be implemented using similar definitions in such a way that the lower and upper endpoints of any calculated interval are equal to the downward and upward rounded endpoints of the exact interval.

If a machine arithmetic does not comply with the IEC/IEEE floating-point standard, *simulated* directed rounding operations implemented by software can be used to obtain suitable machine interval arithmetic operations (see Kearfott et al. [227]). The resulting error in the interval endpoints is generally slightly larger than when using the directed rounding operations supplied by hardware which conforms with the IEC/IEEE standard.

For algebraic and transcendental standard functions f, machine interval functions $\square\mathsf{f}$ retaining the inclusion property (8.19) can be obtained using sophisticated approximation methods (see Kearfott et al. [227]). However, the calculated interval endpoints can be significantly less accurate than those obtained with arithmetic operations.

Application of Interval Analysis to Numerical Integration. Using the interval analytic methods described in this section, it is possible to enclose the exact value of an integration formula

$$\mathrm{Q}_n f = \sum_{i=1}^{n} c_i f(\mathbf{x}_i)$$

in a machine interval provided (i) both the integration abscissas $\mathbf{x}_1,\ldots,\mathbf{x}_n$ and the weights $c_1,\ldots,c_n$ can be enclosed in machine intervals and (ii) a machine interval function $\square\mathsf{f}$ satisfying (8.19) is available. The resulting interval enclosure of $\mathrm{Q}_n f$ provides rigorous bounds on the maximum effect of data and rounding errors on the value of $\mathrm{Q}_n f$.

In order to enclose the exact integration result $\mathrm{I}f$, one then has to bound the remainder term $E(f) = \mathrm{Q}_n f - \mathrm{I}f$. This requires $E(f)$ to be available in a representation that is amenable to analytic interval methods. If, for instance, a mean value representation of the integration error can be given,

$$E(f) = C_k f^{(k)}(\xi), \quad \xi \in [a,b],$$

then

$$E(f) \in C_k \mathsf{f}^{(k)}([a,b]),$$

where $\mathsf{f}^{(k)}$ is an interval evaluation of $f^{(k)}$.

Example (Error Bounds Obtained Using Interval Analysis). The discretization error of the n-point Gauss formula G_n for a function $f \in C^{(2n)}[a,b]$ is given by

$$G_n(f;[a,b]) - \mathrm{I}(f;[a,b]) = C_n f^{(2n)}(\xi)(b-a)^{2n+1},$$

where

$$C_n = \frac{1}{(2n+1)!}\left(\frac{(n!)^2}{(2n)!}\right)^2$$

(see Davis and Rabinowitz [92]). The following table shows the interval enclosures $\mathrm{I}(f;[0,1]) \in [\underline{\mathrm{I}}f, \bar{\mathrm{I}}f]$ obtained by applying the Gauss formula G_7 and the associated remainder term to various integrands. The table also provides the corresponding approximate results $G_7 f$ (computed with double precision accuracy) and the results of exact integration.[91] The computations were based on the portable, standard interval function library INTLIB provided in `TOMS/737`.

	$\exp(x)$	$10^{-4}\exp(10x)$
$G_7 f$	1.7182818284590451	2.2025451640846279
$\mathrm{I}f$	1.7182818284590452	2.2025465794806717
$\underline{\mathrm{I}}f$	1.7182818284590360	2.2025451705769469
$\bar{\mathrm{I}}f$	1.7182818284590538	2.2026881688961071

	$10\sin(10x)$	$(x+1/2)^{-1}$
$G_7 f$	1.8390295221358823	1.0986122727383456
$\mathrm{I}f$	1.8390715290764525	1.0986122886681096
$\underline{\mathrm{I}}f$	1.8389645980517586	1.0986122727512668
$\bar{\mathrm{I}}f$	1.8390944462199907	1.0987977386580814

In all cases the computed intervals $[\underline{\mathrm{I}}f, \bar{\mathrm{I}}f]$ indeed contain the exact results $\mathrm{I}f$. However, the guaranteed error bounds $(\bar{\mathrm{I}}f - \underline{\mathrm{I}}f)/2$ for the "optimal" integral approximations $\mathrm{I}f \approx (\bar{\mathrm{I}}f + \underline{\mathrm{I}}f)/2$ are, except for $f(x) = 10\sin(x)$, significantly larger than the actual integration errors $G_n f - \mathrm{I}f$.

A number of problems arise when the interval analytic approach described in this section is used to bound the error in numerical integration results.

(i) Obtaining appropriate machine interval functions can be difficult if f is not provided in closed form. If, for instance, evaluating f requires solving an algebraic or differential equation, then calculating an interval enclosure $\square\mathrm{f}$ is at least as difficult a problem as calculating an enclosure of $\mathrm{I}f$ for any given $\square\mathrm{f}$. Even if enclosures can be obtained in such cases, they may be too wide to be of any practical value.

(ii) For many kinds of integration formulas, interval extensions of derivatives $f^{(k)}$ are required for bounding the remainder term $E(f)$. Again, this may prove difficult when f is not provided in closed form; additionally, obtainable enclosures of $f^{(k)}$ may not be sufficiently accurate. A possible way around this problem is using *derivative-free* representations of the remainder term $E(f)$, if such representations are available (see, for instance, Eiermann [126]).

(iii) Analytical bounds on the remainder term $E(f)$ may significantly overestimate the actual integration error. The resulting integral enclosures are unnecessarily wide.

[91] All results are rounded to the significant number of digits.

As a result of these and other difficulties, verified numerical integration has been of relatively little relevance to practical computations, although some exceptions exist (see Holzmann, Lang, and Schütt [210] and Storck [422]).

References to Relevant Literature. The foregoing discussion of interval analysis in general and verified integration in particular provides only a very brief introduction into these topics. More comprehensive treatments of analytic interval methods can be found, for instance, in Alefeld and Herzberger [3], Hansen [199], and Neumaier [314]. Fundamental aspects of verified integration are discussed in Caprani, Madsen, and Rall [50]. The verified calculation of Gauss abscissas and weights is dealt with in Storck [420], and a description of verified Romberg integration is given in Storck [421]. *Adaptive* integration algorithms based on the verified calculation of integration formulas are discussed in Corliss [67], Corliss and Krenz [68], and Krämer and Wedner [240].

8.2. Discretization Refinement

Most integration algorithms incorporate specific mechanisms for generating finite or infinite sequences $\{Q_{n_j} f\}$ of approximate integral values of (presumably) increasing accuracy. Such sequences are usually obtained by increasing the number of samples of f, a procedure which is generally referred to as *discretization refinement*.

Sampling Strategies. The efficiency and reliability of numerical integration algorithms depend heavily on the sampling strategy applied. Let

$$S(f) = (f(\mathbf{x}_1), \ldots, f(\mathbf{x}_n))$$

denote the sample information used in a particular integration algorithm. If the location of any sampling point $\mathbf{x}_i$ is chosen independently of any other (previously sampled) value of f, then $S(f)$ conveys *nonadaptive information* about f. Modules implementing integration formulas

$$Q_n(f; B) = \sum_{i=1}^{n} c_i f(\mathbf{x}_i),$$

for instance, are based on nonadaptive sampling because the abscissas and weights are chosen independently of the observed behavior of f (see Section 3.3.2).

According to the definition given, the number of samples taken in nonadaptive sampling strategies does *not* have to be independent of the observed behavior of f. However, if n is *a priori* determined in nonadaptive sampling strategies, then the required function values $f(\mathbf{x}_1), \ldots, f(\mathbf{x}_n)$ can be computed in any order. They can even be evaluated in parallel. A fixed sample of size n is therefore also referred to as *parallel information.*

Otherwise, assume that f is evaluated first at an *a priori* determined set of sampling points $\mathbf{x}_1, \ldots, \mathbf{x}_{n_1}$. The choice of the next set of points $\mathbf{x}_{n_1+1}, \ldots, \mathbf{x}_{n_2}$ is based on knowledge of both the $\mathbf{x}_1, \ldots, \mathbf{x}_{n_1}$ and the $f(\mathbf{x}_1), \ldots, f(\mathbf{x}_{n_1})$. After $f(\mathbf{x}_{n_1+1}), \ldots, f(\mathbf{x}_{n_2})$ are evaluated, $\mathbf{x}_{n_2+1}, \ldots, \mathbf{x}_{n_3}$ are chosen according to $\mathbf{x}_1, \ldots, \mathbf{x}_{n_2}$ and $f(\mathbf{x}_1), \ldots, f(\mathbf{x}_{n_2})$. Generally, the choice of $\mathbf{x}_{n_j+1}, \ldots, \mathbf{x}_{n_{j+1}}$ is based on knowledge already obtained up to this point, i.e., the sampling points $\mathbf{x}_1, \ldots, \mathbf{x}_{n_j}$ and the function values $f(\mathbf{x}_1), \ldots, f(\mathbf{x}_{n_j})$. In this case, $S(f)$ is said to contain *adaptive information.* The sampling points $\mathbf{x}_{n_j+1}, \ldots, \mathbf{x}_{n_{j+1}}$ are undetermined and the function evaluations $f(\mathbf{x}_{n_j+1}), \ldots, f(\mathbf{x}_{n_{j+1}})$ cannot be performed until $f(\mathbf{x}_1), \ldots, f(\mathbf{x}_{n_j})$ have been evaluated. If $n_1 = n_2 = \cdots = 1$, then none of the function evaluations can be done in parallel. In this case the information resulting from an adaptive sampling strategy is said to be *sequential.*

8.2.1. Nonadaptive Algorithms. *Nonadaptive* integration algorithms apply a specific sequence $\{Q_{n_j}\}$ of integration formulas for the given integration region B (and possibly the given weight function w) to the integrand f. A generic description of the structure of nonadaptive integration algorithms is provided by the meta-algorithm in Table 8.1.

Unless the interface of a nonadaptive integration algorithm provides the user with some means of specifying relevant characteristics of f (for instance, the location and/or order of singularities), there is no *a priori* information available about f. Hence, the computational procedure performed in nonadaptive integration algorithms is independent of f (except that the number of integration formulas applied may vary according to the difficulty of the integration problem).

Nonadaptive integration algorithms require an unnecessarily large computational effort if the local behavior of f varies greatly in different parts of the integration region (as a result, for instance, of discontinuities, singularities, or nonlinear oscillations). Therefore, nonadaptive integration algorithms should be used only for smooth, "well-behaved" integrands which do not exhibit any local irregularities.

Software (Nonadaptive Algorithms). The five (similar) subroutines `CMLIB/dqng` ≈ `IMSL/MATH-LIBRARY/dqdng` ≈ `NAG/d01bdf` ≈ `QUADPACK/dqng` ≈ `SLATEC/dqng` employ a nonadaptive integration algorithm based on the Patterson sequence (see Section 5.2.10) of 10-, 21-, 43-, and 87-point formulas for calculating univariate integrals over bounded intervals. `NAG/d01arf` uses the Patterson sequence of 1-, 3-, 7-, 15-, 31-, 63-, 127-, 255-, and 511-point formulas. The same sequence (except for the 511-point formula) is also used in `NAG/d01daf`, which computes two-dimensional, iterated integrals for approximating both the outer and the inner integrals.

TABLE 8.1. *Meta-algorithm for nonadaptive integration.*

```
j := 0;
e_0 := largest_machine_number;
do while e_j > ε
    j := j + 1;
    calculate an approximate value q_j := Q_{n_j}(f; B);
    calculate an error estimate e_j := E_{n_j}(f; B) for q_j
end do
```

8.2.2. Adaptive Algorithms. In adaptive integration algorithms, the integration region B is recursively subdivided into a set of disjoint subregions $\{B_s\}$. Using appropriate (nonadaptive) integration formulas, f is approximately integrated on and corresponding error estimates are obtained for each subregion B_s separately. The subregions generated during the subdivision process (or a subset thereof), and the data associated with them, are stored in special data structures. Adaptive integration algorithms terminate when the overall integral and error estimate, i.e., the sum of all local integral and error estimates on subregions of a partition $\{B_s\}$ of B, satisfy the given accuracy requirement.

The meta-algorithm in Table 8.2 provides an abstract description of the structure of adaptive integration algorithms. It can be used, for instance, as a foundation for developing adaptive integration algorithms based on an appropriate integration formula Q_n and a corresponding error estimator E. The various parts of the meta-algorithm are described in more detail in the remainder of this section.

Basic Integration Module. The central task of calculating approximate integral values and corresponding error estimates on subregions $B_s \subseteq B$ in adaptive algorithms is performed by so-called *basic integration modules*, whose input/output behavior can be specified as follows. For the

$$\textit{input data}: \qquad f, \quad B_s \text{ with } B_s \subseteq B \text{ (and possibly also } w),$$

a basic integration module calculates an approximation

$$Q(f; B_s) \quad \approx \quad I(f; B_s)$$

as well as an error estimate $E(f; B_s)$. The error estimate is supposed (but cannot generally be guaranteed) to satisfy the inequality

$$|Q(f; B_s) - I(f; B_s)| \leq E(f; B_s).$$

In order to obtain integral approximations and corresponding error estimates, basic integration modules apply one, two, or more integration formulas and

TABLE 8.2. *Meta-algorithm for adaptive integration.*

```
q := Q(f; B);
e := E(f; B);
insert (B, q, e) into the data structure;
do while e > ε
    choose an element of the data structure (index s) for partitioning;
    subdivide the chosen region B_s into subregions
        B_l,   l = 1, ..., L;
    calculate approximate values q_l and error estimates e_l :
        q_l := Q(f; B_l)  and  e_l := E(f; B_l),   l = 1, ..., L;
    remove (B_s, q_s, e_s) from the data structure;
    insert (B_1, q_1, e_1), ..., (B_L, q_L, e_L) into the data structure;
    q := q − q_s + (q_1 + ··· + q_L);
    e := e − e_s + (e_1 + ··· + e_L)
end do
```

may additionally employ convergence acceleration techniques. However, basic integration modules use *nonadaptive* information in any case.

Most adaptive integration algorithms use only a single basic integration module. If, however, the subregions $\{B_s\}$ generated are not *affine equivalent* (i.e., they cannot be transformed into each other using affine transformations), then several basic integration modules corresponding to different kinds of integration regions must be used in an adaptive algorithm. This typically happens when B is a nonstandard, multidimensional integration region. Multiple basic integration modules also have to be used if different weight functions are associated with different subregions. This is usually the case when the weight function is nonconstant, $w(\mathbf{x}) \not\equiv 1$.

Subregion Data. Various types of data are associated with the subregions B_s generated during an adaptive integration process:

(i) parameters determining B_s, for instance, the $N+1$ vertices of B_s if B_s is an N-dimensional simplex;

(ii) the approximate integral value $\mathrm{Q}(f; B_s) \approx \mathrm{I}(f; B_s)$;

(iii) the error estimate $\mathrm{E}(f; B_s) \approx |\mathrm{Q}(f; B_s) - \mathrm{I}(f; B_s)|$; and

(iv) miscellaneous auxiliary information, such as function values $\{f(\mathbf{x}_i)\}$ at abscissas $\{\mathbf{x}_i\}$ in B_s to be reused in future calculations.

The data structure used to store the information associated with the different subregions depends on the specific subdivision strategy applied.

Subdivision Decision Processes. Adaptive integration algorithms are designed to minimize the computational effort (the number of f-evaluations) required to meet a given error requirement by placing abscissas predominantly in those parts of the integration region B in which the numerical integration is most difficult. The local density of the abscissa set $\{\mathbf{x}_i\}$ is determined by the local fineness of the partition $\{B_s\}$ of B. In order to select only sufficiently "difficult" regions B_s for further subdivision and to construct appropriate refinements $\{B_1, \ldots, B_L\}$ of such regions B_s, a *dynamic* decision process, which takes information extracted from (potentially all) previously sampled function values into account, is used to control the subdivision refinement. The efficiency and reliability of adaptive algorithms crucially depends on the quality of this dynamic decision process.

Since the "difficulty" of a subregion B_s can be quantified by the (absolute or relative) integration error in B_s, the selection of subregions for further subdivision depends mainly on the (absolute or relative) error estimates in different subregions. The (error) information taken into consideration when selecting a subregion may be *local* or *global*, resulting in local or global subdivision strategies. Local information refers only to the subregion *currently being processed*, while global information refers to data concerning *all* subregions of a partition of B.

Example (Subregion Tree). Adaptive subdivision strategies for univariate integration problems are usually based on a *bisection* of subintervals. If $f : [a, b] \to \mathbb{R}$ is smooth in the interval of integration except for a single difficult spot $x_0 \in [a, b]$ (for instance, a discontinuity or singularity), then the subintervals $[a_s, b_s]$ chosen for further refinement are generally those which contain x_0. Figure 8.2 shows the highly unbalanced tree structure of the resulting subintervals for $[a, b] = [0, 1]$ and $x_0 = 0$.

Local Subdivision Strategies. Local subdivision strategies are distinguished by two characteristics: (i) the subregions in $\{B_s\}$ are classified as being *active* or *inactive*, according to whether or not they are eligible for further subdivision; (ii) the active/inactive classification of subregions is based on *local* error information.

More specifically, local subdivision strategies assign a separate tolerance $\varepsilon(B_s)$ to any subregion $B_s \subseteq B$ by defining

$$\varepsilon(B_s) := \frac{\text{vol}(B_s)}{\text{vol}(B)} \varepsilon; \tag{8.20}$$

i.e., the tolerance is proportional to the volume $\text{vol}(B_s)$.[92] A subregion is rendered *inactive* as soon as the inequality

$$\text{E}(f; B_s) \le \varepsilon(B_s)$$

[92]This definition can easily be extended to weighted integrals.

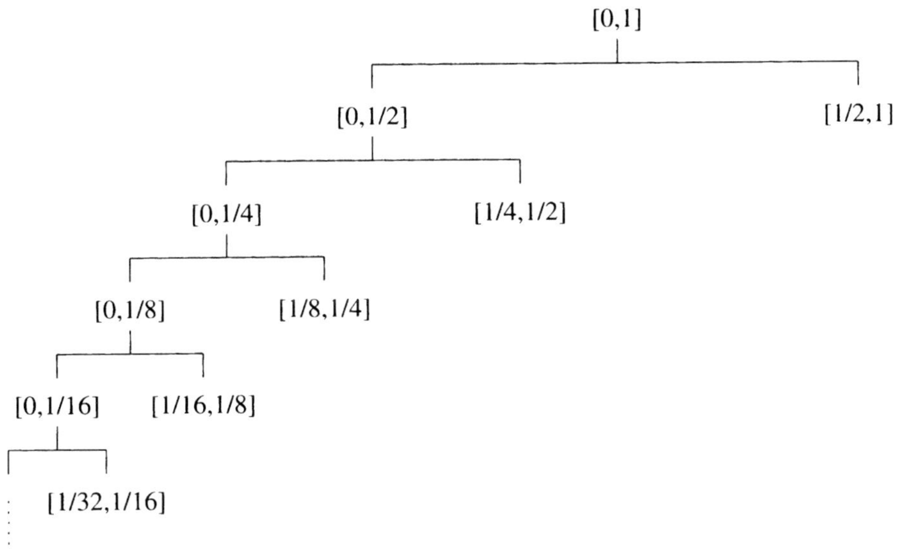

FIG. 8.2. *Subregion tree resulting from an adaptive subdivision strategy.*

is satisfied. Once *all* subregions of an adaptive partition $\{B_s\}$ are rendered inactive, local subdivision strategies do not proceed any further. In this case,

$$\begin{aligned}
\left|\sum_{B_s} \mathrm{Q}(f;B_s) - \mathrm{I}(f;B)\right| &= \left|\sum_{B_s} (\mathrm{Q}(f;B_s) - \mathrm{I}(f;B_s))\right| \\
&\le \sum_{B_s} |\mathrm{Q}(f;B_s) - \mathrm{I}(f;B_s)| \\
&\le \sum_{B_s} \mathrm{E}(f;B_s) \le \sum_{B_s} \varepsilon(B_s) \\
&= \sum_{B_s} \frac{\mathrm{vol}(B_s)}{\mathrm{vol}(B)} \varepsilon \\
&= \varepsilon
\end{aligned}$$

shows that the overall accuracy requirement ε is met, provided the error bound

$$|\mathrm{Q}(f;B_s) - \mathrm{I}(f;B_s)| \le \mathrm{E}(f;B_s)$$

holds in *any* subregion B_s of the partition. In other words, if all local accuracy requirements $\varepsilon(B_s)$ are met, then the overall accuracy requirement ε is also met automatically. In practice, however, the overall accuracy requirement ε is met—and locally adaptive integration algorithms terminate successfully—before all subregions have been rendered inactive.

Storage and Processing Order of Active Subregions. There is usually more than a single active subregion at any given point of the algorithmic procedure. Locally adaptive strategies differ in the way they select the particular subregion to be subdivided from the set of active regions. The set of active subregions is generally processed according to an *a priori* fixed scheme in order to simplify the management of the data structure in which the subregions are stored. As for univariate integrals, active subintervals are often processed according to a depth-first, left-to-right traversal order of the subinterval tree. In this case a *LIFO* (*last in, first out*) *queue* can be used to store active subintervals. If, alternatively, active subintervals are processed according to a breadth-first, left-to-right traversal order of the subinterval tree, a *FIFO* (*first in, first out*) *queue* is employed.

Various locally adaptive subdivision strategies for univariate integration problems are investigated in Rice [365]. It is demonstrated that the order in which active subintervals are processed does not influence the rate of convergence $Q_n f \to If$ as $n \to \infty$.[93] Therefore, the order in which active subregions are processed in locally adaptive algorithms does not have a major impact on the asymptotic efficiency of the algorithms. For practical values of n, however, a breadth-first processing order is often superior to a depth-first processing order (Shapiro [388]).

The data structures needed to store the resulting subregions, LIFO or FIFO queues, in locally adaptive subdivision strategies are simple and require a minimum amount of storage space because only the *active* subregions have to be stored. The computational complexity of storage and retrieval operations in these data structures is independent of the number of subregions stored. For these reasons, in the 1960s and 1970s locally adaptive algorithms were often preferred to globally adaptive algorithms, which require more complex data structures. Since then, however, locally adaptive strategies have been shown to be inferior to globally adaptive strategies on both theoretical and empirical grounds (see Malcolm and Simpson [292] and Rice [365]), and rapidly increasing memory capacities have dispelled concerns regarding the storage requirements of adaptive algorithms.

Furthermore, the simplicity of the data structures used in locally adaptive subdivision strategies hinges on the fact that the error requirement ε is known initially and does not change during the course of computation. If ε is composed of both an absolute and a *relative* error requirement (see Section 8.3.1), then ε is not known initially, but varies as a result of changes in the overall integral approximation Qf. Since the local error requirements $\varepsilon(B_s)$ vary proportionally with ε, subregions which have been classified as inactive at a certain point of the computational procedure may become active again if

[93] The variable n denotes the overall number of function evaluations performed in an adaptive algorithm.

ε decreases. As a result, appropriate data structures would have to store both active *and* inactive subregions B_s in order to access active subregions as well as render inactive regions active again.

For all these reasons, locally adaptive algorithms have been used very little in practical integration programs developed since the late 1970s.

Software (Locally Adaptive Algorithms). The subroutine `NAG/d01ahf` uses a depth-first, right-to-left locally adaptive subdivision strategy to calculate univariate integrals over bounded intervals. The basic integration module implements a Patterson sequence of 1-, 3-, 7-, 15-, 31-, 63-, 127-, and 255-point formulas (see Section 5.2.10) and additionally attempts to accelerate the convergence of these formulas using the ε-algorithm (see Section 3.7.4). `NAG/d01ahf` employs *nonuniform* interval bisection based on the relative gradient of the integrand at the endpoints of the subinterval to be divided.

Global Subdivision Strategies. Global subdivision strategies do *not* classify subregions as being active or inactive; rather, *all* subregions are eligible for further subdivision at any time during the algorithmic procedure. The order in which subregions are processed is determined dynamically based on the error estimates in different subregions. More precisely, the particular subregion B_s chosen for further subdivision is required to possess the largest (possibly weighted) error estimate at the moment of decision:

$$w_s \mathrm{E}(f; B_s) = \max\{w_1 \mathrm{E}(f; B_1), \ldots, w_k \mathrm{E}(f; B_k)\}.$$

Different global subdivision strategies are obtained for different choices of the weights w_s:

Kind of global subdivision strategy	**Weights** w_s
absolute	1
relative	$1/\mathrm{Q}(\lvert f \rvert; B_s)$
region volume	$1/\operatorname{vol}(B_s)$

For a discussion of these and other global subdivision strategies, see Shapiro [388].[94] However, only the *absolute* global subdivision strategy has been employed in practical integration programs; the resulting algorithms are henceforth simply referred to as globally adaptive.

The efficiency of globally adaptive integration algorithms has been confirmed by both theoretical analyses and empirical data (see, for instance, Malcolm and Simpson [292] and Rice [365]). This is why a large number of integration programs developed since the late 1970s are based on globally adaptive algorithms.

[94] Note that the conclusions drawn in Shapiro [388] are at odds with the experiences of most other researchers. This discrepancy is most probably due to Shapiro's rather unusual choice of the basic integration module.

Data Structures for Global Subdivision Strategies. Data structures used in globally adaptive integration programs for storing subregions and information associated with them must support two fundamental operations: insert and delmax. insert stores an element, i.e., a subregion B_s along with the associated information, in the data structure, while delmax retrieves the element with the largest *key* value, i.e., the largest (weighted) error estimate $w_s\mathrm{E}(f; B_s)$, and deletes it from the data structure. Such data structures are generally referred to as *priority queues* (Kruse [249]).

There are different ways of implementing priority queues in integration programs. For a thorough discussion of this topic, see Krommer and Ueberhuber [245], [247]. However, practically all globally adaptive algorithms developed over the last decade are based on (binary) *heaps*. This is mainly due to the fact that the computational complexity of insert and delmax operations in heaps grows proportionally to $\log M$ as the number M of elements stored increases (whereas the computational complexity of insert operations grows proportionally to M if sorted lists are used to implement priority queues). This modest growth is also why the computational cost of storing and retrieving a subregion B_s is generally—even for large values of M—negligible compared to the computational effort required to calculate the corresponding integral approximation $\mathrm{Q}(f; B_s)$ and error estimate $\mathrm{E}(f; B_s)$ (Krommer and Ueberhuber [245]).

Spatial Subdivision of Integration Regions. Once a particular region B_s has been selected for further subdivision, a specific refinement $\{B_1, \ldots, B_L\}$ of B_s must be constructed. In general, the spatial partition $\{B_1, \ldots, B_L\}$ depends on the dimension N and the kind of the region B_s.

In univariate integration programs, (finite) intervals are usually *bisected* into two subintervals of equal length. While theoretical and empirical evidence provided by Hanke [198] and Soerevik [415] shows that in many cases the *trisection* of intervals into subintervals of equal length can be expected to give better results than bisection, the expected improvement, a reduction in the overall number of function evaluations of about 5%, is fairly minor (see, for instance, the example on page 355). *Nonuniform* trisection is proposed and empirically shown to be superior to bisection in Berntsen, Espelid, and Soerevik [30]. In their subdivision scheme the location of the two partition points in $[a_s, b_s]$ is determined based on the fourth-order divided differences of the f-values at the abscissas $x_1, \ldots, x_{n_s}$ in $[a_s, b_s]$ which are used in the basic integration module for calculating the estimates $\mathrm{Q}(f; [a_s, b_s])$ and $\mathrm{E}(f; [a_s, b_s])$.

Multivariate integration programs usually divide an N-dimensional cube C_s along a plane $x_k = \text{const}$, $k \in \{1, \ldots, N\}$, into two subcubes of equal size. The coordinate k is chosen such that the variation of f-samples in C_s

(measured in terms of an appropriate divided difference) in the kth direction attains a maximum value. However, as shown in Berntsen [23], N-dimensional cubes can often be divided along several planes without reducing the efficiency of the adaptive integration scheme.

N-dimensional simplices (triangles, tetrahedra, etc.) are frequently divided into N^2 congruent subsimplices by bisecting the sides of the simplex. *Nonuniform* partitions of triangles are discussed in Espelid and Genz [138].

8.2.3. Software Based on Globally Adaptive Integration Algorithms. This section describes a number of integration programs employing global subdivision strategies. The various subroutines are classified according to the type of integration problem (characterized by the integrand and the integration region) they solve. Software in which the convergence of a globally adaptive integration algorithm is accelerated using extrapolation methods can be found in Section 8.3.3.

General Univariate Integrand. The globally adaptive integration programs `QUADPACK/dqag` $\approx$ `IMSL/MATH-LIBRARY/dqdag` evaluate univariate integrals over finite intervals using basic integration modules based on Gauss–Kronrod formula pairs G_n and K_{2n+1}, where n = 7, 10, 15, 20, 25, 30. The specific Gauss–Kronrod formula pair to be applied can be chosen by the user. The subintervals generated during the subdivision process are stored in a *sorted list*, and the storage space available for this data structure is allocated *statically*. If the preset maximum number of subintervals proves insufficient, then the user has to modify and recompile these routines.

This problem is overcome in the subroutine `QUADPACK/dqage`. This subprogram is functionally identical to the aforementioned routines, but enables the calling (sub)program to pass the storage space available for the sorted list as subroutine arguments. The subroutines `CMLIB/dqag` $\approx$ `CMLIB/dqage` $\approx$ `IMSL/MATH-LIBRARY/dq2ag` $\approx$ `SLATEC/dqag` $\approx$ `SLATEC/dqage` are all identical to `QUADPACK/dqage`, except for minor differences in the user interface.

The subroutine `NAG/d01auf` is also based on `QUADPACK/dqage`, but uses a user-supplied subroutine (rather than a function) to evaluate the integrand at an array of abscissas. This feature can lead to a performance improvement on vector computers, provided that the integrand evaluations can be vectorized (see Section 9.3.2).

The basic integration module of the globally adaptive integration program `TOMS/691/dqxg` is based on an embedded sequence of four quadrature formulas. The *stable, recursive, monotone formulas* employed are described in Favati, Lotti, and Romani [145]. The user must specify the pair of formulas initially used for approximately integrating f on each subinterval $[a_s, b_s]$. The integration algorithm may, however, decide to use additional formulas of higher degrees, provided f is sufficiently smooth on $[a_s, b_s]$.

Oscillatory Univariate Integrand. The subroutine `NAG/d01akf` is based on `QUADPACK/dqage`, but provides only the Gauss–Kronrod formula pair with the maximum number of abscissas, i.e., G_{30} and K_{61}, in its basic integration module. Among all formula pairs in `QUADPACK/dqage`, this formula pair proves generally to be the most efficient one if the integrand is smooth and oscillatory.

Integrand with Algebraic-Logarithmic Weight Function. The globally adaptive programs `QUADPACK/dqaws` ≈ `IMSL/MATH-LIBRARY/dqdaws` evaluate univariate integrals over finite intervals $[a, b]$ with respect to an algebraic weight function

$$w(x) = (x-a)^\alpha (b-x)^\beta$$

or the algebraic weight functions

$$\begin{aligned} w(x) &= (x-a)^\alpha (b-x)^\beta \log(x-a), \\ w(x) &= (x-a)^\alpha (b-x)^\beta \log(b-x), \text{ or} \\ w(x) &= (x-a)^\alpha (b-x)^\beta \log(x-a) \log(b-x). \end{aligned}$$

The exponents α and β of the endpoint singularities must be provided by the user. The integration module uses a pair of 13/25-point modified Clenshaw–Curtis formulas for subintervals $[a_s, b_s]$ which contain at least one of the endpoints a and b. Otherwise, a 7/15-point Gauss–Kronrod formula pair is employed.

The subroutines `QUADPACK/dqawse` ≈ `CMLIB/dqaws` ≈ `CMLIB/dqawse` ≈ `IMSL/MATH-LIBRARY/dq2aws` ≈ `NAG/d01apf` ≈ `SLATEC/dqaws` ≈ `SLATEC/dqawse` are functionally identical to the aforementioned programs, but enable the calling (sub)program to pass the storage space available as subroutine arguments.

Univariate, Cauchy Singular Integrand Function. The two subprograms `QUADPACK/dqawc` ≈ `IMSL/MATH-LIBRARY/dqdawc` calculate the value of the Hilbert transform (1.14) of a given function f for a given pole $y \in (a, b)$. In each subdivision step the subinterval $[a_s, b_s]$ with the largest error estimate is divided into two subintervals at a point $c_s \in (a_s, b_s)$ according to the following rules. (i) If $y \notin [a_s, b_s]$, the division is made at the midpoint $c_s := (a_s + b_s)/2$. (ii) Otherwise, c_s is chosen as the center of the interval $[a_s, c]$, if $[a_s, c]$ is longer than $[c, b_s]$, or as the center of $[c, b_s]$ if the opposite is true. This choice of c_s avoids dividing $[a_s, b_s]$ at the singularity c. If c is contained in or is close to $[a_s, b_s]$, $c \in (a_s - d, b_s + d)$, $d := (b_s - a_s)/20$, then the basic integration module uses a pair of modified Clenshaw–Curtis formulas with 13 and 25 abscissas; otherwise, a 7/15-point Gauss–Kronrod formula pair is applied.

The subroutines QUADPACK/dqawce ≈ CMLIB/dqawc ≈ CMLIB/dqawce ≈ IMSL/MATH-LIBRARY/dq2awc ≈ NAG/d01aqf ≈ SLATEC/dqawc ≈ SLATEC/dqawce are functionally identical to the aforementioned programs, but enable the calling (sub)program to pass the storage space available as subroutine arguments.

Multivariate Integrands on Cubic Regions. The four subprograms CMLIB/dadapt, NAG/d01eaf, NAG/d01fcf, and TOMS/698/dcuhre use a globally adaptive subdivision strategy for evaluating N-dimensional integrals over *cubic regions*. CMLIB/dadapt works for dimensions $N = 2, 3, \ldots, 20$, NAG/d01fcf and TOMS/698/dcuhre for dimensions from $N = 2, 3, \ldots, 15$, and NAG/d01eaf for any dimension $N \geq 1$. At each subdivision step, each of the routines divides the subcube C_s with the largest error estimate along a plane $x_k = \text{const}$ into two subcubes of equal size. The coordinate k is chosen such that the fourth divided difference of the f-samples in C_s in the kth direction attains a maximum value.

The routines differ in the (polynomial) integration formulas and/or the error estimation procedures used in their basic integration modules. For instance, NAG/d01eaf generally (i.e., in any dimension $N \geq 1$) calculates integral approximations using a formula of degree $D = 7$; an additional formula of degree $D = 5$ is used to estimate the discretization error. With TOMS/698/dcuhre, by contrast, the user can generally choose between formulas with degrees of accuracy $D = 7$ and $D = 9$; for $N = 2$ and $N = 3$ additional integration formulas with degrees of accuracy $D = 13$ and $D = 11$ are available. Furthermore, TOMS/698/dcuhre uses null rules for error estimation purposes. A special feature of both NAG/d01eaf and TOMS/698/dcuhre is that they can simultaneously integrate multiple integrands.

Multivariate Integrands on Simplicial Regions. TOMS/584/cubtri evaluates bivariate integrals over triangular regions. At each subdivision step the subtriangle with the largest error estimate is divided into four congruent subtriangles. The basic integration module in TOMS/584/cubtri is based on a formula pair with degrees of accuracy $D = 5$ and $D = 8$.

CMLIB/dtwodq and TOMS/706/dcutri evaluate bivariate integrals over *finite sets* of triangular regions. At each subdivision step CMLIB/dtwodq divides the triangle with the largest error estimate into two subtriangles along the straight line which connects the center of the longest side and the opposite vertex. The user can choose between two different basic integration modules which implement formula pairs with degrees of accuracy $D = 6/8$ and $D = 9/11$, respectively. TOMS/706/dcutri employs the same subdivision strategy as TOMS/584/cubtri, but uses a formula with degree of accuracy $D = 13$ in its basic integration module. Furthermore, TOMS/706/dcutri uses null rules for error estimation purposes. Additionally, `TOMS/706/dcutri` differs from `CMLIB/dtwodq` in that it can integrate multiple integrands simultaneously.

The subroutine `TOMS/720/dcutet` evaluates trivariate integrals over finite sets of tetrahedra. At each subdivision step the subtetrahedron with the largest error estimate is divided into eight congruent subtetrahedra. The basic integration module of `TOMS/720/dcutet` is based on a 43-point cubature formula with degree of accuracy $D = 8$.

8.3. Special Features of Integration Algorithms

In most cases the error estimation procedures and discretization refinement schemes described in Sections 8.1 and 8.2 ensure that numerical integration algorithms calculate sought-after integrals both reliably and efficiently. For some integration problems, however, integration algorithms fail to perform satisfactorily; i.e., they produce inaccurate results or take an excessive amount of time, unless they are enhanced by additional algorithmic measures. For instance, the error estimation procedures described in Section 8.1 (except for those based on interval arithmetic) assume that the numerical error in an integration formula $\mathrm{Q}_n f$ is dominated by the discretization error in $\mathrm{Q}_n f$, i.e., that the effect of rounding errors on the numerical value of $\mathrm{Q}_n f$ is negligible. However, if the given integration problem, and consequently the computation of $\mathrm{Q}_n f$, is badly conditioned (see Section 3.4), then rounding errors may have a significant impact on the calculated value of $\mathrm{Q}_n f$. If this phenomenon occurs, error estimation procedures may become unreliable.

This section describes a number of special features of integration algorithms designed to increase their reliability or efficiency.

8.3.1. Reliability Enhancement

Dealing with Badly Conditioned Problems. Testing for convergence is expressed by the statement "**do while** $e > \varepsilon$" in the meta-algorithms given in Tables 8.1 and 8.2. This statement corresponds to testing the ***absolute*** error if the error requirement $\varepsilon = \varepsilon_{abs}$ remains, independent of the changes in the approximate integral value Qf, constant during the course of computation. If the magnitude of the integral $\mathrm{I}f = \mathrm{I}_w f$ to be calculated is known in advance, then the absolute error requirement ε_{abs} can be chosen in such a way that Qf approximates $\mathrm{I}_w f$ with sufficient accuracy.

Generally, the only information easily available to estimate the magnitude of $\mathrm{I}_w f$ prior to integration are sample values of f. Such sample values can be used to obtain an estimate of $\|f\|_\infty$. An upper bound on $|\mathrm{I}_w f|$ is then given by $K_{f,w}\|f\|_\infty$, where $K_{f,w}$ is the absolute condition number of the integration problem (see Section 3.4). However, the relationship

$$|\mathrm{I}_w f| = \frac{K_{f,w}\|f\|_\infty}{K^{\mathrm{rel}}_{f,w}},$$

which is derived from (3.6), shows that $|\mathrm{I}_w f|$ can be significantly smaller than the bound $K_{f,w}\|f\|_\infty$ when the *relative* conditioning of the integration problem is bad. In this case, choosing the absolute error requirement ε_{abs} based on the estimate $K_{f,w}\|f\|_\infty$ may result in an inaccurate or even meaningless approximate value $\mathrm{Q}f$.

For this reason most numerical integration algorithms enable users to specify a *relative* error requirement ε_{rel} in addition to an absolute error requirement ε_{abs}. Based on ε_{abs} and ε_{rel}, the error tolerance ε imposed on the approximate integration result is obtained in one of the following ways:

1. $\varepsilon := \varepsilon_{abs} + \varepsilon_{rel}\,|\mathrm{Q}f|$,

2. $\varepsilon := \max(\varepsilon_{abs}, \varepsilon_{rel}\,|\mathrm{Q}f|)$, or

3. $\varepsilon := \max(\varepsilon_{abs}, \varepsilon_{rel}\,\mathrm{Q}_{abs}f)$, where $\mathrm{Q}_{abs}f := \mathrm{Q}(|f|; B)$.[95]

The first two definitions simply represent different combinations of the test for convergence with respect to the absolute error,

$$|\mathrm{E}(f;B)| < \varepsilon_{abs},$$

and the test for convergence with respect to the relative error,

$$\frac{|\mathrm{E}(f;B)|}{|\mathrm{Q}(f;B)|} < \varepsilon_{rel}. \tag{8.21}$$

However, the convergence test (8.21) becomes unreliable if the relative conditioning of the linear functional $\mathrm{Q}(f;B)$ is bad (which is usually the result of a bad relative conditioning of the integration problem $\mathrm{I}(f;B)$). If this is the case, then $\mathrm{Q}(f;B)$ is very small compared to the magnitude of the integration weights and integrand values involved in the calculation of $\mathrm{Q}(f;B)$; i.e., heavy cancellation occurs in the calculation of $\mathrm{Q}(f;B)$. If the integration weights in $\mathrm{Q}(f;B)$ are nonnegative (which is usually the case when w is nonnegative), cancellation can occur only if the integrand values are of mixed sign. Therefore, the calculation of $\mathrm{Q}(|f|;B)$ is numerically stable, and the relative L_1-test for convergence,

$$\frac{|\mathrm{E}(f;B)|}{\mathrm{Q}(|f|;B)} < \varepsilon_{rel}, \tag{8.22}$$

[95]In this case the weight function w is assumed to be nonnegative.

is more reliable than the convergence test (8.21). Combining the test (8.22) and the usual test for convergence with respect to the absolute error leads to the third definition of the error tolerance ε.

The additional calculation of $\mathrm{Q}(|f|; B)$ in (8.22) causes only a minor increase in the overall computational effort, as *no* additional f-evaluations are needed. The main disadvantage of (8.22) is that the relative error in $\mathrm{Q}(f; B)$ is *not* necessarily smaller than the relative error requirement ε_{rel}, even if (8.22) is satisfied: the convergence tests (8.21) and (8.22) coincide only if f is strictly positive or negative on B.

Preventing Numerical Instability. Probably the most common operation in numerical integration programs is the simple summation

$$q = y_1 + y_2 + \cdots + y_n. \tag{8.23}$$

Most importantly, this operation occurs when an integration formula $\mathrm{Q}_n f$ is calculated based on weighted integrand values $y_1 = c_1 f(\mathbf{x}_1), \ldots, y_n = c_n f(\mathbf{x}_n)$. It is well known that evaluating (8.23) can be numerically unstable and that the order in which the additions are performed can have a significant impact on the accuracy of the numerical result.

This is why special summation algorithms for the mathematically trivial operation (8.23) have been developed. Among the best known summation techniques are *recursive summation* (see Table 8.3), *pairwise summation* (see Table 8.4), and *compensated summation* or *Kahan summation* (see Table 8.5). While recursive summation has the least computational complexity and storage requirements, it is numerically inferior to the other two algorithms (see Figs. 8.3, 8.4, and 8.5). Note that all three algorithms can be performed using $O(n)$ arithmetic operations.

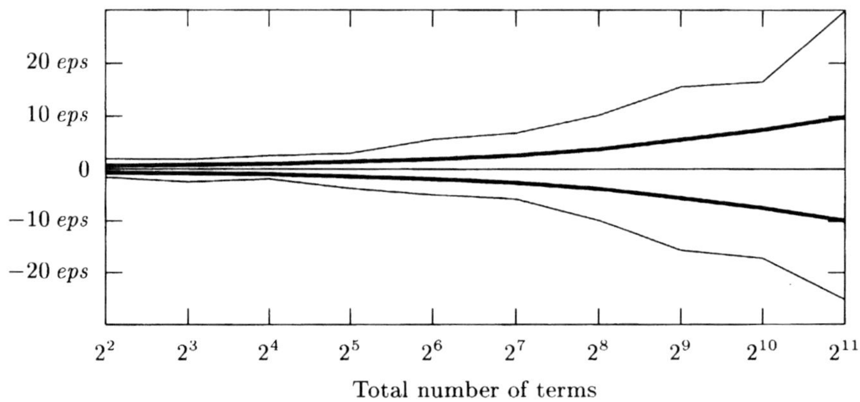

FIG. 8.3. *Extrema of relative rounding errors (—) and the standard deviation of rounding errors (**—**) for the recursive summation of random numbers $x_i \in [0, 1]$.*

TABLE 8.3. *Recursive summation.*

```
q := 0;
do i = 2, 3, ..., n
    q := q + y_i
end do
```

TABLE 8.4. *Pairwise summation.*

```
k := n;
do i = 1, 2, ..., ⌈log_2 n⌉
    l := ⌊k/2⌋;
    do j = 1, 2, ..., l
        y_j := y_{2j-1} + y_{2j}
    end do
    if k ≠ 2l then
        l := l + 1;
        y_l := y_k
    end if
    k := l;
end do
q := y_1;
```

TABLE 8.5. *Compensated (Kahan) summation.*

```
q := y_1;
r := 0;
do i = 2, 3, ..., n
    s := y_i - r;
    t := q + s;
    r := (t - q) - s;
    q := t
end do
```

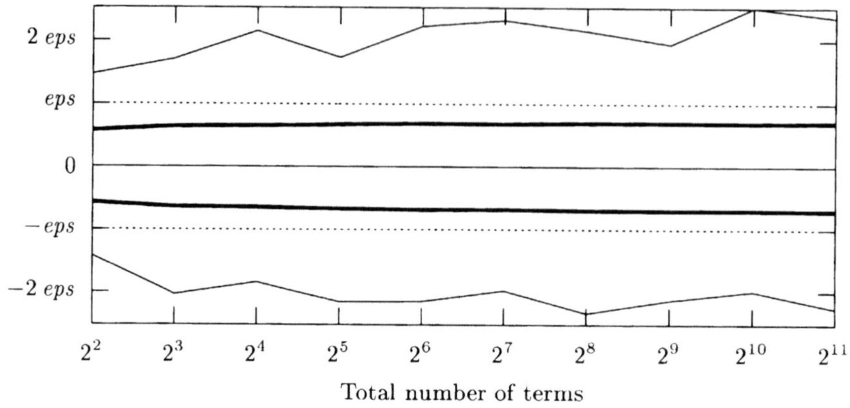

FIG. 8.4. *Extrema of relative rounding errors (—) and the standard deviation of rounding errors (**—**) for the pairwise summation of random numbers* $x_i \in [0, 1]$.

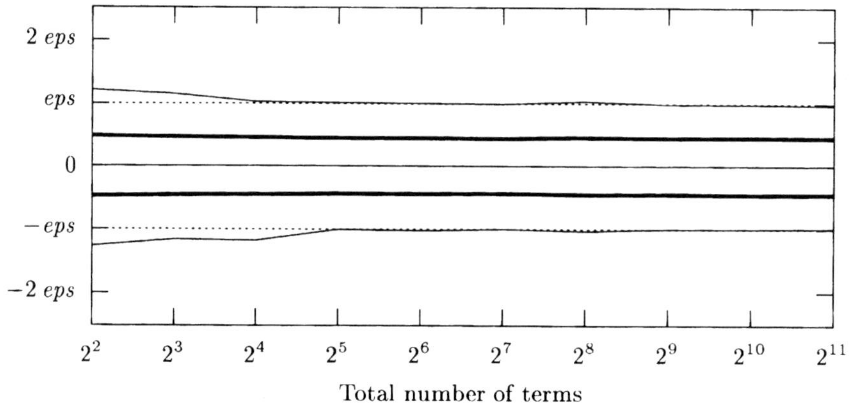

FIG. 8.5. *Extrema of relative rounding errors (—) and the standard deviation of rounding errors (**—**) for the Kahan summation of random numbers* $x_i \in [0, 1]$.

An evaluation of a number of summation algorithms was carried out by Higham [205]. One of the results of Higham's study is that the use of algorithms of higher computational complexity than $O(n)$ (such as summation algorithms based on sorting which require $O(n \log n)$ operations) is not usually justified by a substantially improved accuracy. Furthermore, pairwise and compensated summation are usually comparable in their accuracies, whereas recursive summation performs well only if the summation terms satisfy $|y_1| \leq |y_2| \leq \cdots \leq |y_n|$.

When considering the application of sophisticated summation techniques in numerical integration programs, one should bear in mind that the terms in (8.23) may not all be available at any one time. For instance, the storage space required to store the n numbers $y_1, \ldots, y_n$ may be prohibitive. In this case, certain summation techniques cannot be applied (for example, summation algorithms based on sorting), or they have to be appropriately modified (for example, pairwise summation).

Detecting Noisy Integrands. If an integration problem is badly conditioned with respect to changes in f, then small errors in the sampled function values $f(\mathbf{x}_1), \ldots, f(\mathbf{x}_n)$ may entail large errors in the approximate integration result $Q_n f$. Such errors in the sample values may arise for the following reasons.

(i) When the exact, mathematical function $f : B \subseteq \mathbb{R}^N \to \mathbb{R}$ is implemented on a particular computer system using the floating-point numbers $\mathbb{F}$, the resulting numerical function $\square f$ can assume only values in $\mathbb{F}$: $\square f(B \cap \mathbb{F}^N) \subseteq \mathbb{F}$. The resulting error $\square f - f$ is minimal if the numerical function values $\square f(\mathbf{x})$ for $\mathbf{x} \in B \cap \mathbb{F}^N$ are equal to the optimally rounded exact function values $f(\mathbf{x})$. The deviation of this optimal implementation $\square f(\mathbf{x})$ from the exact function $f(\mathbf{x})$ depends on $\mathbf{x}$ in a highly unsystematic way and is therefore often modeled in terms of stochastic quantities (see Section 8.1.2).

(ii) If f is defined by complicated expressions or in terms of mathematical problems, such as nonlinear algebraic or differential equations, which can be solved only approximately, then the error $\square f - f$ in the implemented integrand is generally significantly larger than the error level introduced by elementary rounding operations.

This situation occurs, for example, when an *iterated*, multidimensional integral is calculated by recursively applying quadrature programs to the constituent one-dimensional integrals. The input data of each of the quadrature routines, except for the quadrature routine used to evaluate the innermost univariate integral, is based on the results of preceding numerical quadratures used to evaluate the corresponding inner integrals. While imposing absolute and/or relative error requirements on the preceding inner numerical quadratures places a limit on the maximum error in the inner integrals, the actual integration errors usually vary unsystematically.

(iii) If the integrand values result from measurements (describing the outcomes of experiments, for instance), then measurement errors usually give rise to a substantial amount of stochastic noise in the data.

If the (pseudo)stochastic errors in the function values $f(\mathbf{x}_1), \ldots, f(\mathbf{x}_n)$ can be modeled as realizations of independent random variables with mean 0 and standard deviation σ, then the resulting mean square error in the integration result $Q_n f$ is given by $\sigma\|Q_n\|_2$ (see the analysis in Section 8.1.2). Therefore, a noise level σ in the sampled function values gives rise to a lower bound $\sigma\|Q_n\|_2$ on the attainable accuracy in numerical integration. The amplifying factor $\|Q_n\|_2$ characterizes the conditioning of the linear functional Q_n with respect to changes in f in a stochastic, *average-case* setting, whereas the condition number $K_{f,Q} = \|Q_n\|_1$ (see Section 3.4.1) measures the *worst-case* effect of data errors.

Robust integration programs should be able to recognize situations in which the magnitude of pseudorandom errors in the sampled function values makes it impossible to meet the user-specified accuracy requirement. As long as the error in $Q_n f$ due to errors in the sample values is significantly smaller than the corresponding discretization error, discretization refinement, i.e., increasing the number of samples, can be expected to *decrease* estimates of the error $Q_n f - If$. In contrast, as soon as the error in $Q_n f$ due to errors in the sample values becomes comparable in size to the discretization error, estimates of $Q_n f - If$ begin to behave unsystematically (sometimes increasing, sometimes decreasing) when n is increased. While phase-factor effects and nonasymptotic error behavior can occasionally produce similar effects (see Section 8.1), the most likely cause of *persistent*, unsystematic error behavior is that the noise level prevents any further improvement of the numerical integration result.

Therefore, *noise detectors* implemented in integration programs count the number of times that discretization refinements have failed to decrease error estimates. Adaptive subdivision algorithms, for instance, count the number of region subdivisions which have *not* lead to improved error estimates:

$$\textbf{if} \quad \sum_{l=1}^{L} e_l > e_s \quad \textbf{then} \quad \text{ncount} := \text{ncount} + 1.$$

When this number reaches an *a priori* determined threshold, the algorithm terminates with an error message saying that the requested accuracy could not be attained due to a "noisy integrand." In this case, the error estimate returned informs the user of the maximum accuracy which can possibly be achieved given the level of noise.

Detecting Nonintegrable Functions. The function f to be integrated by a numerical integration program may *not* be (Riemann) integrable. This is usually caused by f having one or more singularities of order $\alpha \geq N$ in B. It is, in principal, impossible to decide on the basis of a *finite* set of sample values

$$S(f) = (f(\mathbf{x}_1), \ldots, f(\mathbf{x}_n))$$

whether a function f is integrable or not (see Section 10.1.1). Therefore, numerical integration algorithms *cannot* reliably detect nonintegrable functions.

When applied to a nonintegrable, singular function f, an integration algorithm may miss the singular behavior of f altogether by failing to choose the abscissas $\mathbf{x}_1, \ldots, \mathbf{x}_n$ sufficiently close to the singularity(-ies). In such cases the sequence of integral approximations generated by the integration algorithm appears to converge, and the integration algorithm terminates "successfully," i.e., without raising an error, but returns incorrect integral and error estimates.

The probability that the singular behavior of f is missed can often be reduced by increasing the density of the sampling points $\mathbf{x}_1, \ldots, \mathbf{x}_n$. However, since the maximum attainable density is limited by the characteristics of the particular finite precision arithmetic used, it may not be possible to detect a singularity even by sampling f at *all* floating-point numbers.

Example (Nondetectable Singularity). The function

$$f(x) = 3x^2 + \frac{1}{\pi^4} \log[(\pi - x)^2] + 1$$

has a log singularity at the point $x = \pi$,

$$f(x) \to -\infty \quad \text{as} \quad x \to \pi,$$

as well as two zeros close to this point. Figure 8.6 shows the values of an implementation $\square f$ of f for abscissas $x \in \mathbb{F}$ near π, where $\mathbb{F}$ is the system of double precision IEC/IEEE numbers. It can be seen that $\square f$ is a strictly positive function for *all* $x \in \mathbb{F}$. This observation also applies to the restriction $f|_{\mathbb{F}}$ of the *exact* function f, which shows that the positivity of $\square f$ is due to the lack of sampling points $x \in \mathbb{F}$ sufficiently close to π (rather than to the loss of accuracy in $\square f$ caused by rounding and truncation errors). Hence, even by sampling f at *all* floating-point numbers in $\mathbb{F}$, neither the pole at $x = \pi$ nor the two zeros of the function f can be detected.[96]

On the other hand, an integration algorithm may choose one or more sampling points sufficiently close to the singularity(-ies) of f in order for the singular behavior to become apparent. In such cases the sequence of integral approximations generated by refining the discretization of f usually diverges, and the integration algorithm fails to terminate. This is why, in practical integration programs, a finite limit is put on the number of refinement steps using one or several of the following algorithmic parameters.

(i) An *upper limit* on the *computational effort*: This parameter is usually specified in terms of the maximum number of f-evaluations or, in adaptive algorithms, the maximum number of subdivision steps to be performed. This parameter can be set by the user or determined automatically by the integration algorithm.

(ii) A *lower limit* on the *size of subregions*: If the discretization refinement reduces the size of a subregion to the distance between floating-point

[96]This function is integrable, though, so the fact that the singularity is not detectable is not crucial for numerical integration.

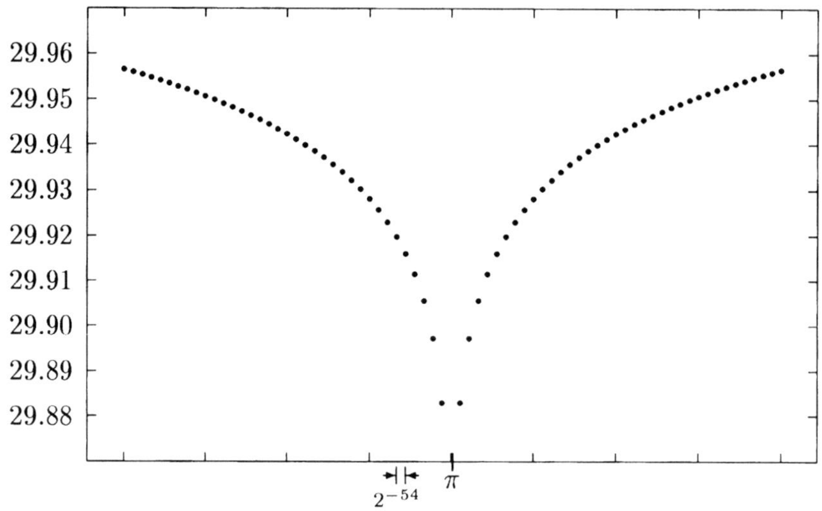

FIG. 8.6. *Numerical values of* $f(x) = 3x^2 + \pi^{-4}\log[(\pi - x)^2] + 1,\ x \in \mathbb{F}$ *(double precision IEC/IEEE floating-point arithmetic).*

numbers, it is not possible to increase the density of samples in this subregion any further. Thus integration programs often have a built-in termination criterion that tests the size of the smallest subregion. If a numerical integration algorithm terminates as a result of one subregion becoming too small, it returns the error condition "nonintegrable function." In practice, this error condition is returned not only for nonintegrable, but also for highly singular, yet integrable functions (e.g., $f(x) := x^{-0.99}$ on the interval $[0, 1]$); i.e., the incorrect diagnosis "nonintegrable function" can be returned for such integrands.

8.3.2. Efficiency Enhancement. Extrapolation methods are designed to improve the efficiency of numerical methods by transforming sequences $\{q_j\}$ of approximate values into sequences $\{\hat{q}_j\}$ which converge faster to the sought-after exact result, provided that the original sequences satisfy certain conditions. In conjunction with numerical integration, the exact result is the integral $\mathrm{I}f$ to be computed, and the approximate results $\{q_j\}$ are obtained using linear algorithms of the form (3.3), $\{q_j := \mathrm{Q}_{n_j} f\}$.

In Section 3.7, general aspects of acceleration methods, as well as a number of specific extrapolation schemes, are described. The applicability

TABLE 8.6. *Meta-algorithm for extrapolation in nonadaptive integration.*

```
j := 0;
ê_j := largest_machine_number;
do while ê_j > ε
    j := j + 1;
    calculate the approximate value q_j = Q_{n_j}(f; B);
    calculate the extrapolated value q̂_j = q^(j)(q_1, ..., q_j);
    calculate an error estimate ê_j for q̂_j
end do
```

of extrapolation methods to particular classes of integration formulas and the properties of the resulting numerical integral approximations are discussed in Sections 5.1.3 and 5.3.6. Extrapolation methods for special integration problems are the subject of Sections 7.2.4 and 7.3.4. This section, by contrast, focuses on how extrapolation methods can be properly embedded into the overall algorithmic structure of numerical integration programs.

Extrapolation in Adaptive and Nonadaptive Algorithms. The approximations $\{q_j\}$ used in the extrapolation process can be the results of either adaptive or nonadaptive algorithms. As can be seen from the meta-algorithm in Table 8.6, incorporating extrapolation methods into *nonadaptive* algorithms is straightforward, at least from a high-level algorithmic point of view.

Software (Extrapolation in Nonadaptive Algorithms). A nonadaptive extrapolation-based algorithm for calculating improper integrals over bounded or unbounded planar regions is described in Robinson and de Doncker [373].

The subroutine `NAG/d01paf` calculates integrals over N-dimensional simplices ($N \geq 2$) using the ε-algorithm for convergence acceleration purposes. The method applied is described in detail in de Doncker [100].

In *adaptive* integration algorithms, extrapolation schemes can be used within the basic integration module(s) to accelerate the convergence of integral approximations for each individual subregion $B_s \subseteq B$; this method is referred to as *local extrapolation*. They can also be used to accelerate the convergence of a sequence $\{q_j\}$ of integral approximations for the whole integration region, as generated by an instance of the meta-algorithm in Table 8.2; this method is referred to as *global extrapolation*.

Local Extrapolation. Basic integration modules calculate approximate integral values $\mathrm{Q}(f; B_s) \approx \mathrm{I}(f; B_s)$ and corresponding error estimates $\mathrm{E}(f; B_s)$ for given eligible input regions $B_s \subseteq B$. $\mathrm{Q}(f; B_s)$ is usually obtained by applying a suitable integration formula to the integrand f. Alternatively, $\mathrm{Q}(f; B_s)$ can be obtained by applying a sequence of integration formulas $\{\mathrm{Q}_{n_j}\}$

along with a corresponding extrapolation scheme to f. In this case, the basic integration module essentially implements an instance of the meta-algorithm in Table 8.6. The only necessary modification is that the statement

do while $\hat{e}_j > \varepsilon$

is replaced with

do while $\hat{e}_j > \varepsilon_s$ **and** $j \leq J$.

The first condition, $\hat{e}_j > \varepsilon_s$, checks for *local* convergence, while the second condition, $j \leq J$, bounds the maximum number of extrapolation steps.

The upper bound J on the maximum number of extrapolation steps is rather low, typically less than 10, mainly for the following two reasons. (i) The number of integrand evaluations usually increases by a constant factor with each additional extrapolation step and thus grows exponentially as a function of j. (ii) If the extrapolation process fails to converge rapidly, then it is likely that the integrand has local irregularities (discontinuities, peaks, etc.), which are dealt with most effectively by local discretization refinement. For both reasons, continuing the extrapolation process beyond a small number of steps generally proves less efficient than proceeding with the adaptive subdivision scheme. Note that adaptive integration algorithms may still converge globally even when local convergence cannot be attained on one or more subregions within J extrapolation steps.

Global Extrapolation. Extrapolation methods are used in numerical integration to accelerate the convergence of sequences $\{(k_j^N \times \mathrm{Q}_n)f \ : \ j = 0, 1, \ldots\}$ of compound integration formulas (see Sections 5.1.3 and 5.3.6). Adaptive subdivision algorithms, on the other hand, choose integration abscissas according to the behavior of each individual integrand. The resulting integral approximations do *not* normally correspond to compound integration formulas. It is therefore not at all clear how the convergence of adaptively calculated sequences of integral approximations can possibly be accelerated using extrapolation methods. Nevertheless, the seemingly incompatible concepts of extrapolation and adaptation can be reconciled as follows.

Consider a sequence of compound integration formulas

$$\{k_j^N \times \mathrm{Q}_n \ : \ k_j = 2^j, \, j = 0, 1, 2, \ldots\}.$$

Each formula $k_j^N \times \mathrm{Q}_n$ is obtained by partitioning B into L^j, $L := 2^N$, congruent subregions and applying a properly transformed copy of Q_n to each of the resulting subregions. Let

$$B^{s_1}, \qquad s_1 = 1, 2, \ldots, L,$$

denote the L subregions obtained by dividing B, and for $j = 2, 3, \ldots$, let

$$B^{s_1,\ldots,s_{j-1},s_j}, \qquad s_j = 1, 2, \ldots, L,$$

be defined as the subregions obtained by dividing $B^{s_1,\ldots,s_{j-1}}$. Now, assume that f is sufficiently smooth on a subregion $B^{s_1,\ldots,s_{j'}}$ for $\mathrm{Q}_n(f; B^{s_1,\ldots,s_{j'}})$ to be a highly accurate approximation to $\mathrm{I}(f; B^{s_1,\ldots,s_{j'}})$. If it is so, then subdividing $B^{s_1,\ldots,s_{j'}}$ and applying Q_n to each of the resulting subregions does not lead to a significant change in the approximate integral value; i.e., for any $j > j'$

$$\mathrm{Q}_n(f; B^{s_1,\ldots,s_{j'}}) \approx \mathrm{I}(f; B^{s_1,\ldots,s_{j'}}) \approx \sum_{s_{j'+1},\ldots,s_j=1}^{L} \mathrm{Q}_n(f; B^{s_1,\ldots,s_{j'},s_{j'+1},\ldots,s_j}).$$

Hence, while the computation of $(L^j \times \mathrm{Q}_n)f$ for $j > j'$ usually requires further subdivisions of $B^{s_1,\ldots,s_{j'}}$, such subdivisions are not necessary if f is sufficiently smooth on $B^{s_1,\ldots,s_{j'}}$.

In order to avoid such superfluous subdivisions during the jth extrapolation step, subregions $B^{s_1,\ldots,s_{j'}}$, $j' < j$, can be processed according to a globally adaptive strategy; i.e., the subregion $B^{s_1,\ldots,s_{j'}}$ chosen for further subdivision is required to possess the largest error estimate at the point of decision. Note that the subregions $B^{s_1,\ldots,s_j}$ resulting from j subdivisions of the original integration region B are *not* eligible for further subdivision as long as the jth extrapolation step has not been completed; they are temporarily rendered *inactive*. After the jth extrapolation step, these subregions are rendered *active* again.[97]

The globally adaptive subdivision process performed during the jth extrapolation step is terminated when the sum of the errors associated with all *active* subregions is less than a threshold ε_{e}. This means that $(L^j \times \mathrm{Q}_n)f$ is approximated with a presumed accuracy of ε_{e} in the jth extrapolation step. The threshold is chosen such that the errors in the extrapolated values, which originate from using approximations of $(L^j \times \mathrm{Q}_n)f$ instead of the exact values, can be expected to be small compared to the user-required tolerance ε. The algorithmic structure of the resulting integration programs is described by the meta-algorithm in Table 8.7 (see, for instance, Kahaner and Stoer [224]).

8.3.3. Software Based on Global Extrapolation. This section describes a number of integration programs which employ globally adaptive subdivision strategies enhanced by global extrapolation methods. The various subroutines are classified according to the type of integration problem (characterized by the integrand and the integration region) they solve.

[97] Note the difference between *inactive* subregions in locally adaptive methods and in adaptive methods using global extrapolation. In locally adaptive methods, subregions which are rendered inactive never become active again.

TABLE 8.7. *Meta-algorithm for global extrapolation in adaptive integration.*

```
j := 0;
q̂_j := q̄_j := Q_n(f; B);
ê_j := E(f; B);
insert (B, q̂_j, ê_j) into the data structure;
do while ê_j > ε
    j := j + 1;
    render all inactive subregions active;
    q̄_j := q̄_{j-1};
    ē_j := Σ_i e_i;
    do while ē_j > ε_e
        choose an active element of the data structure (index s)
        for partitioning;
        subdivide the chosen region B_s into subregions
            B_l,   l = 1, ..., L;
        calculate approximate values q_l :
            q_l := Q_n(f; B_l),   l = 1, ..., L;
        calculate error estimates e_l :
            e_l := E(f; B_l),   l = 1, ..., L;
        remove (B_s, q_s, e_s) from the data structure;
        q̄_j := q̄_j - q_s + (q_1 + ··· + q_L);
        ē_j := ē_j - e_s;
        if B_s is the result of j-1 subdivisions then
            render (B_1, q_1, e_1), ..., (B_L, q_L, e_L) inactive
        else
            render (B_1, q_1, e_1), ..., (B_L, q_L, e_L) active
            ē_j := ē_j + (e_1 + ··· + e_L);
        end if
        insert (B_1, q_1, e_1), ..., (B_L, q_L, e_L) into the data structure;
    end do
    calculate the extrapolated value q̂_j := q^(j)(q̄_0, ..., q̄_j);
    calculate an error estimate ê_j for q̂_j
end do
```

Univariate Integrand with Unknown Singularities. The globally adaptive integration programs QUADPACK/dqags $\approx$ IMSL/MATH-LIBRARY/dqdags evaluate univariate integrals over finite intervals using a basic integration module based on a 10/21-point Gauss–Kronrod formula pair. In addition, the ε-algorithm is used to accelerate the convergence of the sequence of integral approximations (see Section 3.7.4). The subintervals generated during the subdivision process are stored in a *sorted list*, and the storage space available for this data structure is allocated *statically*. If the preset maximum number of subintervals proves insufficient, the user must modify and recompile these routines.

This problem is overcome in the subroutine QUADPACK/dqagse. This subprogram is functionally identical to the aforementioned routines, but enables the calling (sub)program to pass the storage space available for the sorted list as subroutine arguments. CMLIB/dqags $\approx$ CMLIB/dqagse $\approx$ IMSL/MATH-LIBRARY/dq2ags $\approx$ NAG/d01ajf $\approx$ SLATEC/dqags $\approx$ SLATEC/dqagse are all identical to QUADPACK/dqagse, except for minor differences in the user interface.

The subroutine NAG/d01atf is identical to NAG/d01ajf except that it uses a user-supplied subroutine (rather than a function) to evaluate the integrand at an array of abscissas. This feature can lead to a performance improvement on vector computers, provided that the integrand evaluations can be vectorized (see Section 9.3.2).

The basic integration module of the globally adaptive integration program TOMS/691/dqxgs is based on QUADPACK/dqags, but uses an embedded sequence of four quadrature formulas in its basic integration module. The *stable, recursive, monotone formulas* employed are described in Favati, Lotti, and Romani [145].

Univariate Integrand with Singularities of Known Location. The subroutines QUADPACK/dqagp $\approx$ IMSL/MATH-LIBRARY/dqdagp are based on the same integration algorithm as QUADPACK/dqags. Additionally, they enable the user to specify the location of singularities in the interior of the integration interval. The integration interval is initially subdivided at these points, which usually improves the convergence rate of the ε-algorithm.

The subprograms QUADPACK/dqagpe $\approx$ CMLIB/dqagp $\approx$ CMLIB/dqagpe $\approx$ IMSL/MATH-LIBRARY/dq2agp $\approx$ NAG/d01alf $\approx$ SLATEC/dqagp $\approx$ SLATEC/dqagpe are functionally identical to the aforementioned programs, but enable the calling (sub)program to pass the storage space available as subroutine arguments.

Integrand with Trigonometric Weight Function. The globally adaptive integration programs QUADPACK/dqawo $\approx$ IMSL/MATH-LIBRARY/dqdawo evaluate univariate integrals over finite intervals with respect to the trigonometric weight functions $w(x) = \cos(\omega x)$ or $w(x) = \sin(\omega x)$. The parameter ω must be specified by the user. If the length of a subinterval exceeds a certain ω-dependent value, the basic integration module employs a 13/25-point modified Clenshaw–Curtis formula pair. Otherwise, a 7/15-point Gauss–Kronrod formula pair is used. If Gauss–Kronrod formula pairs are applied to one or more subintervals, the ε-algorithm is additionally used to accelerate the convergence of the sequence of integral approximations (see Section 3.7.4).

The subprograms `QUADPACK/dqawoe` $\approx$ `CMLIB/dqawo` $\approx$ `CMLIB/dqawoe` $\approx$ `IMSL/MATH-LIBRARY/dq2awo` $\approx$ `NAG/d01anf` $\approx$ `SLATEC/dqawo` $\approx$ `SLATEC/dqawoe` are functionally identical to the aforementioned programs, but enable the calling (sub)program to pass the storage space available as subroutine arguments.

The subroutines `QUADPACK/dqawf` $\approx$ `IMSL/MATH-LIBRARY/dqdawf` evaluate univariate integrals over semi-infinite intervals $[a, \infty)$ with respect to the trigonometric weight functions $w(x) = \cos(\omega x)$ or $w(x) = \sin(\omega x)$. The globally adaptive subroutine `QUADPACK/dqawoe` (or `IMSL/MATH-LIBRARY/dq2awo`) is applied to each of the finite subintervals $[a_k, b_k] := [a + (k-1)c, a + kc]$, $k = 1, 2, \ldots$, where $c := (2\lfloor|\omega|\rfloor + 1)\pi/|\omega|$. Since c equates to an odd number of half-periods, the integrals

$$\mathrm{I}_k f := \int_{a_k}^{b_k} w(x) f(x)\, dx$$

alternate in sign if f is nonnegative and monotonically decreasing over $[a, \infty)$. In this case, the ε-algorithm is known to accelerate the convergence of the alternating series

$$\mathrm{I}f = \sum_{k=1}^{\infty} \mathrm{I}_k f.$$

Both routines employ the ε-algorithm even if this series is *not* alternating.

The subprograms `QUADPACK/dqawfe` $\approx$ `CMLIB/dqawf` $\approx$ `CMLIB/dqawfe` $\approx$ `IMSL/MATH-LIBRARY/dq2awf` $\approx$ `NAG/d01asf` $\approx$ `SLATEC/dqawf` $\approx$ `SLATEC/dqawfe` are functionally identical to the aforementioned programs, but enable the calling (sub)program to pass the storage space available as subroutine arguments.

Integrals over Unbounded Regions. The subroutines `QUADPACK/dqagi` $\approx$ `IMSL/MATH-LIBRARY/dqdagi` use a globally adaptive subdivision strategy to evaluate univariate integrals over unbounded intervals. Semi-infinite intervals $[a, \infty)$ are first mapped onto the finite interval $[0, 1]$ transformed using the transformation $x = \psi(\overline{x}) = (\overline{x} - a)/(\overline{x} + a)$. The unbounded integration interval B is first transformed into $[0, 1]$: for $B = [a, \infty)$ the transformation $x = \psi(\overline{x}) = (\overline{x} - a)/(\overline{x} + a)$ is used; for $B = (-\infty, \infty)$ the integration interval is divided at the origin, and the aforementioned transformation is applied to both subintervals. The transformed integration problem is then solved with the slightly modified program `QUADPACK/dqags`: the pair of 10/21-point Gauss–Kronrod formulas is replaced with a pair of 7/15-point Gauss–Kronrod formulas. The reduced number of abscissas in this pair of Gauss–Kronrod formulas proves to be more efficient for integrands with singularities, arising, for instance, from the aforementioned transformation.

The subroutines `QUADPACK/dqagie` $\approx$ `CMLIB/dqagi` $\approx$ `NAG/d01amf` $\approx$ `SLATEC/dqagi` $\approx$ `SLATEC/dqagie` $\approx$ `CMLIB/dqagie` $\approx$ `IMSL/MATH-LIBRARY/dq2agi` return additional information about the integration process.

Bivariate Integrands on Triangular Regions. `TOMS/612/triex` uses a globally adaptive subdivision strategy for evaluating bivariate integrals over triangular regions. At each subdivision step, the routine divides the subtriangle with the largest error into four congruent subtriangles. The basic integration module in `TOMS/612/triex` is based on a formula pair with degrees of accuracy $D = 9$ and $D = 11$. In addition, the ε-algorithm is used in order to accelerate the convergence of integral approximations.

Multivariate Integrands on Cubic Regions. A fundamental problem with most multivariate extrapolation schemes is that the number of f-evaluations required to perform a given number of extrapolation steps increases exponentially as the dimension N of the integration problem grows (see Section 7.3.4). This phenomenon usually persists even when extrapolation methods are combined with globally adaptive subdivision schemes because subregions have to be divided *uniformly* into 2^N congruent subsubregions. In order to overcome this problem, Espelid [136], [137] proposes extrapolation-based globally adaptive numerical integration schemes which employ *nonuniform* subdivision strategies. The proposed methods are implemented in `NUMERALGO/na7`, which evaluates N-dimensional integrals over cubic regions for dimensions from $N = 2$ to $N = 15$. Algorithmic details can be found in Espelid and Genz [139].

8.3.4. Integration of Multiple Integrands. Practical applications often require a (possibly large) number m of integrals with the same weight function w and the same integration region B, but with different integrands $f_1, \ldots, f_m$:

$B \to \mathbb{R}$ to be computed (see, for instance, Section 2.1):

$$\begin{aligned} \mathrm{I}_w f_1 &= \int_B w(\mathbf{x}) f_1(\mathbf{x})\, d\mathbf{x}, \\ &\vdots \\ \mathrm{I}_w f_m &= \int_B w(\mathbf{x}) f_m(\mathbf{x})\, d\mathbf{x}. \end{aligned}$$

While such problems can be solved by calculating the m integrals separately, i.e., each in a separate run of an adaptive or nonadaptive integration program, a substantial amount of work can sometimes be saved by calculating the m integrals *simultaneously*, using a *single* program run.

Exploitation of Common Terms. The evaluation of the integrands $f_1, \ldots, f_m$ at fixed abscissas often involves several arithmetic expressions that are common to all functions $f_1, \ldots, f_m$ (see, for instance, Section 2.1.2). Evaluating such *common terms* only once for all m integrands can reduce the computational complexity of integration algorithms significantly.

If the integrands are evaluated *simultaneously* for fixed abscissas, then the exploitation of the common terms is easy: only the common terms belonging to a single abscissa need to be stored temporarily. However, if an integrand is evaluated at *several* integration abscissas $\{\mathbf{x}_i\}$ before the next integrand is evaluated at the same abscissas, then the common terms belonging to *all* $\{\mathbf{x}_i\}$ have to be stored temporarily, which results in an increased storage overhead and in a more complex algorithmic structure. Although the integrand level is conceptually the highest algorithmic level in numerical integration (see Section 9.1), it often forms the lowest algorithmic level in practical integration programs in order to facilitate the exploitation of common terms. In these cases the integrands are usually described by a subprogram which supplies the values $f_1(\mathbf{x}), \ldots, f_m(\mathbf{x})$ of the integrand for any specific input argument $\mathbf{x} \in B$. This algorithmic organization has the additional advantage of facilitating the vectorization of integration algorithms at the integrand level (see Section 9.3.2).

Example (Statistical Integration Problems). In the statistical integration problems described in Section 2.1, different simple polynomial functions $f_1, \ldots, f_m$ have to be integrated with respect to the same density function w. If the integrands are evaluated simultaneously for fixed abscissas, then the exploitation of the common term w is simple:

```
do for i ∈ {1,...,n}
    comterm := w(x_i);
    do for j ∈ {1,...,m}
        compute comterm · f_j(x_i)
    end do
end do
```

However, if each integrand is evaluated for n integration abscissas simultaneously, the exploitation of the common term w results in a more complex algorithmic structure and requires additional storage space of $O(n)$:

do for $j \in \{1, \ldots, m\}$

  do for $i \in \{1, \ldots, n\}$

    if $j = 1$ **then**

      $\text{comterm}(i) := w(\mathbf{x}_i)$

    end if

    compute $\text{comterm}(i) \cdot f_j(\mathbf{x}_i)$

  end do

end do

Adaptive Algorithms for Multiple Integrands. Adaptive integration algorithms choose the sequence of sampling points according to the observed behavior of the integrand. When m integrals are to be calculated simultaneously and the integrands differ significantly in their behavior, abscissas chosen appropriately for one integrand may be unsuitable for others. Hence, the question arises as to how the location of sampling points should be chosen in the presence of multiple integrands.

The approach taken by most adaptive integration algorithms is to tailor the sequence of sampling points to the behavior of the (locally) *most difficult* integrand; i.e., decisions concerning the subdivision of subregions are based on the maximum $\max\{\mathrm{E}(f_1; B_s), \ldots, \mathrm{E}(f_m; B_s)\}$ of the integrands' error estimates over the emerging subregions $\{B_s\}$. Note that in order to avoid unnecessary evaluations of well-behaved integrands, only families of functions exhibiting similar behavior should be integrated simultaneously using a single call to an adaptive integration program.

Chapter 9

Parallel Numerical Integration

The numerical solution of integration problems sometimes requires extensive computations. Substantial effort has therefore been invested into finding ways of exploiting the power of advanced computer systems like vector or parallel computers (SIMD computers and shared memory and distributed memory MIMD machines). Even network-based distributed computing has been used.

This chapter deals with the development of numerical integration software for advanced computer systems, particularly for parallel and distributed computer architectures. A classification of the level and degree of parallelism involved in numerical integration programs is given in Section 9.1. Different parallel integration programs may exploit the available parallelism in totally different ways. Section 9.2 introduces a notation for describing the way in which various sections of numerical integration algorithms are actually parallelized in a particular program. Additionally, the performance-relevant characteristics of tasks arising in parallel numerical integration are discussed. Particular attention is given to the problem of load distribution, i.e., the problem of allocating tasks to computational resources, which is of crucial importance to the performance of parallel numerical integration programs. Section 9.3 reviews the parallelization schemes that have already been implemented and tested on advanced computer systems.

This chapter is supposed to provide a concise introduction to the development of numerical integration software for advanced computer systems. Readers who wish to further extend their knowledge of this subject are referred to the monograph by Krommer and Ueberhuber [247].

9.1. Parallelism in Integration Algorithms

This section provides a classification of the level and degree of potential parallelism inherent in numerical integration programs. In addition, methods used to increase the degree of parallelism are introduced, and their impact on the efficiency of numerical integration algorithms is investigated.

Figure 9.1 gives a graphical representation of the hierarchical structure

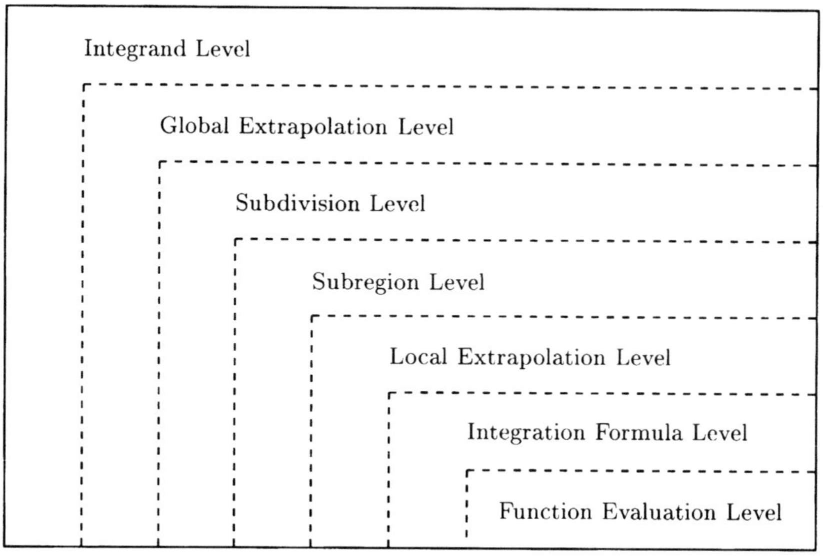

FIG. 9.1. *Hierarchy of algorithmic levels in numerical integration programs.*

TABLE 9.1. *Integrand level meta-algorithm.*

do for $j \in \{1, \ldots, m\}$

compute an approximate value q_j *for* $\mathrm{I}f_j$ *with*
$|q_j - \mathrm{I}f_j| \leq \varepsilon_j$

end do

of numerical integration programs (see Chapter 8). Specific integration algorithms usually incorporate only a subset of these hierarchical levels. In this chapter, the type and degree of parallelism found on each algorithmic level are discussed in a top-down manner.

Note that the main objective when parallelizing *any* algorithmic level is to make it possible to perform integrand evaluations concurrently. This is because the largest portion of computer time in numerical integration programs is spent on evaluating the integrand(s). Therefore, any effective parallelization strategy must aim to minimize the execution time attributable to f-evaluations.

9.1.1. Integrand Level. The integrand level occurs in numerical integration algorithms when m functions $f_1, \ldots, f_m$ are integrated simultaneously on a single integration region B using a single weight function w (see Section 8.3.4). The simple meta-algorithm in Table 9.1 describes the computational structure of this algorithmic level. Since the $f_1, \ldots, f_m$ are generally independent, the computations of $q_1, \ldots, q_m$ can be performed in parallel.

The number m of integrals (i.e., the degree of parallelism at the integrand

TABLE 9.2. *Global extrapolation level meta-algorithm.*

```
r := 0;
ê_{j,r} := largest_machine_number;
do while ê_{j,r} > ε_j
    r := r + 1;
    calculate the approximate value q_{j,r} for If_j;
    calculate the extrapolated value q̂_{j,r} = q^{(r)}(q_{j,1}, ..., q_{j,r});
    calculate an error estimate ê_{j,r} for q̂_{j,r}
end do
q_j := q̂_{j,r}
```

level) is a characteristic property of a given integration *problem* and in general *cannot* be increased by modifying the integration *algorithm*. Large numbers of integrals occur, for instance, in statistical applications (see Section 2.1). On the other hand, it often happens that only a single integral needs to be computed, in which case no parallelism can be obtained at the integrand level at all.

The number m of integrals can only be increased by algorithmic means if the program calling an integration subroutine is itself an integration routine. For instance, iterated, multidimensional integrals can be computed by recursively applying integration programs to lower-dimensional integration problems for the inner integrals. In this case the number m of integrals in a low-dimensional subroutine is equal to the number of integrand values required by the high-dimensional calling program. By raising this number, the degree of parallelism at the integrand level of the low-dimensional subroutine can be increased.

It should be noted that the integrand level, while conceptually being the highest algorithmic level in numerical integration, usually forms the lowest algorithmic level in practical integration programs in order to maximize the computational savings that can be made by treating several integrands simultaneously (see Section 8.3.4). In such cases it is generally impossible or impractical to exploit the inherent parallelism at the integrand level (see Section 9.2).

9.1.2. Global Extrapolation Level. The global extrapolation level in numerical integration algorithms (see Section 8.3.2) can be described by the meta-algorithm in Table 9.2. Here, most of the time is spent on the computation of the approximate values $\{q_{j,r}\}$ using adaptive or nonadaptive algorithms.

The number of extrapolation steps r required to compute an approximation

TABLE 9.3. *Subdivision level meta-algorithm.*

```
e_{j,r} := largest_machine_number;
do while e_{j,r} > ε_{j,r}
    choose a subregion B_{j,r,s} ⊆ B;
    process B_{j,r,s};
    update q_{j,r} and e_{j,r}
end do
```

of the requested accuracy is generally *not* known in advance. For instance, it might happen that one single, *trivial*, extrapolation step $\hat{q}_{j,1} = q_{j,1}$ is sufficient to meet the given error requirement. Therefore, if several approximate values $q_{j,r_1}, \ldots, q_{j,r_2}$, $r_1 < r_2$, are computed concurrently, then $\hat{e}_{j,r_1}$ may possibly satisfy the given error requirement already, i.e., $\hat{e}_{j,r_1} \leq \varepsilon_j$. In such cases, all the additional effort spent on evaluating $q_{j,r_1+1}, \ldots, q_{j,r_2}$ is superfluous. Given this observation, the computation of $q_{j,1}, q_{j,2}, \ldots$ is *inherently* sequential.

The only remedy for such unnecessary computations when concurrently calculating the sequence $q_{j,1}, q_{j,2}, \ldots$ would be a reliable prediction of the number of extrapolation steps still required. For instance, if the error requirement is almost met by $\hat{e}_{j,r_1}$, an efficient algorithm should only evaluate q_{j,r_1+1} in the next step and not several $q_{j,r_1+1}, \ldots, q_{j,r_2}$ concurrently. On the other hand, if $\hat{e}_{j,r_1}$ is significantly larger than ε_j, $\hat{e}_{j,r_1} \gg \varepsilon_j$, then it normally makes sense to compute several $q_{j,r_1+1}, \ldots, q_{j,r_2}$ in parallel. However, these considerations are purely theoretical in nature. To the authors' knowledge the feasibility of this approach has *not* been tested in practice.

The maximum number of extrapolation steps is usually fairly small, hardly ever larger than 10. As a result, the potential parallelism at the global extrapolation level is, apart from being data-dependent, quite limited.

9.1.3. Subdivision Level. The level of subdividing (parts of) the integration region in numerical integration algorithms (see Section 8.2.2) can be described by the meta-algorithm in Table 9.3. The selection of the subregion $B_{j,r,s}$ is governed by an adaptive sampling strategy. Processing the subregion $B_{j,r,s}$ usually consists of subdividing $B_{j,r,s}$ and applying an appropriate basic integration module to the resulting subsubregions (see Table 8.2).

Parallelism at the subdivision level means that several regions may possibly be processed (i.e., subdivided, etc.) at the same time. The extent of the potential parallelism at the subdivision level strongly depends on the nature of the integrand. The region for which the greatest computational effort is required may be very small, for instance, when the integrand has a single peak or singularity. In such cases, only a small number of regions are worth subdividing and

TABLE 9.4. *Meta-algorithm for the main loop in* QUADPACK/dqag.

do while $e > \varepsilon$
choose subinterval I_s *with largest error estimate* e_s *from the data structure*
subdivide I_s *into* I_1 *and* I_2;
calculate approximate values q_l :
$q_l := \mathrm{Q}(f; I_l), \quad l = 1, 2;$
calculate error estimates e_l :
$e_l := \mathrm{E}(f; I_l), \quad l = 1, 2;$
remove (I_s, q_s, e_s) *from the data structure*;
insert (I_1, q_1, e_1) *and* (I_2, q_2, e_2) *into the data structure*;
$q := q - q_s + (q_1 + q_2)$;
$e := e - e_s + (e_1 + e_2)$
end do

processing at the same time. In such cases, the lack of potential parallelism is due to the peculiarities of a given integrand and does *not* depend on the kind of adaptive strategy applied in the algorithm.

However, parallelism at the subdivision level may also depend on the nature of the adaptive strategy. In *locally* adaptive strategies, the order in which active subregions are processed does not usually play a crucial role in the efficiency of the algorithm (see Section 8.2.2). Hence, affluent parallelism may be available at the subdivision level in locally adaptive algorithms if there is a large number of regions worth being subdivided (as in the case of heavy oscillations). On the other hand, traditional *globally* adaptive strategies allow only one region to be processed at a time (even if there is a large number of regions worth being subdivided), which leads to inherently sequential algorithms.

For globally adaptive algorithms, parallelism at the subdivision level can only be achieved by increasing the number of (sub)regions eligible for simultaneous processing. This kind of parallelization technique requires a modification of the underlying adaptive algorithm and thus may increase the total volume of work, i.e., the number of f-evaluations performed.

Example (Parallelism at the Subdivision Level). The algorithmic structure of the globally adaptive integration program QUADPACK/dqag is given in Table 9.4. The parallelization of QUADPACK/dqag (and all other adaptive algorithms) at the subdivision level is relatively easy if a single task pool method (see Section 9.2.2) is applied, i.e., if all subinterval data are stored in one shared data structure: the main loop of the meta-algorithm in Table 9.4 is simply performed concurrently by each node of the parallel system. Accesses to the shared data structure are appropriately synchronized in order to preserve its integrity.

If this parallelization method is applied to a globally adaptive algorithm, each processor selects the subregion B_s which possesses the largest error estimate among all subregions not being processed by any other node at the moment of decision. Thus, the efficiency of the resulting parallel algorithm (measured by the overall number of f-

evaluations performed) can be expected to be optimal in the class of globally adaptive algorithms parallelized at the subdivision level. Nevertheless, the total amount of work may be (substantially) larger than that of its sequential counterpart.

Experiments were carried out on a Sequent Balance 21000 shared memory computer with 28 processors. The basic integration module `QUADPACK/dqk15`, which implements a 7/15-point Gauss–Kronrod formula pair, was used in conjunction with the parallelized version of `QUADPACK/dqag` to numerically compute the integral

$$If := \int_0^1 x^{-0.9}\, dx = 10, \tag{9.1}$$

whose integrand has an integrable, endpoint singularity at $x = 0$. The following table shows the number of function evaluations needed to attain the relative accuracy $\varepsilon_{rel} = 10^{-9}$ for different numbers P of available processors. In addition, the *potential* speed-up and the efficiency (see Section 10.1.2) are given. The potential speed-up indicates the maximum acceleration possible assuming that any overhead due to parallel execution is insignificant.

P	n	*Potential* speed-up	Efficiency
1	8715	1.00	1.00
2	17415	1.00	0.50
3	26025	1.00	0.33

Clearly, there was no potential speed-up at all: since there was only one subinterval worth being subdivided at any stage of the subdivision procedure, only *one* processor could be employed usefully. All other processors were kept busy with unnecessary work.

The following table shows the results of a similar experiment with

$$If := \int_0^1 \cos^2(cx)\, dx \approx 0.5. \tag{9.2}$$

To enable a meaningful comparison, the parameter c was chosen in such a way that the number of required function evaluations in the sequential algorithm was roughly the same as for (9.1).

P	n	*Potential* speed-up	Efficiency
1	8565	1.00	1.00
5	8685	4.93	0.99
10	8805	9.73	0.97
15	8925	14.39	0.96
20	8955	19.13	0.96
25	8985	23.83	0.95

Even when using a large number of processors, the efficiency of the parallel algorithm does not substantially deteriorate. Due to the uniformly oscillatory nature of the integrand in (9.2), a large number of subintervals can usefully be subdivided during most of the computational procedure.

TABLE 9.5. *Subregion level meta-algorithm.*

subdivide the chosen region $B_{j,r,s}$ *into subregions*
$B_{j,r,s,l}, \quad l = 1, \ldots, L;$
do for $l \in \{1, \ldots, L\}$
$q_{j,r,s,l} := \mathrm{Q}(f; B_{j,r,s,l});$
$e_{j,r,s,l} := \mathrm{E}(f; B_{j,r,s,l})$
end do

9.1.4. Subregion Level. The subregion level in numerical integration algorithms (see Section 8.2.2) can be described by the meta-algorithm in Table 9.5. The subregions $B_{j,r,s,1}, \ldots, B_{j,r,s,L}$ can be processed independently; i.e., $(q_{j,r,s,1}, e_{j,r,s,1}), \ldots, (q_{j,r,s,L}, e_{j,r,s,L})$ can be calculated independently, and can therefore also be processed in parallel.

However, the number of subregions created in a single subdivision step is usually quite small. For instance, intervals, and more generally, cubic regions, are normally only split into two subintervals (subcubes). In order to increase the parallelism at the subregion level, the number L of subregions into which a selected region is partitioned can be raised. The impact that such modifications have on the efficiency of adaptive integration algorithms depends on the particular characteristics of the integrand.

Example (Parallelism at the Subregion Level). The subroutine `QUADPACK/dqag` (see Table 9.4) was modified in such a way that the number L of subintervals into which I_s was subdivided could be varied between different program runs. The following table shows the number of function evaluations needed to solve the endpoint singularity problem (9.1) for different values of L. (The relative accuracy requirement was chosen as $\varepsilon_{rel} = 10^{-9}$.) In addition, the number of function evaluations per processor and the potential speed-up are given. It is assumed that, after a chosen interval I_s has been divided, the resulting subintervals $I_1, \ldots, I_L$ are processed by different processors.

L	n	n/L	*Potential* speed-up	Efficiency
2	8715	4357	2.00	1.00
3	8250	2750	3.17	*1.06*
4	8715	2718	4.00	1.00
8	11655	1456	5.99	0.75
16	17535	1096	7.95	0.50
32	27855	870	10.01	0.31

It is interesting to note that *trisection* ($L = 3$) performs better than the *bisection* ($L = 2$) strategy used in `QUADPACK/dqag`. Trisection leads to a reduction in the overall number of required function evaluations, a fact confirming the observations in Hanke [198] and Soerevik [415] (see Section 8.2.2). As a result, linear and even superlinear speed-ups can be attained for small values of $L = P$. However, for larger values of L, the number of function evaluations increases roughly in proportion to

TABLE 9.6. *Local extrapolation level meta-algorithm.*

```
k := 0;
ê_{j,r,s,l,k} := largest_machine_number;
do while ê_{j,r,s,l,k} > ε_s and k ≤ K_s
    k := k + 1;
    q_{j,r,s,l,k} := Q_{n_k}(f; B_{j,r,s,l});
    calculate the extrapolated value
        q̂_{j,r,s,l,k} := q^{(k)}(q_{j,r,s,l,1}, ..., q_{j,r,s,l,k});
    calculate an error estimate ê_{j,r,s,l,k} for q̂_{j,r,s,l,k}
end do
q_{j,r,s,l} := q̂_{j,r,s,l,k};
e_{j,r,s,l} := ê_{j,r,s,l,k}
```

$L/\log L$. Correspondingly, the speed-up curve grows in proportion to $\log L$ only; doubling L only results in an increase of 2 in the potential speed-up. A theoretical explanation for this behavior can be found, for instance, in Minsky and Papert [298].

9.1.5. Local Extrapolation Level. The local extrapolation level in numerical integration algorithms (see Section 8.3.2) can be described by the meta-algorithm in Table 9.6. The approximate values $q_{j,r,s,l,k}$ result from applying integration formulas Q_{n_k} to subregions $B_{j,r,s,l} \subseteq B$. In contrast, the corresponding values $q_{j,r}$ at the global extrapolation level result from applying nonadaptive or adaptive algorithms to the whole integration region B.

Most of the computational effort in the meta-algorithm in Table 9.6 is required for evaluating the integration formulas $Q_{n_1}, Q_{n_2}, \ldots$. All considerations concerning parallelism at the global extrapolation level apply also to the local extrapolation level (see Section 9.1.2).

9.1.6. Integration Formula Level. Integration formulas

$$Q_{n_k} f = \sum_{i=1}^{n_k} c_{k,i} f(\mathbf{x}_i)$$

are usually specified for standard regions (see Section 3.1.1) and have to undergo appropriate (preferably, affine) transformations in order to be applicable to other integration regions (see Section 6.1.3). The integration formula level in numerical integration programs can therefore be described by the meta-algorithm in Table 9.7. Since the sampling points $\mathbf{x}_{k,1}, \ldots, \mathbf{x}_{k,n_k}$ are chosen independently, the quantities

$$f(A_{j,r,s,l}\mathbf{x}_{k,1} + \mathbf{b}_{j,r,s,l}), \ldots, f(A_{j,r,s,l}\mathbf{x}_{k,n_k} + \mathbf{b}_{j,r,s,l})$$

TABLE 9.7. *Integration formula level meta-algorithm.*

$q_{j,r,s,l,k} := 0;$

do for $i \in \{1, \ldots, n_k\}$

$q_{j,r,s,l,k} := q_{j,r,s,l,k} + |\det A_{j,r,s,l}| c_{k,i} f_j(A_{j,r,s,l}\mathbf{x}_{k,i} + \mathbf{b}_{j,r,s,l})$

end do

can be computed in parallel. The number n_k of independent function evaluations depends strongly on the dimension N of the integral. For instance, the largest value found in the univariate integration package QUADPACK is $n_k = 87$, whereas in the multidimensional integration subroutine TOMS/698/cuhre, the default integration rule uses $n_k = 37{,}789$ abscissas for the highest admissible dimension $N_{\max} = 15$.

In order to increase the degree of parallelism, one would like to raise the number n_k of abscissas in integration formulas, i.e., apply integration formulas which use larger numbers of abscissas. There are, however, two fundamental difficulties with this approach.

(i) Efficient formulas with larger numbers of abscissas may not be readily available: the construction of some classes of integration rules, mainly multivariate polynomial rules (see Section 6.2), poses both theoretically and practically substantial problems. The problems become more severe and eventually insurmountable as the number n_k of abscissas increases.

(ii) Changing the number of abscissas used in integration formulas has an impact on the efficiency of integration algorithms. For instance, adaptive algorithms integrate singular functions most efficiently when the number of abscissas used in integration formulas is small. For strongly oscillatory integrands, on the other hand, using integration rules with large numbers of abscissas generally proves to be most efficient.

Example (Parallelism at the Integration Formula Level). The following table shows the number n of function evaluations performed by the globally adaptive integration program QUADPACK/dqag when calculating the singular integral (9.1). The values of n are given as a function of the number of abscissas n_k in the basic integration module QUADPACK/dqkn_k, $n_k = 15, 21, \ldots, 61$, used by QUADPACK/dqag. In addition, the numbers n_b of calls to the basic integration module are given.

n_k	15	21	31	41	51	61
n	8715	12243	18197	24149	30039	36051
n_b	581	583	587	589	589	591

It can be seen that n_b remains practically unchanged when n_k and, hence, the accuracy of the basic integration module increase. This leads to a dramatic decrease in the efficiency of QUADPACK/dqag.

It can generally be observed that adaptive algorithms are most efficient at approximating singular integrals when the numbers of abscissas in their basic integration modules are chosen as low as possible. Note, however, that the *reliability* of the resulting programs declines as n_k is reduced. n_k should therefore not be chosen too small.

The following table shows the corresponding results for the computation of the integral (9.2), which is characterized by a uniformly oscillatory integrand.

n_k	15	21	31	41	51	61
n	8565	5355	3937	5207	3213	3843
n_b	571	255	127	127	63	63

Here, a significant reduction can be observed in n_b when n_k and, hence, the accuracy of the basic integration module are increased. Optimum efficiency is achieved using a relatively large number of abscissas.

9.1.7. Integrand Evaluation Level. Even the evaluation of an integrand f at a single abscissa $\mathbf{x} \in B$ may involve some parallelism. In particular, the evaluation of multidimensional integrands often requires one or several parallelizable loops over the dimension of the integration problem.

Example (Parallelism at the Integrand Evaluation Level). A computational kernel frequently found in multidimensional integration is

$$f(\mathbf{x}) = f(x_1, \ldots, x_N) = \exp\left(\sum_{g=1}^{N} a_g^2 x_g^2\right).$$

The evaluation of the terms $a_1^2 x_1^2, \ldots, a_N^2 x_N^2$ can be performed in parallel.

It is not possible for integration programs to exploit any potential parallelism at the integrand evaluation level because the integrand is provided by the user as an external subprogram (which supplies the value $f(\mathbf{x})$ of the integrand for any given input argument $\mathbf{x} \in B$). Instead, the supplier of the subprogram bears the responsibility for exploiting this kind of parallelism.

9.2. Parallelization Schemes for Integration Algorithms

Section 9.1 classifies the level and quantifies the degree of parallelism involved in numerical integration programs. Parallel integration programs may exploit the available parallelism in totally different ways by employing different *parallelization schemes*. This section provides a general discussion of parallelization schemes for numerical integration programs.

Section 9.2.1 discusses the structure of and introduces a simple notation for identifying various parallelization schemes. In Section 9.2.2, particular attention is given to the problem of distributing the resulting parallel tasks to the available processors. The constituent algorithmic sections and the performance-relevant characteristics of the tasks in parallel numerical integration are described in Sections 9.2.3 and 9.2.4, respectively.

TABLE 9.8. *Template for describing parallel integration algorithms.*

Level	Parallelization
subdivision	$strategy_s$
subregion	$strategy_l$
integration formula	$strategy_i$

TABLE 9.9. *Parallelization at the integration formula level.*

Level	Parallelization
subdivision	sequential
subregion	sequential
integration formula	**parallel**

9.2.1. General Structure of Parallelization Schemes. Krommer and Ueberhuber [247] introduced a concise notation in tabular form to indicate how the various sections of numerical integration algorithms are actually parallelized for specific programs. For the purposes of this book, a simplified version of this notational system, represented by the template in Table 9.8, is sufficient. The parallelization strategy applied at a particular algorithmic level is characterized by one of the attributes *parallel*, *task*, or *sequential*. These key words signify the way in which the iterations of the main loop of this algorithmic level are performed: *parallel* indicates that the iterations are, either individually or in groups, performed concurrently by different processors, whereas *task* and *sequential* both signify that the iterations are performed sequentially, one after the other. Specifically, *task* means that the corresponding iterations are performed *within* the parallelized loop iterations, while *sequential* means that the iterations are performed *outside* the parallelized loops. In other words, the *parallel* loop iterations, with the *task* iterations incorporated into them, constitute the *parallel tasks* into which the overall work in numerical integration programs is partitioned.[98]

The algorithmic levels omitted from Table 9.8 are, for the reasons described in Sections 9.1.1, 9.1.2, and 9.1.7, generally not considered suitable for parallelization. Using the above terminology, the iterations associated with these algorithmic levels have the attributes *task* or *sequential*, depending on whether the corresponding loops are performed within or outside the parallelized loops.

Example (Parallelization at the Integration Formula Level). Table 9.9 indicates that the iterations of the loop

[98]Note that the simple template in Table 9.8 can easily be extended (by including additional attribute columns) to include information on the load distribution scheme, i.e., on the mechanism employed to assign parallel tasks to the available processors. For the purposes of this book, however, the simple template in Table 9.8 proves sufficient.

TABLE 9.10. *Parallelization scheme using loop permutation.*

Level	Parallelization
subdivision	sequential
subregion	task
integration formula	**parallel**

do for $i \in \{1, \ldots, n_k\}$

$$q_{j,r,s,l,k} := q_{j,r,s,l,k} + |\det A_{j,r,s,l}| c_{k,i} f_j(A_{j,r,s,l}\mathbf{x}_{k,i} + \mathbf{b}_{j,r,s,l})$$

end do

of the meta-algorithm in Table 9.7 are performed concurrently by different processors; thus Table 9.9 represents algorithms which are parallelized at the integration formula level.

Most parallel integration programs have the following generic computational structure. The uppermost algorithmic levels are performed *sequentially*, some intermediate levels *concurrently*, and the lowest levels are incorporated into parallel tasks. The parallelization strategy given in Table 9.9, for instance, conforms to this generic structure. Parallel integration programs which adhere to this generic scheme can be obtained from their sequential counterparts quite naturally. Only the loops which correspond to parallelized algorithmic levels have to performed concurrently rather than sequentially; other changes are not required.

In order to obtain a parallel integration program which *cannot* be subsumed under this general model, one has to change the order in which the loops corresponding to different algorithmic levels appear in the program; i.e., one must perform a *loop permutation*. For a number of reasons, performing such loop permutations may be difficult, detrimental, or even impossible.[99] (i) A significant effort may be required to modify the loop order because the whole integration program may have to be restructured. (ii) Loop permutations can result in increased storage and/or communication requirements (see the example on page 367). (iii) The conditional statement **do while** in the loops at the subdivision level and at the extrapolation levels can only be executed when the corresponding loop body, which includes all inner loops, has been performed completely. Hence, the inner loops cannot be moved out of the main **do while** loop. (iv) Moving an outer loop into a **do while** loop may adversely affect the algorithmic efficiency of an integration program.

Example (Parallelization Strategy based on Loop Permutation). Ta-

[99] An exception is the loop permutation required to render the integrand level the lowest algorithmic level in integration programs. This loop permutation is often performed in order to maximize the computational savings that can be made by treating several integrands simultaneously (see Section 8.3.4).

ble 9.10 represents algorithms which are parallelized at the integration formula level. Algorithms represented by Table 9.9 are different in that the l loop (i.e., the loop over the subregions created in a subdivision step) is performed within the concurrent i loop:

do for $i \in \{1, \ldots, n_k\}$

  do for $l \in \{1, \ldots, L\}$

    $q_{j,r,s,l,k} := q_{j,r,s,l,k} + |\det A_{j,r,s,l}|\, c_{k,i}\, f_j(A_{j,r,s,l}\mathbf{x}_{k,i} + \mathbf{b}_{j,r,s,l})$

  end do

end do

Interchanging the i and the l loops requires a significant amount of recoding because the two loops are usually in different program modules: the i loop is part of the basic integration module, whereas the l loop is not, and so interchanging the loops requires a modification of the modular concept used in numerical integration programs.

More recoding effort is required when the values $\{q_{j,r,s,l,k} = \mathrm{Q}_{n_k}(f_j; B_{j,r,s,l})\}$ are to be used in a local extrapolation scheme. According to the meta-algorithm in Table 9.6, the sets of integration formulas $\{\mathrm{Q}_{n_k}\}$ to be applied for the different subregions $B_{j,r,s,1}, \ldots, B_{j,r,s,L}$ may vary, whereas using the above double loop, the same set of integration formulas is evaluated for all subregions $B_{j,r,s,1}, \ldots, B_{j,r,s,L}$. The easiest way to overcome this problem is to use the same integration formula for each subroutine by changing the meta-algorithm in Table 9.6 to

do while $\max\{\hat{e}_{j,r,s,l,k} : l = 1, 2, \ldots, L\} > \varepsilon_s$ **and** $k \leq K_s$

  $k := k + 1$;

  calculate $q_{j,r,s,l,k}$;

  calculate the extrapolated value

    $\hat{q}_{j,r,s,l,k} := q^{(k)}(q_{j,r,s,l,1}, \ldots, q_{j,r,s,l,k})$;

  calculate an error estimate $\hat{e}_{j,r,s,l,k}$ *for* $\hat{q}_{j,r,s,l,k}$

end do

Using this algorithmic organization, the number of extrapolation steps performed for a subregion $B_{j,r,s,l}$ is independent of the index $l \in \{1, 2, \ldots, L\}$ and is equal to the maximum number of extrapolation steps required to attain convergence for any of the subregions $B_{j,r,s,1}, \ldots, B_{j,r,s,L}$. Therefore, the efficiency of the resulting integration algorithm can be significantly worse than the efficiency of the original algorithm given in Table 9.6 if the number of extrapolation steps required to attain convergence depends strongly on the subregion.

9.2.2. Load Distribution. Section 9.2.1 describes the different ways in which the overall workload imposed on a computer system by a numerical integration program can be partitioned into independent tasks. This section deals with the problem of *load balancing*, or *load distribution*, i.e., with the question of how the resulting parallel tasks should be distributed to the processors of a parallel or distributed computer system if a maximum (or at least satisfactory) program performance is to be achieved.[100] Due to space limitations, only a very coarse outline of this subject can be given in this book. For thorough discussions, see Krommer and Ueberhuber [243], [247].

Broadly speaking, any load balancing scheme can be classified as being either *static* or *dynamic*. *Static* load balancing uses only a priori information about the tasks and about the parallel system, i.e., information known prior to any task execution. Static load balancing performs well as long as the main performance-relevant characteristics of the tasks and of the parallel system are known in advance (see Section 9.2.4). *Dynamic* load balancing, on the other hand, reacts to the current system state during the execution of parallel tasks, which is why dynamic load balancing is also called *adaptive* load balancing. By dynamically redistributing work, dynamic load balancing can improve the performance of parallel systems when crucial performance parameters are unpredictable.

Static Load Distribution. The most important static load balancing schemes are based on *block-cyclic* distributions, which can be described as follows. To begin with, suppose the set of tasks to be concurrently performed lends itself to an M-dimensional parameterization:

$$\{\mathrm{Ta}_{i_1,\ldots,i_M} \;:\; i_k = 1,2,\ldots,m_k,\; k=1,2,\ldots,M\}.$$

Example (Lattice Rules). A rank-m lattice rule (see Section 6.5.3) can be represented in the form

$$\mathrm{Q}_n f = \frac{1}{n_1\cdots n_m}\sum_{i_1=0}^{n_1-1}\cdots\sum_{i_m=0}^{n_m-1} f\left(\frac{i_1}{n_1}\mathbf{z}_1+\cdots\frac{i_m}{n_m}\mathbf{z}_m\right).$$

Hence, evaluating $\mathrm{Q}_n f$ consists of the m-dimensional set of tasks

$$\left\{\mathrm{Ta}_{i_1,\ldots,i_m} := \text{compute } f\left(\frac{i_1}{n_1}\mathbf{z}_1+\cdots\frac{i_m}{n_m}\mathbf{z}_m\right) \;:\; i_k=0,\ldots,n_k-1,\; k=1,\ldots,m\right\}.$$

Now, assume that the set of processors is parameterized by a multiindex of the same dimension:

$$\{p_{j_1,\ldots,j_M} \;:\; j_k = 1,2,\ldots,n_k,\; k=1,2,\ldots,M\}.$$

In a $(b_1,\ldots,b_M)$-*block-cyclic* distribution ($b_1,\ldots,b_M\in\mathbb{N}$), tasks $\mathrm{Ta}_{i_1,\ldots,i_M}$ are mapped onto processors $p_{j_1,\ldots,j_M}$ according to

$$j_k = 1 + (((i_k-1)\ \mathrm{div}\ b_k)\ \mathrm{mod}\ n_k),\qquad k=1,2,\ldots,M. \tag{9.3}$$

[100] Performance indices most heavily used in the context of parallel integration are discussed in Section 10.1.2.

TABLE 9.11. $(2,2)$-*block-cyclic distribution of tasks.*

	$\mathrm{Ta}_{*,1}$	$\mathrm{Ta}_{*,2}$	$\mathrm{Ta}_{*,3}$	$\mathrm{Ta}_{*,4}$	$\mathrm{Ta}_{*,5}$
$\mathrm{Ta}_{1,*}$	$p_{1,1}$	$p_{1,1}$	$p_{1,2}$	$p_{1,2}$	$p_{1,1}$
$\mathrm{Ta}_{2,*}$	$p_{1,1}$	$p_{1,1}$	$p_{1,2}$	$p_{1,2}$	$p_{1,1}$
$\mathrm{Ta}_{3,*}$	$p_{2,1}$	$p_{2,1}$	$p_{2,2}$	$p_{2,2}$	$p_{2,1}$
$\mathrm{Ta}_{4,*}$	$p_{2,1}$	$p_{2,1}$	$p_{2,2}$	$p_{2,2}$	$p_{2,1}$
$\mathrm{Ta}_{5,*}$	$p_{1,1}$	$p_{1,1}$	$p_{1,2}$	$p_{1,2}$	$p_{1,1}$

TABLE 9.12. *Cyclic distribution of tasks.*

	$\mathrm{Ta}_{*,1}$	$\mathrm{Ta}_{*,2}$	$\mathrm{Ta}_{*,3}$	$\mathrm{Ta}_{*,4}$	$\mathrm{Ta}_{*,5}$
$\mathrm{Ta}_{1,*}$	$p_{1,1}$	$p_{1,2}$	$p_{1,1}$	$p_{1,2}$	$p_{1,1}$
$\mathrm{Ta}_{2,*}$	$p_{2,1}$	$p_{2,2}$	$p_{2,1}$	$p_{2,2}$	$p_{2,1}$
$\mathrm{Ta}_{3,*}$	$p_{1,1}$	$p_{1,2}$	$p_{1,1}$	$p_{1,2}$	$p_{1,1}$
$\mathrm{Ta}_{4,*}$	$p_{2,1}$	$p_{2,2}$	$p_{2,1}$	$p_{2,2}$	$p_{2,1}$
$\mathrm{Ta}_{5,*}$	$p_{1,1}$	$p_{1,2}$	$p_{1,1}$	$p_{1,2}$	$p_{1,1}$

If $b_1 = b_2 = \cdots = b_M = 1$, then formula (9.3) becomes

$$j_k = 1 + ((i_k - 1) \bmod n_k), \qquad k = 1, 2, \ldots, M,$$

and the corresponding distributions are said to be *cyclic*. Similarly, if $b_k n_k \geq m_k$ for all $k = 1, \ldots, M$, then formula (9.3) becomes

$$j_k = 1 + ((i_k - 1) \text{ div } b_k), \qquad k = 1, 2, \ldots, M,$$

and the corresponding distributions are referred to as *block* distributions.

Example (Block-Cyclic Distributions). Tables 9.11 and 9.12 illustrate the $(2,2)$-block-cyclic distribution and the cyclic distribution of a 5×5, two-dimensional task set onto a 2×2 processor mesh.

The optimal values of the distribution parameters $b_1, \ldots, b_M$ (i.e., the parameter values for which optimal program performance is attained) depend on the interdependencies between the tasks $\{\mathrm{Ta}_{i_1,\ldots,i_M}\}$. If the tasks are mutually independent (which is a valid assumption whenever static load distribution schemes can usefully be employed in parallel numerical integration programs), then cyclic distributions are to be preferred because they ensure that the *number of tasks* per processor is optimally balanced.

Cyclic distributions are also best suited to deal with tasks $\{\mathrm{Ta}_{i_1,\ldots,i_M}\}$ which are *locally correlated* in the following way. If for two tasks $\mathrm{Ta}_{i_1^1,\ldots,i_M^1}$ and $\mathrm{Ta}_{i_1^2,\ldots,i_M^2}$ the index distance

$$\|(i_1^1 - i_1^2, \ldots, i_M^1 - i_M^2)\| \tag{9.4}$$

is small,[101] $Ta_{i_1^1,\ldots,i_M^1}$ and $Ta_{i_1^2,\ldots,i_M^2}$ tend to have similar execution times. Since cyclic distributions map tasks with a small index distance (9.4) onto different processors, the danger of burdening a single processor with all the expensive tasks is avoided. Such local task correlations can occur when the time needed for evaluating f depends on the given abscissa $\mathbf{x}$ because, in such cases, f-evaluations with abscissas close to each other often tend to have similar execution times.

Dynamic Load Distribution. Dynamic load balancing schemes are applied when some or all of the performance-relevant task or system characteristics are initially unknown. Most importantly, this situation arises whenever *adaptive* integration algorithms are to be parallelized (see Section 8.2.2). In such instances, the sequence of subregions $\{B_s\}$ (and consequently the abscissas $\{\mathbf{x}_i\}$ at which f is to be evaluated) are determined at run-time using a *dynamic* decision process. As a result, dynamic load balancing schemes have to use run-time information to enable an appropriate (re)distribution of work.

Dynamic load balancing algorithms can be categorized as follows.

Flat vs. Hierarchical Schemes. In *hierarchical* load balancing algorithms, a divide-and-conquer strategy is employed. The set of processors is divided into *processor clusters*, and the overall task of load balancing is split into several *intracluster* and one *intercluster* load balancing problems. Intracluster load balancing requires that work assigned to a cluster be distributed amongst its nodes; intercluster load balancing deals with the problem of assigning work to the different clusters.

The purpose of subdividing the processor set is to replace a single complex load balancing problem by several less intricate subproblems. The intracluster load balancing problems may be solved in different ways. For instance, the divide-and-conquer strategy can be applied recursively to these subproblems, resulting in further subdivisions of the processor set and a corresponding load balancing hierarchy.

By contrast, *flat* load balancing methods do *not* subdivide the parallel systems to which they are applied.

Single Task Pool vs. Multiple Task Pool Schemes. The computational load can be balanced in a parallel system using a *single task pool* which contains all tasks pending execution. The processors access this task pool, fetch one or more tasks, process the task(s), and store any new tasks they may have created in the task pool. This cycle is repeated until all tasks have been processed or any other termination criterion has

[101] Here, $\|\cdot\|$ denotes a suitable norm.

been met. The task pool resides either in a shared memory area or in the local memory of one processor designated to manage the task pool.

By contrast, *multiple task pool* algorithms allow several or even all processors to store sets of tasks locally.

Synchronous vs. Asynchronous Schemes. *Synchronous* load balancing methods separate the load balancing phase from the task processing phase at a global level: all processors must synchronize before a new phase can be started.

By contrast, *asynchronous* algorithms allow the load balancing phases and the task processing phases of different processors to overlap.

The above classification of dynamic load balancing schemes is sufficient for the purposes of this book. More detailed descriptions and thorough discussions can be found in Krommer and Ueberhuber [243].

Load Indices for Dynamic Load Balancing. Dynamic load balancing schemes attempt to improve the performance of parallel programs (or parallel systems) by dynamically redistributing work according to information about the load on different processors gathered at run-time. In this context, the question arises as to which *load index* or *load metric* should be used specifically to *measure* the processor load. Often, several load indices are conceivable for a given load balancing problem, and the choice of a suitable load metric may crucially influence the results of a load balancing procedure.

Ideally, a properly defined load metric should meet the following requirement. If the costs of redistributing work, i.e., transferring tasks, can be assumed to be insignificant, then the performance of a parallel program should attain its maximum when the load (with respect to this load metric) on different processors is evenly distributed throughout the computations.

However, it is often not possible to find such exact load indices. All one can expect from a practicable load index is that performance and load imbalance are correlated in the sense that a load (im)balance is likely to result in (sub)optimal program performance. A logical equivalence between performance and load balance either does not exist or cannot be demonstrated to exist.

A further complication in dynamic load balancing schemes is that task transfer costs are *not* generally negligible. The load redistribution overhead can indeed *annihilate* any possible performance gain resulting from attaining an optimally balanced load. The issues raised in this section are further discussed in Section 9.3.3 in connection with practical parallelizations of numerical integration programs.

9.2.3. Constituent Parts of Parallel Tasks. Parallelizing integration programs at one or more algorithmic levels makes it possible to execute a number of independent tasks concurrently. These parallel tasks involve the following processing stages.

Acquisition of Input Data. The *input data* of a parallel task consists of all information necessary to identify the set of loop iterations which have to be performed within the task. The precise composition of and the data structures used to store input data depend on the parallelization and load distribution strategy applied. Processors must acquire the input data of the tasks assigned to them before actually performing these tasks.

Computation of Integration Abscissas and Weights. Based on the given input data, processors must compute the integration abscissas

$$\{A_{j,r,s,l}\mathbf{x}_{k,i} + \mathbf{b}_{j,r,s,l}\}$$

(and the corresponding weights $\{|\det A_{j,r,s,l}|c_{k,i}\}$) at which they are supposed to evaluate the integrand(s).

Evaluation of Integrand(s). The assigned integrand evaluations

$$\{f_j(A_{j,r,s,l}\mathbf{x}_{k,i} + \mathbf{b}_{j,r,s,l})\}$$

must be performed.

Local Reduction of Partial Results. Each of the algorithmic levels in Table 9.8 involves a *reduction* operation. Reduction operations reduce the partial results of an algorithmic level to a single result which serves as a partial result for the next highest algorithmic level. For instance, the partial results

$$f_j(A_{j,r,s,l}\mathbf{x}_{k,1} + \mathbf{b}_{j,r,s,l}), \ldots, f_j(A_{j,r,s,l}\mathbf{x}_{k,n_k} + \mathbf{b}_{j,r,s,l})$$

from the integration formula level are combined to give the weighted sum

$$q_{j,r,s,l,k} = \sum_{i=1}^{n_k} |\det A_{j,r,s,l}|c_{k,i}f_j(A_{j,r,s,l}\mathbf{x}_{k,i} + \mathbf{b}_{j,r,s,l}),$$

which is a partial result at the local extrapolation level.

If an algorithmic level is assigned a *task* label, then the corresponding reduction operation constitutes part of each parallel task.

Example (Local Reduction of Partial Results). When an integration algorithm is parallelized at the integration formula level (see Table 9.9), the integrand evaluation level is performed within the resulting parallel tasks. A single integrand evaluation may involve a reduction operation (see, for instance, the example on page 358). Clearly, the processor to which an integrand evaluation is assigned must perform the corresponding reduction operation.

Moreover, if several iterations of a parallelized loop are performed by the same processor, the corresponding local contribution to the global reduction result must be computed. For instance, if several partial results

$$f_j(A_{j,r,s,l}\mathbf{x}_{k,i_1} + \mathbf{b}_{j,r,s,l}), \ldots, f_j(A_{j,r,s,l}\mathbf{x}_{k,i_p} + \mathbf{b}_{j,r,s,l})$$

are computed by a particular processor, then this processor must compute the corresponding contribution

$$\sum_{q=1}^{p} |\det A_{j,r,s,l}| c_{k,i} f_j(A_{j,r,s,l}\mathbf{x}_{k,i_q} + \mathbf{b}_{j,r,s,l})$$

to the overall result $q_{j,r,s,l,k}$.

Global Reduction of Partial Results. If an algorithmic level is parallelized, then a number of processors participate in calculating the corresponding partial results. A joint effort involving all participating processors is then required to compute the global reduction result. Usually, only a single global reduction operation is incorporated into each parallel task. However, if a parallelization scheme has been obtained using loop permutations (see Section 9.2.1), then multiple global reduction operations may have to be performed for each task.

Example (Multiple Global Reduction Operations). In the parallelization scheme described by Table 9.10, the l loop is moved into the parallelized i loop. As a result, each parallelized loop iteration computes partial results for the L entities $q_{j,r,s,1,k}, \ldots, q_{j,r,s,L,k}$, which means that L global reduction operations must be performed at the end of each task.

9.2.4. Performance Characteristics of Parallel Tasks. This section discusses the performance-relevant characteristics of parallel tasks in integration algorithms or programs. Additionally, techniques are described which help to improve and/or to reduce the effects of bad performance-relevant task characteristics.

Algorithmic Efficiency. Parallelizing a sequential numerical integration program may require modifying the underlying algorithm, which may adversely affect the algorithmic efficiency of the resulting parallel integration program (measured by the overall number of f-evaluations required). Examples of this effect are given on pages 353, 355, and 357. For instance, it is shown in Section 9.1 that parallelizing the subdivision level in integration programs may indeed lead to a reduced algorithmic efficiency. At other algorithmic levels, a decline in the algorithmic efficiency may occur when attempts are made to raise the degree of parallelism at the respective level. For instance, increasing the number of abscissas used in integration formulas raises the degree of parallelism at the integration formula level, but often entails an increase in the overall number of f-evaluations. A loss of algorithmic efficiency may diminish or even annihilate any performance gain due to parallel execution.

Degree of Parallelism. Some present-day computer systems provide massive *hardware parallelism*, offering thousands of processors (see Section 9.3.1). To achieve satisfactory performance on such systems, numerical algorithms must be parallelized in such a way that ultimate *software parallelism*—a maximum number of parallel tasks—emerges. Otherwise, only an unacceptably small subset of the available processors can be used in parallel computations.

The degrees of potential parallelism available at the various algorithmic levels are discussed in Section 9.1. If a numerical integration program is parallelized at several algorithmic levels, then the corresponding degrees of parallelism multiply. Such multilevel parallelizations are therefore particularly well suited for systems offering massive hardware parallelism.

Kinds of Parallelism. Parallel computer systems may be restricted in their ability to exploit the parallelism available in numerical integration algorithms. For instance, the operation of different *processor elements* in SIMD systems is controlled by a single instruction stream issued by a *control unit*; all processor elements operate synchronously. As a result, only data parallelism can be exploited by SIMD systems (see the example below). If tasks are not data parallel, only a single processor element can be used at a time.

This problem arises in parallel numerical integration whenever the evaluation of a fixed integrand f_j at different abscissas $\{\mathbf{x}_i\}$ requires totally different operations or when the required number of operations varies considerably. For instance, the integrand may be defined by an iterative process whose convergence behavior depends greatly on the given abscissa $\mathbf{x}_i$. SIMD computers, which can only exploit data parallelism, may not be able to perform such f-evaluations at different abscissas in parallel.

Example (Lack of Data Parallelism). When different SIMD processors concurrently evaluate the integrand

$$f(x_1, \ldots, x_N) := \begin{cases} 0, & x_1 < 0, \\ \exp\left(\sum_{g=1}^{N} a_g^2 x_g^2\right), & x_1 \geq 0, \end{cases}$$

at different abscissas, those processors whose abscissas have coordinate values $x_1 < 0$ stay idle while the others are busy computing their function values.

Predictability of Execution Times. The computational effort required to perform tasks in parallel numerical integration may be highly *data-dependent*. In the example above, for instance, the time needed to evaluate f varies significantly for different abscissas. Unpredictable task execution times may lead to an unevenly distributed load on different processors (and a corresponding performance degradation) if static load distribution schemes are applied (see Section 9.2.2).

Locality of Reference/Access and Transfer Times. When performing parallel tasks, processors may reference data objects which are stored in levels of the memory hierarchy affected by high *access* and/or *transfer times* (see Hennessy and Patterson [203]). This is especially true when data objects are stored in a low level of the local memory hierarchy (for instance, in the swap space on a hard disc) or in *remote* memory, i.e., memory pertaining to another processor. In the latter case, data objects must be transferred over an interconnection network, and the corresponding access and transfer times are usually several orders of magnitude longer than the access time of data objects stored in high levels of the local memory hierarchy. The delays caused by accessing such data objects can result in a significant performance degradation if processors are not able to perform useful computational work at the same time.

The adverse effects of long access and transfer times can be reduced in two ways.

Localizing Memory References. Algorithms can be modified in such a way that references to data objects in slow levels of the memory hierarchy become less frequent and/or that the size of the data objects to be transferred is reduced. This technique is referred to as increasing the *locality of reference* of algorithms.

Increasing Delay Tolerance. Algorithms can *tolerate* long access and transfer times without suffering any significant performance degradation if numerical computations and the transfer of data objects can overlap. However, this is only possible when there is dedicated hardware available which can manage the transfer of data objects autonomously. For remote memory accesses, this means that dedicated *communication hardware* is able to handle incoming and outgoing messages without involving the main computational processor.

In parallel numerical integration algorithms, references to slow layers of the memory hierarchy inevitably occur in global reduction operations (see Section 9.2.3). When the partial results involved in a reduction operation are distributed over several processors, accesses to nonlocal memory are necessary for computing the global reduction result.

If dynamic load balancing methods are used to assign tasks to processors (see Section 9.2.2), then the set of tasks to be performed by a given processor is not initially known. Instead, processors have to acquire the input data of their tasks at run-time in ways determined by the specific load balancing method employed, which usually entails references to remote memory.

Remote memory references can be avoided to a certain extent by using multiple task pool methods for dynamic load balancing purposes. However, multiple task pool methods can only be usefully applied if *new* tasks are generated when tasks are processed. These new tasks can then be used to replenish the local task pool, which would otherwise become depleted (which, in turn, would force processors to make remote memory references in order to fetch new tasks). In parallel numerical integration algorithms, appropriate tasks result only from parallelizations at the *subdivision level*: the subregions $B_{j,r,s,1}, \dots, B_{j,r,s,L}$ created in a subdivision step may be subjected to further subdivisions; i.e., they may become new tasks at the subdivision level.

Since references to remote data objects in parallel numerical integration are made exclusively at the beginning and at the end of task processing, the locality of reference increases when the *size* of the computational work within each task is increased. The task size, on the other hand, is determined by the algorithmic levels incorporated into each task, and thus gets larger when a higher algorithmic level is parallelized. The task size can also be raised by incorporating several parallelized loop iterations—instead of only a single iteration—in a single task.

Locality of Reference/Synchronization Delays. *Synchronization delays* may occur when referencing data objects which are common to several processors (for instance, shared variables or messages). Two forms of synchronization can be distinguished.

Access Control (mutual exclusion) coordinates the competition of different parallel processors for a common data object.

Sequence Control (condition synchronization) imposes a constraint on the order in which different processors may access common data objects.

Clearly, the utilization of a processor drops to zero while it has to wait as the result of a synchronization delay. Again, the effects of synchronization delays can be reduced by increasing the locality of reference, which results in less frequent accesses to common data objects and thus in fewer synchronization delays. Synchronization delays are also tolerable if other useful computational work can be performed while a data object cannot be accessed.

As for parallel integration, synchronization delays occur in global reduction operations. For instance, if the global reduction result is stored in a shared memory area and each participating processor autonomously updates the global result according to the local partial results, accesses of different processors to the global result must be synchronized.

Synchronization delays can also occur in connection with *control dependencies* in numerical integration programs (Dowd [117]). Suppose, for instance, that an adaptive integration program is parallelized at a level *below* the subdivision level. Now, at the end of each subdivision step, it must be decided (i) whether another subdivision step is necessary and (ii) which subregion(s) should be further subdivided. This decision is made by taking into account *all* results obtained within the previous subdivision step. As a result, all processors must synchronize before the decision process and must wait until the decision has been made before they can access the input data pertaining to the new tasks (which describe the subregions to be processed in the new subdivision step). Therefore, synchronization delays in adaptive parallel integration programs during the accessing of input data can only be avoided if the subdivision step is parallelized, i.e., if each processor is allowed to independently subdivide the subregions assigned to it.

9.3. Practical Parallelization of Integration Algorithms

The discussion in Section 9.2 shows that the performance of a parallel numerical integration program essentially depends on three factors:

(i) the choice of the parallelized section(s) of the integration algorithm,

(ii) the characteristics of the underlying parallel or distributed computer system, and

(iii) the load distribution scheme used to map parallel tasks onto given processors.

It is clearly beyond the scope of this book to exhaustively discuss the performance behavior of all algorithmic schemes that may result from the application of different load distribution methods to the various kinds of parallel tasks on different target machines. The following discussion is limited to those particular instances of parallel numerical integration schemes which have already been implemented in practice.

In Section 9.3.1, a concise account is given of present-day computer systems suitable for scientific computations. Sections 9.3.2 and 9.3.3 then deal with integration algorithms which utilize pipelined processors and multiple processors, respectively. First, those schemes in which only a single algorithmic level is vectorized or parallelized are introduced. Subsequently, schemes which exploit parallelism at several algorithmic levels are considered. For each choice of the parallelized sections, system-independent and system-specific parallelization issues are discussed separately. Additionally, this section provides a large number of references to relevant articles in which practical experiences with different parallelization schemes are described. Concise descriptions of these articles can be found in Krommer and Ueberhuber [247].

9.3.1. Computer Systems for Computational Integration. This section describes the main performance-relevant features of various parallel architectures. It is not intended to provide a general, hardware-oriented description of high-performance systems (see Hennessy and Patterson [203] or Dowd [117]). Instead, the focus of attention lies on issues directly related to parallel integration. Specifically, the relevance of the features described is linked to the performance-relevant characteristics of parallel tasks in numerical integration (see Section 9.2.4). For instance, whether or not the degree of parallelism in parallel integration algorithms is sufficient depends on the number of processors available; whether or not memory references are sufficiently local depends on the latencies and transfer rates of the underlying communication system, etc.

Performance-relevant aspects of advanced computer systems, in particular, the influence of local memory hierarchies,[102] which currently play only a minor role in the performance behavior of numerical integration algorithms, are deliberately omitted.[103] Also excluded are features such as the multiple (concurrently operating) arithmetic operation units available in *superscalar* processors, which cannot be exploited in a *portable* way.

Pipelined Processors. Processors with *pipelined* operation units, i.e., *vector processors* and *pipelined scalar processors*, can perform different stages of certain arithmetic and logical operations concurrently, which enables them to *overlap* the execution of sequential operations.[104] In vector processors pipelining can only be performed when *vector operations* (for instance, vector additions) are carried out and special *vector units* are used to perform the operations. Only if the time-consuming sections of a program can be expressed as sequences of vector operations can a reasonable program performance be expected. By contrast, pipelined scalar processors do not contain dedicated vector units, and it is usually the compiler's task to arrange operations in such a way that the processor can overlap the execution of sequential scalar operations. However, proper software coding (for instance, performing *loop unrolling* transformations) can greatly aid the compiler in generating optimized instruction sequences (see Dowd [117]).

Hockney and Jesshope [209] introduced a simple but instructive model for the execution time t of pipelined operations:

$$t(n) = \frac{n + n_{1/2}}{r_\infty}, \tag{9.5}$$

[102]Properties of local memory hierarchies (which include caches, main memory, and swap space) have little impact on the performance of numerical integration programs because evaluating an integrand at a single abscissa often involves a large number of floating-point operations. As a result, the computational intensity (i.e., the number of floating-point operations per memory reference) is high, which mitigates the effects of local memory delays.

[103]See Krommer and Ueberhuber [242], [244] for a general discussion of the performance-relevant characteristics of advanced computer systems.

[104]This may also apply to some intrinsic functions.

where n is the length of the vectors involved. The parameter r_∞ denotes the *maximum (asymptotic) performance*, i.e., the Mflop/s rate for vectors of infinite length. $n_{1/2}$, the *half-performance length*, is the vector length for which half of the maximum performance is achieved. For a thorough discussion of the $(r_\infty, n_{1/2})$ model, see Hennessy and Patterson [203].

Practical values of $(r_\infty, n_{1/2})$ depend on the specific pipelined operation and on the underlying computer system. On *memory-to-memory vector computers*, $n_{1/2}$ values are usually much higher than on *register-to-register vector computers* because of high memory access times. For pipelined scalar processors, $n_{1/2}$ values are generally significantly lower than for vector processors. On machines with large $n_{1/2}$ values, operations with *long* vectors, i.e., high degrees of software parallelism, are required to attain a satisfactory performance.

While the run-time model (9.5) suggests that the execution time increases linearly with n, register-to-register vector computers display a slightly different behavior: there is a more or less significant drop in performance whenever vector lengths exceed multiples of the vector register length.

SIMD Computers. Many of the early parallel computer systems adhered to the *single instruction, multiple data* (SIMD) principle. According to the SIMD paradigm, all *processor elements* operate *synchronously*, each on a *different data set*, and the operations of different processor elements are controlled by a *single instruction* stream issued by a *control unit*. When necessary, different processor elements communicate with each other by passing messages over an interconnection network.

The main idea behind the SIMD design is that of making the individual processor elements as simple as possible in order to keep the production costs per processor element low. For instance, since processor elements do not have to store program instructions, the local memory hierarchy on each processor can be kept small and simple; since I/O operations are performed by the control unit, the processor elements do not require I/O subsystems; etc. Hence, it is economically feasible to build computers which provide massive hardware parallelism: the number P of processor elements in SIMD systems typically ranges between $2^{10} = 1024$ and $2^{16} = 65{,}536$. However, the power of the individual processor elements is usually fairly low in comparison to the power of advanced microprocessors (which, again, helps to keep costs down).

The basic design of SIMD computers has a number of implications: (i) Since processor elements operate synchronously, SIMD systems are very inflexible and may not be able to exploit the parallelism inherent in an algorithm (see the example on page 368). (ii) Since single processor elements are relatively slow, SIMD machines are *fine-grained*; i.e., communication operations are relatively inexpensive. (iii) The control unit clearly has to be significantly more powerful than the individual processor elements. Hence, computational tasks which involve only a minor degree of parallelism should be performed by the control unit rather than by the processor elements because, for such tasks, the overhead resulting from the distribution of work and the collection of results annihilates all performance improvement achieved by executing tasks in parallel.

MIMD Systems. In *multiple instruction, multiple data* (MIMD) systems, each processor operates according to its own, private instruction stream. As a result, MIMD systems are very flexible and can potentially take advantage of any kind of parallelism inherent in an algorithm. MIMD systems can be classified as follows according to the degree of coupling between different processors and the extent of autonomy of each individual processor.

Shared Memory MIMD Computers. In shared memory MIMD computers, all processors have *equal access*, usually through a *bus* or a *crossbar switch*, to a common, shared memory area. Additionally, each processor has exclusive access to its own local memory hierarchy, which consists of one or more cache levels. Processors communicate by reading and writing values of *shared variables*, i.e., variables stored in the shared memory. This communication mechanism is generally several orders of magnitude faster than passing messages over interconnection networks. Hence, shared memory MIMD computers are fine-grained in that communication is relatively inexpensive.

The main problem with shared memory MIMD computers is that the number P of processors they can accommodate is quite limited: typically, shared memory systems contain between 4 and 24 processors.

Distributed Memory MIMD Computers. Distributed memory MIMD computers are characterized by the lack of a shared memory area to which all processors have equal access. Instead, each processor has its own, full-fledged local memory hierarchy, which includes one or more cache levels, main memory, and (possibly also) swap space on a hard disc. Communication between processors is achieved by transferring data between different local memory hierarchies using a dedicated interconnection network. This can be done either explicitly, using a *message passing* system, or transparently, using a *virtual shared memory* system.

The main advantage of distributed memory MIMD computers is that they can accommodate (significantly) larger numbers of processors than their shared memory counterparts. Systems incorporating several thousand processors have been developed in practice. Their main disadvantage is that communication between processors is slower than on shared memory computers, which makes distributed memory MIMD computers *coarse-grained*.

Network-Based MIMD Computing Environments. Network-based MIMD computing environments consist of several completely self-contained computers, typically PCs, workstations, or shared memory computers, which are connected by an interconnection network, typically a local area network (LAN). Since the autonomous systems can communicate with each other using the interconnection network, they can perform computational tasks concurrently. However, communication latencies and transfer times are generally even higher than on distributed memory computers.

In contrast to the parallel systems discussed so far, network-based systems may involve a high degree of *heterogeneity*: (i) the individual processor-memory nodes can vary in their performance characteristics; (ii) the overall interconnection network may consist of several different physical interconnection subnetworks. This heterogeneity makes the exploitation of parallelism (in particular, the development of good load distribution schemes) significantly more difficult.

Network-based computer systems can also be affected by dynamically changing resources. In a workstation cluster, for example, some machines may be temporarily used exclusively by certain applications and thus become unavailable for others. If the utilization of a resource in a distributed computing environment is unpredictable, then static load distribution may result in poor performance. An example of this is a distributed computing environment whose nodes are shared by several users. For such computer systems dynamic load distribution schemes may be required in order to bring about a reasonable performance.

9.3.2. Pipelining in Integration Methods. This section discusses practical issues associated with performing loop iterations of different algorithmic levels in pipeline mode, i.e., of overlapping the execution of sequential loop iterations. Note that pipelining the iterations of a loop usually requires the loop to be at the *innermost* position of the overall loop nest, which may require *loop permutations* to be carried out. (See Section 9.2.1 for a discussion of the possible difficulties associated with performing loop permutations in numerical integration programs.)

Integrand Evaluation Level. Evaluating multidimensional integrands often involves one or several *vectorizable* loops, i.e., loops which can be expressed in terms of vector operations, over the dimension of the integration problem (see the example in Section 9.1.7). Since these loops are at the innermost position of the overall loop nest, their iterations can be performed in pipeline mode without performing any loop permutations. The number of iterations in these loops, i.e., the dimension N of the integral, is only of modest size in most integration problems encountered in practice. (Problems with $N > 20$

rarely occur; they are almost impossible to solve numerically.) Therefore, computer systems with large $n_{1/2}$ values in the run-time model (9.5) do not get sufficient performance improvement from pipelining iterations at the integrand evaluation level. See Berntsen [23].

Integration Formula Level. Evaluating an integrand at a set (a vector) of abscissas, say, the n_k abscissas specified in an integration formula Q_{n_k}, can often be expressed as a sequence of vector operations of length n_k. Since n_k is usually larger than the dimension N of the integration problem, pipelining the iterations at the integration formula level can result in higher performance improvement than pipelining the iterations at the integrand evaluation level.

There are, however, two practical problems with this approach. To begin with, most integration programs require a user-specified function procedure to supply a single integrand value (at a single abscissa) at a time. In such cases, it is not possible to pipeline loop iterations over multiple abscissas simply because there are no such loop iterations in the user-specified function procedure.

And then, integration programs often make the abscissas of integration formulas available one after the other (see Section 9.3.4). Pipelining at the integration formula level, on the other hand, requires all integration abscissas (or a subset of them) to be available at a time, which has certain disadvantages. For instance, an additional workspace of size $O(n_k)$ is required for the simultaneous storage of all needed abscissas. Furthermore, algorithms which make the abscissas simultaneously available may be computationally more expensive than algorithms which make only one abscissa available at a time. This is true, for instance, for the abscissas of *invariant* cubature rules (see Section 6.2.3). The extra cost of abscissa generation may annihilate all performance gained by pipelining. See Berntsen [23] and Gladwell [176].

Integrand Level. Several recently developed numerical integration programs require users to provide function subprograms which *simultaneously* evaluate multiple integrands for a given abscissa (see Section 8.3.4). In such cases, pipelining the loop iterations over different integrands can be an attractive option. See Berntsen [23], Berntsen and Espelid [28], Berntsen, Espelid, and Genz [29], Espelid [135], Genz [166], and Gladwell [176].

Software (Pipelining at the Integrand Level). The three subroutines `NAG/d01eaf`, `TOMS/698/dcuhre`, and `TOMS/706/dcutri` use a user-supplied subroutine to evaluate a vector of integrands at a single abscissa. This enables users (or the compiler) to pipeline the iterations over different integrands and thus to gain in performance.

Multiple Algorithmic Levels. Combining several loop levels into a single loop (a technique referred to as *loop coalescence*) enables one to pipeline iterations of multiple loops, which results in increased pipeline lengths. Such

TABLE 9.13. *Parallelization at the subregion level.*

Level	Parallelization
subdivision	sequential
subregion	**parallel**
integration formula	task

long pipelines are important if high performance is to be achieved on systems with large $n_{1/2}$ values. See Gladwell [176] and Simpson and Yazici [399].

9.3.3. Parallelization of Integration Methods. This section discusses practical issues associated with performing loop iterations of different algorithmic levels in parallel, using multiple processors.

Integration Formula Level. Integration algorithms which are parallelized at the integration formula level are described in Table 9.9. It is important to realize that the task size at this level is usually small because each task consists solely of performing a single integrand evaluation (or, possibly, several evaluations if multiple iterations are assembled in the tasks). As a result, eligible target machines must be *fine-grained*—distributed memory and network-based MIMD systems cannot be used. Furthermore, SIMD systems cannot generally be used efficiently because the number of abscissas in integration formulas is too small.

On shared memory computers, parallelization at the integration formula level can be expected to result in good performance for high-dimensional integrals. In such cases, the number of abscissas is significantly larger than the number of processors (and thus sufficient software parallelism results), and the integrand evaluations are sufficiently expensive (which results in a sufficient locality of reference). See Berntsen [23], Genz [166], and Lyness and Plowman [280].

Software (Integration Algorithms Parallelized at the Formula Level). The routine `NAG/PAR-LIBRARY/d01gcfp` (which is a parallel equivalent of the sequential routine `NAG/d01gcf`) uses rank-1 lattice rules of Korobov form to calculate iterated integrals of any dimension $N \leq 20$ (see Section 6.5.6). Each lattice rule is evaluated in parallel based on a block distribution of the corresponding set of integration abscissas (see Section 9.2.2).

The routine `NAG/PAR-LIBRARY/d01dafp` (the parallel equivalent of the sequential routine `NAG/d01daf`) evaluates bivariate iterated integrals using a Patterson sequence of quadrature formulas (see Section 5.2.10) to perform both the inner and outer integrations. Different function values required for the outer integration (i.e., different inner integrals) are evaluated concurrently based on a single task pool, dynamic load balancing scheme (see Section 9.2.2).

TABLE 9.14. *Parallelization at the subdivision level.*

Level	Parallelization
subdivision	**parallel**
subregion	task
integration formula	task

Subregion Level. Integration algorithms which are parallelized at the subregion level are described in Table 9.13. The number of subregions created in each subdivision step is usually quite small (see Section 9.1.4). By modifying the spatial subdivision scheme (see Section 8.2.2), one can increase the number of subregions generated in each subdivision step and thus obtain a reasonable degree of parallelism. However, as demonstrated in the example on page 355, such algorithmic modifications may entail a significant efficiency degradation if (as is the case for singular integrands) the region requiring the greatest computational effort is very small. Otherwise, if a high sampling density is required over an extended region (as is the case for highly oscillatory integrands), then no significant efficiency degradation is to be expected from modifying the spatial subdivision scheme. Yet, the example on page 353 shows that in this case there is a large amount of parallelism available at the higher *subdivision level.* Parallelization at the subdivision level is usually preferable to parallelization at the subregion level for such integration problems because the resulting tasks have a higher locality of reference.

For the aforementioned reasons, parallelizing integration algorithms at the subregion level does not usually result in a satisfactory performance improvement on computer systems with large numbers of processors and/or high communication delays; satisfactory speed-ups may be attainable on shared memory systems. See Lapegna [251].

Subdivision Level. Adaptive integration algorithms which are parallelized at the subdivision level are described in Table 9.14. As explained in Section 9.2.4, parallelizing adaptive integration algorithms at the subdivision level has the potential to maximize the locality of reference of the resulting tasks and thus to minimize the performance loss due to communication delays. This, in addition to the inherent high algorithmic efficiency of adaptive subdivision strategies, is why a number of researchers have taken a particular interest in this parallelization scheme.

Parallelizing adaptive integration algorithms at the subdivision level means that the meta-algorithm in Table 9.3 is performed concurrently on all participating processors. The tasks at this level correspond to subregions $\{B_{j,r,s}\}$ eligible for further subdivision, which are determined at run-time using a dynamic decision process (see Section 8.2.2). As a result, tasks must be distributed dynamically at run-time if satisfactory performance is to be achieved (see Section 9.2.2).

While implementing proper dynamic load balancing schemes is the main task when parallelizing adaptive integration algorithms at the subdivision level, some changes may also be required in the constituent sequential code sections. For instance, the need to distribute work dynamically at run-time may have an impact on the data structures used for storing the subregions $\{B_{j,r,s}\}$. A thorough discussion of this issue can be found in Krommer and Ueberhuber [245].

It may also be necessary to adapt the noise detection mechanism used (see Section 8.3.1). A noise detector counts, for instance, the number of times that subdividing a region does not improve the corresponding error estimate. In *sequential* globally adaptive integration algorithms, only the subregion with the largest error estimate at the moment of decision is subdivided, whereas in *parallel* globally adaptive integration algorithms, subregions with smaller error estimates are subdivided as well. Their error estimates can even be so small that subdividing them does not lead to improved error estimates, which results in an increase in the noise detection counter. Consequently, the upper threshold on the counter can be reached only because subdividing certain subregions is no longer beneficial, and the algorithm terminates even though it may still be worthwhile to subdivide other subregions (with larger error estimates).

Example (Premature Termination). In the example on page 353, the subroutine `QUADPACK/dqag` is parallelized at the subdivision level. For the computation of the singular integral (9.1), the noise detection mechanism incorporated into `QUADPACK/dqag` had to be removed. Since only one processor can perform useful work, all other processors are busy with intervals whose subdivision does not improve the corresponding error estimates. Without removing the noise detection mechanism, the given error requirement could not be met because the algorithm terminated prematurely.

SIMD Computers. The number of subregions worth subdividing at the same time is usually significantly smaller than the number of processor elements in SIMD systems. Therefore, only a small part of the available processor elements can be kept busy, and the resulting performance is likely to be unsatisfactory. See Lemme and Rice [257].

Shared Memory Computers. On shared memory computers, the subdivision level in adaptive integration programs can easily be parallelized using an asynchronous single task pool method for dynamic load balancing purposes (see Section 9.2.2). For instance, such a method is used in the example on page 353. Since the single task pool can be kept in the rapidly accessible shared memory area, transfer times for retrieving tasks from and storing tasks in the task pool are minimal. A possible cause of performance degradation, however, is the synchronization delays that occur when several processors attempt to access the task pool simultaneously. This problem can be overcome, for instance, by implementing the task pool using *parallel* data structures (see Krommer and Ueberhuber [245]). For more information on integration methods using shared memory computers, see de Doncker and Vakalis [105], [106], Kapenga and de Doncker [102], [226], and Rice [363], [364], [366], [367].

Distributed Memory Computers. Single task pool methods cannot usually be used efficiently on distributed memory computers for dynamically assigning subregions to processors; access to the task pool entails access to remote memory, which results in prohibitively long access and/or transfer times. Only for high-dimensional integrals and/or for computationally expensive integrands may the time required for accessing the task pool be insignificant in comparison to the time required for processing a subregion. As a result, multiple task pool methods (see Section 9.2.2) should be used to dynamically allocate processors to subregions on distributed memory computers.

For similar reasons, the overhead associated with updating the global integral and error estimates $q_{j,r}$ and $e_{j,r}$ in Table 9.3 at each subdivision step is prohibitive. Instead, only the local integral and error contributions, i.e., the contributions originating from the locally stored subregions, are updated at each subdivision step. The $q_{j,r}$ and $e_{j,r}$ are modified significantly less frequently, for instance, only when local contributions to them have changed by a specific amount.

Additionally, in order to keep the communication overhead as low as possible, only a single *master* processor may keep track of the current values of $q_{j,r}$ and $e_{j,r}$; all other *worker* processors merely transfer the accumulated changes in their local contributions to the master processor. This can be done *asynchronously*; i.e., the worker processors can resume their numerical work without waiting for the transfer to be completed. As soon as $q_{j,r}$ and $e_{j,r}$ satisfy the given error requirements, the master processor broadcasts a corresponding *completion* message to the worker processors, which indicates that the current algorithmic stage has been completed successfully.[105]

Using multiple task pools (usually each on a different processor) for dynamic load balancing purposes implies that subregions must be transferred between different task pools at run-time according to information concerning the load on different processors. In this context, the question arises as to how suitable load indices, which can be used to measure processor load, should be defined (see Section 9.2.2).

Clearly, any useful load index must be based on the subregions stored in a processor's task pool. In *locally* adaptive integration algorithms, the *number*

[105]Similar messages are broadcast by the master processor when error conditions occur.

of active subregions assigned to a processor is an obvious choice for the load index: if the load is balanced with respect to this load metric, then none of the P processors is idle as long as at least P active subintervals are available at the same time.

In *globally* adaptive subdivision strategies, subregions should be evenly distributed to the available processors with respect to the size of their error estimates in order to avoid superfluous subdivisions. Therefore, the load on a processor should depend on the error estimates for the subregions assigned to it. However, to the authors' knowledge there is no mathematical model describing exactly how the efficiency of a parallel globally adaptive algorithm depends on the distribution of error estimates to different processors. As a result, an exact load metric for globally adaptive integration algorithms cannot be defined.

Nevertheless, various useful practical load indices can be devised, each with its specific advantages and disadvantages. For instance, the load on a processor can be measured by the local sum of error estimates from all of its subregions. This is an obvious choice for a load index because the overall difficulty of a numerical integration problem can be characterized by the global sum of all error estimates. Yet, with respect to this load metric, a processor with a large number of "easy" subregions may be just as loaded as a processor with a small number of "difficult" subregions. The first processor will subdivide its "easy" regions, although there are still more "difficult" regions assigned to the second processor. This problem does not occur when the load on a processor is defined, for instance, by the *largest* error estimate from any subregion assigned to it. However, when this load index is used, a processor with a large number of "difficult" subregions appears just as loaded as a processor with only one similarly "difficult" subregion, which does not seem to be appropriate either.

For more information on this topic, see Berntsen and Espelid [26], [27], Burrage [47], de Doncker and Kapenga [103], de Doncker, Kapenga, and Zuidema [104], Genz [166], Lapegna and D'Alessio [252], Maierhöfer [288], [289], Maierhöfer and Skorobohatyj [290], and Miller and Davis [297].

Software (Adaptive Integrations Parallelized at the Subdivision Level). The routines `NAG/PAR-LIBRARY/d01atfp` and `NAG/PAR-LIBRARY/d01aufp` are parallel versions of the sequential routines `NAG/d01atf` and `NAG/d01auf`, respectively. They both employ the same multiple task pool load distribution scheme in order to distribute subintervals dynamically. This load distribution scheme is also used in `NAG/PAR-LIBRARY/d01axfp`, which is a parallel variant of `NAG/d01akf`.

The subroutine `NAG/PAR-LIBRARY/d01fafp` uses a globally adaptive subdivision strategy for evaluating N-dimensional integrals over cubic regions. Invariant polynomial integration formulas (see Section 6.2.3) are employed to approximate the given integral on each subregion, and sequences of null rules (see Section 8.1.2) are used to derive estimates of the corresponding integration errors. Subregions are distributed dynamically to the participating processors based on a multiple task pool load distribution scheme (see D'Apuzzo, Lapegna, and Murli [84] for details).

TABLE 9.15. *Parallelization at the subdivision and subregion levels.*

Level	Parallelization
subdivision	**parallel**
subregion	**parallel**
integration formula	task

Subdivision and Subregion Levels. Table 9.15 describes adaptive integration algorithms which are parallelized at the subdivision and subregion levels. As demonstrated in the example on page 353, only a small degree of parallelism may be available at the subdivision level when dealing with an integrand with a single peak or singularity. The example on page 355 indicates that in this case at least a logarithmic speed-up can be obtained by parallelizing the subregion level. As a result, it is advantageous to try to exploit parallelism at both levels; as long as there is a sufficient number of subregions worth being subdivided, the number L of subsubregions into which a selected subregion is divided is chosen as in the underlying sequential integration algorithm. As soon as the number of subregions worth being subdivided becomes insufficient, L is raised in order to increase the degree of utilizable parallelism.

Shared Memory Computers. On shared memory computers, the subdivision level in adaptive integration programs can easily be parallelized using single task pool methods for dynamic load balancing purposes (see Section 9.2.2). For parallelizations at the subdivision level only, *asynchronous* methods are used, as there is no inherent reason for all processors to synchronize when storing and retrieving tasks. When the subregion level is being parallelized as well, the situation is quite different because the subdivision factor L has to be determined dynamically. When using asynchronous single task pool methods, L must be determined based on partly obsolete information because decision making and task processing proceed simultaneously. By contrast, synchronous methods do not suffer from this deficiency. This is why all load distribution methods used for parallelizing the subdivision and subregion levels on shared memory systems are based on *synchronous* single task pool methods. The various algorithms differ in the heuristics used to select subregions suitable for further subdivision and to determine the subdivision factor L.

For more information on this topic, see Berntsen [23], Berntsen and Espelid [28], Berntsen, Espelid, and Genz [29], Bull and Freeman [45], Freeman and Bull [152], Genz [166], and Napierala and Gladwell [312].

Distributed Memory Computers. Parallelizing the subdivision and subregion levels of adaptive integration programs is difficult on distributed memory computers; because of long communication delays, maintaining global, up-to-

TABLE 9.16. *Parallelization at the subdivision, subregion, and integration formula levels.*

Level	Parallelization
subdivision	**parallel**
subregion	**parallel**
integration formula	**parallel**

date error information, as is required for effective subdivision heuristics, is prohibitively expensive. See Freeman and Bull [152] and Genz [166].

Subdivision, Subregion, and Integration Formula Levels. Table 9.16 describes adaptive integration algorithms which are parallelized at the subdivision, subregion, and integration levels, resulting in ultimate software parallelism. Since each task consists of performing a single integrand evaluation only (or, possibly, several evaluations if multiple iterations are assembled into tasks), suitable target machines must be fine-grained. Consequently, this parallelization scheme is best suited for SIMD computers. See Genz [164].

9.3.4. Pipelining and Parallelizing Abscissa Computations.

The abscissas and weights of integration formulas can either be stored at *compile-time* or generated at *run-time*. The method chosen should be economical with respect to both the required storage space and the computational complexity of the abscissa generation process.

In general, storing integration abscissas and weights at compile-time is preferred if an integration formula involves only a small number of abscissas and/or if the abscissas and weights are difficult to compute. Most importantly, this method is used for simple interpolatory quadrature formulas (see Section 5.2). On the other hand, generating integration abscissas and weights at run-time is usually preferred if an integration formula involves a large number of abscissas and/or if the abscissas and weights are easy to compute. Most importantly, this method is used for compound and product formulas (see Sections 5.3 and 6.1.4), invariant formulas (see Section 6.2.3), and pseudorandom and number-theoretic formulas (see Sections 6.3, 6.4, and 6.5).

If evaluating the integrand(s) is inexpensive, then both the abscissa generation procedure and the affine transformations,

$$\bar{x}_{j,r,s,l,k,i} := A_{j,r,s,l}\mathbf{x}_{k,i} + \mathbf{b}_{j,r,s,l}, \qquad i = 1, 2, \ldots, n_k, \tag{9.6}$$

needed to make integration formulas applicable to different integration regions $\{B_{j,r,s,l}\}$ may consume a considerable portion of the overall computing time.

Pipelining. For each index $i \in \{1, \ldots, n_k\}$, the transformation (9.6) represents a BLAS-2 operation (Dongarra et al. [114]). If all abscissas $\mathbf{x}_{k,1}, \ldots, \mathbf{x}_{k,n_k}$

are assembled in a matrix $X_k \in \mathbb{R}^{N \times n_k}$, then (9.6) can be written as

$$A_{j,r,s,l}X_k + B_{j,r,s,l}, \tag{9.7}$$

where the matrix B consists of the columns $\mathbf{b}_{j,r,s,l}$. The transformation (9.7) is a BLAS-3 operation (Dongarra et al. [113]). While machine-optimized versions of these BLAS subroutines are readily available on virtually all current high-performance microprocessors, they are not actually used in integration programs for the following reasons. (i) Often, integration programs require very specific affine transformations (for instance, scaling or shift operations) which can be computed more efficiently than by calculating (9.6). (ii) The vector and matrix orders in (9.6) are equal to the dimension N of the given integration problem. Since N is small, $N \leq 20$, little performance can be gained from calling optimized BLAS routines. (iii) The BLAS-3 formulation (9.7) requires multiple abscissas to be available simultaneously, which gives rise to the problems described in Section 9.3.2.

Pipelining the generation of abscissas is straightforward and advantageous for most pseudorandom and number-theoretic formulas, but not for other kinds of integration formulas.

Parallelization. When parallelizing the computation of integration abscissas and/or the associated affine transformations, each abscissa $\overline{x}_{j,r,s,l,k,i}$ should be computed by the processor assigned to evaluate the integrand(s) at $\overline{x}_{j,r,s,l,k,i}$. Since these computations can be performed in parallel without any communication between processors, near-linear speed-ups (for the corresponding algorithmic sections) can be expected. In any case, affine transformations (9.6) can be performed in parallel without any problem. This also applies to generating the integration abscissas as long as each integration formula is evaluated by a separate processor.

Problems may arise, however, when several processors cooperate in evaluating a single integration formula. For instance, generating the abscissas of an invariant integration formula is *not* a data parallel procedure and thus cannot be performed concurrently on SIMD computers. In addition, whenever the abscissas of an integration formula are defined using a *recursive* procedure (which makes the abscissas available one after the other), special algorithmic modifications, which depend on the specific kind of integration formula, are required in order to enable multiple recursions to proceed independently on different processors. See Genz [166] and Simpson and Yazici [399].

9.3.5. Pipelining and Parallelizing Reduction Operations. In numerical integration algorithms, reduction operations (see Section 9.2.3) are mostly required for the evaluation of integration formulas

$$Q_n f_j = \sum_{i=1}^{n} c_i f_j(\mathbf{x}_i), \quad j = 1, 2, \ldots, m,$$

for given weights $c_1, \ldots, c_n$ and function values $f_j(\mathbf{x}_1), \ldots, f_j(\mathbf{x}_n)$, $j = 1, \ldots, m$. Depending on the loop order, these calculations can easily be seen to consist of m DOT-products of length n or of n AXPY operations of length m (see Golub and Van Loan [179]).

Machine-optimized versions of these BLAS routines (Lawson et al. [255]) are readily available for virtually all current high-performance microprocessors. They can be used to perform the *local* part of reduction operations in numerical integration algorithms. However, the resulting performance gain is usually minor, since the cost of reduction operations is generally insignificant in comparison to the computational effort required for the evaluation of integrands. Moreover, algorithmic modifications (in particular, making multiple function values available simultaneously rather than one at a time) may be required to make BLAS routines applicable.

Global reduction operations additionally require *global sums* to be calculated. Global sum operations are defined in most parallel programming languages and environments (see Section 10.1.3), and machine-optimized implementations are widely available. Using these optimized, global summation routines can significantly reduce the effects of communication delays in global reduction operations and is thus highly recommendable. See Gladwell [176].

Chapter 10

Assessment of Numerical Integration Software

The preceding chapters describe a large number of discretization methods, error estimation procedures, and discretization refinement schemes used in numerical integration programs on both sequential and parallel computers. Given the bewildering variety of algorithmic components and targeted computer systems, the question arises as to how the quality of numerical integration software can be assessed objectively.[106] Answering this question requires both clarifying the relevant quality *criteria*, which determine the usefulness of a software product (see Section 10.1), and developing appropriate *evaluation techniques*, which provide objective (quantitative) ratings of software products in terms of these criteria (see Section 10.2).

10.1. Assessment Criteria

This section describes the most relevant criteria for assessing the quality of numerical integration software, specifically, *reliability*, *performance*, and *portability*, and discusses problems that may prevent numerical integration software from performing satisfactorily with respect to these criteria.[107] While the aforementioned criteria are important to all numerical integration programs, the relative weights assigned to them usually depend on the specific requirements of a given integration problem. High-dimensional integration problems, for instance, are often at the brink of computational tractability, and this gives prominence to the performance aspect of numerical integration programs. When integrating functions with singularities or peaks, on the other hand, the "difficult" spots can be missed altogether, resulting in completely

[106]The term *software* is generally used to denote both programs *and* their accompanying documentation. While the following discussion focuses on the assessment of numerical integration *programs*, it should be made clear that good documentation is absolutely *crucial* to the usefulness of any software product.

[107]For general discussions of software quality criteria, the reader should refer to the large number of publications in the field of software engineering, for instance, Boehm et al. [36], Bauer [16], Berzins [32], etc.

incorrect integration results. In such cases, reliability is of the utmost importance.

10.1.1. Reliability. A particular software product is *reliable* if it performs its functions in accordance with the given program *specification.* The reliability of a computer program can only be assessed on the basis of a sound program specification which completely and unequivocally spells out

(i) the meaning of the program's input and output data,

(ii) the format in which input data must be provided on entry to the program and the format in which output data are returned on exit from the program,

(iii) the constraints imposed on eligible input data,

(iv) the (mathematical) properties of correct output data, and

(v) the behavior of the program for ineligible input data.

Computer programs which are *not* accompanied by appropriate specifications should be considered as being *inherently unreliable* since their behavior is not clearly described and is thus innately unpredictable.

The main aspects of reliability are *correctness*, *robustness*, and *accuracy*. A program is *correct* if it produces correct output data, i.e., results which conform to the program specification, for *all eligible* input data. The *robustness* of a program characterizes its ability to deal with unexpected or inappropriate input data. The *accuracy* of a program indicates the degree to which its numerically calculated results coincide with mathematically exact results.

Issues concerning the reliability of numerical integration programs are addressed in various contexts in this book; the kind of input data for numerical integration algorithms, the format in which these input data must be provided, and the various (accuracy) requirements imposed on numerical integration results are discussed in Section 3.2. The *inherent incorrectness* of *any* sampling method, i.e., any method which uses only *discrete* sample information $S(f) = (f(\mathbf{x}_1), \ldots, f(\mathbf{x}_n))$, when it comes to estimating the error of numerical integration results, is demonstrated in Section 8.1. Section 8.3.1 deals with the effects of using machine rather than exact arithmetic and of using noisy rather than accurate input data on the robustness and accuracy of numerical integration procedures. Section 8.3.1 also describes specific algorithmic measures which can be used to improve these quality indices.

In Section 8.1.5 it is shown that, for a restricted set of integration problems, correct numerical methods can be developed using *interval arithmetic*. However, it is also explained why the *guaranteed* accuracy of verified integration results can be unacceptably low for practical purposes.

10.1.2. Performance. A number of indices have been proposed and are used to characterize the performance of scientific programs on sequential and parallel computer systems.

Absolute Performance Indices are based on measured, simulated, or analytically predicted run-times and/or numbers of floating-point operations (Addison et al. [2]).

Temporal Performance is defined as the reciprocal run-time of a program. This performance index attains a maximum when the program's run-time is at a minimum, irrespective of the number of floating-point operations carried out.

Benchmark Performance is the ratio between the number of floating-point operations needed for the execution of a *sequential* algorithm and the run-time of the *equivalent parallel* algorithm. The unit of this metric is *flop/s* (or Mflop/s, Gflop/s).

Hardware Performance is the ratio between the required number of floating-point operations of a (sequential or parallel) algorithm and its run-time. Again, this metric is measured in *flop/s* (or Mflop/s, Gflop/s).

Complexity Performance is defined as the reciprocal computational complexity, measured by the number of floating-point operations required, of an algorithm.

When considering the suitability of these indices for measuring the performance of numerical integration programs, it is important to remember that most computer time in numerical integration programs is spent in the subprogram which evaluates the integrand(s), rather than in the actual integration algorithm. Hence, benchmark and hardware performance indices are not relevant to numerical integration because they measure the performance of the user-provided integrand evaluation subprogram rather than the performance of the actual integration algorithm.

Furthermore, on uniprocessor computers the overall run-time and the computational complexity of numerical integration programs are usually (approximately) proportional to the required number of integrand evaluations. This number is therefore a valid indicator of the absolute performance of numerical integration algorithms on these systems.

Relative Performance Indices are derived by comparing the performance of (usually parallel) algorithms to certain performance standards.

Speed-up is the ratio, $S_P := t_1/t_P$, between the run-time t_1 of a sequential algorithm and the run-time t_P of a corresponding parallelized algorithm using P processors.

Efficiency is the ratio, S_P/P, between the speed-up of a parallel algorithm and the number of processors. Efficiency measures the concurrent utilization of a multiprocessor's capacity: the closer S_P/P is to 1, the better the use that is made of the potentially P-fold power of a P processor system.

Effectivity is the ratio, $S_P/(Pt_P)$, between the speed-up of, and the overall work (measured by Pt_P) performed in, a parallel algorithm. It measures both the absolute increase in speed *and* the relative utilization of the computer's capacity.

System Efficiency is obtained by relating benchmark or hardware performance indices to the peak performance of a machine, i.e., to the maximum rate of execution possible for the given hardware characteristics (Dongarra [112]).

As for numerical integration, system efficiency measures the performance of the user-provided integrand evaluation subprogram and is therefore not relevant to assessing the performance of numerical integration algorithms. The remaining relative performance indices can be used to assess various performance aspects of parallel integration programs. However, relative performance indices can be highly misleading if the sequential programs they are based on perform badly. In such cases, high relative performance indices do *not* necessarily indicate a good absolute performance, which should be the ultimate goal of any kind of parallelization.

In order to prevent this problem, the performance of any parallel algorithm should be compared to that of a commonly used and accepted sequential program of proven quality. It is a methodological deficiency frequently observed in literature on parallel numerical integration that no such comparisons are carried out. This drastically diminishes the significance of the reported relative performance indices.

10.1.3. Portability. A substantial software recoding effort may be required to port programs from the computer platforms they were originally developed for to other platforms. A program is *portable* if no such effort is required, and *highly portable* if any recoding effort is insignificant to the original software development costs. Portability secures software development investments by ensuring that software remains usable when the underlying hardware and/or system software change.

In order to be portable, numerical programs must be written entirely using *standardized*, *commonly available* programming languages and environments. Sequential programs should therefore be coded in Fortran 77, Fortran 90 [4], or ANSI C [230]. Mixing Fortran and C software modules can lead to difficulties, as interlanguage calling mechanisms are not standardized. Even mixing Fortran 77 and Fortran 90 modules can be problematic if different compilers are used for the two kinds of modules. For instance, some Fortran compilers append underscore characters to the names of all referenced routines. In these cases, underscore characters have to be added to names of all C routines called from Fortran modules.[108]

Portable parallel numerical programs should make use of standard message passing systems such as those provided by the *parallel virtual machine* (PVM) environment (Geist et al. [162]) or implementations of the *message passing interface* (MPI) standard (Gropp, Lusk, and Skjellum [183], Snir et al. [412]). Even higher portability, albeit with significantly restricted functionality, is provided by the *basic linear algebra communication subroutines* (BLACS) (Dongarra and Whaley [116]). *High performance Fortran* (HPF) (Koelbel et al. [238]) provides an alternative platform for writing parallel numerical programs based on the data distribution paradigm.

When assessing the suitability of different parallel programming systems for numerical integration purposes in terms of portability, flexibility, and efficiency, MPI currently emerges as being the preferable tool: BLACS and HPF lack the flexibility necessary to implement some of the dynamic load balancing schemes applied in parallel numerical integration (see Section 9.2.2); PVM, while currently in widespread use, is no longer being developed, whereas MPI will be extended and further developed in future.

Portable numerical programs must also accommodate different hardware implementations of standard data types and operations. Most importantly, floating-point numbers and operations can be implemented using different radices, mantissa lengths, maximum/minimum exponents, and/or rounding mechanisms on different computer systems. Even IEC/IEEE-conformant implementations of floating-point numbers and operations may vary because, among other things, the IEC/IEEE floating-point standard only specifies *minimum* requirements for standard conformant implementations. Nevertheless, numerical programs can achieve a high degree of portability despite the presence of different floating-point arithmetics by using portable language elements, available in both Fortran 90 and ANSI C, to extract information about the given floating-point arithmetic and then adapting their algorithmic behavior accordingly (Naur [313]).

Portability and Efficiency. Sequential portable numerical integration programs usually achieve a reasonable overall efficiency due to the power of optimizing compilers and the high computational intensity of integrand evaluations (see Section 9.3.1). When using vector processors, significantly more

[108]Portability of mixed-language programs can be achieved by using appropriate platform-dependent macro definitions which can be invoked by setting specific compiler flags.

effort has to be made if the capacity of target machines is to be fully utilized. With parallel systems, the situation is even worse. In order to fully utilize the potential performance of parallel computers, crucial algorithmic parameters, like the granularity of parallel tasks, the communication patterns between parallel tasks, the load distribution, etc., must be adapted to the characteristics of the underlying computer system (see Section 9.2.4). The resulting programs are (even if they are portable) implicitly machine-dependent in the sense that they run efficiently only if a given computer system exhibits appropriate performance characteristics (see Section 9.3.1).

A possible solution to the problem of inefficient portable algorithms is the *architecture adaptive algorithms* (AAA) methodology proposed by Krommer and Ueberhuber [242], [244]. Using portable language elements, algorithms can acquire information about the performance-relevant characteristics of the underlying computer system and adapt themselves accordingly. A (from the algorithm developer's point of view) less demanding alternative is enabling users to specify performance-relevant algorithmic parameters themselves, thereby giving them the means to tune portable parallel algorithms to the characteristics of the target computer.

Software (User-Provided Algorithmic Parameters). `NAG/PAR-LIBRARY/d01dafp` enables users to specify an algorithmic parameter which determines the granularity of the applied dynamic load balancing scheme. This subroutine argument can be chosen to fit the granularity of the given parallel or distributed computer system.

10.2. Assessment Techniques

This section describes techniques which can be used to evaluate different numerical integration programs in terms of the quality criteria introduced in Section 10.1. Section 10.2.1 looks at the fundamental approaches to evaluating programs and algorithms and their applicability for numerical integration purposes. In Section 10.2.2, theoretical performance analyses of numerical integration algorithms based on *complexity-theoretic* methods are discussed, whereas in Section 10.2.3, implementation-based evaluation methods using *test batteries* and *performance profiles* are dealt with.

10.2.1. Approaches to Assessing Numerical Integration Software.

Three fundamentally different approaches to evaluating numerical algorithms and programs can be distinguished. *Theoretical analyses* are based on mathematically derived theorems about (certain aspects of) the behavior of numerical algorithms; *implementation and testing analyses* are based on performing carefully planned experiments on specific implementations of numerical algorithms on particular computer systems; *simulation analyses* perform experiments on *mathematical models* of (certain aspects of) numerical algorithms and/or computer systems.

Theoretical analyses are used extensively throughout this book, mostly in connection with integration formulas, to establish theorems about the error, the convergence order, the numerical stability, etc. of numerical integration schemes. When it comes to comparing different integration methods, however, theoretical analyses are usually only suitable for assessing different variants of a particular class of integration schemes, but not for assessing methods which belong to totally different classes of integration schemes.

Example (Cubature Formulas). The degree of accuracy of polynomial cubature formulas is a valuable, analytical indicator of the accuracy of these formulas for smooth integrands (see Section 6.2). Hence, polynomial cubature formulas with high degrees of accuracy are generally preferable (as far as accuracy is concerned) to formulas with low degrees of accuracy. Similarly, the figure of merit provides a valuable, analytical indicator of the accuracy of lattice rules (see Section 6.5). However, neither the degree of accuracy nor the figure of merit can be used to compare the accuracies of polynomial cubature formulas to those of lattice rules.

Theoretical analyses of complex integration algorithms and programs often prove intractable, and simplifying assumptions about the systems under consideration may be needed to make theoretical analyses feasible (see Section 10.2.2). This is why implementation and testing analyses are generally used for assessing practical integration algorithms and programs (see Section 10.2.3). On uniprocessor computers, this approach can be performed with a reasonable effort and gives satisfactory results for the following reasons.

(i) A comprehensive characterization of the behavior of any integration algorithm can be obtained by testing the algorithm against a standardized, representative sample of integration problems.

(ii) Since sequential numerical integration algorithms are *deterministic*,[109] only a single test run is needed for each integration problem.

(iii) Test results are machine-independent expect for (usually minor) variations due to the use of different implementations of floating-point numbers and operations. Insight into the possible effects of using different machine arithmetics can be gained by performing tests on a small sample of uniprocessor computers which represent the various commonly used kinds of machine arithmetic.

Unfortunately, with advanced computer systems—in particular, with parallel and distributed computers—the situation is quite different.

(i) As explained in Section 10.1.3, the performance of a particular numerical integration program on an advanced computer system may depend

[109]Numerical integration methods based on pseudorandom formulas may be an exception (see Section 6.4).

heavily on the characteristics (the communication speed, processor power, etc.) of the given system. In order to obtain a comprehensive view of the performance of a parallel numerical integration algorithm, the algorithm must be implemented on a broad range of advanced computer systems. This is cumbersome and may even pose insurmountable practical problems, in particular, when an advanced computer system to be tested is not physically available. Furthermore, using implementation and testing analyses, it is *not* usually possible to assess the performance of parallel algorithms on *future* computer systems, whose performance-relevant characteristics may differ from the characteristics of present-day computers.

(ii) Parallel numerical integration algorithms may be indeterministic; i.e., the results of different program runs may differ. Indeterminism is usually introduced into parallel numerical algorithms in order to reduce the performance loss that results from synchronizing multiple paths of execution. Indeterministic algorithms are not necessarily unreliable, but their behavior may depend, among other things, on the characteristics of the underlying computer system. Hence, the reliability of an indeterministic, parallel integration algorithm for a particular integration problem cannot be established using just a single test run. Instead, extensive test runs on a broad range of platforms may be required to substantiate the reliability of an indeterministic scheme. This gives rise to the previously mentioned implementation and testing problems.

For these reasons, simulation analyses, which, so far, have hardly been used in connection with numerical integration, should be applied to complement theoretical and practical analyses in assessing the quality of numerical integration algorithms for advanced computer systems. By running numerical integration programs on simulated parallel computer systems, for instance, it is possible to obtain performance results for systems with characteristics not found in present-day computers. The authors' experience with the simulation of parallel numerical integration programs is summarized in Krommer and Ueberhuber [246].

10.2.2. Theoretical Performance Assessment. This section introduces the framework of *information-based complexity theory* for analyzing numerical integration algorithms. Due to space constraints only a very coarse outline of this subject can be given. For thorough discussions, see Traub and Wozniakowski [432] and Traub, Wasilkowski, and Wozniakowski [430], [431].

The key problem addressed in complexity-theoretic analyses of numerical integration algorithms can phrased (with different variations) as follows. Given a class $\mathcal{F}$ of functions $f : B \subseteq \mathbb{R}^N \to \mathbb{R}$ and a number $n \in \mathbb{N}$, how should the integration abscissas $\mathbf{x}_1, \ldots, \mathbf{x}_n$ and the weights $c_1, \ldots, c_n$ be chosen if the error in the resulting integration formula Q_n for integrands $f \in \mathcal{F}$ is to be minimized?

This general problem formulation raises a number of questions that need to be answered if a precise mathematical problem definition is to be obtained.

(i) Which class $\mathcal{F}$ of functions is considered?

(ii) How exactly is the error in Q_n for functions $f \in \mathcal{F}$ measured?

(iii) In which sense must Q_n minimize the integration errors for integrands $f \in \mathcal{F}$?

These questions can be answered in different ways, giving rise to different *problem settings*.

(i) A typical choice for $\mathcal{F}$ is

$$\mathcal{F} := \{ f \in C^k[a,b] \ : \ \|f^{(k)}\|_\infty \leq M_k \},$$

but spaces of $(b - a)$-periodic or singular functions (and corresponding spaces of multivariate functions) have also been studied.

(ii) The error in an n-point rule Q_n, applied to a specific integrand f, is usually measured by its *absolute error*

$$E_{\mathrm{abs}}(f; \mathrm{Q}_n) := |\mathrm{Q}_n f - \mathrm{I} f|$$

or its *relative error*

$$E_{\mathrm{rel}}(f; \mathrm{Q}_n) := \left| \frac{\mathrm{Q}_n f - \mathrm{I} f}{\mathrm{I} f} \right|.$$

(iii) The aggregate integration error $E(\mathcal{F}; \mathrm{Q}_n)$ may be defined in fundamentally different ways.

Worst-Case Error. The *worst-case error* E^{wor} is defined by

$$E^{\mathrm{wor}}(\mathcal{F}; \mathrm{Q}_n) := \sup\{ E(f; \mathrm{Q}_n) \ : \ f \in \mathcal{F} \}$$

as the largest error in Q_n for any element $f \in \mathcal{F}$.

Average-Case Error. Let P be a probability distribution on $\mathcal{F}$. The *average-case error* E^{avg} is defined by

$$E^{\mathrm{avg}}(\mathcal{F}; \mathrm{Q}_n) := \sqrt{\int_{\mathcal{F}} E(f; \mathrm{Q}_n)^2 \, dP(f)}.$$

In other words, $E^{\mathrm{avg}}(\mathcal{F}; \mathrm{Q}_n)$ is the quadratic mean of the error $E(f; \mathrm{Q}_n)$ with respect to the probability distribution P.

Probabilistic Error. Let $\delta \in [0,1]$. The *probabilistic error* E_δ^{prob} is defined by

$$E_\delta^{\text{prob}}(\mathcal{F}; \mathrm{Q}_n) := \inf \{ E^{\text{wor}}(\mathcal{F} \setminus \mathcal{A}; \mathrm{Q}_n) \ : \ \mathcal{A} \subseteq \mathcal{F},\ P(\mathcal{A}) \leq \delta \} .$$

In other words, E_δ^{prob} is given by the smallest worst-case error on a subset $\mathcal{F} \backslash \mathcal{A} \subseteq \mathcal{F}$ with measure $P(\mathcal{F} \backslash \mathcal{A}) \geq 1 - \delta$.

Moreover, the probability measure P used in the definitions of the average-case error and the probabilistic error can be defined in different ways.

The main problem with complexity-theoretic analyses of integration algorithms is that the seemingly objective results of these investigations may depend on the choice of a specific problem setting. For instance, *worst-case analyses* of integration algorithms for classes $\mathcal{F}$ of well-behaved functions show that sampling the abscissas $\mathbf{x}_1, \ldots, \mathbf{x}_n$ adaptively is *not* superior to nonadaptive sampling (Wozniakowski [448]). On the other hand, Wasilkowski and Gao [444] show that for spaces which contain singular functions, adaptive sampling is indeed superior to nonadaptive sampling in a probabilistic problem setting, provided that an appropriate probability distribution P is chosen. For other choices of P, however, it has been demonstrated that adaptive sampling does *not* help (Wasilkowski [443]). The apparently contradictory outcomes of these investigations can be explained by the different problem settings they are based on, in particular, the different probability weights assigned to singular integrands.

As long as the question as to which particular problem setting is relevant for a specific range of applications remains open, the results of complexity-theoretic analyses are of limited value for the practical assessment of numerical integration algorithms. Hence, for the time being, the implementation and testing of numerical integration schemes using test batteries and performance profiles (as described in the following section) remains the prevalent assessment technique.

10.2.3. Test Batteries and Performance Profiles. When a new method is introduced for numerically calculating integrals, its usefulness is usually demonstrated by solving a small number of selected test problems. Only limited information is given as to the practical range of application of the new numerical integration scheme, and often the following phenomenon can be observed: while accurate results can be efficiently produced for problems similar to the selected test problems, results for other kinds of integrals are either inaccurate or require an excessive computational effort (Genz [167]).

Clearly, this way of testing new algorithms and software does *not* provide a sound basis for assessing the quality of numerical integration algorithms.

Instead, standard testing schemes for numerical integration software rely on extensive *test batteries* and *performance profiles*. But even these standard testing schemes may have to be modified and extended if parallel numerical integration software is to be evaluated.

Test Batteries. A *test battery* is a collection of integration problems in which each problem is distinguished by a particular, individual characteristic. The integration problems are chosen such that their exact analytical solutions are known. Once a test battery has been devised, the quality of a numerical integration scheme can be assessed by giving it the task of evaluating all integrals contained in the test battery and by recording the relevant output data. These data usually include, for each test problem, (i) the approximate integral value, (ii) the estimated error in the approximate result, (iii) the required number of f-evaluations, and (iv) the value of the error parameter on exit from the program.

Based on this information, both the reliability and the performance of the tested integration program can be assessed for the different problems in the test battery.

Correctness. A tested integration program is correct for a particular integration problem if the actual error, i.e., the difference between the approximate and the exact integral values, is smaller than the estimated error in the approximate result.

Accuracy. A correct numerical integration program is sufficiently accurate for a particular integration problem if the estimated error in the approximate result is less than the problem's error requirement.

Robustness. The robustness of an integration program can be assessed by including integrands in the test battery (for instance, nonintegrable functions) which are beyond the program's specified range of applicability. For such integrands, an appropriate error parameter should be set by the tested routine.

Performance. A numerical integration program solves a particular integration problem efficiently if the number of required f-evaluations is "small." The observed numbers of f-evaluations can be assessed, for instance, by comparing them to the corresponding results of well-established programs or to an appropriate index of difficulty for the test integrand.

Example (Quadpack). The documentation in QUADPACK [339] contains numerical examples which demonstrate the range of applicability of its various subroutines. The numerical examples involve 14 kinds of integration problems with different characteristics. The following table shows a selection of these test integrands for the interval [0, 1] along with a short description of their distinctive features:

$f(x)$	*Characteristics of Integration Problem*
$x^{-0.9}\log(1/x)$	algebraic-logarithmic, endpoint singularity
$4^{-10}/((x-\pi/4)^2+4^{-20})$	interior peak
$\cos(2^{10}\sin x)$	heavy, nonlinear oscillations
$\|x-1/3\|^{-0.8}$	interior, algebraic singularity

Performance Profiles. The results of test battery analyses have often been observed to be unreliable. The main problem is that the suitability of a numerical integration program for a certain class of integration problems cannot be evaluated accurately by testing the program against a *single* particular problem of this class.

Example (Integrand with Peak). Suppose the suitability of numerical quadrature programs for integrands with interior peaks is to be evaluated. A suitable test integrand (for the interval [0, 1]) would be

$$f_{u,a}(x) := \frac{4^{-a}}{(x-u)^2+4^{-2a}}, \quad u \in (0,1), \quad a \geq 1. \tag{10.1}$$

$f_{u,a}$ has a pronounced peak at $x = u$ if a is chosen sufficiently large. It has been demonstrated by Lyness and Kaganove [277] that the integration error in Gauss-Legendre quadrature formulas for $f_{u,a}$ depends heavily on the location of the peak, i.e., on the selection of the parameter u.[110] This shows that the quality of Gauss-Legendre quadrature formulas (and any numerical integration algorithm based on them) for integrands with peaks *cannot* be assessed using only a single test integrand.

The performance profile method is an attempt to tackle this problem by using *parameterized families* of test integrands, instead of single test integrands, for representing different classes of integration problems. By testing a numerical integration program against a representative sample of test integrands from such a family, the suitability of the numerical integration program for the corresponding type of integration problem can be accurately assessed.

Test families are usually parameterized using ordered pairs $(\mathbf{u}, \mathbf{a})$, $\mathbf{u}, \mathbf{a} \in \mathbb{R}^N$. The value of the parameter $\mathbf{a}$ determines the *difficulty* of the integration problem, whereas by varying $\mathbf{u}$ (while keeping $\mathbf{a}$ constant) different integration problems of the same kind and the same difficulty can be generated. The parameter $\mathbf{u}$ does *not* affect the difficulty of the resulting integration problem and is usually varied randomly to obtain a representative sample of similar integration problems.

Example (Integrand with Peak). Consider the family of peaked integrands $f_{u,a}$ in (10.1). The parameter u describes the location of the peak, while the parameter

[110] This phenomenon is discussed in more detail in connection with error estimation procedures in Section 8.1.1.

a determines the height, 4^a, of the peak at $x = u$. The difficulty of $f_{u,a}$ depends primarily on the height of the peak, i.e., on the value of the parameter a. Thus, by choosing a random sample of parameters $u_1, \ldots, u_L \in (0, 1)$ while keeping a constant, a sample of integration problems of the same kind and the same difficulty can be generated. By varying a, the difficulty of these integration problems can be changed.

The performance profile method was originally introduced for testing the quality of one-dimensional integration programs (Lyness and Kaganove [277]). For instance, the documentation from the software package QUADPACK [339] contains extensive performance profiles of its constituent programs. Subsequently, this test method has also been applied to multidimensional integration software (Genz [167]). For instance, the performance profile method was used by Berntsen [22] for evaluating the quality of the various NAG integration subroutines for N-dimensional cubes. Berntsen et al. [25] and [29] used the performance profile method to assess the quality of numerical integration software for tetrahedra and the N-dimensional cubes, respectively, whereas Berntsen [24] tested the reliability of error estimation methods based on performance profiles.

Testing the Performance of Parallel Numerical Integration Software. When it comes to testing the performance of parallel numerical integration software, the performance profile method in its current form has two major shortcomings.

(i) The only performance data recorded in current performance profile methods is the number of f-evaluations required for the solution of the different test problems. While the number of required f-evaluations is a valid indicator of the performance of *sequential* numerical integration programs (see Section 10.1.2), it is not necessarily a useful indicator for *parallel* numerical integration programs because there may not be a direct relationship between the number of required f-evaluations and the run-time of such algorithms.

(ii) The integrand families used in current performance profile methods are designed in such a way that the suitability of numerical integration software for integrands with different *analytical* characteristics can be assessed. The performance of parallel numerical integration programs may, however, also depend on *computational* characteristics of a given integrand. For instance, whether or not a parallelization method is effective may depend on the computational complexity of f-evaluations (see Section 9.2).

The currently used test families incorporate only integrands with very specific computational characteristics:

— The integrands in all test families are computationally cheap: an

f-evaluation usually requires only a small number of rational operations plus the evaluation of one or two transcendental functions.

— Integrands belonging to the same family generally have the same computational complexity.

— The computational complexity of f-evaluations for a given integrand is essentially independent of the chosen abscissa $\mathbf{x} \in B$.

The computational characteristics of practical integration problems may be quite different. For instance, evaluating integrands originating from integral transform methods (see Section 2.2) may require the solution of algebraic or differential equations. Evaluating such integrands usually takes significantly more computational effort than simply performing a small number of rational operations and transcendental functions. Therefore, the currently used test families do not allow an accurate assessment of the performance of parallel numerical integration programs.

Problem (i) can be easily overcome by incorporating additional performance indices (see Section 10.1.2) into performance profile methods, whereas problem (ii) can be dealt with by including additional parameters in test families. These parameters should make it possible to specify *computational*, rather than analytical, characteristics of the test integrands.

Example (Computational Characteristics). The computational characteristics of test integrands may, for instance, be described by two parameters $(F_\mu(f), F_\sigma(f))$. Let $F_\mathbf{x}(f)$ denote the number of floating-point operations required for the evaluation of an integrand $f: B \subseteq \mathbb{R}^N \to \mathbb{R}$ at an abscissa $\mathbf{x} \in B$. Then,

$$F_\mu(f) := \frac{1}{\text{vol}(B)} \int_B F_\mathbf{x}(f)\, d\mathbf{x}$$

is defined as the average number of floating-point operations required for evaluating the integrand f. $F_\sigma(f)$ denotes the corresponding standard deviation,

$$F_\sigma(f) := \sqrt{\frac{1}{\text{vol}(B)} \int_B \left(F_\mathbf{x}(f) - F_\mu(f)\right)^2 d\mathbf{x}}.$$

While $F_\mu(f)$ determines the overall computational complexity of f, $F_\sigma(f)$ measures the degree of local variations in $F_\mathbf{x}(f)$.

The value of $F_\mu(f)$ may have a profound impact on the optimal grain size of the applied parallelization strategy, while $F_\sigma(f)$ may crucially influence the optimal load balancing scheme. For small values of $F_\sigma(f)$, $F_\sigma(f) \ll F_\mu(f)$, the computational complexities $F_\mathbf{x}(f) \approx F_\mu(f)$ are essentially constant and can thus be predicted in advance with high accuracy. This fact indicates that static load balancing can be used to assign a given set of f-evaluations to the available processors; for larger values of $F_\sigma(f)$, the computational complexities $F_\mathbf{x}(f)$ become increasingly unpredictable, which may indicate that dynamic load balancing should be used for this purpose (see Section 9.2.2).

Bibliography

[1] M. Abramowitz, I. A. Stegun, *Handbook of Mathematical Functions*, Vol. 55 of Appl. Math. Ser., 10th edition, National Bureau of Standards, U.S. Government Printing Office, Washington, D.C., 1972.

[2] C. Addison, J. Allwright, N. Binsted, N. Bishop, B. Carpenter, P. Dalloz, D. Gee, V. Getov, T. Hey, R. Hockney, M. Lemke, J. Merlin, M. Pinches, C. Scott, I. Wolton, *The* GENESIS *distributed memory benchmarks—Part 1: Methodology and general relativity benchmark with results for the* SUPRENUM *computer*, Concurrency: Practice and Experience 5 (1993), pp. 1–22.

[3] G. Alefeld, J. Herzberger, *Introduction to Interval Computations*, Academic Press, New York, 1983.

[4] American National Standards Institute, *Fortran, ANSI X3.198-1992*, New York, 1992.

[5] S. L. Anderson, *Random number generators on vector supercomputers and other advanced architectures*, SIAM Rev. 32 (1990), pp. 221–251.

[6] A. B. Andreev, V. A. Kascieva, M. Vanmaele, *Some results in lumped mass finite element approximation of eigenvalue problems using numerical quadrature formulas*, J. Comp. Appl. Math. 43 (1992), pp. 291–311.

[7] D. D. Ang, J. Lund, F. Stenger, *Complex variable and regularization methods of inversion of the Laplace transform*, Math. Comp. 53 (1989), pp. 589–608.

[8] K. E. Atkinson, *A survey of boundary integral equation methods for the numerical solution of Laplace's equation in three dimensions*, in Numerical Solution of Integral Equations, M. A. Golberg, Ed., Plenum Press, New York, 1990, pp. 1–34.

[9] K. E. Atkinson, E. Venturino, *Numerical integration of line integrals*, SIAM J. Numer. Anal. 30 (1993), pp. 882–888.

[10] J. Ax, *On Schanuel's conjectures*, Ann. Math. 93 (1971), pp. 252–268.

[11] O. Axelsson, V. A. Barker, *Finite Element Solution of Boundary Value Problems*, Academic Press, New York, 1984.

[12] C. T. H. Baker, *On the nature of certain quadrature formulas and their errors*, SIAM J. Numer. Anal. 5 (1968), pp. 783–804.

[13] C. T. H. Baker, *Convergence and stability of quadrature methods applied to Volterra equations with delay*, IMA J. Numer. Anal. 13 (1993), pp. 67–91.

[14] U. Banerjee, *A note on the effect of numerical quadrature in finite element eigenvalue approximation*, Numer. Math. 61 (1992), pp. 145–152.

[15] U. Banerjee, J.E. Osborn, *Estimation of the effect of numerical integration in finite element eigenvalue approximation*, Numer. Math. 56 (1990), pp. 735–762.

[16] F.L. Bauer (Ed.), *Software Engineering—An Advanced Course*, Springer-Verlag, Berlin, Heidelberg, New York, Tokyo, 1977.

[17] F.L. Bauer, H. Rutishauser, E. Stiefel, *New aspects in numerical quadrature*, in Proceedings of Symposia in Applied Mathematics, Amer. Math. Soc., Providence, RI, 1963, pp. 199–219.

[18] M. Beckers, R. Cools, *A Relation between Cubature Formulae of Trigonometric Degree and Lattice Rules*, Report TW 181, Department of Computer Science, Katholieke Universiteit Leuven, Belgium, 1992.

[19] M. Beckers, R. Cools, *A relation between cubature formulae of trigonometric degree and lattice rules*, in Numerical Integration IV—Proceedings of the Conference at the Mathematical Institute, Oberwolfach, 1992, H. Braß, G. Hämmerlin, Eds., Birkhäuser, Basel, 1993, pp. 13–24.

[20] M. Beckers, A. Haegemans, *Transformation of integrands for lattice rules*, in Numerical Integration—Recent Developments, Software and Applications, T.O. Espelid, A. Genz, Eds., Kluwer, Dordrecht, the Netherlands, 1992, pp. 329–340.

[21] J.O. Berger, *Statistical Decision Theory—Foundations, Concepts, and Methods*, Springer-Verlag, Berlin, Heidelberg, New York, Tokyo, 1980.

[22] J. Berntsen, *A Test of NAG Software for Automatic Numerical Integration over the Cube*, Reports in Informatics 15, Department of Informatics, University of Bergen, Norway, 1985.

[23] J. Berntsen, *Adaptive Multidimensional Quadrature Routines on Shared Memory Parallel Computers*, Reports in Informatics 29, Department of Informatics, University of Bergen, Norway, 1987.

[24] J. Berntsen, *Practical Error Estimation in Adaptive Multidimensional Quadrature Routines*, Reports in Informatics 30, Department of Informatics, University of Bergen, Norway, 1988.

[25] J. Berntsen, R. Cools, T.O. Espelid, *A Test of* DCUTET, Reports in Informatics 46, Department of Informatics, University of Bergen, Norway, 1990.

[26] J. Berntsen, T.O. Espelid, *A Parallel Global Adaptive Quadrature Algorithm for Machines with Hypercube Structure*, Reports in Informatics 27, Department of Informatics, University of Bergen, Norway, 1987.

[27] J. Berntsen, T.O. Espelid, *A parallel global adaptive quadrature algorithm for hypercubes*, Parallel Comput. 8 (1988), pp. 313–323.

[28] J. Berntsen, T.O. Espelid, DCUTRI*: An Algorithm for Adaptive Cubature over a Collection of Triangles*, ACM Trans. Math. Software 18 (1992), pp. 329–342.

[29] J. Berntsen, T.O. Espelid, A. Genz, *An adaptive algorithm for the approximate calculation of multiple integrals*, ACM Trans. Math. Software 17 (1991), pp. 437–451.

[30] J. Berntsen, T.O. Espelid, T. Soerevik, *On the subdivision strategy in adaptive quadrature algorithms*, J. Comp. Appl. Math. 35 (1991), pp. 119–132.

[31] B. Bertram, *Numerical implementation of a Brakhage-like technique for weakly singular integral equations using product integration*, Comput. Math. Appl. 26 (1993), pp. 21–26.

[32] V.A. Berzins, *Software Engineering with Abstractions*, Addison-Wesley, Reading, MA, 1991.

[33] B. Bialecki, *A sinc-Hunter quadrature rule for Cauchy principal value integrals*, Math. Comp. 55 (1990), pp. 665–681.

[34] B. Bialecki, *A sinc quadrature rule for Hadamard finite-part integrals*, Numer. Math. 57 (1990), pp. 263–269.

[35] B. Bialecki, *Sinc quadratures for Cauchy principal value integrals*, in Numerical Integration—Recent Developments, Software and Applications, T. O. Espelid, A. Genz, Eds., Kluwer, Dordrecht, the Netherlands, 1992, pp. 81–92.

[36] B.W. Boehm, J.R. Brown, H. Kaspar, H. Lipow, G.J. MacLeod, M.J. Merrit, *Characteristics of Software Quality*, North-Holland, Amsterdam, New York, Oxford, 1977.

[37] R.F. Boisvert, S.E. Howe, D.K. Kahaner, GAMS—*a framework for the management of scientific software*, ACM Trans. Math. Software 11 (1985), pp. 313–356.

[38] M. Bourdeau, A. Pitre, *Tables of good lattices in four and five dimensions*, Numer. Math. 47 (1985), pp. 39–43.

[39] H. Braß, *Quadraturverfahren*, Vandenhoeck und Ruprecht, Göttingen, Germany, 1977.

[40] C. Brezinski, *A subroutine for the general interpolation and extrapolation problems*, ACM Trans. Math. Software 8 (1982), pp. 290–301.

[41] C. Brezinski, M. Redivo Zaglia, *Extrapolation Methods: Theory and Practice*, Elsevier, Amsterdam, New York, 1991.

[42] I.N. Bronstein, K.A. Semendjajew, *Taschenbuch der Mathematik*, 25th edition, Teubner, Stuttgart, 1991.

[43] M. Bronstein, *Integration of elementary functions*, J. Symbol. Comput. 9 (1990), pp. 117–173.

[44] M. Bronstein, *Symbolic Integration I—Transcendental Functions*, Springer-Verlag, Berlin, Heidelberg, New York, Tokyo, 1996.

[45] J.M. Bull, T.L. Freeman, *Parallel globally adaptive quadrature on the KSR-1*, Adv. Comput. Math. 2 (1994), pp. 357–373.

[46] B. Buning, N. Wittels, D.S. Fussell, *Non-uniform patch luminance for global illumination*, in Proceedings of Graphics Interface '92, Canadian Information Processing Society, Toronto, 1992, pp. 310–318.

[47] K. Burrage, *An adaptive numerical integration code for a chain of transputers*, Parallel Comput. 16 (1990), pp. 305–312.

[48] F. Calio, E. Marchetti, *Derivation and implementation of an algorithm for singular integrands*, Computing 38 (1987), pp. 235–245.

[49] F. Calio, E. Marchetti, *Complex Gauss Kronrod integration rules for certain Cauchy principal value integrals*, Computing 50 (1993), pp. 165–173.

[50] O. Caprani, K. Madsen, L.B. Rall, *Integration of interval functions*, SIAM J. Math. Anal. 12 (1981), pp. 321–341.

[51] R. Carino, *Numerical Integration over Finite Regions Using Extrapolation by Nonlinear Sequence Transformations*, Ph.D. Thesis, La Trobe University, Bundoora, Victoria, Australia, 1992.

[52] R. Carino, I. Robinson, E. de Doncker, *An algebraic study of the Levin transform in numerical integration*, in Numerical Integration—Recent Developments, Software and Applications, T. O. Espelid, A. Genz, Eds., Kluwer, Dordrecht, the Netherlands, 1992, pp. 175–186.

[53] R. Carino, I. Robinson, E. de Doncker, *An algorithm for automatic integration of certain singular functions over a triangle*, in Numerical Integration—Recent

Developments, Software and Applications, T. O. Espelid, A. Genz, Eds., Kluwer, Dordrecht, the Netherlands, 1992, pp. 295–304.

[54] B.W. Char, K.O. Geddes, G.H. Gonnet, B.L. Leong, M.B. Monagan, S. Watt, *First Leaves: A Tutorial Introduction to Maple V*, Springer-Verlag, Berlin, Heidelberg, New York, Tokyo, 1992.

[55] M.M. Chawla, S. Kumar, *Convergence of quadratures for Cauchy principal value integrals*, Computing 23 (1979), pp. 67–72.

[56] E.W. Cheney, *Introduction to Approximation Theory*, McGraw-Hill, New York, 1966.

[57] G.W. Cherry, *Integration in finite terms with special functions: The logarithmic integral*, SIAM J. Comput. 15 (1986), pp. 1–21.

[58] C. Chiodo, G. Criscuolo, *On the convergence of a rule by Monegato for the numerical evaluation of Cauchy principal value integrals*, Computing 40 (1988), pp. 67–74.

[59] P.G. Ciarlet, *The Finite Element Method for Elliptic Problems*, Vol. 4 of Stud. Math. Appl., North-Holland, Amsterdam, New York, Oxford, 1978.

[60] T.F. Coleman, C. Van Loan, *Handbook for Matrix Computations*, SIAM, Philadelphia, 1988.

[61] R. Cools, *A survey of methods for constructing cubature formulae*, in Numerical Integration—Recent Developments, Software and Applications, T. O. Espelid, A. Genz, Eds., Kluwer, Dordrecht, the Netherlands, 1992, pp. 1–24.

[62] R. Cools, D.P. Laurie, L. Pluym, *Algorithm 764: Cubpack++—a C++ package for automatic two-dimensional cubature*, ACM Trans. Math. Software 23 (1997), pp. 1–15.

[63] R. Cools, D.P. Laurie, L. Pluym, *A User Manual for Cubpack++—Version 1.1*, Reports TW 255, Computer Science, Katholieke Universiteit Leuven, Belgium, 1997.

[64] R. Cools, P. Rabinowitz, *Monomial Cubature Rules Since "Stroud"—A Compilation*, Report TW 161, Department of Computer Science, Katholieke Universiteit Leuven, Belgium, 1991.

[65] R. Cools, H.J. Schmid, *Minimal cubature formulae of degree $2k - 1$ for two classical functionals*, Computing 43 (1989), pp. 141–157.

[66] R. Cools, I.H. Sloan, *Minimal Cubature Formulae of Trigonometric Degree*, Preprint AMR93/5, Department of Applied Mathematics, University of New South Wales, Australia, 1993.

[67] G.F. Corliss, *Performance of self-validating adaptive quadrature*, in Numerical Integration—Recent Developments, Software and Applications, P. Keast, G. Fairweather, Eds., D. Reidel, Dordrecht, the Netherlands, 1987, pp. 239–259.

[68] G.F. Corliss, G. Krenz, *Indefinite integration with validation*, ACM Trans. Math. Software 15 (1989), pp. 375–393.

[69] G. Criscuolo, G. Mastroianni, *On the convergence of an interpolatory product rule for evaluating Cauchy principal value integrals*, Math. Comp. 48 (1987), pp. 725–735.

[70] G. Criscuolo, G. Mastroianni, *On the convergence of product formulas for the numerical evaluation of derivatives of Cauchy principal value integrals*, SIAM J. Numer. Anal. 25 (1988), pp. 713–727.

[71] G. Criscuolo, G. Mastroianni, *On the uniform convergence of Gaussian quadra-*

ture rules for Cauchy principal value integrals, Numer. Math. 54 (1989), pp. 445–461.

[72] G. Criscuolo, G. Mastroianni, G. Monegato, *Convergence properties of a class of product formulas for weakly singular integral equations*, Math. Comp. 55 (1990), pp. 213–230.

[73] K.S. Crump, *Numerical inversion of Laplace transforms using a Fourier series approximation*, J. Assoc. Comput. Mach. 23 (1976), pp. 89–96.

[74] C. Dagnino, *Extended product integration rules*, BIT 23 (1983), pp. 488–499.

[75] C. Dagnino, *Extensions of some results for interpolatory product integration rules to rules not necessarily of interpolatory type*, SIAM J. Numer. Anal. 23 (1986), pp. 1284–1289.

[76] C. Dagnino, *Product integration of singular integrands based on cubic spline interpolation at equally spaced nodes*, Numer. Math. 57 (1990), pp. 97–104.

[77] C. Dagnino, V. Demichelis, E. Santi, *An algorithm for numerical integration based on quasi-interpolating splines*, Numer. Algorithms 5 (1993), pp. 443–452.

[78] C. Dagnino, V. Demichelis, E. Santi, *Numerical integration based on quasi-interpolating splines*, Computing 50 (1993), pp. 149–163.

[79] C. Dagnino, C. Fiorentino, *Computation of nodes and weights of extended Gaussian rules*, Computing 32 (1984), pp. 271–278.

[80] C. Dagnino, A. Palamara Orsi, *Product integration of piecewise continuous integrands based on cubic spline interpolation at equally spaced nodes*, Numer. Math. 52 (1988), pp. 459–466.

[81] C. Dagnino, E. Santi, *On the evaluation of one-dimensional Cauchy principal value integrals by rules based on cubic spline interpolation*, Computing 43 (1990), pp. 267–276.

[82] C. Dagnino, E. Santi, *On the convergence of spline product quadratures for Cauchy principal value integrals*, J. Comput. Appl. Math. 36 (1991), pp. 181–187.

[83] J. Dagpunar, *Principles of Random Variate Generation*, Clarendon Press, Oxford, UK, 1988.

[84] M. D'Apuzzo, M. Lapegna, A. Murli, *Scalability and Load Balancing in Adaptive Algorithms for Multi-Dimensional Integration*, Technical Report 2/95, Centro di Ricerche per il Calcolo Parallelo e i Supercalcolatori, Naples, Italy, 1995.

[85] J.H. Davenport, *On the Integration of Algebraic Functions*, Vol. 102 of Lecture Notes in Comput. Sci., Springer-Verlag, Berlin, Heidelberg, New York, Tokyo, 1981.

[86] J.H. Davenport, *Integration formelle*, Report 375, IMAG, Grenoble, France, 1983.

[87] J.H. Davenport, *Integration: Formal and numeric approaches*, in Tools, Methods and Languages for Scientific and Engineering Computation, B. Ford, J.C. Rault, F. Thomasset, Eds., North-Holland, Amsterdam, New York, Oxford, 1984, pp. 417–426.

[88] J.H. Davenport, Y. Siret, E. Tournier, *Computer Algebra—Systems and Algorithms for Algebraic Computation*, 2nd edition, Academic Press, New York, 1993.

[89] B. Davies, B. Martin, *Numerical inversion of the Laplace transform: A survey and comparison of methods*, J. Comp. Phys. 33 (1979), pp. 1–32.

[90] P.J. Davis, *Interpolation and Approximation*, Blaisdell, New York, 1963.

[91] P.J. Davis, *A construction of nonnegative approximate quadratures*, Math. Comp. 21 (1967), pp. 578–582.

[92] P. J. Davis, P. Rabinowitz, *Methods of Numerical Integration*, 2nd edition, Academic Press, New York, 1984.

[93] C. de Boor, CADRE: *An algorithm for numerical quadrature*, in Mathematical Software, J. R. Rice, Ed., Academic Press, New York, 1971, pp. 417–449.

[94] C. de Boor, *On the error in multivariate polynomial interpolation*, Appl. Numer. Math. 10 (1992), pp. 297–305.

[95] C. de Boor, *Gauss elimination by segments and multivariate polynomial interpolation*, in Approximation and Computation: A Festschrift in Honor of Walter Gautschi, R. V. M. Zahar, Ed., Vol. 119 of Internat. Ser. Numer. Math., Birkhäuser, Boston, 1994, pp. 1–22.

[96] C. de Boor, A. Ron, *On multivariate polynomial interpolation*, Construct. Approx. 6 (1990), pp. 287–302.

[97] C. de Boor, A. Ron, *On polynomial ideals of finite codimension with applications to box spline theory*, J. Math. Anal. Appl. 158 (1991), pp. 168–193.

[98] C. de Boor, A. Ron, *Computational aspects of polynomial interpolation in several variables*, Math. Comp. 58 (1992), pp. 705–727.

[99] C. de Boor, A. Ron, *The least solution for the polynomial interpolation problem*, Math. Z. 210 (1992), pp. 347–378.

[100] E. de Doncker, *New Euler–Maclaurin expansions and their application to quadrature over the s-dimensional simplex*, Math. Comp. 33 (1979), pp. 1003–1018.

[101] E. de Doncker, *Asymptotic expansions and their application in numerical integration*, in Numerical Integration—Recent Developments, Software and Applications, P. Keast, G. Fairweather, Eds., D. Reidel, Dordrecht, the Netherlands, 1987, pp. 141–151.

[102] E. de Doncker, J. A. Kapenga, *A portable parallel algorithm for multivariate numerical integration and its performance analysis*, in Third SIAM Conference on Parallel Processing for Scientific Computing, Los Angeles, SIAM, Philadelphia, 1988, pp. 109–113.

[103] E. de Doncker, J. A. Kapenga, *Parallel cubature on loosely coupled systems*, in Numerical Integration—Recent Developments, Software and Applications, T. O. Espelid, A. Genz, Eds., Kluwer, Dordrecht, the Netherlands, 1992, pp. 317–327.

[104] E. de Doncker, J. A. Kapenga, S. Zuidema, *Automatic integration using* DIME, paper presented at the conference Numerical Integration—Recent Developments, Software and Applications, Bergen, Norway, 1991.

[105] E. de Doncker, I. Vakalis, *A parallel global adaptive algorithm for integration over the n-dimensional simplex on shared memory machines*, paper presented at the conference Numerical Integration—Recent Developments, Software and Applications, Bergen, Norway, 1991.

[106] E. de Doncker, I. Vakalis, *Convergence results and speedup of parallel numerical integration algorithms*, in Proceedings of the Sixth SIAM Conference on Parallel Processing, R. F. Sincovec, D. E. Keyes, M. R. Leuze, L. R. Petzold, D. A. Reed, Eds., SIAM, Philadelphia, 1993, pp. 539–545.

[107] D. Delbourgo, D. Elliott, *On the approximate evaluation of Hadamard finite-part integrals*, IMA J. Numer. Anal. 14 (1994), pp. 485–500.

[108] L. Devroye, *Non-Uniform Random Variate Generation*, Springer-Verlag, Berlin, Heidelberg, New York, Tokyo, 1986.

[109] K. Diethelm, *Modified compound quadrature rules for strongly singular integrals*, Computing 52 (1994), pp. 337–354.

[110] S. Disney, I. H. Sloan, *Lattice integration rules of maximal rank*, SIAM J. Numer. Anal. 29 (1992), pp. 566–577.

[111] V. A. Ditkin, A. P. Prudnikov, *Integral Transforms and Operational Calculus*, Pergamon Press, New York, Oxford, 1965.

[112] J. J. Dongarra, *The* LINPACK *benchmark: An explanation*, in Evaluating Supercomputers, A. J. von der Steen, Ed., Chapman and Hall, London, 1990.

[113] J. J. Dongarra, J. du Croz, S. Hammarling, I. Duff, *A set of level 3 basic linear algebra subprograms*, ACM Trans. Math. Software 16 (1990), pp. 1–17.

[114] J. J. Dongarra, J. du Croz, S. Hammarling, R. J. Hanson, *An extended set of Fortran basic linear algebra subprograms*, ACM Trans. Math. Software 14 (1988), pp. 1–17.

[115] J. J. Dongarra, E. Grosse, *Distribution of mathematical software via electronic mail*, Comm. ACM 30 (1987), pp. 403–407.

[116] J. J. Dongarra, R. C. Whaley, *A Users' Guide to the* BLACS *v1.0.* LAPACK *Working Note No.94*, Technical Report CS-95-281, Department of Computer Science, University of Tennessee, Knoxville, TN, 1995, (http://www.netlib.org/lapack/lawns/lawn94.ps).

[117] K. Dowd, *High Performance Computing*, O'Reilly, Sebastopol, CA, 1993.

[118] Z. Drezner, *Computation of the multivariate normal integral*, ACM Trans. Math. Software 18 (1992), pp. 470–480.

[119] D. G. Duffy, *On the numerical inversion of Laplace transforms: Comparison of three new methods on characteristic problems from applications*, ACM Trans. Math. Software 19 (1993), pp. 333–359.

[120] M. G. Duffy, *Quadrature over a pyramid or cube with a singularity at a vertex*, SIAM J. Numer. Anal. 19 (1982), pp. 1260–1262.

[121] C. H. Edwards, *Advanced Calculus in Several Variables*, Academic Press, New York, 1973.

[122] P. P. B. Eggermont, C. Lubich, *Operational quadrature methods for Wiener-Hopf integral equations*, Math. Comp. 60 (1993), pp. 699–718.

[123] U. T. Ehrenmark, *Far field asymptotics of the two-dimensional linearized sloping beach problem*, SIAM J. Appl. Math. 47 (1987), pp. 965–981.

[124] U. T. Ehrenmark, *A three-point formula for numerical quadrature of oscillatory integrals with variable frequency*, J. Comput. Appl. Math. 21 (1988), pp. 87–99.

[125] J. Eichenauer-Herrmann, H. Niederreiter, *On the discrepancy of quadratic congruential pseudorandom numbers*, J. Comp. Appl. Math. 34 (1991), pp. 243–249.

[126] M. C. Eiermann, *Automatic, guaranteed integration of analytic functions*, BIT 29 (1989), pp. 270–282.

[127] D. Elliott, *An asymptotic analysis of two algorithms for certain Hadamard finite-part integrals*, IMA J. Numer. Anal. 13 (1993), pp. 445–462.

[128] D. Elliott, D. F. Paget, *On the convergence of a quadrature rule for evaluating certain Cauchy principal value integrals*, Numer. Math. 23 (1975), pp. 311–319.

[129] D. Elliott, D. F. Paget, *On the convergence of a quadrature rule for evaluating certain Cauchy principal value integrals—an addendum*, Numer. Math. 25 (1976), pp. 287–289.

[130] D. Elliott, D. F. Paget, *Gauss type quadrature rules for Cauchy principal value integrals*, Math. Comp. 33 (1979), pp. 301–309.

[131] H. Engels, *Numerical Quadrature and Cubature*, Academic Press, New York, 1980.

[132] A. Erdelyi, W. Magnus, F. Oberhettinger, F. Tricomi, *Tables of Integral Transforms*, McGraw-Hill, New York, 1954.

[133] T. O. Espelid, *On the construction of good symmetric integration rules*, SIAM J. Numer. Anal. 24 (1987), pp. 855–881.

[134] T. O. Espelid, *Integration Rules, Null Rules and Error Estimation*, Reports in Informatics 33, Department of Informatics, University of Bergen, Norway, 1988.

[135] T. O. Espelid, DQAINT*: An algorithm for adaptive quadrature (of a vector function) over a collection of finite intervals*, in Numerical Integration—Recent Developments, Software and Applications, T. O. Espelid, A. Genz, Eds., Kluwer, Dordrecht, the Netherlands, 1992, pp. 341–342.

[136] T. O. Espelid, *Integrating singularities using non-uniform subdivision and extrapolation*, in Numerical Integration IV, H. Braß, G. Hämmerlin, Eds., Vol. 112 of Internat. Ser. Numer. Math., Birkhäuser, Basel, 1993, pp. 77–89.

[137] T. O. Espelid, *On integrating vertex singularities using extrapolation*, BIT 34 (1994), pp. 62–79.

[138] T. O. Espelid, A. Genz, *On the Subdivision Strategy in Adaptive Cubature Algorithms for Triangular Regions*, Reports in Informatics 74, Department of Informatics, University of Bergen, Norway, 1992.

[139] T. O. Espelid, A. Genz, DECUHR*: An algorithm for automatic integration of singular functions over a hyperrectangular region*, Numer. Algorithms 8 (1994), pp. 201–220.

[140] T. O. Espelid, K. J. Overholt, DQAINF*: An algorithm for automatic integration of infinite oscillating tails*, Numer. Algorithms 8 (1994), pp. 83–101.

[141] T. O. Espelid, T. Soerevik, *A discussion of a new error estimate for adaptive quadrature*, BIT 29 (1989), pp. 283–294.

[142] G. A. Evans, *Practical Numerical Integration*, Wiley, Chichester, 1993.

[143] G. A. Evans, *Two robust methods for irregular oscillatory integrals over a finite range*, Appl. Numer. Math. 14 (1994), pp. 383–395.

[144] H. Faure, *Discrépances de suites associées à un système de numération (en dimension un)*, Bull. Soc. Math. France 109 (1981), pp. 143–182.

[145] P. Favati, G. Lotti, F. Romani, *Interpolatory integration formulas for optimal composition*, ACM Trans. Math. Software 17 (1991), pp. 207–217.

[146] P. Favati, G. Lotti, F. Romani, *Local error estimates in quadrature*, BIT 31 (1991), pp. 102–111.

[147] L. Fejér, *Mechanische Quadraturen mit positiven Cotes'schen Zahlen*, Math. Z. 37 (1933), pp. 287–310.

[148] T. Fessler, W. F. Ford, D. A. Smith, HURRY*: An acceleration algorithm for scalar sequences and series*, ACM Trans. Math. Software 9 (1983), pp. 346–354.

[149] L. N. G. Filon, *On a quadrature formula for trigonometric integrals*, Proc. Roy. Soc. Edinburgh 49 (1928), pp. 38–47.

[150] W. F. Ford, A. Sidi, *An algorithm for a generalization of the Richardson extrapolation process*, SIAM J. Numer. Anal. 24 (1987), pp. 1212–1232.

[151] B. L. Fox, *Implementation and relative efficiency of quasirandom sequence generators*, ACM Trans. Math. Software 12 (1986), pp. 362–376.

[152] T. L. Freeman, J. M. Bull, *Shared memory and message passing implementations of parallel algorithms for numerical integration*, in Parallel Scientific Computing, J. Dongarra, J. Wasniewski, Eds., Vol. 879 of *Lecture Notes in Comput. Sci.*, Springer-Verlag, Berlin, Heidelberg, New York, Tokyo, 1994, pp. 219–228.

[153] F. N. Fritsch, D. K. Kahaner, J. N. Lyness, *Double integration using one-dimensional adaptive quadrature routines: A software interface problem*, ACM Trans. Math. Software 7 (1981), pp. 46–75.

[154] B. S. Garbov, G. Giunta, J. N. Lyness, A. Murli, *A Fortran software package for the numerical inversion of the Laplace transform based on Weeks' method*, ACM Trans. Math. Software 14 (1988), pp. 171–176.

[155] B. S. Garbov, G. Giunta, J. N. Lyness, A. Murli, *Software for an implementation of Weeks' method for the inverse Laplace transform problem*, ACM Trans. Math. Software 14 (1988), pp. 163–170.

[156] K. Gatermann, *The construction of symmetric cubature formulas for the square and the triangle*, Computing 40 (1988), pp. 229–240.

[157] W. Gautschi, *Quadrature formulae on half-infinite intervals*, BIT 31 (1991), pp. 438–446.

[158] W. Gautschi, *On the computation of generalized Fermi–Dirac and Bose–Einstein integrals*, Comput. Phys. Comm. 74 (1993), pp. 233–238.

[159] W. Gautschi, M. A. Kovacevic, G. V. Milovanovic, *The numerical evaluation of singular integrals with coth-kernel*, BIT 27 (1987), pp. 389–402.

[160] W. Gautschi, J. Wimp, *Computing the Hilbert transform of a Jacobi weight function*, BIT 27 (1987), pp. 203–215.

[161] K. O. Geddes, S. R. Czapor, G. Labahn, *Algorithms for Computer Algebra*, Kluwer, Dordrecht, the Netherlands, 1992.

[162] A. Geist, A. Beguelin, J. J. Dongarra, W. Jiang, R. Manchek, V. Sunderam, *PVM: Parallel Virtual Machine. A Users' Guide and Tutorial for Networked Parallel Computing*, MIT Press, Cambridge, MA, 1994.

[163] W. M. Gentleman, *Implementing Clenshaw–Curtis quadrature*, Comm. ACM 15 (1972), pp. 337–342, 343–346.

[164] A. Genz, *Numerical multiple integration on parallel computers*, Comput. Phys. Comm. 26 (1982), pp. 349–352.

[165] A. Genz, *Fully symmetric interpolatory rules for multiple integrals*, SIAM J. Numer. Anal. 23 (1986), pp. 1273–1283.

[166] A. Genz, *The numerical evaluation of multiple integrals on parallel computers*, in Numerical Integration—Recent Developments, Software and Applications, P. Keast, G. Fairweather, Eds., D. Reidel, Dordrecht, the Netherlands, 1987, pp. 219–229.

[167] A. Genz, *A package for testing multiple integration subroutines*, in Numerical Integration—Recent Developments, Software and Applications, P. Keast, G. Fairweather, Eds., D. Reidel, Dordrecht, the Netherlands, 1987, pp. 337–340.

[168] A. Genz, *Numerical computation of multivariate normal probabilities*, J. Comp. Graph Stat. 1 (1992), pp. 141–149.

[169] A. Genz, *Statistics applications of subregion adaptive multiple numerical integration*, in Numerical Integration—Recent Developments, Software and Applications, T. O. Espelid, A. Genz, Eds., Kluwer, Dordrecht, the Netherlands, 1992, pp. 267–280.

[170] A. Genz, *Comparison of methods for the computation of multivariate normal probabilities*, Comput. Sci. Statist. 25 (1993).

[171] A. Genz, R. E. Kass, *Subregion adaptive integration of functions having a dominant peak*, J. Comp. Graph. Stat. 6 (1997), pp. 1–20.

[172] A. Genz, A. A. Malik, *An imbedded family of fully symmetric numerical integration rules*, SIAM J. Numer. Anal. 20 (1983), pp. 580–588.

[173] A. Gerasoulis, *Piecewise-polynomial quadratures for Cauchy singular integrals*, SIAM J. Numer. Anal. 23 (1986), pp. 891–902.

[174] G. Giunta, G. Laccetti, M. R. Rizzardi, *More on the Weeks' method for the numerical inversion of the Laplace transform*, Numer. Math. 54 (1988), pp. 193–200.

[175] G. Giunta, A. Murli, *Algorithm 649: A package for computing trigonometric Fourier coefficients based on Lyness's algorithm*, ACM Trans. Math. Software 13 (1987), pp. 97–107.

[176] I. Gladwell, *Vectorization of one-dimensional quadrature codes*, in Numerical Integration—Recent Developments, Software and Applications, P. Keast, G. Fairweather, Eds., D. Reidel, Dordrecht, the Netherlands, 1987, pp. 231–238.

[177] M. A. Golberg, *Introduction to the numerical solution of Cauchy singular integral equations*, in Numerical Solution of Integral Equations, M. A. Golberg, Ed., Plenum Press, New York, 1990, pp. 183–308.

[178] G. H. Golub, *Some modified matrix eigenvalue problems*, SIAM Rev. 15 (1973), pp. 318–334.

[179] G. H. Golub, C. F. Van Loan, *Matrix Computations*, 3rd edition, The Johns Hopkins University Press, Baltimore, 1996.

[180] G. H. Golub, J. H. Welsch, *Calculation of Gauss quadrature rules*, Math. Comp. 23 (1969), pp. 221–230.

[181] H. L. Gray, S. Wang, *A new method for approximating improper integrals*, SIAM J. Numer. Anal. 29 (1992), pp. 271–283.

[182] W. H. Greub, *Linear Algebra*, 3rd edition, Springer-Verlag, Berlin, Heidelberg, New York, Tokyo, 1963.

[183] W. Gropp, E. Lusk, A. Skjellum, *Using MPI: Portable Programming with the Message-Passing Interface*, MIT Press, Cambridge, MA, 1994.

[184] C. Großmann, H. G. Roos, *Numerik partieller Differentialgleichungen*, Teubner, Stuttgart, 1992.

[185] H. Grothe, *Matrixgeneratoren zur Erzeugung gleichverteilter Zufallsvektoren*, in Zufallszahlen und Simulationen, L. Afflerbach, J. Lehn, Eds., Teubner, Stuttgart, 1986, pp. 29–34.

[186] M. Guiggiani, *The evaluation of Cauchy principal value integrals in the boundary element method—A review*, Math. Comput. Modelling 15 (1991), pp. 175–184.

[187] M. Guiggiani, G. Krishnasamy, T. J. Rudolphi, F. J. Rizzo, *A general algorithm for the numerical solution of hypersingular boundary integral equations*, J. Appl. Mech. 59 (1992), pp. 604–614.

[188] S. Haber, *A modified Monte Carlo quadrature*, Math. Comp. 20 (1966), pp. 361–368.

[189] S. Haber, *A modified Monte Carlo quadrature II*, Math. Comp. 21 (1967), pp. 388–397.

[190] W. Hackbusch, *Multi-Grid Methods and Applications*, Vol. 4 of Springer Ser. Comput. Math., Springer-Verlag, Berlin, Heidelberg, New York, Tokyo, 1985.

[191] W. Hackbusch, *Integralgleichungen—Theorie und Numerik*, Teubner, Stuttgart, 1989.

[192] W. Hackbusch, *Iterative Lösung großer schwachbesetzter Gleichungssysteme*, Teubner, Stuttgart, 1991.

[193] W. Hackbusch, *Elliptic Differential Equations—Theory and Numerical Treatment*, Springer-Verlag, Berlin, Heidelberg, New York, Tokyo, 1992.

[194] W. Hackbusch, *Integral Equations: Theory and Numerical Treatment*, Birkhäuser, Basel, 1995.

[195] W. Hackbusch, S. A. Sauter, *On the efficient use of the Galerkin method to solve Fredholm integral equations*, Appl. Math. 38 (1993), pp. 301–322.

[196] J. Hadamard, *Lectures on Cauchy's Problem in Linear Partial Differential Equations*, Yale University Press, New Haven, CT, 1923.

[197] A. Haegemans, R. Cools, *Construction of three-dimensional cubature formulae with points on regular and semi-regular polytopes*, in Numerical Integration—Recent Developments, Software and Applications, P. Keast, G. Fairweather, Eds., D. Reidel, Dordrecht, the Netherlands, 1987, pp. 153–163.

[198] W. Hanke, *Die optimale Sektion bei adaptiven Integrationsverfahren mit globaler Strategie*, Z. Angew. Math. Mech. 62 (1982), pp. T327–T329.

[199] E. R. Hansen, *Global Optimization Using Interval Analysis*, Marcel Dekker, New York, 1992.

[200] J. F. Hart et al., *Computer Approximations*, Wiley, New York, 1968.

[201] T. Hasegawa, T. Torii, *An automatic quadrature for Cauchy principal value integrals*, Math. Comp. 56 (1991), pp. 741–754.

[202] T. Hasegawa, T. Torii, *Hilbert and Hadamard transforms by generalized Chebyshev expansion*, J. Comput. Appl. Math. 51 (1994), pp. 71–83.

[203] J. L. Hennessy, D. A. Patterson, *Computer Architecture—A Quantitative Approach*, 2nd edition, Morgan Kaufmann, San Mateo, CA, 1995.

[204] E. Hermite, *Sur l'integration des fractions rationelles*, Nouvelles Annales de Mathématiques (1872), pp. 145–148.

[205] N. Higham, *The accuracy of floating point summation*, SIAM J. Sci. Comput. 14 (1993), pp. 783–799.

[206] D. R. Hill, C. B. Moler, *Experiments in Computational Matrix Algebra*, Birkhäuser, Basel, 1988.

[207] E. Hlawka, *Funktionen von beschränkter Variation in der Theorie der Gleichverteilung*, Ann. Math. Pura Appl. 54 (1961), pp. 325–333.

[208] E. Hlawka, J. Schoißengeier, R. Taschner, *Geometrische und analytische Zahlentheorie*, Manz Verlag, Vienna, 1986.

[209] R. W. Hockney, C. R. Jesshope, *Parallel Computers 2*, Adam Hilger, Bristol, Philadelphia, 1988.

[210] O. Holzmann, B. Lang, H. Schütt, *Newton's constant of gravitation and verified numerical quadrature*, Reliable Computing 2 (1996), pp. 229–240.

[211] E. Horowitz, *Algorithms for partial fraction decomposition and rational integration*, in Proceeding of SYMSAM '71, S. R. Petrick, Ed., ACM, New York, 1971, pp. 441–457.

[212] L. K. Hua, Y. Wang, *Applications of Number Theory to Numerical Analysis*, Springer-Verlag, Berlin, Heidelberg, New York, Tokyo, 1981.

[213] D. B. Hunter, *Some Gauss-type formulae for the evaluation of Cauchy principal values of integrals*, Numer. Math. 27 (1972), pp. 419–424.

[214] J. P. Imhof, *On the method for numerical integration of Clenshaw and Curtis*, Numer. Math. 5 (1963), pp. 138–141.

[215] N. I. Ioakimidis, *On the uniform convergence of Gaussian quadrature rules for*

Cauchy principal value integrals and their derivatives, Math. Comp. 44 (1985), pp. 191–198.

[216] M. Iri, S. Moriguti, Y. Takasawa, *On a certain quadrature formula*, Kokyuroku of the Research Institute for Mathematical Sciences, Kyoto University, 91 (1970), pp. 82–118.

[217] E. Isaacson, H. B. Keller, *Analysis of Numerical Methods*, Wiley, New York, 1966.

[218] R. D. Jenks, R. S. Sutor, *AXIOM: The Scientific Computation System*, Springer-Verlag, Berlin, Heidelberg, New York, Tokyo, 1992.

[219] A. J. Jerri, *Integral and Discrete Transforms with Applications and Error Analysis*, Vol. 162 of Monogr. Textbooks Pure Appl. Math., Marcel Dekker, New York, 1992.

[220] S. Joe, D. C. Hunt, *The number of lattice rules having given invariants*, Bull. Austral. Math. Soc. 46 (1992), pp. 479–495.

[221] S. Joe, I. H. Sloan, *Imbedded lattice rules for multidimensional integration*, SIAM J. Numer. Anal. 29 (1992), pp. 1119–1135.

[222] S. Joe, I. H. Sloan, *Implementation of a lattice method for numerical multiple integration*, ACM Trans. Math. Software 19 (1993), pp. 523–545.

[223] D. K. Kahaner, *Numerical quadrature by the ε-algorithm*, Math. Comp. 26 (1972), pp. 689–693.

[224] D. K. Kahaner, J. Stoer, *Extrapolated adaptive quadrature*, SIAM J. Sci. Comput. 4 (1983), pp. 31–44.

[225] M. H. Kalos, P. A. Whitlock, *Monte Carlo Methods*, Wiley, New York, 1986.

[226] J. A. Kapenga, E. de Doncker, *A parallelization of adaptive task partitioning algorithms*, Parallel Comput. 7 (1988), pp. 211–225.

[227] R. B. Kearfott, M. Dawande, K. Du, C. Hu, *Algorithm 737: INTLIB: A portable Fortran 77 interval standard-function library*, ACM Trans. Math. Software 20 (1994), pp. 447–459.

[228] P. Keast, J. C. Diaz, *Fully symmetric integration formulas for the surface of the sphere in s dimensions*, SIAM J. Numer. Anal. 20 (1983), pp. 406–419.

[229] G. Kedem, S. K. Zaremba, *A table of good lattice points in three dimensions*, Numer. Math. 23 (1974), pp. 175–180.

[230] B. W. Kernighan, D. M. Ritchie, *The C Programming Language*, 2nd edition, Prentice-Hall, Englewood Cliffs, NJ, 1988.

[231] B. D. Khan, *A computation for tomographic reconstruction from projections*, Appl. Math. Comput. 64 (1994), pp. 79–91.

[232] R. Kieser, *Über hypersinguläre Integrale und einseitige Sprungrelationen in der Methode der Randelemente*, Ph.D. Thesis, University of Stuttgart, 1991.

[233] R. Kieser, C. Schwab, W. L. Wendland, *Numerical evaluation of singular and finite-part integrals on curved surfaces using symbolic manipulation*, Computing 49 (1992), pp. 279–301.

[234] M. Klerer, F. Grossman, *Error rates in tables of indefinite integrals*, Indust. Math. 18 (1968), pp. 31–62.

[235] P. H. Knowles, *Integration of a class of transcendental Liouvillian functions with error-functions I*, J. Symbol. Comput. 13 (1992), pp. 525–543.

[236] P. H. Knowles, *Integration of a class of transcendental Liouvillian functions with error-functions II*, J. Symbol. Comput. 16 (1993), pp. 227–241.

[237] D. E. Knuth, *Seminumerical Algorithms*, Vol. 2 of The Art of Computer Programming, 2nd edition, Addison-Wesley, Reading, MA, 1981.

[238] C. H. Koelbel, D. B. Loveman, R. S. Schreiber, G. L. Steele Jr., M. E. Zosel, *The High Performance Fortran Handbook*, The MIT Press, Cambridge, MA, 1994.

[239] M. Kolar, S. F. O'Shea, *Fast, portable, and reliable algorithm for the calculation of Halton numbers*, Comput. Math. Appl. 25 (1993), pp. 3–13.

[240] W. Krämer, S. Wedner, *Two adaptive Gauss-Legendre type algorithms for the verified computation of definite integrals*, Reliable Computing 2 (1996), pp. 241–254.

[241] E. Krol, *The Whole Internet—User's Guide and Catalog*, O'Reilly, Sebastopol, CA, 1992.

[242] A. R. Krommer, C. W. Ueberhuber, *Architecture Adaptive Algorithms*, Technical Report ACPC/TR 92-2, Austrian Center for Parallel Computation, Vienna, 1992.

[243] A. R. Krommer, C. W. Ueberhuber, *Dynamic Load Balancing—An Overview*, Technical Report ACPC/TR 92-18, Austrian Center for Parallel Computation, Vienna, 1992.

[244] A. R. Krommer, C. W. Ueberhuber, *Architecture adaptive algorithms*, Parallel Comput. 19 (1993), pp. 409–435.

[245] A. R. Krommer, C. W. Ueberhuber, *Data Management for Parallel Integration*, Technical Report SciPaC/TR 93-5, Scientific Parallel Computation Group, Technical University of Vienna, 1993.

[246] A. R. Krommer, C. W. Ueberhuber, *A Simulation System for* AAA *Software Development*, Report 104/93, Institute for Applied and Numerical Mathematics, Technical University of Vienna, 1993.

[247] A. R. Krommer, C. W. Ueberhuber, *Numerical Integration on Advanced Computer Systems*, Vol. 848 of Lecture Notes in Comput. Sci., Springer-Verlag, Berlin, Heidelberg, New York, Tokyo, 1994.

[248] A. S. Kronrod, *Nodes and Weights of Quadrature Formulas*, Consultants Bureau, New York, 1965.

[249] R. L. Kruse, *Data Structures and Program Design*, Prentice-Hall, Englewood Cliffs, NJ, 1984.

[250] V. I. Krylov, *Approximate Calculation of Integrals*, Macmillan, New York, London, 1962.

[251] M. Lapegna, *Global adaptive quadrature for the approximate computation of multidimensional integrals on a distributed-memory multiprocessor*, Concurrency: Practice and Experience 4 (1992), pp. 413–426.

[252] M. Lapegna, A. D'Alessio, *A scalable parallel algorithm for the adaptive multidimensional quadrature*, in Proceedings of the Sixth SIAM Conference on Parallel Processing, R. F. Sincovec, D. E. Keyes, M. R. Leuze, L. R. Petzold, D. A. Reed, Eds., SIAM, Philadelphia, 1993, pp. 933–936.

[253] D. P. Laurie, *Sharper error estimates in adaptive quadrature*, BIT 23 (1983), pp. 258–261.

[254] D. P. Laurie, *Null rules and orthogonal expansions*, in Approximation and Computation: A Festschrift in Honor of Walter Gautschi, R. V. M. Zahar, Ed., Vol. 119 of Internat. Ser. Numer. Math., Birkhäuser, Boston, 1994, pp. 359–370.

[255] C. L. Lawson, R. J. Hanson, D. R. Kincaid, F. T. Krogh, *Basic linear algebra subprograms for Fortran usage*, ACM Trans. Math. Software 5 (1979), pp. 303–323.

[256] D. Lazard, R. Rioboo, *Integration of rational functions—Rational computation of the logarithmic part*, J. Symbol. Comput. 9 (1990), pp. 113–115.

[257] J.M. Lemme, J.R. Rice, *Adaptive quadrature algorithms for the ILLIAC IV*, Internat. J. Comput. Inform. Sci. 9 (1980), pp. 63–72.

[258] D. Levin, *Development of non-linear transformations for improving convergence of sequences*, Internat. J. Comput. Math 3 (1973), pp. 371–388.

[259] D. Levin, *Procedures for computing one and two dimensional integrals of functions with rapid irregular oscillations*, Math. Comp. 38 (1982), pp. 531–538.

[260] D. Levin, A. Sidi, *Two new classes of non-linear transformations for accelerating the convergence of infinite integrals and series*, Appl. Math. Comp. 9 (1981), pp. 175–215.

[261] S. Lewanowicz, *Evaluation of Bessel function integrals with algebraic singularities*, J. Comp. Appl. Math. 37 (1991), pp. 101–112.

[262] T.G. Lewis, W.H. Payne, *Generalized feedback shift register pseudorandom number algorithm*, J. Assoc. Comput. Mach. 20 (1973), pp. 456–468.

[263] P.R. Lipow, F. Stenger, *How slowly can quadrature formulas converge*, Math. Comp. 26 (1972), pp. 917–922.

[264] D.S. Lubinsky, P. Rabinowitz, *Rates of convergence of Gaussian quadrature for singular integrands*, Math. Comp. 43 (1984), pp. 219–242.

[265] D.S. Lubinsky, A. Sidi, *Convergence of product integration rules for functions with interior and endpoint singularities*, Math. Comp. 46 (1986), pp. 229–245.

[266] J. Lund, *Sinc function quadrature rules for the Fourier integral*, Math. Comp. 41 (1983), pp. 103–113.

[267] J. Lund, K.L. Bowers, *Sinc methods for quadrature and differential equations*, SIAM, Philadelphia, 1992.

[268] J.N. Lyness, *The calculation of trigonometric Fourier coefficients*, J. Comput. Phys. 54 (1984), pp. 57–73.

[269] J.N. Lyness, *An error functional expansion for n-dimensional quadrature with an integrand function singular at a point*, Math. Comp. 30 (1976), pp. 1–23.

[270] J.N. Lyness, *The Euler–Maclaurin expansion for the Cauchy principal value integral*, Numer. Math. 46 (1985), pp. 611–622.

[271] J.N. Lyness, *An introduction to lattice rules and their generator matrices*, IMA J. Numer. Anal. 9 (1989), pp. 405–419.

[272] J.N. Lyness, *The canonical forms of a lattice rule*, in Numerical Integration IV—Proceedings of the Conference at the Mathematical Institute, Oberwolfach, 1992, H. Braß, G. Hämmerlin, Eds., Birkhäuser, Basel, 1993, pp. 225–240.

[273] J.N. Lyness, *Finite-part integrals and the Euler–Maclaurin expansion*, in Approximation and Computation: A Festschrift in Honor of Walter Gautschi, R.V.M. Zahar, Ed., Vol. 119 of Internat. Ser. Numer. Math., Birkhäuser, Boston, 1994, pp. 397–407.

[274] J.N. Lyness, E. de Doncker, *On quadrature error expansions I*, J. Comp. Appl. Math. 17 (1987), pp. 131–149.

[275] J.N. Lyness, E. de Doncker, *Quadrature error expansions II—The full corner singularity*, Numer. Math. 64 (1993), pp. 355–370.

[276] J.N. Lyness, G. Giunta, *A modification of the Weeks' method for numerical inversion of the Laplace transform*, Math. Comp. 47 (1986), pp. 313–322.

[277] J.N. Lyness, J.J. Kaganove, *Comments on the nature of automatic quadrature routines*, ACM Trans. Math. Software 2 (1976), pp. 65–81.

[278] J.N. Lyness, G. Monegato, *Quadrature error functional expansion for the simplex when the integrand has singularities at vertices*, Math. Comp. 34 (1980), pp. 213–225.

[279] J.N. Lyness, B.W. Ninham, *Numerical quadrature and asymptotic expansions*, Math. Comp. 21 (1967), pp. 162–178.

[280] J.N. Lyness, S.E. Plowman, *Trapezoidal Rule Quadrature Algorithms for MIMD Distributed Memory Computers*, 1992, preprint.

[281] J.N. Lyness, I.H. Sloan, *Some properties of rank-2 lattice rules*, Math. Comp. 53 (1989), pp. 627–637.

[282] J.N. Lyness, T. Soerevik, *Lattice rules by component scaling*, Math. Comp., 61 (1993), pp. 799–820.

[283] J.N. Lyness, T. Soerevik, *The number of lattice rules*, BIT 29 (1989), pp. 527–534.

[284] J.N. Lyness, T. Soerevik, *A search program for finding optimal integration lattices*, Computing 47 (1991), pp. 103–120.

[285] J.N. Lyness, T. Soerevik, *An algorithm for finding optimal integration lattices of composite order*, BIT 32 (1992), pp. 665–675.

[286] M.A.H. MacCallum, F.J. Wright, *Algebraic computing with* REDUCE, in Lecture Notes from the First Brazilian School on Computer Algebra, M.J. Reboucas, Ed., Oxford University Press, Oxford, UK, 1991.

[287] J.I. Maeztu, *On symmetric cubature formulae for planar regions*, IMA J. Numer. Anal. 9 (1989), pp. 167–183.

[288] G. Maierhöfer, *Ein paralleler adaptiver Algorithmus für die numerische Integration*, Progress Report SC-89-8, Konrad-Zuse-Zentrum für Informationstechnik, Berlin, 1989.

[289] G. Maierhöfer, *A parallel adaptive algorithm for the numerical integration—Ein paralleler adaptiver Algorithmus für die Numerische Integration*, in Parallele Datenverarbeitung mit dem Transputer 1, Transputer-Anwender-Treffen, TAT '89, Springer-Verlag, Berlin, Heidelberg, New York, Tokyo, 1990, pp. 43–53.

[290] G. Maierhöfer, G. Skorobohatyj, *Ein paralleler adaptiver Algorithmus zur Numerischen Integration—Seine Implementierung für* SUPRENUM*-artige Architekturen mit* SUSI, Technical Report TR 88-5, Konrad-Zuse-Zentrum für Informationstechnik, Berlin, 1988.

[291] D. Maisonneuve, *Recherche et utilisation des "bons treillis"*, in Applications of Number Theory to Numerical Analysis, S.K. Zaremba, Ed., Academic Press, New York, 1972, pp. 121–201.

[292] M.A. Malcolm, R.B. Simpson, *Local versus global strategies for adaptive quadrature*, ACM Trans. Math. Software 1 (1975), pp. 129–146.

[293] K.W. Mangler, *Improper Integrals in Theoretical Aerodynamics*, Report 2424, Royal Aircraft Establishment, Farnborough, 1951.

[294] F. Mantel, P. Rabinowitz, *The application of integer programming to the computation of fully symmetric integration formulas in two and three dimensions*, SIAM J. Numer. Anal. 14 (1977), pp. 391–425.

[295] G. Mastroianni, G. Monegato, *Polynomial approximations of functions with endpoint singularities and product integration formulas*, Math. Comp. 62 (1994), pp. 725–738.

[296] G. Mastroianni, G. Monegato, *Convergence of product integration rules over*

$(0, \infty)$ *for functions with weak singularities at the origin*, Math. Comp. 64 (1995), pp. 237–249.

[297] V. A. Miller, G. J. Davis, *Adaptive quadrature on a message-passing multiprocessor*, J. Parallel Distrib. Comput. 14 (1992), pp. 417–425.

[298] M. Minsky, S. Papert, *On some associative, parallel, and analog computations*, in Associative Information Techniques, E. J. Jacks, Ed., Elsevier, Amsterdam, New York, 1971.

[299] R. A. Mitchell, *Error estimates arising from certain pseudorandom sequences in a quasi-random search method*, Math. Comp. 55 (1990), pp. 289–297.

[300] H. M. Möller, *Polynomideale und Kubaturformeln*, Ph.D. Thesis, University of Dortmund, Germany, 1973.

[301] H. M. Möller, *Kubaturformeln mit minimaler Knotenanzahl*, Numer. Math. 25 (1976), pp. 185–200.

[302] H. M. Möller, *On the construction of cubature formulae with few nodes using Groebner bases*, in Numerical Integration—Recent Developments, Software and Applications, P. Keast, G. Fairweather, Eds., D. Reidel, Dordrecht, the Netherlands, 1987, pp. 177–192.

[303] G. Monegato, *The numerical evaluation of one-dimensional Cauchy principal value integrals*, Computing 29 (1982), pp. 337–354.

[304] G. Monegato, *Stieltjes polynomials and related quadrature rules*, SIAM Rev. 24 (1982), pp. 137–158.

[305] G. Monegato, *On the weights of certain quadratures for the numerical evaluation of Cauchy principal value integrals and their derivatives*, Numer. Math. 50 (1987), pp. 273–281.

[306] G. Monegato, *The numerical evaluation of a 2-D Cauchy principal value integral arising in boundary integral equation methods*, Math. Comp. 62 (1994), pp. 765–777.

[307] G. Monegato, *Numerical evaluation of hypersingular integrals*, J. Comput. Appl. Math. 50 (1994), pp. 9–31.

[308] G. Mühlbach, *Neville-Aitken algorithms for interpolating by functions of Chebyshev-systems in the sense of Newton and in a generalized sense of Hermite*, in Theory of Approximations with Applications, A. G. Law, B. N. Sahney, Eds., Academic Press, New York, 1976, pp. 200–212.

[309] A. Murli, M. Rizzardi, *Talbot's method for the Laplace inversion problem*, ACM Trans. Math. Software 16 (1990), pp. 158–168.

[310] G. Myerson, *On ignoring the singularity*, SIAM J. Numer. Anal. 28 (1991), pp. 1803–1807.

[311] NAG Ltd., *NAG Fortran Library Manual—Mark 17*, Oxford, UK, 1996.

[312] M. Napierala, I. Gladwell, *Reducing ranking effects in parallel adaptive quadrature*, in Seventh SIAM Conference on Parallel Processing for Scientific Computing, D. H. Bailey, J. R. Gilbert, M. V. Mascagni, R. S. Schreiber, H. D. Simon, V. J. Torczon, L. T. Watson, Eds., SIAM, Philadelphia, 1995, pp. 261–266.

[313] P. Naur, *Machine dependent programming in common languages*, BIT 7 (1967), pp. 123–131.

[314] A. Neumaier, *Interval Methods for Systems of Equations*, Cambridge University Press, Cambridge, UK, 1990.

[315] H. Niederreiter, *On the distribution of pseudorandom numbers generated by the linear congruential method III*, Math. Comp. 30 (1976), pp. 571–597.

[316] H. Niederreiter, *Quasi-Monte Carlo methods and pseudorandom numbers*, Bull. Amer. Math. Soc. 84 (1978), pp. 957–1041.

[317] H. Niederreiter, *Point sets and sequences with small discrepancy*, Monatsh. Math. 104 (1987), pp. 273–337.

[318] H. Niederreiter, *Lower bounds for the discrepancy of inverse congruential pseudorandom numbers*, Math. Comput. 55 (1990), pp. 277–287.

[319] H. Niederreiter, *The existence of efficient lattice rules for multidimensional numerical integration*, Math. Comp. 58 (1992), pp. 305–314, S7–S16.

[320] H. Niederreiter, *Existence theorems for efficient lattice rules*, in Numerical Integration—Recent Developments, Software and Applications, T. O. Espelid, A. Genz, Eds., Kluwer, Dordrecht, the Netherlands, 1992, pp. 71–80.

[321] H. Niederreiter, *Random Number Generation and Quasi-Monte Carlo Methods*, SIAM, Philadelphia, 1992.

[322] H. Niederreiter, I. H. Sloan, *Lattice rules for multiple integration and discrepancy*, Math. Comp. 54 (1990), pp. 303–312.

[323] H. Niederreiter, I. H. Sloan, *Quasi-Monte Carlo methods with modified vertex weights*, in Numerical Integration IV—Proceedings of the Conference at the Mathematical Institute, Oberwolfach, 1992, H. Braß, G. Hämmerlin, Eds., Birkhäuser, Basel, 1993, pp. 253–265.

[324] H. Niederreiter, I. H. Sloan, *Integration of nonperiodic functions of two variables by Fibonacci lattice rules*, J. Comput. Appl. Math. 51 (1994), pp. 57–70.

[325] S. E. Notaris, *An overview of results on the existence or nonexistence and the error term of Gauss–Kronrod quadrature formulae*, in Approximation and Computation: A Festschrift in Honor of Walter Gautschi, R. V. M. Zahar, Ed., Vol. 119 of Internat. Ser. Numer. Math., Birkhäuser, Boston, 1994, pp. 485–496.

[326] T. Ooura, M. Mori, *The double exponential formula for oscillatory functions over the half infinite interval*, J. Comput. Appl. Math. 38 (1991), pp. 353–360.

[327] D. F. Paget, *The numerical evaluation of Hadamard finite-part integrals*, Numer. Math. 36 (1981), pp. 447–453.

[328] A. Palamara Orsi, *Two algorithms for the construction of product formulas*, Computing 49 (1993), pp. 363–372.

[329] A. Papoulis, *Probability, Random Variables, Stochastic Processes*, 3rd edition, McGraw-Hill, New York, 1991.

[330] T. N. L. Patterson, *The optimum addition of points to quadrature formulae*, Math. Comp. 22 (1968), pp. 847–856.

[331] T. N. L. Patterson, *An algorithm for generating interpolatory quadrature rules of the highest degree of precision with preassigned nodes for general weight functions*, ACM Trans. Math. Software 15 (1989), pp. 123–136.

[332] B. Pichon, *Numerical calculation of the generalized Fermi–Dirac integrals*, Comput. Phys. Comm. 55 (1989), pp. 127–136.

[333] R. Piessens, *Modified Clenshaw–Curtis integration and applications to numerical computation of integral transforms*, in Numerical Integration—Recent Developments, Software and Applications, P. Keast, G. Fairweather, Eds., D. Reidel, Dordrecht, the Netherlands, 1987, pp. 35–41.

[334] R. Piessens, M. Branders, *A note on the optimal addition of abscissas to quadrature formulas of Gauss and Lobatto type*, Math. Comp. 28 (1974), pp. 135–140, 344–347.

[335] R. Piessens, M. Branders, *Computation of oscillating integrals*, J. Comput. Appl. Math. 1 (1975), pp. 153–164.

[336] R. Piessens, M. Branders, *Modified Clenshaw–Curtis method for the computation of Bessel function integrals*, BIT 23 (1983), pp. 370–381.

[337] R. Piessens, M. Branders, *Computation of Fourier transform integrals using Chebyshev series expansions*, Computing 32 (1984), pp. 177–186.

[338] R. Piessens, M. Branders, *On the computation of Fourier transforms of singular functions*, J. Comput. Appl. Math. 43 (1992), pp. 159–169.

[339] R. Piessens, E. de Doncker, C. W. Ueberhuber, D. K. Kahaner, QUADPACK—*A Subroutine Package for Automatic Integration*, Springer-Verlag, Berlin, Heidelberg, New York, Tokyo, 1983.

[340] J. C. Piquette, *A method for symbolic evaluation of indefinite integrals containing special functions or their products*, J. Symbol. Comput. 11 (1991), pp. 231–249.

[341] F. A. Potra, E. Venturino, *Low-order methods for Cauchy principal value integrals with endpoint singularities*, IMA J. Numer. Anal. 14 (1994), pp. 295–310.

[342] W. H. Press, B. P. Flannery, S. A. Teukolsky, W. T. Vetterling, *Numerical Recipes in C—The Art of Scientific Computing*, 2nd edition, Cambridge University Press, Cambridge, UK, 1992.

[343] W. H. Press, B. P. Flannery, S. A. Teukolsky, W. T. Vetterling, *Numerical Recipes in Fortran—The Art of Scientific Computing*, 2nd edition, Cambridge University Press, Cambridge, UK, 1992.

[344] J. F. Price, I. H. Sloan, *Pointwise convergence of multiple Fourier series: Sufficient conditions and an application to numerical integration*, J. Math. Anal. Comput. 169 (1992), pp. 140–156.

[345] J. S. Quarterman, S. Carl-Mitchell, *The Internet Connection—System Connectivity and Configuration*, Addison-Wesley, Reading, MA, 1994.

[346] P. Rabinowitz, *Gauss–Kronrod integration rules for Cauchy principal value integrals*, Math. Comp. 41 (1983), pp. 63–78.

[347] P. Rabinowitz, *The convergence of interpolatory product integration rules*, BIT 26 (1986), pp. 131–134.

[348] P. Rabinowitz, *Rates of convergence of Gauss, Lobatto, and Radau integration rules for singular integrands*, Math. Comp. 47 (1986), pp. 625–638.

[349] P. Rabinowitz, *The convergence of noninterpolatory product integration rules*, in Numerical Integration—Recent Developments, Software and Applications, P. Keast, G. Fairweather, Eds., D. Reidel, Dordrecht, the Netherlands, 1987, pp. 1–16.

[350] P. Rabinowitz, *Convergence results for piecewise linear quadratures for Cauchy principal value integrals*, Math. Comp. 51 (1988), pp. 741–747.

[351] P. Rabinowitz, *On an interpolatory product rule for evaluating Cauchy principal value integrals*, BIT 29 (1989), pp. 347–355.

[352] P. Rabinowitz, *Generalized noninterpolatory rules for Cauchy principal value integrals*, Math. Comp. 54 (1990), pp. 271–279.

[353] P. Rabinowitz, *Numerical evaluation of Cauchy principal value integrals with singular integrands*, Math. Comp. 55 (1990), pp. 265–276.

[354] P. Rabinowitz, *Numerical integration based on approximating splines*, J. Comput. Appl. Math. 33 (1990), pp. 73–83.

[355] P. Rabinowitz, *Product integration of singular integrands using Hermite–Fejer interpolation*, BIT 31 (1991), pp. 321–327.

[356] P. Rabinowitz, *Uniform convergence results for Cauchy principal value integrals*, Math. Comp. 56 (1991), pp. 731–740.

[357] P. Rabinowitz, D.S. Lubinsky, *Noninterpolatory integration rules for Cauchy principal value integrals*, Math. Comp. 53 (1989), pp. 279–295.

[358] P. Rabinowitz, N. Richter, *Perfectly symmetric two-dimensional integration formulas with minimal number of points*, Math. Comp. 23 (1969), pp. 953–976.

[359] P. Rabinowitz, E. Santi, *On the uniform convergence of Cauchy principal values of quasi-interpolating splines*, BIT 35 (1995), pp. 277–290.

[360] P. Rabinowitz, W.E. Smith, *Interpolatory product integration in the presence of singularities: L_2 theory*, J. Comput. Appl. Math. 39 (1992), pp. 79–87.

[361] P. Rabinowitz, W.E. Smith, *Interpolatory product integration in the presence of singularities: L_P theory*, in Numerical Integration—Recent Developments, Software and Applications, T.O. Espelid, A. Genz, Eds., Kluwer, Dordrecht, the Netherlands, 1992, pp. 93–109.

[362] G.G. Rasputin, *Cubature formulas with a minimal number of nodes for centrally symmetric n-dimensional integrals*, Izv. Vyssh. Uchebn. Zaved Mat. 34 (1990), pp. 50–57.

[363] J.R. Rice, *Adaptive quadrature: Convergence of parallel and sequential algorithms*, Bull. Amer. Math. Soc. 80 (1974), pp. 1250–1254.

[364] J.R. Rice, *Parallel algorithms for adaptive quadrature—Convergence*, in Information Processing '74, J.L. Rosenfeld, Ed., North-Holland, Amsterdam, New York, Oxford, 1974, pp. 600–604.

[365] J.R. Rice, *A metalgorithm for adaptive quadrature*, J. Assoc. Comput. Mach. 22 (1975), pp. 61–82.

[366] J.R. Rice, *Parallel algorithms for adaptive quadrature II—Metalgorithm correctness*, Acta Inform. 5 (1975), pp. 273–285.

[367] J.R. Rice, *Parallel algorithms for adaptive quadrature III—Program correctness*, ACM Trans. Math. Software 2 (1976), pp. 1–30.

[368] A. Rich, J. Rich, D. Stoutemyer, DERIVE *User Manual*, Soft Warehouse, Honolulu, HA, 1997.

[369] R.H. Risch, *The problem of integration in finite terms*, Trans. Amer. Math. Soc. 38 (1969), pp. 167–189.

[370] R.H. Risch, *Algebraic properties of the elementary functions of analysis*, Amer. J. Math. 101 (1979), pp. 743–759.

[371] J.F. Ritt, *Integration in Finite Terms*, Columbia University Press, New York, 1948.

[372] T.J. Rivlin, *The Chebyshev Polynomials*, Wiley, New York, 1974.

[373] I. Robinson, E. de Doncker, *Automatic computation of improper integrals over a bounded or unbounded planar region*, Computing 27 (1981), pp. 253–284.

[374] G. Rodriguez, S. Seatzu, *On the numerical inversion of the Laplace transform in reproducing kernel Hilbert spaces*, IMA J. Numer. Anal. 13 (1993), pp. 463–475.

[375] W. Romberg, *Vereinfachte numerische Integration*, Det. Kong. Norske Videnskabers Selskab Forhandlinger 28 (1955), pp. 30–36.

[376] D. Rosen, D.E. Cormack, *Singular and near singular integrals in the BEM: A global approach*, SIAM J. Appl. Math. 53 (1993), pp. 340–357.

[377] M. Rosenlicht, *Integration in finite terms*, Amer. Math. Monthly 79 (1972), pp. 963–972.

[378] K.F. Roth, *On irregularities of distribution*, Mathematika 1 (1954), pp. 73–79.

[379] M. Rothstein, *Aspects of Symbolic Integration and Simplification of Exponential and Primitive Functions*, Ph.D. Thesis, University of Wisconsin, Madison, 1976.

[380] T. W. Sag, G. Szekeres, *Numerical evaluation of high-dimensional integrals*, Math. Comp. 18 (1964), pp. 245–253.

[381] R. P. Sagar, *A Gaussian quadrature for the calculation of generalized Fermi-Dirac integrals*, Comput. Phys. Comm. 66 (1991), pp. 271–275.

[382] R. P. Sagar, *On the evaluation of the Fermi-Dirac integrals*, Astrophys. J. 276 (1991), pp. 364–366.

[383] H. J. Schmid, *On cubature formulae with a minimal number of knots*, Numer. Math. 31 (1978), pp. 281–297.

[384] H. J. Schmid, *Interpolatorische Kubaturformeln und reelle Ideale*, Math. Z. 170 (1980), pp. 267–282.

[385] W. M. Schmidt, *Irregularities of distribution*, Acta Arith. 21 (1972), pp. 45–50.

[386] C. Schwab, W. L. Wendland, *Kernel properties and representation of boundary integral operators*, Math. Nachr. (1992), pp. 187–218.

[387] C. Schwab, W. L. Wendland, *On numerical cubatures of singular surface integrals in boundary element methods*, Numer. Math. 62 (1992), pp. 343–369.

[388] H. D. Shapiro, *Increasing robustness in global adaptive quadrature through interval heuristics*, ACM Trans. Math. Software 10 (1984), pp. 117–139.

[389] A. Sidi, *Some properties of a generalization of the Richardson extrapolation process*, J. Inst. Math. Appl. 24 (1979), pp. 327–346.

[390] A. Sidi, *Extrapolation methods for oscillatory infinite integrals*, J. Inst. Math. Appl. 26 (1980), pp. 1–20.

[391] A. Sidi, *An algorithm for a special case of a generalization of the Richardson extrapolation process*, Numer. Math. 38 (1982), pp. 299–307.

[392] A. Sidi, *The numerical evaluation of very oscillatory infinite integrals by extrapolation*, Math. Comp. 38 (1982), pp. 517–529.

[393] A. Sidi, *Generalizations of Richardson extrapolation with applications to numerical integration*, in Numerical Integration III—Proceedings of the Conference at the Mathematical Institute, Oberwolfach, 1988, H. Braß, G. Hämmerlin, Eds., Birkhäuser, Basel, 1988, pp. 237–250.

[394] A. Sidi, *A user-friendly extrapolation method for oscillatory infinite integrals*, Math. Comp. 51 (1988), pp. 249–266.

[395] A. Sidi, *On rates of acceleration of extrapolation methods for oscillatory infinite integrals*, BIT 30 (1990), pp. 347–357.

[396] A. Sidi, *A new variable transformation for numerical integration*, in Numerical Integration IV—Proceedings of the Conference at the Mathematical Institute, Oberwolfach, 1992, H. Braß, G. Hämmerlin, Eds., Birkhäuser, Basel, 1993, pp. 359–373.

[397] K. Sikorski, F. Stenger, *A Fortran subroutine for numerical integration in H_p*, ACM Trans. Math. Software 10 (1984), pp. 152–160.

[398] K. Sikorski, F. Stenger, *Optimal quadratures in H_p spaces*, ACM Trans. Math. Software 10 (1984), pp. 140–151.

[399] R. B. Simpson, A. Yazici, *An organization of the extrapolation method for vector processing*, Parallel Comput. 4 (1987), pp. 175–188.

[400] M. Singer, B. Saunders, B. F. Caviness, *An extension of Liouville's theorem on integration in finite terms*, SIAM J. Computing 14 (1985), pp. 966–990.

[401] I. H. Sloan, *Numerical integration in high dimensions—The lattice rule approach,*

in Numerical Integration—Recent Developments, Software and Applications, T.O. Espelid, A. Genz, Eds., Kluwer, Dordrecht, the Netherlands, 1992, pp. 55–69.

[402] I.H. Sloan, S. Joe, *Lattice Methods for Multiple Integration*, Clarendon Press, Oxford, UK, 1994.

[403] I.H. Sloan, P.J. Kachoyan, *Lattice methods for multiple integration: Theory, error analysis and examples*, SIAM J. Numer. Anal. 24 (1987), pp. 116–128.

[404] I.H. Sloan, J.N. Lyness, *The representation of lattice quadrature rules as multiple sums*, Math. Comp. 52 (1989), pp. 81–94.

[405] I.H. Sloan, J.N. Lyness, *Lattice rules: Projection regularity and unique representations*, Math. Comp. 54 (1990), pp. 649–660.

[406] I.H. Sloan, W.E. Smith, *Properties of interpolatory product integration rules*, SIAM J. Numer. Anal. 19 (1982), pp. 427–442.

[407] I.H. Sloan, L. Walsh, *A computer search of rank-2 lattice rules for multidimensional quadrature*, Math. Comp. 54 (1990), pp. 281–302.

[408] D.A. Smith, W.F. Ford, *Acceleration of linear and logarithmic convergence*, SIAM J. Numer. Anal. 16 (1979), pp. 223–240.

[409] D.A. Smith, W.F. Ford, *Numerical comparison of nonlinear convergence accelerators*, Math. Comp. 38 (1982), pp. 481–499.

[410] W.E. Smith, D.F. Paget, *Optimal nodes for interpolatory product integration*, SIAM J. Numer. Anal. 29 (1992), pp. 586–600.

[411] W.E. Smith, I.H. Sloan, A.H. Opie, *Product integration over infinite intervals—Rules based on the zeros of Hermite polynomials*, Math. Comp. 40 (1983), pp. 519–535.

[412] M. Snir, S.W. Otto, S. Huss-Lederman, D.W. Walker, J.J. Dongarra, *MPI: The Complete Reference*, MIT Press, Cambridge, MA, 1996.

[413] I.M. Sobol, *The distribution of points in a cube and the approximate evaluation of integrals*, Zh. Vychisl. Mat. i Mat. Fiz. 7 (1967), pp. 784–802.

[414] S.L. Sobolev, *Cubature formulas on the sphere invariant under finite groups of rotation*, Soviet Math. 3 (1962), pp. 1307–1310.

[415] T. Soerevik, *Reliable and Efficient Algorithms for Adaptive Quadrature*, Ph.D. Thesis, University of Bergen, Norway, 1988.

[416] T. Soerevik, T.O. Espelid, *Fully symmetric integration rules for the 4-cube*, BIT 29 (1989), pp. 148–153.

[417] F. Stenger, *Numerical Methods Based on Sinc and Analytic Functions*, Vol. 20 of Springer Ser. Comput. Math., Springer-Verlag, Berlin, Heidelberg, New York, Tokyo, 1993.

[418] H.J. Stetter, *Verification in computer algebra systems*, Computing Suppl. 9 (1993), pp. 247–263.

[419] H.W. Stolle, R. Strauss, *On the numerical integration of certain singular integrals*, Computing 48 (1992), pp. 177–189.

[420] U. Storck, *Verified calculation of the nodes and weights for Gaussian quadrature formulas*, Interval Comput. 4 (1993), pp. 114–124.

[421] U. Storck, *Zweidimensionale Integration mit automatischer Ergebnisverifikation (Two-dimensional integration with automatic result verification)*, Z. Angew. Math. Mech. 73 (1993), pp. T 897–T 899.

[422] U. Storck, *Verifizierte Berechnung mehrfach geschachtelter singulärer Integrale der Gaskinetik*, Ph.D. Thesis, University of Karlsruhe, Germany, 1995.

[423] A. H. Stroud, *Approximate Calculation of Multiple Integrals*, Prentice-Hall, Englewood Cliffs, NJ, 1971.

[424] A. H. Stroud, *Numerical Quadrature and Solution of Ordinary Differential Equations*, Springer-Verlag, Berlin, Heidelberg, New York, Tokyo, 1974.

[425] A. Talbot, *The accurate numerical inversion of Laplace transforms*, J. Inst. Math. Appl. 23 (1979), pp. 97–120.

[426] R. C. Tausworthe, *Random numbers generated by linear recurrence modulo two*, Math. Comp. 19 (1965), pp. 201–209.

[427] B. M. Trager, *Algebraic factoring and rational function integration*, in Proc. SYMSAC '76, R. D. Jenkins, Ed., ACM, New York, 1976.

[428] B. M. Trager, *Integration of simple radical extensions*, in Proc. EUROCAM '79, Vol. 72 of Lecture Notes in Comput. Sci., Springer-Verlag, Berlin, Heidelberg, New York, Tokyo, 1979, pp. 408–414.

[429] B. M. Trager, *Integration of Algebraic Functions*, Ph.D. Thesis, MIT, Cambridge, MA, 1984.

[430] J. F. Traub, G. W. Wasilkowski, H. Wozniakowski, *Information, Uncertainty, Complexity*, Addison-Wesley, Reading, MA, 1983.

[431] J. F. Traub, G. W. Wasilkowski, H. Wozniakowski, *Information-Based Complexity*, Academic Press, New York, 1988.

[432] J. F. Traub, H. Wozniakowski, *A General Theory of Optimal Algorithms*, Academic Press, New York, 1980.

[433] W. F. Trench, *Advanced Calculus*, Harper and Row, New York, Hagerstown, San Francisco, London, 1978.

[434] G. Tsamasphyros, P. S. Theocaris, *On the convergence of some quadrature rules for Cauchy principal-value and finite-part integrals*, Computing 31 (1983), pp. 105–114.

[435] C. W. Ueberhuber, *Numerical Computation*, Springer-Verlag, Berlin, Heidelberg, New York, Tokyo, 1997.

[436] B. L. van der Waerden, *Modern Algebra*, Ungar, New York, 1970.

[437] C. F. Van Loan, *Computational Frameworks for the Fast Fourier Transform*, SIAM, Philadelphia, 1992.

[438] E. Venturino, *On solving singular integral equations via a hyperbolic tangent quadrature rule*, Math. Comp. 47 (1986), pp. 159–167.

[439] P. Verlinden, R. Cools, D. Roose, A. Haegemans, *The construction of cubature formulae for a family of integrals: A bifurcation problem*, Computing 40 (1988), pp. 337–346.

[440] P. Verlinden, A. Haegemans, *The construction of cubature formulae by continuation*, Computing 45 (1990), pp. 145–155.

[441] Visual Numerics Inc., *IMSL MATH/LIBRARY: User's Manual—Version 3.0*, Houston, TX, 1994.

[442] Visual Numerics Inc., *IMSL STAT/LIBRARY: User's Manual—Version 3.0*, Houston, TX, 1994.

[443] G. W. Wasilkowski, *Information of varying cardinality*, J. Complexity 2 (1986), pp. 204–228.

[444] G. W. Wasilkowski, F. Gao, *On the power of adaptive information for functions with singularities*, Math. Comp. 58 (1992), pp. 285–304.

[445] W. T. Weeks, *Numerical inversion of Laplace transforms using Laguerre functions*, J. Assoc. Comput. Mach. 3 (1966), pp. 419–429.

[446] M. W. Wilson, *A general algorithm for nonnegative quadrature formulas*, Math. Comp. 23 (1969), pp. 253–258.

[447] S. Wolfram, MATHEMATICA, Addison-Wesley, Reading, MA, 1991.

[448] H. Wozniakowski, *A survey of information-based complexity*, J. Complexity 1 (1985), pp. 11–44.

[449] P. Wynn, *On a device for computing the* $e_m(S_n)$ *transformation*, Mathematical Tables and Aids to Computing 10 (1956), pp. 91–96.

[450] Y. Xu, *Common Zeros of Polynomials in Several Variables and Higher Dimensional Quadrature*, Vol. 312 of *Pitman Res. Notes Math. Ser.*, Longman, Harlow, UK, 1994.

[451] F. Zhang, *A hyperbolic tangent quadrature rule for solving singular integral equations with Hadamard finite part integrals*, Comput. Math. Appl. 22 (1991), pp. 59–73.

Author Index

Software Index

Subject Index